Physicochemical Treatment of Hazardous Wastes

WALTER Z. TANG

LEWIS PUBLISHERS

A CRC Press Company
Boca Raton   London   New York   Washington, D.C.

Library of Congress Cataloging-in-Publication Data

Tang, Walter Z.
    Physicochemical treatment of hazardous wastes / Walter Z. Tang
        p.   cm.
    Includes bibliographical references and index.
    ISBN 1-56676-927-2 (alk. paper)
    1. Hazardous wastes—Purification. I. Title.

TD1060.T35 2003
628.4′2—dc21   2003055435

Visit the CRC Press Web site at www.crcpress.com

© 2004 by CRC Press LLC
Lewis Publishers is an imprint of CRC Press LLC

No claim to original U.S. Government works
International Standard Book Number 1-56676-927-2
Library of Congress Card Number 2003055435
Printed in the United States of America 1 2 3 4 5 6 7 8 9 0
Printed on acid-free paper

Dedication

In memory of my father, Yuxiang Tang

To my mother, Yongcui Hu, and

To my children, William and Elizabeth,

with love.

pollutants are explained in terms of elementary reactions. These elementary reactions not only are important in assessing the treatability of organic pollutants using QSAR but are also critical in properly designing AOP processes. Finally, QSAR models are discussed to demonstrate the effect of molecular structure on their degradation kinetics and to rank the treatability of each organic compound.

This book is intended for graduates, engineers, and scientists affiliated with universities, consulting firms, or national laboratories and who are dealing with the remediation of hazardous wastes in water, groundwater, and industrial wastewater. Due to the in-depth discussion of organic chemistry, graduate students in environmental engineering and upper-level undergraduates in chemistry, chemical engineering, or environmental sciences who intend to enter environmental engineering should find it useful in their professional development. Students will learn a systematic approach to applying various sciences to the search for effective treatment technologies in terms of thermokinetic principles. Engineers will find the QSAR models extremely useful in selecting treatment processes for hazardous wastes according to the molecular structures of organic pollutants. Scientists in industrial and governmental laboratories, as well as designers and reviewers in remediation projects, will also find the book helpful in their efforts to restore our environment and keep it clean.

During the 1970s, the U.S. Environmental Protection Agency designated phase-transfer technologies such as air stripping and activated carbon adsorption as the best available technologies. The search for mineralizing organic pollutants shifted the focus from phase-transfer technologies to oxidative technologies after the Hazardous Waste Amendment in 1984. As a result, AOPs were developed in laboratories, extended to pilot sites, and finally applied in the field from the 1980s to the present. The concept of an AOP includes any process that uses hydroxyl radicals as the predominant species; however, the concept failed to provide fundamental theories such as transition state theory to guide research communities in their search for the most effective oxidation processes. In a strict sense, then, AOP should be defined as a Catalytic Oxidation Process (COP), which would provide sound scientific footing for the search for innovative technologies. It is well documented that oxidants such as oxygen, ozone, and hydrogen peroxide oxidize organic pollutants slowly. It is only when they are catalytically decomposed into other active species such as hydroxyl radicals that the activation barrier of the activated complex can be significantly lowered. The catalysts normally used are ultraviolet photons, transition metals or their ions, ultrasound, and electrons. Increasing temperature and pressure can further enhance the catalytic effect.

My research on AOPs began over 12 years ago at the University of Delaware. When I worked on the degradation of phenols by a visible photon/CdS system, I had to wake up at midnight in order to take samples from a photocatalytic reactor because the reaction half time in degrading 0.001-M phenol is about 1 day. After 1 day I found that Fenton's reagent was an extremely

Preface

On average, one ton of hazardous waste per person is generated annually by industries in the United States. Before the Resource Conservation and Recovery Act of 1984, hazardous wastes were improperly disposed of into the environment without any regulation. As a result, remediation of these contaminated sites and management of the ongoing hazardous waste sources are two major tasks to be achieved by treatment technologies. Due to the complex nature of the contaminated media and of the pollutants, environmental professionals are facing a host of questions, such as: What are the contaminated media? What is the nature of the pollutants? What are the concentrations of each pollutant? Among biological, physicochemical, or thermal technologies, if physicochemical processes are to be the solution, the treatability of various pollutants must be assessed before a process can be properly designed. This book systematically examines the treatability of hazardous wastes by various physicochemical treatment processes according to the Quantitative Structure–Activity Relationships (QSARs) between kinetic rate constants and molecular descriptors.

I have attempted to achieve five major goals in this book: (1) fundamental theories of thermokinetics such as the transition state theory are used to integrate research findings in Advanced Oxidation Process (AOP) research; (2) reaction kinetics and mechanisms for each AOP are explained in terms of elementary reactions and the reactive center; (3) QSARs are introduced as methodologies to assess the treatability of organic compounds; (4) computational molecular descriptors such as the E_{HOMO} and E_{LUMO} are used extensively in the QSAR analysis; (5) the kinetics of various AOPs are compared so that the most effective process can be selected for a given class of organic pollutants.

This book is divided into five parts. Chapter 1 to Chapter 4 define the hazardous waste problems and physicochemical approaches to solve these problems. Chapter 5 explains QSAR theory and its application to predicting molecular descriptors and hydroxyl radical reactions. Chapter 6 to Chapter 12 focus on each of the eight most important AOPs. Chapter 13 presents a major reductive treatment technology, zero-valence iron, and Chapter 14 compares each AOP according to its oxidation kinetics for specific classes of organic compounds. Each chapter begins with an introduction of the process and its historical development. The intention is to demonstrate how fundamental sciences guide the search for these innovative technologies. Also, such introductions provide the information necessary for readers to delve into the literature for current research topics. Then, the principles of the process and the degradation kinetics, along with mechanisms of organic fast process, I added hydrogen peroxide and ferrous ion separately to the reactor. The reaction half time reduced from one day to a few hours. When I added hydrogen peroxide first and then the ferrous sulfate, the reaction half time was reduced to a few minutes. It became clear to me during my investigation of the oxidation kinetics and mechanisms of chlorinated phenols by Fenton's reagent that the efficiency of AOPs depends upon both the rate and the amount of hydroxyl radical generated and the molecular structure of organic compounds.

It has long been recognized that the treatability of different classes of organic compounds differs significantly. Furthermore, the treatability of chlorinated compounds within a given class of organic pollutants decreases as the chlorine content in a molecule increases. Indeed, the carbon in tetrachloride has been oxidized by chlorine so much that it is even insensitive to hydroxyl radical attack. Therefore, elementary iron may be a more economical way to reduce these pollutants rather than to oxidize them. To quantify the effect of chlorine, QSAR models are used to assess the effect of chlorine on molecular descriptors such as E_{HOMO} and E_{LUMO}. The treatability of organic compounds by each AOP, then, can be evaluated using QSAR models of the oxidation kinetic rate constants and molecular descriptors.

Thermokinetics, group theory, and computational QSARs should find broad application in future research effort on AOPs for several reasons: (1) thermokinetics bridges thermodynamics and kinetics, which serve as the foundation for QSAR analysis; (2) group theory may offer kinetic calculations of activated complex for a given class of compounds, and the resulting degradation rate constants can be more accurately estimated; and (3) as more data regarding operational costs become available for each technology, QSARs may be incorporated into the calculations to estimate the operational cost of a specific compound. In addition, nanotechnology will become another research focus in the next decade to develop nanoparticles such as elementary iron, TiO_2, nanofiltration, and electromembranes in the physicochemical treatment of hazardous wastes.

About the Author

 **Walter Z. Tang** (B.S., Sanitary Engineering, Chongqing University, Chongqing, China, 1983; M.S., Environmental Engineering, Tsinghua University, Beijing, China, 1986; M.S., Environmental Engineering, University of Missouri-Rolla, 1988; Ph.D., Environmental Engineering, University of Delaware, 1993) is an Associate Professor and Graduate Director for Environmental Engineering in the Department of Civil and Environmental Engineering at Florida International University (FIU), Miami, FL. He has been a registered Professional Engineer in Florida since 1993. Dr. Tang has had extensive research experience over the past 14 years in the area of physicochemical treatment processes; environmental applications of aquatic, organic, catalytic, and colloidal chemistry; advanced oxidation processes; environmental molecular structure–activity relationships (QSARs); and methodology in environmental impact assessment.

Dr. Tang is the principal investigator for 14 research projects supported by the U.S. Environmental Protection Agency, the National Institutes of Health, and the National Science Foundation. He has published 24 peer-reviewed papers and 41 conference papers, co-authored one book, and contributed one chapter to a book. Also, he has written graduate teaching manuals for three different graduate courses. He has been a referee for 12 journals and has served as a proposal reviewer for the NSF and the National Research Council. Dr. Tang has organized and presided over 11 sessions at various national and international conferences on advanced oxidation processes (AOPs) and was the invited speaker at Florida Atlantic University in 2001.

Dr. Tang has supervised three post doctors, three visiting professors, and 35 graduate students in environmental engineering, and he has taught six undergraduate courses and nine graduate courses in the Department of Civil and Environmental Engineering at FIU. Dr. Tang received FIU's Faculty Research Award in 1997, Faculty Teaching Award in 1998, and Departmental Teacher of the Year Award in 1998. He is a member of Chi Epsilon and is listed in *Who's Who in the World*, *Who's Who in America*, *Who's Who in Science and Engineering*, and *Who's Who Among America's Teachers*.

Since 1994, Dr. Tang has been a co-principal investigator in joint research projects on AOPs with professors at Tsinghua University, Chongqing University, and the Third Medical University of Chinese Military in Chongqing, China. As a research fellow in the China–Cornell Fellowship Program supported by the Rockefeller Foundation, Dr. Tang offered six seminars at Tsinghua University. As a co-principal investigator from 1998 to 2002 of the Two-Bases Program sponsored by the China National Science Foundation, he advised a Ph.D. student at Tsinghua University on his dissertation: QSARs in the Anaerobic Degradability of Organic Pollutants. Chongqing University and Chongqing Jianzhu University granted the visiting professorship to Dr. Tang in 1999. He won six joint research projects sponsored by the Chinese Ministry of Education for Chongqing University. He was the invited speaker at Nankai University and Gansu Industry University in 2002 and at Wuhan University in 2003. He was named the Outstanding Chinese Scholar in the southern region of the United States and served as a Foreign Expert in the State Sunshine Program of China. The Chinese Ministry of Education invited Dr. Tang to Beijing as a state guest for the 50th anniversary of China National Day in 1999.

Acknowledgments

I would like to acknowledge the contributions to this book made by my former graduate students: Tzai-Shian Jung, Angela Pierotti, Sangeeta Dulashia, Todd Hendrix, Ricardo Martinez, Lucero Vaca, Stephanie Tassos, Rena Chen, Taweeporn Fongtong, Jiun-Jia Hsu, Kenneth Morris, Jose Polar, Carlos Hernandez, and Jeffrey Czajkowski. I thank Jiashun Huang, Dennis Maddox, and Pia Hansson Nunoo for their many hours devoted to typing and drawing of the figures. I would like to thank all the students since 1991 at Florida International University (FIU) who took the graduate course, Advanced Treatment System, upon which the book is based. Students who assisted in this book include Bernine Khan, Lillian Costa-Mayoral, Christopher Wilson, and Oscar Carmona. A special acknowledgment goes to Georgio Tachiev of the Hemisphere Center for Environmental Technology at FIU for his constructive proofreading.

I am grateful to Dr. C.P. Huang at the University of Delaware for introducing me to the research of AOPs. Many QSAR models were developed through financial support from the U.S. Environmental Protection Agency, National Science Foundation, and National Institutes of Health, and their support is greatly appreciated. Thanks go to Mrs. Virginia Broadway at the USEPA for supporting and administrating five EPA fellowships to my students over the last decade. Dr. William Cooper and his colleagues are acknowledged for their work on high-energy electron beams. I would like to thank Dean Vish Prasad and Associate Dean David Shen of the College of Engineering at FIU for allowing me to complete the book. I am in debt to Gail Renard and Sara Kreisman, my book editors at CRC Press LLC, who provided excellent professional guidance and spent numerous days editing and proofreading the manuscript.

Table of Contents

1

Environmental Laws

1.1 Introduction

In the early 1960s, public interest in environmental matters provoked the rapid proliferation of environmental laws and regulations in the United States, as illustrated in Figure 1.1; however, the effects of hazardous waste on public health and the environment did not attract the attention of the government until the 1970s. Since 1970, the government has been taking an active role in protecting the people of the United States from various forms of hazardous waste, and, as a result, a regulatory-driven industry known as *environmental remediation* was generated (LaGrega et al., 1994). Ten major pieces of legislation were passed to provide this protection:

- National Environmental Policy Act (NEPA)
- Occupational Safety and Health Act (OSHA)
- Clean Water Act (CWA)
- Safe Drinking Water Act (SDWA)
- Toxic Substances Control Act (TSCA)
- Resource Conservation and Recovery Act (RCRA)
- Comprehensive Environmental Response, Compensation and Liability Act (CERCLA)
- Hazardous and Solid Waste Amendments (HSWA)
- Superfund Amendments and Reauthorization Act (SARA)
- Clean Air Act (CAA)

Each law attempts to achieve specific goals by setting environmental standards for different classes of hazardous waste. Treatment technologies are researched by universities and companies and implemented by the environmental remediation industry.

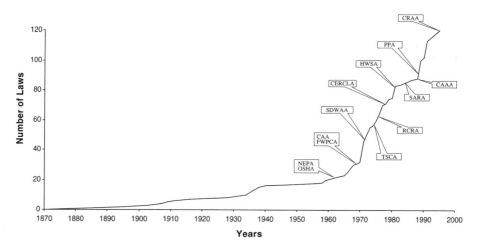

FIGURE 1.1
Cumulative growth in federal environmental laws and amendments in the United States.

1.2   Environmental Laws

1.2.1   National Environmental Policy Act (NEPA)

> The objective of this act is to declare a national policy which will encour-
> age productive and enjoyable harmony between man and his environ-
> ment; to promote efforts which will prevent or eliminate damage to the
> environment and biosphere and stimulate the health and welfare of man;
> to enrich the understanding of the ecological systems and natural re-
> sources important to the Nation; and to establish a Council on Environ-
> mental Quality. (NEPA)

The National Environmental Policy Act (NEPA) requires federal agencies to
assess the environmental impact early in the planning process of any project.
For those projects that are expected to have an effect on the quality of the
environment, the proposing agency is required to file a formal environmental
impact statement. In addition, the NEPA has had an impact in certain areas;
for example, it has:

- Provided a systematic means of dealing with environmental concerns and has included environmental costs in the decision-making
  process.
- Opened governmental activities and projects to public scrutiny and
  public participation.

- Delayed many projects because of the time required to comply with NEPA regulations.
- Caused many projects to be modified or abandoned that were intended to balance environmental costs with other benefits.

1.2.2   Occupational Safety and Health Act (OSHA)

> To assure safe and healthful working conditions for working men and women; by authorizing enforcement of the standards developed under the Act; by assisting and encouraging the States in their efforts to assure safe and healthful working conditions; by providing for research, information, education, and training in the field of occupational safety and health; and for other purposes. (OSHA)

To achieve its vision, the Occupational Safety and Health Act (OSHA) has established three interdependent and complementary strategic goals to guide the development of programs and activities for the agency. The successful accomplishment of any one of the strategic goals would not be possible without parallel successes in achieving the other goals. For example, the focus on reducing hazards, exposures, injuries, illnesses, and deaths in the workplace would be difficult to achieve without realizing the goals that call for the engagement of workers and employers in this effort, as well as the development of strong public confidence and support for its activities.

1.2.3   Clean Water Act (CWA)

The Clean Water Act (CWA) focuses on keeping streams, rivers, lakes, and bays clean and safe. The origin of the CWA can be traced back to the Rivers and Harbors Act of 1899, which prohibited the discharge of all refuse other than sewage (ironically) to any navigable water of the United States. It also vaguely included permitting authority for the Army Corps of Engineers. The goal of this act was more likely to keep the waterways free from obstruction than to keep them clean. Over the next 50 years, very little attention was given to our waters, and as a result water pollution rose to crisis proportions by the 1970s. A clear example of this occurred in Cleveland, OH, in June of 1969, when the Cuyahoga River caught fire due to the massive amounts of floating chemicals and oil. In 1971, the nation was moved by Ralph Nader's book, *Water Wasteland*, which outlined serious water quality problems throughout the country (Adler et al., 1993). These events caused the government to pass the Federal Water Pollution Control Act Amendments of 1972, which created what we now know as the CWA. Earlier versions of the act were the primary vehicles that provided federal money for local water pollution control construction.

The CWA is based on a simple idea: No discharges to water are allowed by anyone without a permit. In regard to the issue of private property vs. federal government intervention, the CWA establishes regulations only for discharges to the waters of the United States. This includes all navigable waters, their tributaries, and areas that flow to waters of the United States at some point during the year (e.g., dry riverbeds). Subsequent legal decisions have included the groundwater that eventually flows to a body of water in the United States. The CWA has three levels of goals (Adler et al., 1993):

- Zero discharge is allowed.
- All waters should be fishable and allow swimming.
- No toxicities are allowed in any waters of the United States.

The CWA established a National Pollutant Discharge Elimination System (NPDES) program to be developed and overseen by the U.S. Environmental Protection Agency (USEPA). It requires that all dischargers of any type of pollutant apply for an NPDES permit. These permits set specific limits for pollutant discharges and attempt to stop pollution at its source. The two main groups that must have an NPDES permit are industrial facilities and wastewater utilities. The effluent limitations vary for each NPDES permit; however, most limitations are taken from technology-based effluent limitation guidelines issued by the USEPA. As a result, different levels of effluent limitations have been established depending on the treatment efficiency of different technologies, including:

- Best practicable control technology (BPT) is the minimum level set for all pollutants.
- Best conventional technology (BCT) applies to conventional pollutants.
- Best available technology (BAT) applies to toxic and unconventional pollutants.
- Best available demonstrated technology (BADT) is used for new dischargers.

Another factor that must be considered in the NPDES program is the receiving water body. The CWA requires states to classify each water body based on its actual or potential use. Water quality standards are then developed for each classification. Once this process is complete, water quality management plans are written to keep dischargers within these limits.

To assist the states, the USEPA has created a general set of water quality standards that includes criteria for 137 pollutants. This information is provided in the USEPA report *Quality Criteria for Water 1986*, also known as the *Gold Book*. When applying standards to source dischargers, the NPDES permit includes these water-quality-based effluent limitations (WQBELs). Rather than focus solely on individual pollutants, whole effluent toxicity tests have become more commonplace (Moore and Vicory, 1998). These tests reflect the overall toxicity of an effluent on various organisms so as to obtain a clearer picture of the quality of the water. A somewhat controversial requirement of the CWA is the process for establishing total maximum daily loads (TMDLs) for each water body. Each state plan must include these TMDLs, which set the maximum amount of each pollutant that should be found in a water body. The TMDLs are then divided among the NPDES permit holders that discharge to that water body. Unfortunately, this task has been found to be much more complex and costly than originally thought. In fact, the USEPA is being sued by many different organizations over the strict deadlines. It has been estimated that TMDL studies alone will cost billions of dollars throughout the United States, and that they will not be completed for another 8 to 13 years (Hun, 1998b). To help the states, the USEPA organized a TMDL Advisory Committee in 1996 to coordinate the development of TMDLs and to ensure that the proper stakeholders are involved.

The CWA also addresses oil and hazardous substances. It expressly prohibits the discharge of oil or hazardous substances in harmful quantities to the waters of the United States. The USEPA has classified 300 hazardous substances according to the levels of danger they present to health and the environment. These substances are listed by their Chemical Abstract Number (CAS#) and are provided by the USEPA. If a discharge of any of these hazardous substances or oil occurs that is above the reportable quantity (RQ) level, the National Response Center (NRC) must be notified. Severe penalties, including jail and fines of up to $10,000, can be issued for failing to notify the NRC. Also, the CWA places the responsibility of all costs of cleanup with the dischargers.

One final aspect of the CWA focuses primarily on pretreatment programs. The majority of CWA regulations eliminate direct discharges into U.S. waters; however, indirect pollution sources that discharge into the collection system of a downstream treatment plant are also regulated through pretreatment requirements. Because these sources of pollution do not require an NPDES permit, certain standards have been created to prevent industrial pollutants from passing untreated through the downstream treatment process. These standards are typically regulated on the local level. Any violation of the CWA is subject to enforcement by the USEPA, including civil fines and/or criminal penalties. The CWA also gives citizens the right to sue dischargers which helps prevent unauthorized releases.

1.2.4 Safe Drinking Water Act (SDWA)

The Safe Drinking Water Act (SDWA) was passed in 1974 and was significantly amended in 1996. The goal of the SDWA is very straightforward: to protect the public from harmful substances in their drinking water. The regulation gives the USEPA the authority to develop drinking water standards for each substance. To date, rules for over 80 contaminants (i.e., toluene, PCBs, lead, cadmium, etc.) have been promulgated. National Primary Drinking Water Regulations (NPDWRs) are set for each contaminant. In order for a contaminant to be regulated by the SDWA, it must satisfy the following conditions:

- The contaminant must have adverse effects on human health.
- The pollutants must be likely to occur in public drinking water systems.
- Regulations will have positive impacts on reducing the amount of contaminants found in the drinking water supply.

The USEPA is required to develop maximum contaminant level goals (MCLGs) for each contaminant and then set the maximum contaminant levels (MCLs) in the NPDWR as close to the MCLGs as feasibly possible, based primarily on cost and available technology. The SDWA does state, however, that every NPDWR must present different technologies that will meet the MCLs. If testing is not easily available for a particular contaminant, the USEPA can establish a specific requirement that a certain treatment technology be used. For example, because *Cryptosporidium* is not easily detected in water, filtering is often required for utilities that use surface water as a source.

The SDWA states that the USEPA must decide whether or not to regulate at least five different contaminants every 5 years, and that every 5 years they must publish a list of chemicals from which these contaminants are to be selected. This list is known as the Drinking Water Contaminant Candidate List and is subject to public review and comment (Pontius, 1998). To validate the analysis of the USEPA, the SDWA requires that they consult with the Science Advisory Board. Even though the USEPA must first select chemicals to review that pose the greatest threat to health, the SDWA allows the regulation of a chemical without scientific proof of danger should there be a valid health threat. However, every analysis must consider a benefit/cost study before regulation of a contaminant occurs.

1.2.5 Toxic Substances Control Act (TSCA)

The Toxic Substances Control Act (TSCA) was passed in 1976 after much debate in Congress, despite intense lobbying by chemical manufacturers (Davis, 1993). It was the first environmental law that tried to look at the other end of the pipe. In other words, it tries to prevent the manufacture of toxic substances rather than try to control their discharge. Prior to the TSCA, the USEPA was solely in a reactive stance and able to regulate chemicals only after the damage was already done.

The role of the TSCA is to identify all commercially manufactured chemicals that come in contact with the general population and then to regulate the chemicals, based on their characteristics. All newly created chemicals are reviewed and approved before they can be manufactured in or imported into the United States. Although over a million chemicals are known to exist, less than 100,000 are commercially produced in or imported into the United States. Pesticides, tobacco, drugs, food products, cosmetics, and certain nuclear materials are excluded from the TSCA and are regulated under separate laws.

One of the first tasks required by the TSCA in 1976 was for the USEPA to assemble a list of all commercially produced chemicals already in existence in the United States. This information was largely provided by the various chemical manufacturing companies and included over 60,000 substances. Since development of the initial list, more than 25,000 chemicals have been added. Although only new chemicals are required to go through the review and approval processes, the TSCA authorizes complete testing of existing chemicals.

Before a company or organization can manufacture or import a new chemical, it must provide the USEPA with a Pre-Manufacture Notice (PMN) and must first identify whether the chemical can be considered new by comparing it to the USEPA's chemical inventory. The PMN must include a chemical description, plan for production, list of intended uses, and effects on health and the environment. Substances used only in research and development or for export do not require a PMN.

It is the responsibility of the USEPA to determine the hazards to health and environment of all existing chemicals, and the agency can also impose restrictions on the manufacture of certain chemicals; however, the burden is on the USEPA to show any possible harm, thereby protecting industries from unnecessary regulations. The TSCA does authorize the USEPA to require that a manufacturer test its chemicals if it feels that a concern for human health or the environment exists. Again, to protect industry, the USEPA must devise the testing procedure. If the USEPA determines that an existing chemical is being used in new ways, a Significant New Use Rule (SNUR) can be issued. The SNUR requires a PMN to be written by the manufacturer and submitted to the USEPA.

With so many chemicals in existence, the USEPA has had to set priorities for testing chemicals. First, low-volume chemicals, which are produced or imported at a rate of less than 10,000 pounds per year, were excluded. Approximately 25,000 chemicals are in this category. Second, polymers, which typically are not toxic, were dropped from the list, which left approximately 15,000 nonpolymer chemicals produced or imported in quantities from 10,000 pounds per year up to 1 million pounds per year. As a result, the USEPA has focused its testing on the 3000 to 4000 high-production-volume (HPV) and nonpolymer chemicals (USEPA, 1998b).

The testing of just one chemical can take a team of scientists 2 to 3 years and require up to 300 laboratory animals and as much as $300,000 or more;

therefore, the TSCA also created the Interagency Testing Committee (ITC) to determine which potentially toxic chemicals to analyze first. The ITC is made up of representatives from the Council on Environmental Quality, the Department of Commerce, the National Science Foundation, the National Institute of Environmental Health Sciences, and the National Institute for Occupational Safety and Health. If existing or new chemicals "present unreasonable risk of injury to health or the environment," then the USEPA can:

- Prohibit manufacture of the chemical.
- Prohibit or limit certain uses.
- Dictate quantity and concentration limits for its manufacture.
- Require quality control measures during its manufacture.
- Require tests to show compliance with regulations.
- Establish recordkeeping requirements.
- Require labeling and public disclosure.

The burden to show unreasonable risk is on the USEPA, although it has been very difficult for them to enforce these regulations. Included in the Regulation of Unreasonable Risk section of the TSCA is specific mention of polychlorinated biphenyls (PCBs). These chemicals were extensively used in commercial and industrial manufacturing beginning in 1929. PCBs have various applications in cooling electrical equipment, heat transfer systems, and hydraulic fluid. Not until the 1970s did the effects of PCBs on health and the environment begin to be understood. PCBs are of significant concern because of their property of bioaccumulation. The chemical is absorbed into the tissue of humans and wildlife and is not released. As it accumulates, the risk for cancer, birth defects, skin lesions, and liver problems increases.

By 1979, the production of PCBs had been phased out because of the strict regulations of the TSCA. Only totally enclosed products that include PCBs are exempted. Also, a threshold of 50 parts per million (ppm) was established by the USEPA, as concentrations under that level do not cause unreasonable risk. Since the inception of these regulations, high levels of PCB concentrations have dropped from 12% of the American population in 1979 to nearly 0% in the late 1980s (Rosenbaum, 1995). In addition to PCBs, the TSCA has also been used to help reduce the unreasonable risk of asbestos in the environment. In 1986, the TSCA was amended to include the Asbestos Hazard Emergency Response Act. The primary function of this act was to remove asbestos from the nation's schools, and it was passed in response to studies that found asbestos to be an airborne carcinogen. In 1989, the USEPA issued regulations that would phase out all uses of asbestos in commercial products by 1997.

The TSCA contains important reporting and recordkeeping requirements. Any manufacturer or importer that finds a chemical substance to be a hazard to health or the environment must send immediate notification to the USEPA. Also, they must keep all records pertaining to adverse reactions to health or the environment for 5 years. Additionally, all employee allegations regarding adverse affects must be kept for 30 years.

Penalties associated with the TSCA can be very steep. Civil penalties of up to $25,000 per day can be issued, and criminal prosecution may also be pursued. The greater the damage to the environment, the greater the fines, which increase accordingly. The TSCA Civil Penalty Policy has been created to determine appropriate penalties for each violation. Proposed penalties have been as high as $17 million.

1.2.6 Resource Conservation and Recovery Act (RCRA)

The Resource Conservation and Recovery Act (RCRA) was originally passed in 1976 with little fanfare while the Congress was debating the TSCA. It was amended in 1978 and again in 1984 (the Hazardous and Solid Waste Amendments). RCRA is the broadest federal law covering the management of solid waste, and it has established a "cradle to grave" ideology. It regulates waste through all aspects of its life, from waste generators to storage facilities, transportation, treatment, and finally disposal. An important goal of the RCRA is to reduce or eliminate the generation of hazardous waste. Although no regulatory requirements for hazardous waste reduction have been created, the USEPA is counting on the high cost of compliance with the RCRA to provide economic incentive for such a reduction (Duke and Masek, 1997). This regulation focuses on two types of solid waste: *nonhazardous municipal waste* and *hazardous waste*, where solid waste is defined as any "garbage, refuse, sludge, and other discarded material, including solid, semisolid, liquid, or contained gaseous material."

Subtitle D of RCRA covers nonhazardous solid waste generators, transporters, treatment facilities, storage facilities, and disposal sites. Extensive rules governing municipal solid waste landfills (MSWLs) have been developed, including how to design, construct, operate, monitor, and close a landfill, as well as leachate systems and caps. The nonhazardous waste regulation is left for the states.

Subtitle C of RCRA is the primary vehicle for managing hazardous waste in the United States. Hazardous waste is defined as any solid waste, or a combination of solid wastes, that because of its quantity, concentration, or physical, chemical, or infectious characteristics may:

- Cause or significantly contribute to an increase in mortality or an increase in serious irreversible or incapacitating reversible illness.
- Pose a substantial present or potential hazard to human health or the environment when improperly treated, stored, transported, disposed of, or otherwise managed.

The two classes of hazardous waste as defined by RCRA are *characteristic waste* and *"listed" waste*. Characteristic waste is defined by the properties it exhibits. The four characteristic properties are as follows:

- *Ignitability*, which is determined by the flash point of the substance
- *Corrosivity*, which occurs when the waste has a pH less than 2 or greater than 12.5.
- *Reactivity*, which is based on the ability of the waste to rapidly change its state
- *Toxicity*, which is determined using the toxicity characteristic leaching procedure (TCLP)

"Listed" waste is any waste that contains a substance that is "listed" by the USEPA as hazardous. This type of waste has been listed based on the waste's "toxicity, persistence, and degradability in nature, potential for accumulation in tissue, and other related factors such as flammability, corrosiveness, and other hazard characteristic." Rules have been developed by the USEPA to ensure proper disposal of these types of hazardous waste. The mixture rule states that any substance mixed with a "listed" hazardous waste becomes a hazardous waste. If it is not a "listed" waste, but instead a characteristic waste, and the mixture does not exhibit any of the characteristics, the mixture is not considered hazardous. The "derived from" rule states that any waste derived from the treatment of a "listed" hazardous waste remains a hazardous waste. Similar to the mixture rule, if the by-product of a characteristic waste does not exhibit any of the hazardous characteristics, it is not considered hazardous.

The USEPA established a Hazardous Waste Identification Rule (HWIR), which allows certain types of low-risk waste listed as hazardous by the USEPA to be exempt from hazardous waste regulations, as long as they can be safely handled as solid waste (USEPA, 1998a). The agency established risk level criteria according to affects of the waste on health and the environment. The HWIR has saved a significant amount of money on treatment, storage, and disposal and encourages pollution prevention and treatment. In the past, there was no reason to treat a "listed" hazardous waste if the by-products would still be considered hazardous.

Another aspect of the RCRA covers treatment, storage, and disposal (TSD) permits. Every TSD business is required to obtain a TSD permit and abide by certain minimum standards. These detailed standards are provided in the RCRA and include rules for design, construction, operation, and closure of TSD facilities. Each TSD must also develop a Preparedness and Prevention report, as well as a Contingency Planning and Emergency Preparedness report. Typically, states handle the TSD permitting program; however, their standards can be no less stringent than those of the USEPA.

Because RCRA regulations begin at the "cradle," or generation, of hazardous waste, any business or organization that generates more than 100 kg of hazardous waste in any one month qualifies as a hazardous waste generator and requires a USEPA generator identification number. As the constant removal of hazardous waste from a generation facility could become incredibly expensive, the RCRA allows hazardous wastes to be stored for up to 90 days on-site before it must be removed. If it remains on-site for longer than 90 days, the facility automatically becomes a hazardous waste storage facility and must comply with the stricter TSD rules (Davenport, 1992).

The RCRA has also outlined a set of systematic rules governing the transport of hazardous waste. A detailed manifest system was established, where a manifest is to be prepared for each shipment of hazardous waste. The manifest includes information on the generator, the nature of the waste, and the quantity. Each transporter of the waste is required to sign and verify the manifest and keep a copy. When the waste reaches its destination, a copy signed by all parties is returned to the origination point to verify arrival of the waste. This system ensures that no waste is lost or disposed of improperly.

A final section of the RCRA concerns underground storage tanks (USTs). Under the RCRA, the USEPA requires that all USTs be protected against spills, corrosion, and overfills within 10 years of the promulgation date. Over 1 million USTs are known to exist in the United States, and many leak hazardous substances (e.g., petroleum products) into the ground and groundwater (Danner et al., 1998). The RCRA requires that all facilities with USTs must do one of the following:

- Upgrade existing USTs to meet the new standards.
- Replace the UST completely.
- Close the UST.

Many small tanks and wastewater treatment plant tanks have been exempt from these rules. Penalties for noncompliance with the UST guidelines can be up to $10,000 per day. All other violations of the RCRA can carry similar penalties, including steep fines and jail time. Civil penalties can be assessed up to $25,000 per day per violation of the RCRA. Criminal penalties up to $250,000 and 15 years in jail can be imposed for knowingly putting someone in imminent danger by violating the RCRA. Similar to other environmental laws, the RCRA authorizes citizen suits in the event that the USEPA fails to implement the RCRA.

1.2.7 Hazardous and Solid Waste Amendments (HSWA)

In 1984, Congress passed amendments to the RCRA that are known as the Hazardous and Solid Waste Amendments (HSWA). The HSWA were filled with specific deadlines and requirements to ensure that the USEPA implemented the RCRA (Rosenbaum, 1995). In the early 1980s, the Reagan administration cut USEPA spending and effectively slowed the work of the agency down to a crawl. For example, as mentioned previously, the RCRA required the USEPA to develop a list of harmful wastes to be regulated. Although over 1000 chemicals were likely to qualify for the list and 450 were identified in 1980, only five new types of waste were identified in the next 6 years (Rosenbaum, 1995).

One of the main components of the HSWA is the land disposal ban for hazardous wastes. This land ban states that no hazardous waste can be disposed of on land until it has been treated to have concentrations of chemicals under a certain level. The USEPA was given the responsibility to create these levels and provide a proper treatment method for each waste. The universe of hazardous waste was broken down into three categories; these groups of waste were evaluated, and specific treatment methods and standards were developed. The treatment standards have been based primarily on available technology rather than on potential risks (Hendrichs, 1991). If, after treatment, the waste no longer meets any of the criteria under which the waste was listed, it can be unlisted. This process requires an extensive petition to be filed with the USEPA and can take several years to be approved.

In addition to the land ban, the 1984 HSWA also established two other regulations:

- A regulatory framework for corrective action of waste releases from RCRA-regulated sites; the program is similar to CERCLA, except that it is typically handled by states instead of the federal government (Martino and Kaszynski, 1991).

- The prohibition of exporting hazardous waste without express permission from the government of the receiving country.

1.2.7.1   Comprehensive Environmental Response, Compensation and Liability Act (CERCLA)

The Comprehensive Environmental Response, Compensation and Liability Act (CERCLA, or Superfund) was passed in 1980. It was the first law to focus on the necessary environmental response to uncontrolled hazardous waste sites or potential hazardous waste releases throughout the country. The Superfund law was significantly amended in 1986 by the Superfund Amendments and Reauthorization Act (SARA), which is discussed in the next section, and was last reauthorized in 1990. It has been scheduled for reauthorization for the past several years, but agreement on changes to the act has not yet been reached.

In 1980, President Carter pushed hard for the CERCLA to be passed — mainly in response to various environmental tragedies in the news, such as the Love Canal toxic waste site near Niagara Falls and the valley of drums in Kentucky. People wanted someone to be held accountable for the toxic waste sites that littered the country. The CERCLA focuses on two issues: responding to existing uncontrolled hazardous waste sites and preventing any potential areas. An important distinction must be made, however. The goal is not necessarily to remove the waste, but to remove the *harm*.

First of all, the CERCLA requires all hazardous waste releases over a prescribed threshold, known as reportable quantities (RQs), to be reported to National Response Center. Action is taken from that point to determine if it will be a CERCLA site. The CERCLA also established development of a National Contingency Plan. This plan includes all procedures for handling hazardous waste in the United States. The act also requires the creation of an uncontrolled hazardous waste site ranking system (HRS). The HRS determines if a site should be placed on the National Priorities List (NPL), which is a list of all the Superfund sites.

The HRS system is based on risk to health and the environment. The criteria examined include the groundwater migration pathway, surface water migration pathway, soil exposure pathway, and air migration pathway (Hendrichs, 1991). The ranking attempts to quantify the risk each site poses on a relative scale. Only those sites placed on the NPL will receive CERCLA funds; however, regulations in the CERCLA can still be applied to non-NPL sites.

The actions at NPL sites are coordinated through a National Response Team to work with other agencies, such as the Federal Emergency Management Agency (FEMA) and the Nuclear Regulatory Agency (NRC), so that no overlapping takes place. The main effectiveness of the CERCLA is in the assignment of financial obligation to the potentially responsible parties (PRPs). A PRP is anyone who generates, stores, transports, treats, or disposes of hazardous waste on the affected site. Another important part of the CERCLA establishes a fund to finance costs and expenses incurred in cleanup that will eventually be assigned to the PRPs. The government wanted to immediately assign the financial responsibility of the cleanups to the PRPs; however, it knew that very few organizations would have the money available to begin the remediation process. Therefore, these funds cover costs until they can be reimbursed. Also, it was clear from the beginning that the federal government would have to carry some of the financial burden due to insolvent or nonexistent companies. The preliminary goal was an 80/20 split, where the PRPs would finance 80% of the costs and the federal government would cover the rest.

The fund is supported by four taxes: a petroleum tax, hazardous chemicals tax, imported substances tax, and environmental tax. The original 1980 CERCLA Act was created with a trust fund of $1.6 billion by Congress. The 1990 reauthorization increased the aggregate cap on Superfund revenue to $11.97 billion (Hendrichs, 1991). In the next reauthorization, the Clinton administration was pushing for "orphan share" funding from a separate account to cover contributions for insolvent or defunct parties (USEPA, 1997). The goal of the CERCLA, however, is to clean up the hazardous waste sites. It authorizes the USEPA to order PRPs to remediate waste sites or remove hazardous substances. The USEPA or PRPs can develop a preliminary nonbinding allocation of responsibility (NBAR), which divides the costs among the PRPs. As the name implies, the allocation is nonbinding and can be shifted at any time.

Relatively small PRPs, known as *de minimis* parties, are allowed early release from CERCLA requirements. The USEPA has completed settlements with 15,000 small-volume contributors at hundreds of Superfund sites. By settling with the USEPA, small polluters are not dragged into the problems of bigger polluters.

Hazardous wastes must be either removed or remediated through long-term remedial action. Removal is merely the elimination of any further release of the hazardous waste. The three types of removal actions are:

- Classic emergency removal actions of a waste that poses an immediate threat to health or the environment (e.g., truck spill on highway)
- Time-critical removal actions, which are taken when a response must be made within 6 months of an action memo (e.g., spill in an industrial site in a residential area)
- Non-time-critical removal actions, where no significant change is expected at the waste site (e.g., remote industrial site)

If it deems it necessary, the USEPA can require an expedited response action (ERA) at a site that would require an immediate engineering evaluation/cost analysis (EE/CA). After a removal action is completed or determined unnecessary, the long-term remedial action must be undertaken. First, a remedial investigation/feasibility study (RI/FS) is performed to determine a proper course of action. This can involve:

- Assessing the site conditions
- Evaluating alternatives to select a remedy
- Performing pilot treatability studies

The USEPA established a National Remedy Review Board in 1995 to review all remedies. This board has saved an estimated $31 million, and future cost reductions of more than $725 million are expected (USEPA, 1998b). The remedial investigation (RI) will also include identification of "applicable or relevant and appropriate requirements" (ARARs). These are remediation standards, standards of control, or other criteria or limitations developed by federal or state law. Applicable requirements are those that have been previously used at a CERCLA site for the same waste. Relevant and appropriate requirements are those not formerly used for waste at a CERCLA site but which address the problem. Advisories and guidance to be considered can also be issued, but they are not as binding.

Among the many different types of ARARs are ambient or chemical-specific requirements, which can be levels set by other laws, such as MCLs, National Ambient Air Quality Standards (NAAQS), or CWA, CAA, and TSCA regulations, and the long-term remedial action would have to meet those goals. Because not that many ambient or chemical-specific requirements have been established, other types of ARARs must usually be identified. An alternative is for the USEPA to use carcinogenic potency factors or reference doses to set the proper level of treatment. It must be remembered, though, that each ARAR is specific to the remedial activity and not the pollutant.

A feasibility study is necessary after the remedial investigation to make sure the proper remedial action has been selected. Once that has been completed, the USEPA publishes a Superfund Record of Decision (ROD), which describes the remedial action selected. The next step is the remedial design/remedial action phase, which can include:

- Storage and confinement
- Perimeter protection using dikes, trenches, ditches, and clay cover
- Cleanup of released hazardous substances
- Recycling, reuse, diversion, destruction, or segregation of reactive wastes
- Dredging or excavations
- Collection of leachate and runoff
- On-site treatment
- Any monitoring reasonably required to assure protection of public health and the environment

Extreme care must be taken in the remedial design so that air emissions and water discharges (point or non-point) from CERCLA sites remain within the limits of the CAA, CWA, and SDWA regulations. Often, fugitive emissions (non-point source) must be considered that are not directly tied to a particular manufacturing or other process. The remedial action is typically the lengthiest stage of the process, and it involves not only remediation of the site, but also the cost of relocating residents, businesses, etc.

Over 1400 Superfund sites are on the NPL list. Table 1.1 compares the progress of the program from January 1993 through the end of fiscal year 1997. The program had completed the cleanup of over 900 sites by 2001 and reported a 20% reduction in cleanup duration, or 2 years on average. PRPs have paid 75% of long-term Superfund costs — over $12 billion. Legal decisions, such as holding previous owners and operators liable for past pollution prior to the CERCLA, will continue to ensure the success of the program.

TABLE 1.1

Comparison of Superfund Progress (1993 vs. 1997)

	1993	1997
Remedial assessment not begun	73	55
Study under way	367	180
Remedy selected	92	63
Design under way	213	124
Construction under way	380	477
Construction completed	155	498

Source: USEPA, 1998b.

1.2.7.2   Superfund Amendments Reauthorization Act (SARA)

The Superfund Amendments Reauthorization Act (SARA) was passed in 1986 and made two major changes to the original CERCLA. First, it established the Emergency Planning and Community Right to Know Act (EPCRA) in Title III; second, it increased spending for Superfund sites to $8.5 billion and provided new cleanup standards that use the best available technologies (Rosenbaum, 1995).

The primary goal of the EPCRA was to develop proper emergency planning for hazardous waste releases and give communities the right to know what hazardous substances are being manufactured or processed in their area. To make sure that communities know what chemical substances are being manufactured, processed, or stored in their vicinity, the EPCRA requires businesses and organizations to submit Material Safety Data Sheets (MSDSs) for all OSHA-regulated substances at its facility. The MSDSs can include:

- Facility information
- Identity of the chemical and its hazardous components
- Physical and chemical characteristics of hazardous substances
- Fire and explosion hazard data
- Reactivity and health hazard data
- Precautions for safe use and control measures

Also, Form R must be submitted to report the release of any toxic chemicals. The EPCRA requires the USEPA to track releases and make the information available to the public. In addition, facilities must submit yearly Toxic Release Inventory (TRI) reports for all permitted and non-permitted discharges (Davenport, 1992). The second goal of the EPCRA is emergency planning, primarily to improve preparedness on the state and local levels in the event of a toxic chemical release. SARA requires every state to create a State Emergency Response Commission (SERC) and also to identify local districts. These local districts are usually county governments, and they are responsible for having a Local Emergency Planning Committee (LEPC). Each LEPC must develop an emergency plan and take into account all facilities in its district that manage toxic chemicals.

Section 302 of the EPCRA establishes threshold-planning quantities (TPQs) for extremely hazardous substances (EHSs). Any facility that goes over the TPQ must submit a report to the SERC and LEPC. Section 304 requires that facilities must immediately report any release of an EHS or CERCLA hazardous substance over the reportable quantity levels. These reports must be submitted to the SERCs and LEPCs for all areas possibly affected. An immediate phone notification is required with a follow-up written report. Also, the EPCRA requires that all facilities that produce MSDSs under OSHA must submit a list of chemicals to the SERC, LEPC, and local fire departments. Reports are only required if the substance is found over normal threshold quantity levels.

A similar report is required of any facility with MSDS requirements under OSHA for hazardous chemicals. It must submit one of two different emergency and hazardous chemical inventory forms. Tier I provides aggregate information on each hazardous chemical with the type of health and physical hazards it presents and the estimates of daily quantities at the facility (maximum and average daily amounts). Tier II is the most complete report and contains all information provided in Tier I. It is commonly preferred by government agencies and includes chemical-specific information, such as the following:

- Chemical or common name
- Physical and health hazards
- Estimates of daily maximum and average amount present on-site
- Number of days the chemical is stored on-site
- Manner of storage and specific location on-site

The EPCRA is enforceable with fines of up to $25,000 per day and 2 years in prison depending on the violation. The focus of the USEPA to date has been on violations of the immediate notification of release rule.

1.2.7.3   Clean Air Act (CAA)

The Clean Air Act was originally passed in 1963 and was primarily a source of government funding for air pollution control. A follow-up law in 1965, the Motor Vehicle Air Pollution Control Act, regulated the emissions from new vehicles and was the first action taken by the government to control air pollution. The Clean Air Act Amendments of 1970 and 1977 established what we now consider to be the true goals of the CAA.

The CAA created a system for the federal government to set goals for the air quality of the country and identify a means of achieving them. Passage of this law showed the first need for a federal Environmental Protection

Agency. The CAA required the government to establish the National Ambient Air Quality Standards (NAAQS), which would protect the public health from air pollution. These standards would be identified by the USEPA and set the maximum concentrations for each pollutant in the ambient, or outside, air. The CAA identified six criteria pollutants to be regulated:

- Carbon monoxide
- Sulfur dioxide
- Nitrogen dioxide
- Ozone
- Lead
- Total suspended particulates (later amendments modified the regulated particulates to those with a diameter less than 10 μ)

The CAA was amended again in 1990 to include the following:

- More stringent motor vehicle standards
- Additional toxic air pollution regulations
- A new system to reduce sulfur dioxide and nitrogen dioxide emissions
- A unified operating permit program
- Phase-out of chlorofluorocarbons (CFCs)
- Additional penalties for violations of the CAA

The toxic air pollution regulations in the 1990 amendments contain requirements expressly for hazardous waste emissions. The CAA established the National Emission Standards for Hazardous Air Pollutants (NESHAPs), which applies to those substances that are harmful to public health. A list of 188 substances has been developed that the USEPA must regulate under the CAA. Similar to nonhazardous pollutants, a source is designated as major if it discharges over 10 tons per year of any one of the 188 listed substances, or over 25 tons per year of any combination of substances. All other stationary sources of hazardous air pollutants are considered area sources.

The USEPA is responsible for creating and enforcing the NESHAPs for all hazardous air pollutant sources. The CAA states that new or existing major sources must have emission standards based on the maximum available control technology (MACT) to reduce hazardous air pollutant emissions. The MACT standards are based on the performance of the best 12% of the control devices in the same source category. These MACT emissions requirements were extended in 1997 to cover wastewater biosolid incinerators at publicly owned treatment works (POTWs) that have the potential to discharge cadmium, lead, and mercury (Richman, 1997).

Area sources may be assigned emissions standards based on generally available control technologies (GACTs); however, the USEPA can require MACT standards, depending on the circumstances. In order to ensure that the USEPA does not hamper an industry, the CAA included a MACT Hammer clause. This provision states that the USEPA must review permits on a case-by-case basis if they have not created a standard for a particular hazardous air pollutant. This would likely result in standards that are more overprotective than they would be after a normal review process. The CAA required MACT standards to be set for all major source categories on a schedule to be completed in a 10-year period. A list of 173 major source categories that emit one or more of the 188 listed pollutants has been published.

The CAA also directed the USEPA to develop a list of 100 substances that are dangerous to human health if accidentally released. Using this list, the CAA requires every stationary source using any of these substances in excess of the threshold level to have a risk management plan to reduce the chances and/or severity of an accidental release. The CAA also established the Chemical Safety Board to investigate and report the accidental releases of hazardous pollutants, in much the same way that the National Transportation Safety Board (NTSB) investigates highway and airplane crashes. Great strides have been made in the implementation of the Clean Air Act. Major pollutant reductions were achieved with the removal of lead and other additives from automobile fuel, as well as requiring catalytic converters on vehicles to destroy the remaining hydrocarbons that are not burned in the engine. Unfortunately, at least 15 "pollutants of concern" are still being deposited in major U.S. waters through air pollution (Hun, 1998a). This list includes cadmium, dichlorodiphenyltrichloroethane (DDT), PCBs, and mercury.

1.3 Summary

The environmental laws governing hazardous waste are a relatively recent effort by the U.S. government. They are still undergoing fine-tuning, and some aspects still require major adjustments. Most of these environmental regulations are less than 20 years old, and we have already seen major improvements. Air pollutant levels have been reduced, hazardous waste sites are being remediated, and rivers and bays once too polluted to swim or fish in are now clean. Most importantly, the population is becoming more aware of the vital role that the environment plays in our lives. The philosophy of the 1970s was that all potential problems imaginable had to be prevented. The priority of the 1980s was renewal of the economy and a trend toward cost-effective regulations, but now it is recognized that the possibilities to be safeguarded against are too numerous for this approach to be affordable.

The common theme of the environmental movement is that good environmental quality is good for the economy in the long run. The short-run economic dislocation problems were largely ignored in the 1970s. Municipalities and corporations were expected to pay whatever was needed to correct past environmental problems and to provide future environmental protection, no matter the price. The balancing of economic and environmental goals is likely to take the form of moderation in achieving some environmental goals that adversely affect economic activities. To the extent possible, regulations will move away from the command-and-control type of approach generally used today. Also, more practical emission standards with built-in economic incentives will be established so that cost-effective pollution control technologies can be used that provide overall lower pollutants over the life of the equipment.

References

Adler, R. et al., *The Clean Water Act: 20 Years Later*, Island Press, Washington, D.C., 1993.

Corbitt, R.A., *Standard Handbook of Environmental Engineering*, McGraw-Hill, New York, 1990.

Danner, B. et al. Don't get buried alive, *Environ. Protect.*, July, 1998.

Davenport, G., The ABCs of hazardous waste legislation, *Chem. Eng. Progr.*, May, 45–50, 1992.

Davis, C., *The Politics of Hazardous Waste*, Prentice Hall, Englewood Cliffs, NJ, 1993.

Duke, L.D. and Masek, B., Evaluating progress in toxic pollution prevention for two industrial sectors, 1987–1993, *Environ. Eng. Sci.*, 14, 81–95, 1997.

Findley, R. and Farber, D., *1997 Supplement to Environmental Law*, West Publishing, Minneapolis, MN, 1997.

Hendrichs, R., Law. Part II. RCRA, CERCLA, and the Clean Air Act, *Res. J. WPCF*, June, 1991.

Hun, T., Air pollutants harm U.S. waters, USEPA study says, *Water Environ. Technol.*, February, 1998a.

Hun, T., Clinton plan earmarks $7.3 billion to address problems by watershed, *Water Environ. Technol.*, April, 1998b.

LaGrega, M.D., Buckingham, P.L., and Evans, J.C., *Hazardous Waste Management*, McGraw-Hill, New York, 1994.

Martino, L. and Kaszynski, G., A comparison of the CERCLA response program and the RCRA corrective action program, *Hazardous Waste Hazardous Mater.*, 8(2), 1991.

Millano, E., Hazardous wastes, *Water Environ. Res.*, Literature Review, 1997.

Moore, R. and Vicory, A., NWQS review presents opportunity for wastewater treatment industry, *Water Environ. Technol.*, March, 1998.

Pontius, F., New horizons in federal regulations, *J. AWWA*, March, 38–50, 1998.

Richman, M., EPA may hold biosolids incinerators to strict CAA emissions standards, *Water Environ. Techno.*, May, 1997.

Rosenbaum, W., Environmental politics and policy, *Congr. Q.*, 1995.

Sears, T., MACT attack, *Environ. Protect.*, August, 1998.

Simonsen, C., *Essentials of Environmental Law*, Pearson Publications, Dallas, TX, 1996.

Sutherland, D., Superfund awakens in state supreme courts, *Environ. Manager*, January, 1998.

U.S. Congress, *Clean Water Act*, 42 U.S.C. 1251 et seq. (1948–1987).

U.S. Congress, *Clean Air Act*, 42 U.S.C. 7401 et seq. (1963–1990).

U.S. Congress, *National Environmental Policy Act*, 42 U.S.C. 4321 et seq. (1970–1975).

U.S. Congress, *Safe Drinking Water Act*, 42 U.S.C. 300f et seq. (1974–1996).

U.S. Congress, *Resource Conservation and Recovery Act*, 42 U.S.C. 6901 et seq. (1976–1986).

U.S. Congress, *Toxic Substances Control Act*, 15 U.S.C. 2601 et seq. (1976–1986).

U.S. Congress, *Comprehensive Environmental Response, Compensation and Liability Act*, 42 U.S.C. 9601 et seq. (1980–1990).

U.S. Congress, *Superfund Amendments and Reauthorization Act*, 42 U.S.C. 11001 et seq. (1986).

USEPA, *The Clinton Administration's Superfund Legislative Reform Principles*, U.S. Environmental Protection Agency website, May 7, 1997. www.epa.gov.

USEPA, Superfund Cleanup Figures, U.S. Environmental Protection Agency website, April, 1998a. www.epa.gov/oerrpage/superfund/action/process/mgmtrpt.htm.

USEPA, Statement of Carol Browner, Administrator of the USEPA, Before the Subcommittee on Finance and Hazardous Materials, U.S. House of Representatives, U.S. Environmental Protection Agency website, March 5, 1998b. www.epa.gov.

2

Environmental Hazardous Wastes

2.1 Introduction

As a result of worldwide urbanization and industrialization, numerous toxic and hazardous organic compounds have found their way into surface and ground waters, both of which are the source of drinking water. The disposal of over 40,000 organic chemicals used by various industries in the United States has resulted in a broad range of hazardous waste problems. The amount of industrial wastewater containing these recalcitrant pollutants is increasing significantly. Many of these chemicals are resistant to degradation and pose a potential health threat to the human population. In addition, many hazardous waste materials are recalcitrant, normally nonbiodegradable, and even toxic to microorganisms; therefore, physicochemical treatment techniques are better alternatives than biological treatment.

Since 1970, one of the most widespread industrial practices has been the halogenation of organics, which produces organic solvents, pesticides, chlorofluorocarbons (CFCs), and polychlorinated biphenyls (PCBs). The halogenation of hydrocarbons yields compounds of lower flammability, higher density, higher viscosity, and improved solvent properties compared to nonhalogenated solvents. For example, 46.5% of chlorine gas used in the United States was for the production of chlorinated organic compounds. Classified as derivatives of aliphatic hydrocarbons, halogenated solvents have been used extensively in a number of industrial processes, and over 400,000 tons of halogenated solvents are used annually for metal cleaning. Due to their higher density, high water solubility, and low degradability, chlorinated solvents are extremely mobile in groundwater.

Chlorinated solvents are commonly used in the manufacturing of pesticides. Carbon tetrachloride is a commercial product widely used in the United States for dry cleaning, metal degreasing, and fire extinguishers. It is also used for the production of CFCs and grain fumigation. Methylene chloride is commonly used in paint stripping, for which it is mixed with alcohols, acids, and amines. Methylene chloride is also used in the extraction of caffeine from coffee and other beverages. As a result, chlorinated organic pollutants can still be found at most contaminated sites because of high resistance to biodegradation under natural environmental conditions.

The most common nonhalogenated solvents are a group of petroleum distillates, aliphatic and aromatic hydrocarbons, alcohols, ketones, esters, and ethers. Nonhalogenated solvents have a number of industrial uses, such as cold cleaning, which includes metal degreasing, parts cleaning, and paint stripping. They have also been used as carriers for paints, varnishes, and printing inks. Prior to the RCRA, waste solvents were disposed of in landfills, sewers, and soil pits. After the Resource Conservation and Recovery Act (RCRA), they had to be disposed of through solvent recycling, fuel blending, and incineration. Hydrocarbons such as benzene, toluene, xylenes, and a number of alkylbenzenes and low-molecular-weight ketones are important classes of nonhalogenated solvents. Ketones consist of R and R' alkyl groups linked to a keto or carbonyl group. Other miscellaneous nonhalogenated solvents include glycols, such as ethylene glycol and propylene glycol; ethers, such as dimethyl ethers; and amines, such as isopropylamine.

2.2   Classification of Hazardous Pollutants

In general, hazardous wastes are classified as organics, heavy metals, and radioactive. Organic pollutants are further classified as halogenated/nonhalogenated volatile organic compounds (VOCs) and halogenated/nonhalogenated semivolatile organic compounds (SVOCs). These classifications help facilitate the selection of remediation technologies according to the treatability of each class in a specific contaminated media:

*   *Halogenated volatile organic compounds (HVOCs)* — Halogenated organic compounds contain molecules of chlorine, fluorine, bromine, and/or iodine. The nature of the halogen bond and the halogen itself can significantly affect the performance of a specific treatment technology. HVOCs are difficult to treat.

*   *Halogenated semivolatile organic compounds (SVOCs)* — Halogenated SVOCs may also contain molecules of chlorine, bromine, iodine, and/or fluorine. The degree of volatilization from halogenated SVOCs is much less than for HVOCs. The most common types of halogenated SVOCs include polychlorinated biphenyl (PCBs), pentachlorophenol (PCP), and hexachlorobenzene.

*   *Nonhalogenated volatile organic compounds (nonhalogenated VOCs)* — Nonhalogenated compounds do not have a halogen attached to them. Common types of nonhalogenated VOCs include acetone, styrene, and methanol.

- *Nonhalogenated semivolatile organic compounds (nonhalogenated SVOCs)* — Nonhalogenated SVOCs do not contain halogens. The degree of volatilization from nonhalogenated SVOCs is much less than for nonhalogenated VOCs. The most common types of nonhalogenated SVOCs include pyrene, fluorene, and dibenzofuran.

Each class of the aforementioned organic pollutants may include hundreds of substituted compounds. For example, chlorinated benzenes may include one hexachlorobenzene, a pentachlorobenzene, three dichlorobenzenes, and three trichlorobenzenes. Table 2.1 lists 100 priority pollutants classified by the USEPA.

Halogenated and nonhalogenated VOCs are found in many products such as gasoline, paints, paint thinners, and solvents that are used for dry cleaning and metal degreasing. Furthermore, halogenated and nonhalogenated SVOCs also have the same properties and behaviors as VOCs. These compounds are typically used in liquid form and readily evaporate. They may cause adverse effects on the environment and human health through contaminated soil and/or groundwater.

Table 2.2 lists the hydrophobic and electronic properties of substituents. The most common substituents include a wide variety of electron-withdrawing substituents, such as $-F$, $-Cl$, $-Br$, $-I$, $-NO_2$, $-SO_3H$, $-CN$, and $-COOH$, and electron-donating substituents, such as $-CH_3$, $-C_2H_5$, $-NH_2$, $-OH$, and $-OCH_3$. These substituents may form thousands of halogenated/nonhalogenated VOCs and SVOCs or non-VOCs.

2.3 Sources of Hazardous Waste

According to Toxic Release Inventory (TRI) reports, a total of 2.58 billion pounds of releases occurred in 1997, with two industries reporting more than half of that total. The chemical manufacturing industry was responsible for 797.5 million pounds of the total releases, and the primary metal industry reported a use of 694.7 million pounds. These amounts represented 30.9% and 27.0% the TRI, as illustrated in Figure 2.1.

The chemical manufacturing industry ranked as the number one source of hazardous waste, with 742.6 million pounds of on-site releases, 342.2 million pounds of air emissions, 106 million pounds of surface water discharges, and 215.8 million pounds of underground injection. The chemicals accounting for the largest amounts of underground injection by chemical manufacturing facilities were nitrate (40.6 million pounds) and ammonia (29.0 million pounds) compounds. The primary metal industry ranked second, with on-site releases of 405.9 million pounds. The paper product sector ranked third, with on-site releases of 228.8 million pounds, consisting mostly of air emissions of 193.8 million pounds.

TABLE 2.1

USEPA Priority Pollutants

1. Acenaphthlene	**Dichloropropane and dichloropropylene**	**Phthalate Esters**
2. Acrolein	32. 1,2-Dichloropropane	66. *bis*(2-Ethylhexyl) phthalate
3. Acrylonitriline	33. 1,2-Dichloropropylene (1,3-dichloropropene)	67. Butyl benzyl phthalate
4. Benzene	34. 2,4-Dimethylphenol	68. Di-*n*-butyl phthalate
5. Benzidine	**Dinitrotoluene**	69. Di-*n*-octyl phthalate
6. Carbon tetrachloride	35. 2,4-Dinitrotoluene	70. Diethyl phthalate
Chlorinated benzenes (other than dichlorobenzenes)	36. 2,6-Dinitrotoluene	71. Dimethyl phthalate
7. Chlorobenzene	37. 1,2-Diphenylhydrazine	**Polynuclear aromatic hydrocarbons**
8. 1,2,4-Trichlorobenzene	38. Ethylbenzene	72. Benzo[*a*]anthracene
9. Hexachlorobenzene	39. Fluoranthene	73. Benzo[*a*]pyrene
Chlorinated ethanes	**Haloethers (others than those listed elsewhere)**	74. 3,4-Benzofluoranthene
10. 1,2-Dichloroethane	40. 4-Chlorophenyl phenyl ether	75.Benz[*k*]fluoranthene
11. 1,1,1-Trichloroethane	41. 4-Bromophenyl phenyl ether	76.Chrysene
12. Hexachloroethane	42. *bis*(2-Chloroisopropyl) ether	77. Acenaphthylene
13. 1,1-Dichloroethane	43. *bis*(2-Chloroethoxy) methane	78. Anthracene
14. 1,1,2-Trichloroethane	**Halomethanes (others than those listed elsewhere)**	79. Benzo[*g,h,i*]perylene
15. 1,1,2,2-Tetrachloroethane	44. Methylene chloride	80. Fluorene
16. Chloroethane	45. Methyl chloride	81. Phenanthrene
Chloroalkyl ethers	46. Methyl bromide	82. Dibenzo[*a,h*]anthracene
17. *bis*(Chloroethyl)ether	47. Bromoform	83. Indeno[1,2,3-*c,d*]pyrene
18. *bis*(2-Chloroethyl) ether	48. Dichlorobromomethane	84. Pyrene
19. 2-Chloroethyl vinyl ether	49. Trichlorofluoromethane	85. Tetrachloroethylene
Chlorinated naphthalene	50. Dichlorodifluoromethane	86. Toluene
20.2-Chloronaphthalene	51. Chlorodibromomethane	87. Trichloroethylene
Chlorinated phenols	52. Hexachlorobutadiene	88. Vinyl chloride
21.2,4,6-Trichlorophenol	53.Hexachlorocyclopentadiene	**Pesticides and metabolites**
22. *p*-Chloro-*m*-cresol	54. Isophorone	89. Aldrin
23. Chloroform (trichloromethane)	55. Naphthalene	90. Dieldrin
24. 2-Chlorophenol	56. Nitrobenzene	91. Chlordane
Dichlorobenzenes	**Nitrophenols**	92. 4,4'-DDT

25. 1,2-Dichlorobenzene
26. 1,3-Dichlorobenzene
27. 1,4-Dichlorobenzene
Dicholorobenzidine
28. 3,3-Dichlorobenzidine
Dichloroethylenes
29. 1,1-Dichloroethylene
30. 1,2-*trans*-Dichloroethylene
31. 2,4-Dichlorophenol

57. 2-Nitrophenol
58. 4-Nitrophenol
59. 2,4-Dinitrophenol
60. 4,6-Dinitro-*o*-cresol
Nitrosamines
61. *N*-Nitrosodimethylamine
62. *N*-Nitrosodiphenylamine
63. *N*-Nitrosodi-*n*-propyl-amine
64. Pentachlorophenol
65. Phenol

93. 4,4′-DDE
94. 4,4′-DDD
Endosulfan and Metabolites
95. α-Endosulfan
96. β-Endosulfan
97. Endosulfan sulfate
Endosulfan and Metabolites
98. Endrin
99. Endrin aldehyde
Heptachlor and metabolites
100. Heptachlor
101. Heptachlor epoxide
100. Heptachlor

TABLE 2.2

Hydrophobic and Electronic Properties of Substituents

Electron Withdrawing	Log P	σ_m	σ_p	F	R	Electron Donating	Log P	σ_m	σ_p	F	R
Br	2.34	0.4	0.2	0.44	-0.17	NH_2	0.04	-0.2	-0.7	0.02	-0.68
Cl	2.35	0.4	0.2	0.41	-0.15	OH	0.59	0.12	-0.4	0.29	-0.64
F	1.62	0.3	0.1	0.43	-0.34	OCH_3	1.34	0.12	-0.3	0.26	-0.51
I	2.60	0.4	0.2	0.40	-0.19	CH_3	1.94	-0.1	-0.2	-0.04	-0.13
NO_2	1.91	0.7	0.8	0.67	0.16	$N(CH_3)_2$	1.66	-0.15	-0.83	0.1	-0.92
COC_6H_5	3.07	0.34	0.43	0.30	0.16	C_6H_5	3.20	0.06	-0.01	0.08	-0.08
COOH	1.58	0.4	0.5	0.33	0.15	OC_6H_5	3.5	0.25	-0.03	0.34	-0.35
CN	1.60	0.6	0.7	0.51	0.19	CH_2OH	0.25	0.00	0.00	0.00	0.00
CH_2Cl	1.66	0.11	0.12	0.10	0.03	C_2H_5	2.50	-0.1	-0.2	-0.05	-0.10
CHO	1.35	0.4	0.4	0.31	0.13	NHC_2H_5	1.56	-0.24	-0.61	-0.11	-0.51
$COCH_3$	1.35	0.4	0.5	0.32	0.20	$NHCH_3$	1.01	-0.30	-0.84	-0.11	-0.74
CF_3	2.36	0.43	0.54	0.38	0.19	OC_2H_5	1.86	0.10	-0.24	0.22	-0.44
SCN	—	0.4	0.5	0.36	0.19	C_3H_7	2.13	-0.07	-0.13	-0.06	-0.08
$CONH_2$	-0.01	0.28	0.36	0.24	0.14	$C(CH_3)_3$	3.56	-0.10	-0.20	-0.07	-0.13
CO_2CH_3	1.47	0.37	0.45	0.33	0.15	NHC_6H_5	2.85	-0.12	-0.40	-0.02	-0.38
$N(CH_3)_3$	4.48	0.88	0.82	0.89	0	OC_3H_7	2.53	0.10	-0.25	0.22	-0.45
H	1.48	0	0	0	0	H	1.48	0	0	0	0

Note: Log P is the partition coefficient; σ_m and σ_p are Hammett sigma constants at the *meta* and *para* positions, respectively; F and R are polar and resonance constants, respectively, proposed by Swain and Lupton (1968).

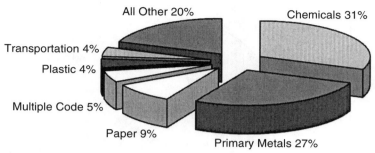

All Other 20%　　　　　Chemicals 31%

Transportation 4%

Plastic 4%

Multiple Code 5%

Paper 9%　　　Primary Metals 27%

FIGURE 2.1
TRI on-site releases.

The five different pathways for the release of chemical waste products in the environment are:

- *Fugitive air* refers to the release of chemicals into the air from on-site equipment leaks, evaporative losses from surface impoundments and spills, and building ventilation systems.
- *Stack air* refers to the release of chemicals into the air through on-site stacks, vents, ducts, pipes, or any confined air stream.
- *Water* refers to the release of chemicals into rivers, lakes, streams, oceans, and other bodies of surface water from all discharge points at the facility. This category includes the release from on-site wastewater treatment systems, open trenches, and stormwater runoff.
- *Underground* refers to underground releases and is defined as the injection of fluids into on-site subsurface wells for the purpose of waste disposal.
- *Land* refers to the release of chemicals onto the land. Land releases include landfills, land treatment, and surface impoundment.

Prior to establishment of the RCRA in 1976, most disposals occurred through methods that were considered easiest for maintenance and industrial personnel. As a result, organic pollutants found their way into the environment through different pathways. For example, liquid wastes containing lubricating oils and other petroleum residues were commonly discarded onto soil and unpaved roads by soil spreading. Substances such as gasoline, heating oil, and jet fuels leak out of rusted and corroded underground storage tanks (USTs). Based on a 1991 survey covering 1.6 million active and closed tanks, gasoline and diesel tanks represented 62 and 20%, respectively, of the total USTs, respectively. The distribution of USTs has probably changed somewhat, as approximately 600,000 tanks have been sealed between 1991 to 1995. The substances stored in RCRA-regulated tanks are shown in Figure 2.2. Because gasoline and diesel fuels account for the majority of the USTs, these substances pose the greatest threat to groundwater due to leakage of the tanks.

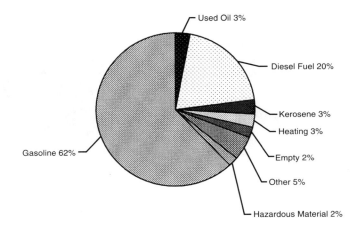

FIGURE 2.2
Distribution of underground storage tanks. (From USEPA, *National Survey of Underground Storage Tanks*, U.S. Environmental Agency, Office of Underground Storage Tanks, Washington, D.C., Spring, 1991.

Before 1976, waste was usually poured into surface storage areas such as pits, ponds, and lagoons. The waste then disappeared by seeping through the soil in the pits. At sanitary landfills, which were designed to accept newspapers, cans, bottles, and other household waste, liquid hazardous wastes were often disposed of in drums and, in some cases, poured directly into the landfills, and the waste often migrated to surface and ground waters. Drum storage areas, where waste chemicals were stored in 55-gallon drums, were often located on loading docks, concrete pads, or at other temporary storage areas until the wastes could be disposed of. The drums that were stored in this manner eventually corroded and leaked, causing chemical releases that seeped into the underlying soil and groundwater. Uncontrolled incineration, in which hazardous waste such as chlorophenols and PCBs combusted, has sometimes resulted in incomplete combustion, formation of more toxic products in the ash, and emission of hazardous air pollutants.

Typical sources of contamination from the contaminated sites of the Department of Defense (DOD) are shown in Figure 2.3. The most prevalent contaminants in groundwater are VOCs and metals, which appear in 74 and 59% of the DOD groundwater sites, respectively. SVOCs and metals were more consistent across different media than were VOCs. SVOCs were found in 31 to 43% of the sites, and metals were found in 59 to 80% of the sites. Fuels were found at fewer than 22% of all sites, a figure that may reflect the reporting of BTEX (benzene, toluene, ethylbenzene, and xylene) constituents of fuels and petroleum under VOCs. Figure 2.3 also shows the major contaminant groups by media and DOD component. The most frequently occurring group, metals, is found at 69% of all sites, followed by VOCs at 65% and SVOCs at 43%. VOCs and metals are found at most sites of all branches of the service, except at Army sites, where VOCs account for only 41% of the sites.

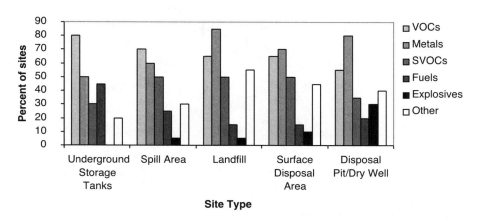

FIGURE 2.3
Source of contamination at different DOD sites. (From DOD, *Restoration Management Information System*, Department of Defense, Washington, D.C., November, 1995.)

2.4 Contaminated Media of Hazardous Wastes

Because hazardous wastes have different physicochemical properties (vapor pressure and solubility), contaminated media may contain inconsistently distributed pollutants. The distribution depends on environmental conditions and is a function of the properties of the chemical. In general, contaminated media are classified as: (1) soil, (2) air, (3) sludge and sediments, and (4) groundwater, and each type may contain pollutants in different phases:

- *Gaseous phase* — Contaminants are present as vapors in saturated zones.

- *Solid phase* — Contaminants in liquid form are adsorbed in soil particles in both saturated and unsaturated zones.

- *Aqueous phase* — Contaminants are dissolved into pore water according to their solubility in both saturated and unsaturated zones.

- *Immiscible phase* — Contaminants are present as nonaqueous-phase liquids (NAPLs), primarily in unsaturated zones.

Quantitative distributions of organic pollutants in different phases largely depend on their physical properties: the solubility constants in water (K_{ow}) and soil (K_{oc}) and Henry's constant. Figure 2.4 illustrates their distribution in these phases.

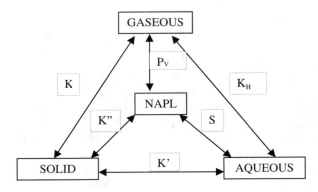

FIGURE 2.4
Schematic of contamination phases of NAPL in soil.

2.4.1 Groundwater

Groundwater can be contaminated with water-soluble substances found in overlying soil. BTEX and fuel oxygenates, such as methyltertiary butyl ether (MTBE), can leak from USTs and contaminate groundwater (Suidan et al., 2002). These organic chemicals are known to be carcinogenic in nature and pose a great threat to human health, specifically to the 53% of the U.S. population who depend on groundwater as their drinking water resource. Concentrations of organic pollutants in groundwater largely depend upon their solubility in water. Water solubility is the maximal concentration of a compound that can be dissolved in water at a given temperature. The dissolution of an organic compound into water mainly depends upon both the structure and size of an organic pollutant, and three factors are involved: Van der Waals forces, hydrogen bonding, and dipole–dipole interactions. Because water is a highly polar solvent, polar organic pollutants tend to have higher water solubility. In addition, substituents will also affect the solubility of a given class of organic compounds:

- The presence of substituents such as hydroxyl (–OH) and amino (–NH$_2$) will increase the solubility of a compound due to hydrogen bonding.
- Because halogens increase molecular volume, the more halogens a molecule contains, the lower the water solubility.
- Due to the higher polar characteristics of nonsubstituted hydrocarbons and less hydrogen bonding, organic acids, amines, alcohols, ethers, and ketones have higher solubilities than their corresponding hydrocarbons.
- As the number of carbons increases in any given organic class, the solubility decreases due to high polarity.

- Double bonds have little effect on the water solubility of their corresponding compounds.
- The larger a polycyclic aromatic hydrocarbon (PAH) is, the less solubility the PAH will have due to the large and nonpolar planar molecular structure.

Table 2.3 lists the solubility constants of typical organic pollutants and their ranges.

2.4.2   Soil

Contaminants in soil can partition between the soil and air, soil and water, and soil and solids; however, the physical structure and chemical composition of surface and subsurface soil are highly variable. Compared to aqueous systems, soil is a complex heterogeneous media composed of solid, liquid, and gaseous phases. The four major components of soil are the inorganic (mineral) fraction, organic matter, water, and air. Soil consists of 50% pore space, which is occupied by air and water; 45% minerals; and 5% organic matter. Figure 2.5 illustrates a typical soil structure that is important to consider for remediation.

The available treatment data show large variations in treatment efficiencies. Some of the important considerations relevant to the selection of treatment technologies for VOCs include the following:

- *Particle size distribution* refers to the distribution of particles in the soil matrix. In general, the three types of soil are sand, clay, and loam. Sand is soil composed of at least 70% sand; clay is soil consisting of at least 35% clay; and loam soil contains equal weights of sand, clay, and silt. Particular size or soil texture can affect the treatability of contaminated soil in two ways. The potential reaction sites are primarily limited to the surface of particles. The surface-to-volume ratio has a major impact on the nature and rate of reactions between the particle and the contaminant; therefore, larger sand-sized particles are less reactive than smaller clay-sized particles, particularly if reactions may occur between the sheets of clay minerals.
- *Cation exchange capacity* is defined as the quantity of cations sorbed per mass of soil and is expressed as milliequivalents of positive charge per 100 g of soil. The high cation-exchange capacity of clay minerals enables them to be more reactive with many contaminants than the minerals' characteristic of organic matter comprised of sand-sized particles. Thus, the relatively large surface area and the high cation exchange capacity make clays and organic matter more difficult to treat than sands and silts.

TABLE 2.3

Water Solubilities of Common Hazardous Compounds

Compound	Mean Water Solubility (mg/L) at 25°C	Range
Aliphatic hydrocarbons		
Cyclohexane	59.3	58.4–66.5
Cyclohexene	213	—
Isooctane	1.25	0.56–2.46
n-Decane	0.021	0.020–0.022
n-Heptane	2.91	2.24–3.57
n-Octane	1.39	0.431–3.37
Monocyclic aromatic hydrocarbons		
Benzene	1770	1740–1800
Ethylbenzene	181	152–208
m-Xylene	163	146–173
o-Xylene	221	171–297
p-Xylene	189	156–214
Toluene	546	515–627
Polycyclic aromatic hydrocarbons		
Anthracene	0.059	0.045–0.073
Benzo[a] anthracene	0.012	0.0094–0.014
Benzo[a]pyrene	0.0038	—
Chrysene	0. 0033	0.0018–0.006
Fluorene	1.84	1.69–1.98
Naphthalene	31.9	30.0–34.4
Phenanthrene	1.09	0.994–1.29
Pyrene	0.134	0.132–0.135
Nonhalogenated solvents		
Acetone	Miscible	—
Methyl butyl ketone	35,000	—
Methyl ethyl ketone	256,000	—
Methyl isobutyl ketone	19,100	—
Halogenated solvents		
Carbon tetrachloride	911	757–1160
Chloroform	7870	7100–9300
Methylene chloride	18,400	16,700–19,400
Perchloroethylene	275	150–400
1,1,1-Trichloroethane	657	—
Trichloroethylene	1310	1100–1470
Insecticides		
Aldrin	0.011	—
Carbaryl	76.5	70–83
Dieldrin	0.198	0.195–0.200
DDT	0.0024	0.0012–0.020
Hexachlorobenzene	0.005	—
Lindane	7.30	6.80–7.80
Malathion	145	—
Methyl parathion	57	—
Parathion	24.0	—

-- *continued*

TABLE 2.3 (CONTINUED)

Water Solubilities of Common Hazardous Compounds

Compound	Mean Water Solubility (mg/L) at 25°C	Range
Herbicides		
Atrazine	33.0	—
2,4-D	697	522–890
2,4,5-T	273	268–278
Trifluralin	1.0	—
Fungicides		
Pentachlorophenol	14.0 at 20°C	—
Industrial intermediates		
Chlorobenzene	460	295–503
2,4-Dichlorophenol	9750	4500–15,000
Dimethyl phthalate	4160	4000–4320
D-*n*-octyl phthalate	3	—
Hexachlorocyclopentadiene	1.80	—
Phenol	77,900	67,000–84,700
2,4,6-Trichlorophenol	800	—
Explosives		
Picric acid	14,000 at 20°C	—
24,6-Trinitrotoluene	130 at 20°C	—
Polychlorinated biphenyls		
PCB 1016	0.42	—
PCB 1232	1.45	—
PCB 1248	0.036	0.017–0.054
PCB 1254	0.011	0.010–0.012
PCB 1260	0.0027	—
Chlorinated dioxins		
2,3,7,8-Tetrachlorodibenzo-*p*-dioxin	0.00032	—

Source: Watts, R.J., *Hazardous Wastes: Sources, Pathways, Receptors,* John Wiley & Sons, New York, 1997. With permission.

- *Porosity* is defined as the percentage of soil occupied by pore space and is an important parameter related to the transport, retardation, and mass transfer of pollutants in groundwater.

- *Volumetric water content* is the fraction of the soil pores that are filled with water.

- *Bulk density of soil* is the soil weight per unit volume, including water and voids. It is used in converting weight to volume in the mass balance calculations.

- *Particle density* is the specific gravity of a soil, which is important in soil washing and in the determination of the setting velocity of suspended soil in flocculation and sedimentation processes.

A SOIL PROFILE

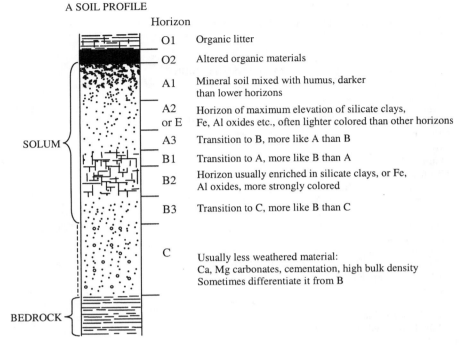

FIGURE 2.5

A soil profile, illustrating the traditional O, A, B, and C horizons and some further subdivisions of them. (From Greenland, D.J. and Hayes M.H.B., Eds., *The Chemistry of Soil Constituents,* John Wiley & Sons, New York, 1978. With permission.)

- *Permeability and hydraulic conductivity* are the controlling factors in the effectiveness of *in situ* treatment technologies. The ability of soil flushing fluids to contact and remove contaminants can be reduced by low soil permeability. Low permeability can lessen the volatilization of VOCs in soil vapor extraction or limit the effectiveness of *in situ* vitrification by slowing vapor releases.

- *Organic content* is usually divided into two categories: humic and nonhumic. Nonhumic chemicals are unaltered amino acids, carbohydrates, fats, and other chemicals that are present in soil as a consequence of living organisms. Humic materials are formed by microbial mediated reactions. They contain polymerized phenols with carboxic, carbonyl, ester, and methoxy groups. It is one of the most important properties influencing the transport of hazardous compounds.

Among these factors, organic content determines the soil distribution coefficient and is an important parameter that quantifies the distribution of organic pollutants in soil. The soil distribution coefficient (K_d) is the ratio between the mass of contaminant sorbed by soil (mg/g) to the mass of contaminant in the aqueous phase (mg/mL), or:

$$K_d = \frac{\text{Mass of contaminant sorbed by soil}}{\text{Mass of contaminant in aqueous phase}} \qquad (2.1)$$

It can be experimentally determined by dividing the equilibrium concentration of organic pollutant adsorbed on soil by its equilibrium concentration in water after 24 hours of mixing. Because the sorption of organic pollutants onto soil is almost exclusively onto the organic fraction due to the hydrophobicity of the organic pollutant, K_{oc}, the soil adsorption coefficient of organic pollutants due to organic content in soil is defined as follows:

$$K_{oc}(\text{mL} / \text{g}) = \frac{\text{Mass of contaminant sorbed by soil (mg / g)}}{\text{Mass of contaminant in aqueous phase (mg / mL)}} \qquad (2.2)$$

Therefore, K_d can be calculated by normalizing K_{oc} according to the following equation:

$$K_d = K_{oc} \times f_{oc} \qquad (2.3)$$

where K_d is the soil distribution coefficient; K_{oc} is the soil adsorption coefficient of organic pollutants due to organic content in soil; and f_{oc} is the weight fraction of organic content in soil (%).

In practice, the above equation is valid only if the organic carbon content is the primary sorbent, for organic pollutants with molecular weights less than 400, and when the organic pollutant does not have special-function groups that could promote ion exchange or complexation. In other words, when hydrophobic partition is the sole adsorption mechanism, K_{oc} is strongly correlated with the hydrophobic partition coefficient (K_{ow}). The octanol/water partition coefficient is defined as follows:

$$K_{ow} = \frac{\text{Concentration in octanol phase (mg / L)}}{\text{Concentration in aqueous phase (mg / L)}} \qquad (2.4)$$

Table 2.4 lists the K_{ow} for different hazardous organic pollutants. To avoid expensive experimental tests when determining K_d, K_{oc} can be estimated by K_{ow} using correlations equations reported in the literature, as shown in Table 2.5.

TABLE 2.4

Octanol/Water Partition Coefficients for Common Hazardous Compounds

Compound	Mean Log K_{ow}	Range
Aliphatic hydrocarbons		
Cyclohexane	3.44	—
Cyclohexene	2.86	—
n-Decane	6.69	—
n-Heptane	4.66	—
Isooctane	5.83	—
n-Octane	5.18	—
Monocyclic aromatic hydrocarbons		
Benzene	2.05	1.95–2.15
Toluene	2.58	2.21–2.79
Ethylbenzene	3.11	3.05–3.15
o-Xylene	3.11	2.77–3.13
m-Xylene	3.20	—
p-Xylene	3.17	3.15–3.18
Polycyclic aromatic hydrocarbons		
Anthracene	4.34	—
Benzo[a]anthracene	5.91	5.90–5.91
Benzo[a]pyrene	6.06	—
Chrysene	5.71	5.60–5.91
Fluorene	4.38	—
Naphthalene	3.51	3.01–4.70
Phenanthrene	4.52	—
Pyrene	5.32	—
Nonhalogenated solvents		
Acetone	0.24	—
Methyl butyl ketone	1.38	—
Methyl ethyl ketone	0.28	0.26–0.29
Methyl isobutyl ketone	1.09	—
Halogenated solvents		
Carbon tetrachloride	2.73	—
Chloroform	1.94	1.90–1.97
Methylene chloride	1.28	1.25–1.30
Perchloroethylene	2.79	2.53–2.88
1,1,1-Trichloroethane	2.33	2.18–2.47
Trichloroethylene	2.33	2.29–2.42
Insecticides		
Aldrin	5.17	—
Carbaryl	2.38	2.31–2.56
Dieldrin	5.16	—
DDT	6.11	5.76–6.36
Endrin	5.02	4.56–5.34
Hexachlorobenzene	5.65	5.45–6.18
Lindane	3.76	3.66–3.85
Malathion	2.84	—
Parathion	3.43	2.15–3.93
Herbicides		
Atrazine	2.68	—
2,4-D	2.94	1.57–4.88
2,4,5-T	3.40	—
Trifluralin	5.31	5.28–5.01

-- *continued*

TABLE 2.4 (CONTINUED)

Octanol/Water Partition Coefficients for Common Hazardous Compounds

Compound	Mean Log K_{ow}	Range
Fungicides		
Pentachlorophenol	4.41	3.81–5.01
Industrial intermediates		
Chlorobenzene	2.84	2.71–2.98
2,4-Dichlorophenol	3.08	—
Dimethyl phthalate	1.70	1.47–2.00
Di-*n*-octyl phthalate	9.54	9.20–9.87
Hexachlorocyclopentadiene	4.52	4.00–5.04
Phenol	1.47	1.46–1.48
2,4,6-Trichlorophenol	3.24	
Explosives		
Picric acid	1.69	1.34–2.03
2,4,6-Trinitrotoluene	2.25	—
Polychlorinated biphenyls	—	—
Aroclor 1016	5.58	—
Aroclor 1232	3.87	3.20–4.54
Aroclor 1248	6.11	—
Aroclor 1254	6.31	—
Aroclor 1260	6.91	—
Chlorinated dioxins		
2,3,7,8-Tetrachlorodibenzo	5.77	5.38–6.15

Source: Watts, R.J., *Hazardous Wastes: Sources, Pathways, Receptors,* John Wiley & Sons, New York, 1997. With permission.

2.4.3 Air

Volatile organic compounds and SVOCs are usually released from contaminated sites through volatilization, which is the transfer process of VOCs or SVOCs from liquid or solid phases to gaseous phase. Vapor pressure and Henry's constants govern the degree of volatilization. In relatively pure forms of organic compounds, volatility is a function of vapor pressure at a given temperature. Vapor pressure represents the tendency of a compound to evaporate in a gas. Raoult's law states that the partial pressure of the vapor in gas (P_{org}) is proportional to the mole fraction of the organic solvent (χ_{org}) and the vapor pressure of the pure organic solvent (P_{org}^{o}) as follows:

$$P_{org} = \chi_{org} \times P_{org}^{o} \tag{2.5}$$

Henry's law describes the tendency for an organic compound to evaporate from water to a gas. It states that the concentration of an organic compound in the aqueous phase is proportional to the concentration in the corresponding vapor phase as follows:

$$P_{org} = H \times C_{org} \tag{2.6}$$

TABLE 2.5

Regression Equation	Organic Carbon	Applicable Compounds
$\log K_{oc} = 1.029 \log K_{ow} - 0.18$	Wide range	Broad range of herbicides and insecticides (atrazine, bromacil, carbofuran, 2-4-D, DDT, diuron, malathion, methyl parathion, simazine)
$\log K_{oc} = 0.94 \log K_{ow} + 0.22$	Wide range	s-Triazine herbicides
$\log K_{oc} = 0.524 \log K_{ow} + 0.855$	1.0–4.0%	Phenyl urea herbicides (e.g., 3-[3-chlorophenyl]1,1-urea dimethyl-(3,3,4- dichlorophenyl]urea)
$\log K_{oc} = 0.52 \log K_{ow} + 0.64$	1.0–5.92% (organic matter)	Phenyl ureas, carbamates, organochlorine insecticides, 4-bromophenol, captan, bromo- and chloroanilines, methylanilines, bromo- and chloro-nitrobenzenes, diphenyl amino folpet, hexachlorobenzene, naphthalene, phenol, simazine
$\log K_{oc} = 0.544 \log K_{ow} + 1.377$	Wide range	PAHS, organochlorine insecticides, benzene, uracil herbicides, carbamates, acid amides, phenoxy herbicides, pentachlorophenol, penzene, DCPB, organophosphate esters, PCBs, ethylene dibromide
$\log K_{oc} = 0.72 \log K_{ow} + 0.49$	<0.01–33%	Chlorobenzene, chlorobenzenes, methylbenzenes, toluene, PCE, n-butylbenzene
$\log K_{oc} = 0.989 \log K_{ow} - 0.346$	0.66–2.38%	Benzene, PAHs
$\log K_{oc} = \log K_{ow} - 0.21$	0.09–3.29%	PAHs
$\log K_{oc} = -0.82 \log S + 4.07$ $S = $ water solubility (mg/L)	0.11–2.38%	PAHs
$\log K_{oc} = -0.62 \log K_{ow} + 2.04$	3.51%	Triazines, pyrimidines, pyridazines
$\log K_{oc} = 0.601 \log K_{ow} + 1.991$	Wide range	Phenylsulfynil and phenylsulfonyl, acetates

Source: Watts, R.J., *Hazardous Wastes: Sources, Pathways, Receptors,* John Wiley & Sons, New York, 1997. With permission.

where P_{org} is the partial pressure of the vapor in gas; H is Henry's constant (atm-m^3/mol); and C_{org} is the concentration of organic in the aqueous phase. Table 2.6 lists Henry's constants for some common organic pollutants.

2.4.4 Sludge and Sediments

Sludge is defined as an aggregate of oil and other types of matter in any form, other than dredged soil, having a combined specific gravity equivalent to or greater than water. An estimated 0.7 billion tons of sewage sludge is generated annually in sewage treatment plants worldwide. The sewage sludge is buried in landfills, incinerated, applied to agricultural land, and dumped into the ocean. The main problem with sludge, especially sludge from urban sewage treatment plants, is that it may contain moderate to high concentrations of hazardous pollutants such as pesticides, PAHs, PCBs, and nonylphenols.

2.5 Distribution of Hazardous Pollutants in Contaminated Sites

The distribution of toxic pollutants in all major remediation programs and contamination sites in the United States corresponds to the following seven programs:

- National Priorities List (Superfund)
- Resource Conservation and Recovery Act (RCRA)
- Underground storage tanks (USTs)
- Department of Defense (DOD)
- Department of Energy (DOE)
- Other federal agencies
- States and private parties

Many of the sites to be remediated under these programs contain similar types of contamination. In most of the programs, both the soil and groundwater are contaminated by VOCs, metals, and SVOCs. Over the past 15 years, almost half a million sites with potential contamination have been reported to state or federal authorities. Over the next decade, federal, state, and local governments and private industries will continue their efforts in the remediation of contaminated sites. This continuing demand for site remediation services and for the technology to treat the different waste streams makes physicochemical treatment processes important for environmental remediation.

TABLE 2.6

Vapor Pressure and Henry's Law Constants for Common Hazardous Chemicals

Compound	Vapor Pressure (mmHg) at 20°C	Henry's Law Constant (atm-m³/mol) at 25°C
Aliphatic hydrocarbons		
Cyclohexane	95	0.194
Cyclohexene	67	0.046
n-Decane	2.7	0.187
n-Heptane	40	2.04
Isooctane	40	3.01
n-Octane	10.4	3.23
Monocyclic aromatic hydrocarbons		
Benzene	76	5.48×10^{-3}
Ethylbenzene	7.08	8.68×10^{-3}
Toluene	22	6.74×10^{-3}
Polycyclic aromatic hydrocarbons		
Anthracene	1.7×10^{-5}	1.77×10^{-5}
Benzo[a]anthracene	2.2×10^{-8}	6.6×10^{-7}
Benzo[a]pyrene	5.0×10^{-7}	$<2.4 \times 10^{-6}$
Chrysene	6.3×10^{-7}	7.26×10^{-20}
Fluorene	0.005	2.1×10^{-4}
Naphthalene	0.054	4.6×10^{-4}
Phenanthrene	2.1×10^{-4}	2.56×10^{-4}
Pyrene	2.5×10^{-6} at 25°C	1.87×10^{-5}
Chlorinated solvents		
Carbon tetrachloride	90	0.0302
Chloroform	160	3.2×10^{-3}
trans-1,2-Dichloroethene	265	0.0067
Hexachloroethane	0.18	2.5×10^{-3}
Methylene chloride	455 at 25°C	2.69×10^{-3}
Perchloroethylene (PCE)	14	0.0153
1,1,1-Trichloroethane (TCA)	96	0.018
Trichloroethylene (TCE)	58	9.1×10^{-3}
Insecticides		
Aldrin	6.7×10^{-6}	1.4×10^{-6}
DDT	7.26×10^{-7} at 30°C	5.20×10^{-5}
Dieldrin	1.78×10^{-7}	5.8×10^{-5}
4,6-Dinitro-o-cresol	3.2×10^{-4}	4.3×10^{-4}
Endrin	7×10^{-7}	5.0×10^{-7}
Hexachlorobenzene	1.09×10^{-5}	1.7×10^{-3}
Lindane	9.04×10^{-6}	3.25×10^{-6}
Herbicides		
2,4-D	4.7×10^{-3}	1.95×10^{-2} at 20°C
Fungicides		
Pentachlorophenol	1.4×10^{-4}	3.4×10^{-6}
Industrial intermediates		
Acrylonitrile	83	1.1×10^{-4}
Chlorobenzene	90	3.7×10^{-3}
o-Chlorophenol	1.42 at 25°C	8.28×10^{-6}
Dibutyl phthalate	1×10^{-5} at 25°C	6.3×10^{-5}
1,2-Dichlorobenzene	1	1.9×10^{-3}
1,3-Dichlorobenzene	2.30 at 25°C	3.60×10^{-3}

-- continued

TABLE 2.6 (CONTINUED)

Vapor Pressure and Henry's Law Constants for Common Hazardous Chemicals

Compound	Vapor Pressure (mmHg) at 20°C	Henry's Law Constant (atm-m³/mol) at 25°C
Diethyl phthalate	0.05 at 70°C	8.46×10^{-7}
Di-*n*-octyl phthalate	1.4×10^{-4} at 25°C	1.41×10^{-12}
Hexachlorocyclopentadiene	0.081 at 25°C	0.016
p-Nitrophenol	1×10^{-4}	3×10^{-5} at 20°C
Phenol	0.2	3.97×10^{-7}
1,2,4-Trichlorobenzene	0.29 at 25°C	2.32×10^{-3}
2,4,6-Trichlorophenol	0.017 at 25°C	9.07×10^{-8}
Polychlorinated biphenyls		
Aroclor 1016	4×10^{-4}	3.3×10^{-4}
Aroclor 1221	6.7×10^{-3}	3.24×10^{-4}
Aroclor 1232	4.60×10^{-3}	8.64×10^{-4}
Aroclor 1248	4.9×10^{-4} at 25°C	3.5×10^{-3}
Aroclor 1260	4.1×10^{-5}	7.1×10^{-3}
Chlorinated dioxins		
2,3,7,8-Tetrachlorodibenzo-*p*-dioxin (TCDD)	6.4×10^{-10}	5.40×10^{-23} at 18 – 22°C

Source: Watts, R.J., *Hazardous Wastes: Sources, Pathways, Receptors,* John Wiley & Sons, New York, 1997. With permission.

2.5.1 National Priorities List Sites

Superfund is a term used for the federal program that is in charge of the remediation of hazardous waste sites. The program is administered by the U.S. Environmental Protection Agency (USEPA) and is under the authority of the Comprehensive Environmental Response, Compensation and Liability Act (CERCLA). In addition to establishing enforcement authority, CERCLA created a trust fund to be used for site identification and remediation, known as the Superfund. The procedures for CERCLA implementation are described in the National Oil and Hazardous Substance Pollution Contingency Plan (NCP). This plan outlines the steps that USEPA and other federal agencies must follow in responding to the release of hazardous substances or oil into the environment. The goal described in the NCP is to select remedies that protect human health and the environment, maintain protection over time, and minimize untreated waste. The sites with the worst contamination problems are included on the National Priority List (NPL). As of 1996, 547 proposed and final NPL sites not owned by the federal government still required further remedial action. The data reflect the industrialized nature of these regions and the number of abandoned industrial and commercial facilities. An additional 124 NPL sites that require remedial action are located at federal facilities. Among these NPL sites, the percentages of sites that require groundwater, soil, sediment, and sludge remediation are 76, 73, 22, and 12%, respectively. The total volume of soil, sludge, and sediment to be remediated at these sites is estimated at 33 million cubic yards.

2.5.1.1   Contaminants

Volatile organic compounds are the contaminant group most frequently found at NPL sites and are the most frequently treated with innovative technologies. VOCs are to be remediated at 71% of the sites, followed by metals (65%) and SVOCs (61%). Figure 2.6 shows the frequency of the major contaminants. All three groups — VOCs, SVOCs and metals — are to be remediated at 25% of the sites, but not necessarily with the same media.

Based on data available for 944 NPL sites for FY1982 to FY1994 Records of Decision (RODs), Figure 2.6 shows the number of sites that contain a specific class of pollutants. It is important to note that a site may contain one or more of the contaminants shown here. As can be seen in the figure, VOCs, SVOCs, and metals are the dominant subgroups for these sites. The 90 sites listed as "other" contain only radioactive elements, nonmetallic inorganic, or unspecified organics or inorganics. VOCs and SVOCs were subdivided into more specific subgroups that better coincide with the application of different technologies, such as bioremediation. The subgroups are described below and are assembled according to the three major contaminant groups:

- *VOCs* — This group includes halogenated VOCs, BTEX, and other nonhalogenated VOCs (ketones and alcohols). The most prevalent class of organics, halogenated VOCs, are widely used as solvents and are being remediated at 64% of the sites. Although many of the BTEX compounds are derived from petroleum products, CERCLA prohibits the listing of sites on the NPL that are contaminated with petroleum products alone.
- *SVOCs* — This group includes PCBs, PAHs, pesticides, phenols, and other SVOCs, such as chlorobenzene and phthalates. The most common SVOCs are PAHs and pesticides.
- *Metals* — This group includes lead, arsenic, chromium, cadmium, zinc, nickel, and other less frequently found metals.

Figure 2.7 shows additional data available for the 944 NPL sites for FY1982 to FY1994 RODs. The following organic pollutants are the ones most frequently found in these sites:

- Trichloroethylene (50%)
- Benzene (43%)
- Toluene (36%)
- Tetrachloroethylene (36%)
- Vinyl chloride (29%)
- PCBs (29%)

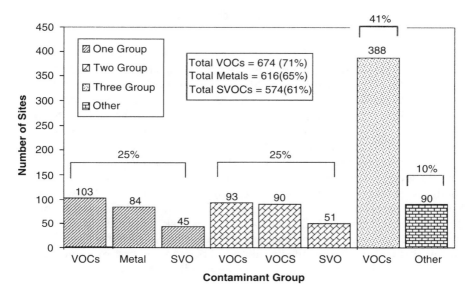

FIGURE 2.6
Distribution of contaminant group. (From USEPA, *POD Information Directory*, U.S. Environmental Protection Agency, Office of Emergency and Remedial Response, Washington, D.C., December, 1995.)

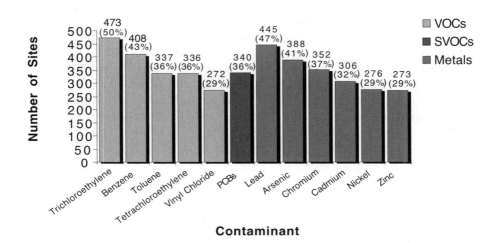

FIGURE 2.7
The distribution of the most frequent individual contaminant. (From USEPA, *POD Information Directory*, U.S. Environmental Protection Agency, Office of Emergency and Remedial Response, Washington, D.C., December, 1995.)

2.5.2 Resource Conservation and Recovery Act

The USEPA estimates that over 6000 facilities are currently operated as treatment, storage, or disposal facilities (TSDFs) regulated under the Resource Conservation and Recovery Act (RCRA), which assigns the responsibility of corrective action to facility owners and operators and authorizes the USEPA to oversee corrective actions. Unlike the Superfund, RCRA responsibility is delegated to states. The USEPA and authorized states have completed initial assessment of potential environmental contamination at over 70% of RCRA facilities, as required by statute to address corrective action. Environmental contamination at many RCRA facilities is expected to be less severe than at Superfund sites; however, the total number of RCRA facilities exceeds the number of Superfund sites. The USEPA developed a computer-based system known as the RCRA National Corrective Action Prioritization System (NCAPS) to help establish priorities for corrective action activities. Among the factors considered in NCAPS are the history of hazardous waste releases, the likelihood of human and environmental exposure, and the type and quantity of waste handle at the facility.

2.5.2.1 *Contaminated Media*

Two separate studies provide an indication of the nature of contaminants at RCRA corrective action sites. In one study, the USEPA Technology Innovation Office obtained information about 275 sites for the purpose of identifying relationships between site characteristics and the use of innovative technologies at RCRA corrective action sites. Groundwater and soil are the most commonly reported contaminated media at the 256 sites for which information was available. The second study is the Regulatory Impact Analysis (RIA), which was developed to support the 1993 Corrective Action Rule. This study analyzed a sample of 79 TSDFs to estimate contamination that is likely to be present in soil and groundwater. The USEPA used a long-term modeling approach to simulate contaminant concentrations likely to be encountered from 1992 to 2119. The RIA suggests that as many as 2600 TSDFs require corrective action, of which 2100 (80%) might have significant releases to on-site groundwater and 780 (30%) will have significant off-site groundwater contamination.

2.5.2.2 *Contaminants*

Of the 214 sites for which contamination data are available (Figure 2.8), HVOCs were most prevalent of all the contaminated groups, being reported at 60% of the sites, followed by heavy metals at 46% and nonhalogenated VOCs at 32%. Figure 2.9 presents the frequency of the most common contaminant groups. Because some groups could be present at more than one site, the total percentage may exceed 100%.

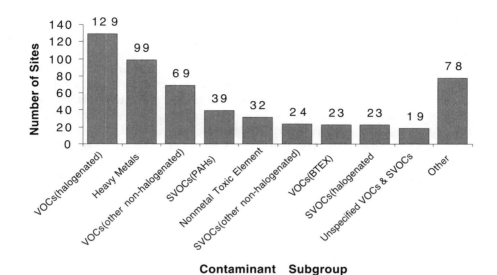

FIGURE 2.8
Distribution of contaminant subgroups at 214 sites. PAH = polynuclear aromatic hydrocarbons; BTEX = benzene, toluene, ethylbenzene, and xylene. (From USEPA, *Analysis of Facility of Corrective Action Data*, U.S. Environmental Protection Agency, Office of Solid Waste and Emergency Response, Technology Innovation Office, Washington, D.C., January, 1994.)

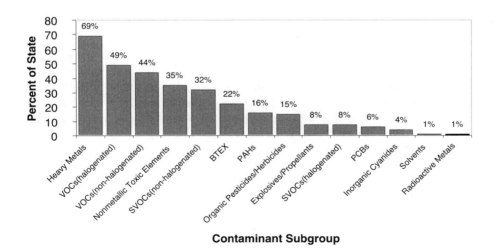

FIGURE 2.9
Distribution of contaminant subgroup. (From DOD, *Restoration Management Information System*, Department of Defense, Office of Deputy Under Secretary of Defense (Environmental Security), Washington, D.C., November, 1995.)

2.5.3 Underground Storage Tanks Sites

Contamination resulting from leaks and spills from USTs is addressed primarily by the tank owners under state UST programs, established according to Subtitle I of the 1984 Hazardous and Solid Waste Amendments to RCRA. All states and territories have passed legislation for UST remediation, and 45 states possess trust funds. Some states have more active enforcement programs than others, and some have promulgated UST requirements for double-lined tanks, more stringent monitoring procedures, or earlier upgrading compliance dates. The USEPA reports that as of 1996, 1,064,478 active tanks and 1,074,022 closed tanks have been registered in the United States. Using the USEPA estimate of an average of 2.7 tanks per sites, approximately 792,037 sites with USTs are subject to corrective action regulation.

2.5.3.1 *Contaminated Media*

The USEPA has estimated that the number of confirmed releases is 418,000. Remedial design or remedial actions have been initiated at almost 25,300 of these sites. The volume of soil (per site) to be remediated ranges from 9 to 800 cubic yards. No information is available regarding the magnitude of surface and ground waters in need of remediation.

2.5.3.2 *Contaminants*

Most USTs contain petroleum products that are mixtures of four types of hydrocarbons: paraffins, olefins, naphthalenes, and aromatics.

2.5.4 Department of Defense

The Department of Defense (DOD) has taken on the task of remediating waste resulting from numerous industrial, commercial, training, and weapon-testing activities, as well as remediation of military bases that are closing so that the properties can be transferred to local communities. The DOD has begun investigating more than 8300 sites located at more than 1500 installations or formerly used defense sites (FUDS) that will ultimately require remediation. The CERCLA and RCRA are the primary federal laws governing the investigation and remediation of DOD contaminated sites. In Executive Order 12580 (signed in January 1987), the President directed the Secretary of Defense to implement investigations and remedial measures, in consultation with the USEPA, for releases of hazardous substances from facilities under the jurisdiction of the Secretary. DOD remediation efforts must also consider the requirements of the state law and the Base Closure and Realignment Act of 1988 and 1990 (BRAC). As of 1995, the DOD had identified 22,089 sites located at 1705 installations, in addition to the FUDS properties with potential hazardous waste contamination involving soil, groundwater, and other media.

2.5.4.1   Contaminated Media

The data on media and contaminants are from the Restoration Management Information System. The types of media identified are groundwater, soil, surface water, and sediments; 71% of the contaminated media is groundwater, and 67% is contaminated soil, which indicates that many sites contain both. Contaminated surface water and sediments account for 19 and 6%, respectively, of the media. The totals add up to more than the total number of the sites, because a site may contain more than one type of contaminated media.

2.5.4.2   Contaminants

The contaminants have been grouped into six categories at the DOD sites:

- Volatile organic compounds (VOCs)
- Semivolatile organic compound (SVOCs)
- Metals
- Fuels
- Explosives
- Other (inorganic elements and compounds such as asbestos, arsenic, cyanides, corrosives, pesticides, and herbicides)

Some sites contain contaminants that are found less frequently in industries and that present unique problems for selecting remediation approaches. For example, over 8% of DOD sites contain explosives, and an unspecified number contain low-level radiation. The occurrence frequency of contaminants also varies from site to site. Contaminants present at the DOD sites are further divided into 19 subgroups, such as HVOCs, nonhalogenated VOCs, and BTEX. Figure 2.9 shows subgroups based on 3212 sites requiring cleanup at 480 installations as of 1994. Subgroups found at less than 1% of the sites include dioxins/furans, organic corrosives, inorganic corrosives, and organic cyanides. Those classified as "other" primarily include inorganic elements and compounds such as asbestos, arsenic, inorganic cyanides, pesticides, and herbicides. Totals may exceed 100% because more than one contaminant may occur at a site (DOD, 1995). Heavy metals are the most prevalent subgroup.

2.5.5   Department of Energy

One of the most serious and costly environmental remediation tasks facing the federal government is the remediation and restoration of more than 100 major installation and other locations that are the responsibility of the U.S. Department of Energy (DOE). The environmental problems associated with DOE properties, unlike those of other industries, include unique radiation hazards, unprecedented volumes of contaminated soil and water, and a large number of contaminated structures, ranging from nuclear reactors to chemical plants, for the extraction of nuclear material to evaporation ponds.

The Office of Environmental Restoration and Waste Management (EM) manages the DOE environmental programs. Remedial action at sites throughout the DOE complex involves treatment, disposal, and in some cases transfer to the Waste Management Program of various types of waste. These types of waste are categorized as:

- Hazardous waste that contains hazardous constituents but no radionuclides
- Mixed waste that contains both hazardous and radioactive materials
- Low-level waste that contains a small amount of radioactive substances in large volumes of material
- Uranium by-product materials that contain very low concentrations of naturally occurring α-emitting radionuclides in large volumes of soil-like materials
- Transuranic waste that contains plutonium, americium, and other elements with atomic numbers higher than uranium
- High-level waste that contains highly radioactive material, including fission products, traces of uranium and plutonium, and other transuranic elements resulting from the chemical reprocessing of spent fuel

2.5.5.1 Contaminants

Groundwater and soil are the predominant types of contaminated media found at DOE sites. The DOE has made substantial progress in identifying specific contaminants of concern for many individual sites. Figure 2.10 shows the frequency of major contaminants and categories of contaminants that have been identified at the DOE installations and other locations where characterization and assessment (C/A) have not been completed. Organics are among the contaminants found at about 38% of the DOE installations that have not begun remediation. Among these are PCBs, petroleum/fuel hydrocarbons, solvents, trichloroethylene (TCE), unspecified VOCs, and unspecified SVOCs. Metals are listed as being present at 35% of the sites. Those contaminants cited most often include lead, beryllium, mercury, arsenic, and chromium. Radioactive contaminants are present at most of the DOE installations. The most frequently cited are uranium, tritium, thorium, and plutonium. Mixed waste, which contains both radioactive and hazardous contaminants, is of particular concern to the DOE because of the lack of acceptable treatment technology.

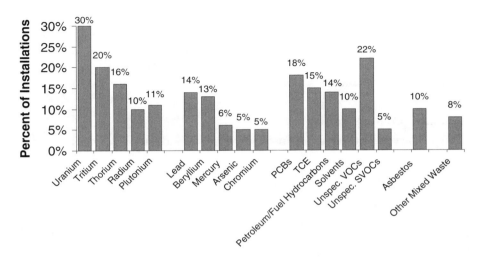

FIGURE 2.10

The most frequent contaminants at DOE sites. Figure includes all contaminants in a group, not only those indicated by the bars. A site may contain more than one contaminant. *Other mixed waste* indicates installations and other locations with mixed waste for which specific contaminants have not been delineated. (From *Estimating the Cold War Mortgage: The 1995 Baseline Environmental Report*, DOE/EM-2032, U.S. Department of Energy, Washington, D.C., March 1995; *Contaminated Media/Waste Database as of March 3, 1995*, DOE/EM-40, U.S. Department of Energy, Washington, D.C., 1995; *Draft Programmatic Environmental Impact Statement for the Uranium Mill Tailings Remedial Action Ground Water Project*, DOE/EIS-0198, U.S. Department of Energy, UMTRA Project Office, Washington, D.C., April, 1995.)

2.5.6 Waste Sites Managed by Other Federal Agencies

With the exception of the DOE and DOD, other federal agencies (OFAs) include agencies responsible for the ownership, operation, and remediation of contaminated waste sites. The OFA listing of sites includes the following:

- U.S. Department of Interior (DOI), which is responsible for a large number of potentially contaminated sites on more than 440 million acres of federal land that it manages
- U.S. Department of Agriculture (USDA), which estimates that it currently has 3000 potentially contaminated sites on land under its management
- National Aeronautics and Space Administration (NASA), which has identified 730 potentially contaminated sites

The federal government must comply as a private party with the provisions of CERCLA and RCRA. These statutes make federal agencies liable for the remediation of contaminated waste at currently or formerly owned facilities.

2.5.6.1   Contaminated Media

Contaminated media at these sites result primarily from sedimentation in surface waters, acid mine drainage, and household chemical waste. As with other contaminated sites throughout the United States, the predominant contaminated media are soil and groundwater.

2.5.6.2   Contaminants

The primary contaminants are fuels, solvents, and industrial waste constituents.

2.5.7   Sites Managed by States and Private Companies

All non-federal agency sites that are not being remediated under CERCLA, RCRA, or UST programs become the responsibility of the state remediation programs. Private parties are individuals not affiliated with federal or state governments. By using data supplied by the states, the USEPA has identified over 79,000 non-NPL sites. Of these sites, 29,000 will require some action beyond a preliminary assessment; however, the actual number of sites that will require remediation, as well as the extent of contamination at these sites, is largely unknown. Most states have enforcement authority and state Superfunds to finance the remediation of abandoned waste sites. In addition, many state sites are remediated by private parties in accordance with state remediation standards. The federal government has actively encouraged and assisted states in the remediation of their contaminated properties. For example, the USEPA has a program dedicated to helping states address brownfield sites. The USEPA defines brownfield sites as abandoned, idle, or under-used industrial and commercial facilities where expansion or redevelopment is complicated by real or perceived environmental contamination.

2.5.7.1   Contaminated Media

A central source of information that characterizes the types of media affected was not available.

2.5.7.2   Contaminants

Some studies show that the most prevalent waste found at these sites includes organic chemicals, metals, solvents, and oily waste.

2.6   Conclusion

In summary, many of the contaminated sites in the federal or state programs contain similar types of contamination. In most programs, about two thirds of the sites have contaminated soil or groundwater, or both. VOCs, the most frequently occurring contaminant type, are present at more than two thirds of Superfund, RCRA, and DOD sites and at almost half of the DOE sites. In addition, VOCs, primarily in the form of BTEX, are also the primary contaminants at UST sites. Metals and SVOCs are most prevalent at Superfund and DOD sites, although they are also present at many sites in the other programs. The DOE must characterize, treat, and dispose of mixed waste, as well as stabilize landfills and deactivate facilities. The DOD is concerned with remediation soil contaminated with explosives and unexploded ordnance. For each program, the contamination characteristics are summarized as follows:

- *NPL* — VOCs are the most frequent contaminant group found and are the most frequently treated with innovative technologies. VOCs are to be remediated at 71% of the sites, followed by metals (65%) and SVOCs (61%). All three groups (VOCs, SVOCs, and metals) are to be remediated at 25% of the sites, but not necessarily with the same media.

- *RCRA* — Halogenated VOCs, the most prevalent of all contaminated groups, were reported at 60% of the sites, followed by heavy metals at 46% and nonhalogenated VOCs at 32%.

- *USTs* — Underground storage tanks contain petroleum products that are mixtures of four types of hydrocarbons: paraffin, olefins, naphthalenes, and aromatics.

- *DOD* — The most prevalent contaminant groups in groundwater are VOCs and metals, which appear in 74 and 59% of DOD groundwater sites, respectively. SVOCs and metals were more consistent across different media than VOCs. SVOCs were found in 31 to 43% of the sites, and metals were found in 59 to 80% of the sites. Fuels were found at fewer than 22% of all sites. The most frequently occurring group, metals, was found at 69% of all sites, followed by VOCs at 65% and SVOCs at 43%.

- *DOE* — Organics, PCBs, petroleum/fuel hydrocarbons, solvents, TCE, unspecified VOCs, and unspecified SVOCs are among the contaminants found. Metals cited most often include lead, beryllium, mercury, arsenic, and chromium. Radioactive contaminants are present at most installations; the most frequently cited are uranium, tritium, thorium, and plutonium. In addition, mixed waste containing both radioactive and hazardous contaminants is of particular concern to the DOE because of the lack of an acceptable treatment technology.

- *Other federal agencies* — The primary contaminants are fuels, solvents, and industrial waste constituents.

- *States and private parties* — Some studies show that the most prevalent waste at these sites is organic chemicals, metals, solvents, and oily waste.

References

Cha, D.K., Song, J.S., and Sarr, D., Treatment technologies, *Water Environ. Res.*, 69, 676–689, 1997a.

Cha, D.K., Song, J.S., Sarr, D., and Kim, B.J., Hazardous waste treatment technologies, *Water Environ. Res.*, 68(4), 575–586, 1997b.

Chang, M.L., Wu, S.C., and Chen, C.Y., Diffusion of volatile organic compounds in pressed humid acid disks, *Environ. Sci. Technol.*, 31(8), 2307–2312, 1997.

Charles, A.W., Ed., *Hazardous Waste Management*, McGraw-Hill, New York, 1989.

Cheremisinoff, P.C., Ed., *Encyclopedia of Environmental Control Technology*, Vol. 1: *Thermal Treatment of Hazardous Waste*, Gulf Publishing, Houston, TX, 1989.

Cheremisinoff, P.C., Ed., *Encyclopedia of Environmental Control Technology*, Vol. 4: *Hazardous Waste Containment and Treatment*, Gulf Publishing, Houston, TX, 1990a.

Cheremisinoff, P.C., Ed., *Encyclopedia of Environmental Control Technology*, Vol. 6: *Pollution Reduction and Contaminant Control*, Gulf Publishing, Houston, TX, 1990b.

DOD, *Restoration Management Information System*, Department of Defense, Office of the Deputy Under Secretary of Defense (Environmental Security), Washington, D.C., November, 1995.

Greenland, D.J. and Hayes M.H.B., Eds., *The Chemistry of Soil Constituents,* John Wiley & Sons, New York, 1978.

Ground-Water Remediation Technologies Analysis Center, *Remediation Technologies*, www.gwrtac.org/htm/tech_topic.htm.

Suidan, M.T. et al., Why MTBE and gasoline oxygenates? *J. Environ. Eng.*, Sept., 772, 2002.

Swain, C.G. and Lupton, E.C. Jr., *J. Am. Chem. Soc.* 90, 4328, 1968.

U.S. Department of Energy, *Contaminated Media/Waste Database as of March 3, 1995*, DOE/EM-40, Washington, D.C., 1995.

U.S. Department of Energy, *Draft Programmatic Environmental Impact Statement for the Uranium Mill Tailings Remedial Action Ground Water Project*, DOE/EIS-0198, UMTRA Project Office, Washington, D.C., April, 1995.

U.S. Department of Energy, *Estimating the Cold War Mortgage: The 1995 Baseline Environmental Report*, DOE/EM-2032, Washington, D.C., March 1995.

USEPA, *Innovative Treatment Technologies: Annual Status Report*, 8th ed., EPA-542-R-96-010, U.S. Environmental Protection Agency, Office of Solid Waste and Emergency Response, Technology Innovation Office, Washington, D.C., November, 1996.

USEPA, *Innovative Treatment Technologies: Annual Status Report Database (ITT Database)*, EPA-542-C-96-002, U.S. Environmental Protection Agency, Office of Solid Waste and Emergency Response, Technology Innovation Office, Washington, D.C., January, 1997a.

USEPA, *Clean Up the Nation's Waste Sites: 1996*, EPA 542-R-96-005, U.S. Environmental Protection Agency, Office of Solid Waste and Emergency Response, Technology Innovation Office, Washington, D.C., April 1997b.

USEPA, Hazardous waste clean-up information at www.clu-in.com, 2003.

Watts, R.J., *Hazardous Wastes: Sources, Pathways, Receptors,* John Wiley & Sons, New York, 1997.

3

Physicochemical Treatment Processes

3.1 Introduction

Treatment technologies can be classified many different ways. For example, for achieving cleanup standards, the Clear Water Act (CWA) classifies treatment technologies as follows:

- Best practicable control technology (BPCT) applies to all pollutants.
- Best conventional technology (BCT) applies to conventional pollutants.
- Best available technology (BAT) applies to toxic and unconventional pollutants.
- Best available demonstrated technology (BADT) is used for new dischargers.

According to treatment mechanisms, however, treatment technologies are classified as biological, physicochemical, and thermal processes. In terms of the place where the actual treatment takes place, the issue of *in situ* vs. *ex situ* comes into play as far as selecting the most cost-effective remediation processes. For example, most bioremediation processes are *in situ*, while physicochemical processes may be implemented both *in situ* and *ex situ*, according to the following:

- *In situ physical/chemical treatment* — The advantage of *in situ* treatment is that it allows soil to be treated without being excavated and transported; however, it generally requires longer time periods, and treatment uniformity is less certain because of the variability in soil and aquifer characteristics. Physical/chemical treatment processes destroy, separate, or contain pollutants by using physicochemical approaches. Physical/chemical technologies for volatile organic compounds (VOCs) include soil flushing, soil vapor extraction, and solidification/stabilization (*in situ* vitrification).

- *Ex situ physical/chemical treatment* — *Ex situ* treatment generally requires excavation of soil or pumping of groundwater — processes that may increase costs but which are less time consuming than *in situ* treatment. In addition, *ex situ* treatment is more reliable due to being able to homogenize the contaminants. The available *ex situ* physicochemical technologies for VOCs consist of soil washing, ultraviolet oxidation, air stripping, liquid-phase carbon adsorption, and dehalogenation.

- *In situ thermal treatment* — *In situ* thermal treatment allows pollutants to be treated without being excavated and transported. Thermal treatment requires a shorter clean-up time; however, high costs are usually associated with the amount of energy and equipment required. For example, enhanced soil vapor extraction is usually an energy-intensive process.

- *Ex situ thermal treatment* — *Ex situ* thermal treatment generally requires excavation of soil or pumping of groundwater, processes that may increase costs; however, much shorter time is required for *ex situ* than *in situ* treatment. Moreover, it can achieve the designed efficiency due to the controlled reaction environments. Thermal processes typically use heat to increase the volatility (separation), detonate (destruction), or melt (immobilization). Innovative *ex situ* thermal treatment technologies for VOCs include thermal desorption (separation technology) and incineration (destruction technology).

The history of physicochemical treatment processes is summarized in Table 3.1.

3.2   Treatment Technologies

3.2.1   Phase Transfer Technologies for Halogenated VOCs and Nonhalogenated VOCs

3.2.1.1   *Air Stripping*

Air stripping can be used to separate a broad range of VOCs from water; however, it is effective only for contaminated water with VOC or semivolatile concentrations with a dimensionless Henry's constant greater than 0.01. Henry's law constant is used to determine whether air stripping will be effective. Generally, organic compounds with Henry's constants greater than 0.01 atm-m^3/mol, such as chloroethane, trichloroethylene (TCE), dichloroethylene (DCE), and perchloroethylene (PCE), are amenable to air stripping. Organic pollutants with low volatility at cold temperatures may require preheating of groundwater. Figure 3.1 shows the relative range of Henry's

TABLE 3.1

Historical Aspects of Treatment Technologies for VOCS and SVOCS

Time	Technologies/Events	Contaminant	Organization
1874	First incinerator (Cheremisinoff, 1989)	Hazardous wastes	British government
1885	First U.S. incinerators (Cheremisinoff, 1989)	Hazardous wastes	U.S. government
1947– 1977	Hot spot: PCBs from two electric plants were discharged into the Hudson River	PCBs	U.S. government
1974	*Vulcanus* incinerator ship test in Gulf of Mexico	Halogenated VOCs	USEPA
1976– 1983	Love Canal, New York	Dioxins, TCP, and hazardous wastes	Hooker Chemical Company, New York
1982	Times Beach, Missouri	Dioxins, and hazardous wastes	U.S. government
1983	*Vulcanus II* in North Sea	Halogenated VOCs	USEPA
1983	Mobile rotary kiln with afterburn facility	HCB, 1,2-4 trichlorobenzenes in toluene	USEPA
1987	Terra Vac, vacuum extraction system (Cheremisinoff, 1989)	VOCs and TCE	Valleys Products Co., Ltd., Montana
1987	Organic extraction using solvents at the general refining Superfund site in Savannah, GA (Cheremisinoff, 1989)	Oily sludges, hydrocarbon-contaminated soil, and triethylamine (TEA)	Resources Conservation Company, Georgia
1988	Organic extraction using solvents at New Bedford Harbor, MA (Cheremisinoff, 1989)	PCBs (sediments)	CF Systems Corporation, Arvada, CO
1989	Debris washing system (Cheremisinoff, 1989)	PCBs, pesticides, and metals	IT Corporation / USEPA
1989	Soil washing system (water-based volume reduction) at MacGillis & Gibbs Superfund site in Minnesota (Cheremisinoff, 1989)	PAHs, PCBs, and PCP	BioTrol, Inc., Minnesota
1990	AquaDetox, integrated vapor extraction and stream vacuum stripping (Cheremisinoff, 1989)	VOCs, PCE, and TCE	Dow Chemical Co., Ltd.
1990s	Detoxifier for *in situ* stream/air stripping (Cheremisinoff, 1989)	VOCs, PCBs, and SVOCs	Toxic Treatment (USA) Inc., California
1991	Membrane microfiltration at the Palmerton Zinc Superfund site in Pennsylvania	VOCs (liquid), metals, and oily, inorganic, organic wastes	E.I. DuPont de Nemours and Oberlin Filter Company
1990s	Udell Technologies' stream injection and vacuum extraction (Cheremisinoff, 1989)	VOCs and SVOCs	McClellan Air Force Base, California
1993	*In situ* stabilization/solidification proprietary binder (Cheremisinoff, 1989)	SVOCs in soil	S.M.W. Seiko, Inc., California

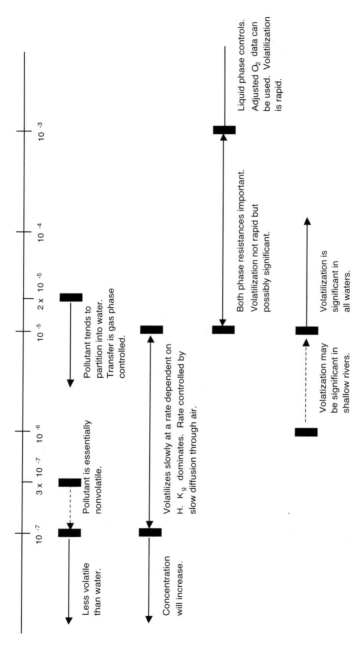

FIGURE 3.1

Ranges and relative values. (From Thomas, R.G., Volatilization from soil, in *Handbook of Chemical Property Estimation Methods*, Lyman, W. J., Reed, W.F., and Rosenblatt, D.H., Eds., McGraw-Hill, New York, 1982, chap. 16. With permission.)

constants of organic pollutants and their volatility in terms of treatability using air-stripping technology.

Typical design considerations include feed water flow rates, water and air temperatures, tower feed and discharge systems (gravity feed or type and location of pumps), and influent contaminant concentrations. In addition, requirements for effluent water contaminant concentrations and restrictions on air emission should be considered according to the various federal and state regulations.

Air stripping involves the mass transfer of volatile contaminants from water to air. An air-stripping tower usually packs a number of trays in a very small chamber to maximize air–water contact while minimizing space. Because of the significant vertical and horizontal space savings, these units are increasingly being used for groundwater treatment. A major operating cost of air strippers is the amount of electricity required for the groundwater pump, the sump discharge pump, and the air blower. Fouling of packing may cause a reduction of the airflow rate, which is caused by the oxidation of minerals (such as iron and manganese) in the feed water, by precipitation of calcium, and by biological growth on the packing material.

In-well stripping technology utilizes the application of VOCs for the treatment of contaminated groundwater. Typically, a blower introduces air or an inert gas (e.g., nitrogen) into the water column within an extraction well. Bubbles produced by the gas create enough pressure to suck the VOCs in the contaminated water into the bubbles by volatilization. Wells are usually equipped with two screened sections and a deflector plate. As the contaminated water encounters the deflector plate, the bubbles break and combine. Water then flows from the screen section, re-infiltrating the vadose zone until all the remediation goals are met. Peng et al. (1994) investigated the volatilization of benzene, toluene, trichloroethene, and tetrachloroethene from quiescent water. The volatilization rate was found to be inversely proportional to the square of the water depth.

3.2.1.2 Soil Vapor Extraction (SVE)

In situ soil vapor extraction (SVE) is a process for removing and venting VOCs from the unsaturated (vadose) zone of soil (Cheremisinoff, 1989). A vacuum is applied to the soil to induce a flow of clean air from the atmosphere into the subsurface. The process resembles continuous flushing of the soil with clean air and continues until volatilization and desorption of contaminants are complete. The gas that leaves the soil can be processed for VOC removal (e.g., via activated carbon adsorption) and then discharged to the atmosphere or injected into the subsurface. It can also be destroyed by combustion in existing boilers that are operating on a continuous basis, depending on local and state air discharge regulations. Vertical extraction vents are typically used at depths of 1.5 m, although they have been successfully applied to a depth of 91 m. Horizontal extraction vents can be used depending on the contaminant zone geometry, drill rig access, or other site-specific factors.

Vapor extraction and air injection wells differ from conventional ground-water monitoring wells in that they are installed in the vadose zone with the screened sections extending above the water table surface. A geo-membrane is placed over the soil surface to prevent short circuiting and the loss of the injected air to the ground surface. In case of a high ground-water table or when the contamination is confined to shallow depths, perforated pipes buried in horizontal trenches would be used in lieu of wells.

Compounds with a Henry's law constant greater than 0.01 or that exhibit vapor pressure greater than 0.5 mmHg (0.02 in Hg) can be successfully treated by *in situ* SVE. For example, solvents such as perchloroethylene (PCE) and trichloroethylene (TCE) and most constituents in gasoline can be effectively removed by this technology. On the other hand, SVE will not remove heavy oils, metals, polychlorinated biphenyls (PCBs), or diox-ins.

The technical factors that must be considered include: (1) volatility of pollutants, (2) lateral and vertical concentration of VOCs, (3) soil types and properties (e.g., structure, texture, permeability, stratification, and moisture content), (4) emission control requirement, (5) schedule for cleanup, (6) position and length of screened interval, (7) blower type, and (8) site factors (e.g., depth and contaminated extension area, depth to water table).

The cost of an SVE system varies with the size of the site, nature of the contaminants, and hydrological setting. Cost estimates for *in situ* SVE range from $10 to $50 per cubic meter of soil or from $10 to $40 per cubic yard (USEPA, 1989).

The advantages of the SVE system are (1) relatively low cost, (2) low maintenance and operating costs, (3) treatment of VOC-contaminated soil in the vadose zone, (4) effective source control, and (5) no further release of contaminants to groundwater. Therefore, it is ideal to use at sites in highly developed areas and/or where contamination has spread to adjacent prop-erties, or underneath buildings.

The volatility of pollutants, type of soil, and groundwater hydrology will affect the treatment efficiency of the process; thus, SVE does have some disadvantages. For example, it is limited to volatile compounds, low groundwater table, and loose sandy formations. It is not recommended for low-hydraulic-conductivity soils (required conductivity exceeding 0.001 to 0.01 cm/sec). Soil that has a high percentage of fines and a high degree of saturation will require higher vacuums (increasing costs). As a result, the technology is not effective in saturated zones. Also, channeling treatments may result from nonhomogeneity of the subsurface; therefore, good knowl-edge of subsurface characteristics is required for the proper location and design of wells. Finally, soil of high organic content or that is extremely dry and has a high sorption capacity of VOCs may result in reduced removal rates.

3.2.2 Phase Transfer Technologies for Halogenated SVOCs, Nonhalogenated SVOCs, and Non-VOCs

3.2.2.1 Activated Carbon Adsorption

Groundwater is pumped through a series of canisters or columns containing activated carbon to which dissolved organic contaminants adsorb. Hydrocarbons, semivolatile organic compounds (SVOCs), nonvolatile organic compounds (non-VOCs), and explosives are all easily removed by GAC; however, because it is a phase transfer technology organic pollutants are not destroyed. All of the used carbon eventually needs to be properly discarded. Periodic replacement or regeneration of saturated carbon is required. Type, pore size, quality of the carbon, and operating temperature will impact performance of the process. Streams with high suspended solids (>50 mg/L), oil, and grease (>10 mg/L) may cause fouling of the carbon and may require frequent treatment. Water-soluble compounds and small molecules are not adsorbed well.

When the concentration of contaminants in the effluent from the bed exceeds a certain level, the carbon can be regenerated at the site. The two most common reactor configurations for carbon adsorption systems are fixed beds or moving beds. The fixed-bed configuration is the most commonly used for the adsorption of non-VOCs from contaminated waters. Liquid-phase carbon adsorption is effective for removing contaminants at low concentrations (<10 mg/L) from water at nearly any flow rate. It is also effective for removing higher concentrations of contaminants from water at low flow rates (typically 2 to 4 L/min or 0.5 to 1 ppm). The removal of suspended solids from contaminated water is critical for effective treatment. The major design variables for liquid-phase carbon applications are empty-bed contact time (EBCT), flow rate, and system configuration, all of which will have an impact on carbon usage. Particle size and hydraulic loading are often chosen to minimize pressure drop and reduce or eliminate backwashing. For a single adsorbent, the EBCT is normally chosen to be large enough to minimize the carbon usage rate. Alternatively, multiple beds in series may be used to decrease carbon usage rate.

The theory and engineering aspects of the technology are well documented by performance data. Carbon adsorption is a relatively nonspecific adsorbent and is effective for removing many organic, explosive, and some inorganic contaminants from liquid and gaseous streams. Costs will depend on waste flow rate, type of contaminant, concentration of contaminant, mass loading, required effluent concentration, and site and timing requirements. Costs are lower at lower concentration levels of a contaminant. Costs are also lower at higher flow rates. At flow rates of 0.4 million L/day (0.1 million gal/day), costs increase from $0.32 to $1.70 per 1000 liters treated water ($1.20 to $6.30 per 1000 gallons treated water).

3.2.2.2 Soil Washing

Contaminants are flushed out from soil by using suitable reagents such as surfactants. The contaminants are dissolved in washing solution or concentrated in a smaller volume of soil through size separation and attrition scrubbing. Soil washing would probably find its greatest site remediation where the soil is contaminated with a single contaminant or a class of contaminants. While certain contaminants such as simple phenols can be readily extracted from both organic and inorganic soils, other contaminants such as PCBs or arsenic cling tenaciously to the soils and resist release into the washing fluids. SVOCs, non-VOCs, and fuels are usually removed through soil washing. Heavy metals can be washed away from soil by using ethylenediamine tetraacetic acid (EDTA), which is recycled in the process. Soil washing can clean a wide range of organic and inorganic contaminants from coarse-grained soil and can recover chemicals from accidental chemical spills; however, the process is not feasible when a mixture containing a range of contaminants with different solubilities is present. The recovered solutions can be very dilute and large in volume, hence costly to treat.

Washing processes that separate the clay silt from the coarser particles can effectively separate and concentrate the contaminants into a smaller volume of soil for further treatment. Sequential washing or different washing solutions may be required for a mixture of contaminants. In designing a soil washing system, soil size distribution (0.24 to 2 mm) and the characteristics, nature, and concentration of contaminants should be evaluated using a bench-scale treatability study. The average cost of $170 per ton depends on the target contaminant.

3.2.3   Thermal Treatment Processes

3.2.3.1   *Thermal Desorption*

Thermal desorption is a physical separation process. Waste is heated to volatilize water and organic contaminants. A carrier gas or vacuum system transports volatilized water and organics to the gas treatment system. The bed temperatures and residence times designed for these systems will volatilize selected contaminants without oxidation. Three types of thermal desorption are available:

1. *Direct fired process* — Fire is applied directly to the surface of contaminated media. The main purpose of the fire is to desorb contaminants from the soil though some contaminants that may be thermally oxidized.

2. *Indirect fired process* — A direct-fired rotary dryer heats an air stream, which, by direct contact, desorbs water and organic contaminants from the soil.

3. *Indirectly heated process* — An externally fired rotary dryer volatilizes the water and organics from the contaminated media into an inert carrier gas stream. The carrier gas is later treated to remove or recover the contaminants.

Two common thermal desorption designs are the rotary dryer and thermal screw. Rotary dryers are horizontal cylinders that can be directly or indirectly fired. The dryer is normally inclined and rotated. For the thermal screw units, screw conveyors or hollow augers are used to transport the medium through an enclosed trough. Hot oil or steam circulates through the auger to indirectly heat the medium. All thermal desorption systems require the treatment of the off-gas to remove particulates and contaminants. Conventional particulate removal equipment, such as wet scrubbers, remove particulates or fabric filters. Contaminants are removed through condensation followed by carbon adsorption, or they are destroyed in a secondary combustion chamber or a catalytic oxidizer. Most of these units are transportable.

Based on the operating temperature of the desorbent, thermal desorption processes can be categorized into two types. The first is high-temperature thermal desorption (HTTD), in which waste is heated to temperatures from 320 to 560°C (600 to 1000°F). This process is frequently used in combination with incineration, solidification/stabilization, or dechlorination of SVOCs, PAHs, PCBs, and pesticides. The second type is low-temperature thermal desorption (LTTD), in which waste is heated to temperatures from 90 to 320°C (200 to 600°F). Nonhalogenated VOCs, SVOCs, and fuels can be effectively treated by LTTD.

All *ex situ* soil thermal treatment systems employ similar feed systems consisting of a screening device to separate and remove materials, a belt conveyor to move the screened soil from the screen to the first thermal treatment chamber, and a weight belt to measure soil mass. The size reduction equipment can be incorporated into the feed system, but its installation is usually avoided to minimize shutdown as a result of equipment failure. Soil storage piles and feed equipment are generally covered to protect them from rain to minimize soil moisture content and material handling problems. Soils and sediments with water contents greater than 20 to 25% may require the installation of a dryer. Some volatilization of contaminants occurs in the dryer, and the gases are routed to a thermal treatment chamber. The cost for this process ranges from $45 to $330 per metric ton ($40 to $300 per ton) of soil.

The advantage of thermal desorption is that it is effective for the full spectrum of organic contaminants; however, dewatering may be necessary to achieve acceptable soil moisture content levels. Highly abrasive feed can potentially damage the processor unit, and heavy metals in the feed may produce a treated solid residue that requires stabilization. Clay, silty-type soil, and soil with a high humic content may increase reaction time because of the strong binding of contaminants between pollutants and soils. Maxymillian Technologies (1994) presented a full-scale test conducted by Clean Berkshires, Inc. They demonstrated that the thermal desorption system could treat soil contaminated with VOCs and PAHs and achieve 99.99% removal.

3.2.3.2   Dehalogenation at High Temperature

Dehalogenation occurs by either the replacement of halogen molecules or the destruction of the contaminant. Soil and sediment that are contaminated with chlorinated organic compounds, especially PCBs, dioxins, and furans, can be remediated through dehalogenation. The contaminated soil is screened, processed with a crusher and pug mill, and mixed with sodium bicarbonate. The mixture is heated to above 330°C (630°F) in a reactor to partially decompose and volatilize the contaminants. The volatilized contaminants are captured, condensed, and treated separately.

Treatability tests should be conducted to identify parameters such as water, alkaline metals, humus contents in the soils, the presence of multiple phases, and total organic halides that could affect processing time and costs. This technology uses standard equipment. The reaction vessel must be equipped to mix and heat the soil and reagents. The cost for the operation of a full-scale facility is estimated to range from $220 to $550 per metric ton ($200 to $500 per ton). The cost does not include excavation, refilling, residue disposal, or analytical costs. Factors such as high clay or moisture content may slightly increase the treatment cost. As opposed to air stripping or activated carbon adsorption, the contaminant is partially degraded rather than being transferred to another medium. The target contaminant groups for dehalogenation treatment are halogenated SVOCs, PCBs, and pesticides; however, high clay and moisture content will increase costs. Concentrations of chlorinated organics greater than 5% may require large volumes of chemical reagents.

The dehalogenation process has been approved by the EPA's Office of Toxic Substances for PCB treatment and has been experimentally implemented for the cleanup of PCB-contaminated soil at the following three Superfund sites: Wide Beach in Erie County, New York (1985); Re-Solve in Massachusetts (1987); and Sol Lynn in Texas (1988). The glycolate process has been used to successfully treat contaminant concentrations of PCBs from less than 2 ppm to reportedly as high as 45,000 ppm. Using this technology, Helland et al. (1995) investigated reductive dechlorination of carbon tetrachloride with elemental iron and found that the rate of dechlorination to chloroform and methylene chloride was a fast first-order process.

3.2.3.3   Incineration

Incineration is required to have a destruction and removal efficiency (DRE) for hazardous wastes of greater than 99.99%. High temperatures ranging from 1600 to 2200°F are used to combust halogenated and other refractory organics. Common types of incinerations are liquid injection incineration, rotary kiln incineration, infrared combustion, fluid-bed incineration, bubbling fluid-bed incineration, and circulating fluid-bed incineration. Under extremely high temperatures, organic pollutants are completely oxidized to carbon dioxide and water. Design factors are threshold temperature, pressure, residence time, mixing intensity, air supply, and materials of construction. Thermodynamics such as heat capacity (in Btu), heat transmission (conduction, convection, and radiation), net heating value (NHV), and DRE are needed. Soil treatment costs range from $200 to $1000 per ton, except for soil contaminated with PCBs or dioxins, in which case costs range from $1500 to $6000 per ton.

Incineration can be applied to all contaminants that have high heat content. It eliminates odor and leachate problems, which are usually associated with landfills, but metals can react with other elements in the feed stream, such as chlorine or sulfur, and form more volatile and toxic compounds. The process has high maintenance and operation costs. Straitz et al. (1995) reported that incineration is a reliable, cost-effective approach to treat VOCs and a wide range of objectionable gas. Chaouki et al. (1995) conducted experiments that examined the effect of fluidized bed combustion on the properties and characteristics of a soil contaminated with PCBs. Particle size distribution and PCB contents were determined before and after a 30-minute incineration. The incineration promoted agglomeration soil, increases in soil pH, and decreases in the PCB content to undetectable levels.

3.2.4 Solidification/Stabilization (Vitrification)

Vitrification is one of the *in situ* solidification/stabilization processes. It melts contaminated soils to destroy organics pollutants by pyrolysis and immobilizes inorganics with high temperatures from 1600 to 2000°C (2900 to 3650°F) by using electric current (USEPA, 1994). Vitrification products are chemically stable, leach-resistant, glass, and crystalline materials. An electric current is passed between electrodes placed in the ground. As the vitrified zone grows, the pyrolysis products migrate to the surface of the vitrified zone and are destroyed by combustion. Off-gas is collected in a hood. Several vitrification processes have been operated on a pilot scale and at full scale, such as DOE's Oak Ridge National Laboratory and Hanford Nuclear Reservation. More than 170 tests have been performed for contaminated soils and sludges. Costs will vary widely, depending on the project size, reagents, and soil properties.

For vitrification to work effectively, the moisture content of soil should be less than 25%. Organic contents are limited to 10%. A minimum content of alkali such as sodium and potassium oxides in soil should be 1.4% by weight, in order to form glass. In addition, particle size, microstructure, and hydrogeologic characteristics should be measured for moderate to high permeability soils. This technology can be applied for destroying or removing organics and immobilizing most inorganics in soil and sludge. The process has been tested on a broad range of VOCs and SVOCs, including dioxins and PCBs. However, reagent delivery and effective mixing are more difficult than that of the *ex situ* process. If contaminants are present below the water table, the soil may require dewatering.

3.2.5 Advanced Oxidation Processes (AOPs)

The *ex situ* treatment process combines the use of ultraviolet (UV) light and chemical oxidants such as ozone and/or hydrogen peroxide to destroy organic contaminants in groundwater. Typically, high-intensity UV radiation reacts with the oxidant through photolysis to generate highly reactive hydroxyl radicals, which attack the organic molecule. UV photolysis is the process in which chemical bonds of the contaminants are broken by UV light, resulting in the production of carbon dioxide and water, etc. Single-lamp, bench-top reactors that can be operated in batch or continuous modes are available for treatability studies.

Design considerations include the existence of free-radical scavengers such as carbonate and bicarbonate species, oxidizer influent dosages, catalysts, UV lamp intensity, retention time, and water turbidity. Costs are between $0.03 and $3.00 per 1000 L ($0.10 to $10.00 per 1000 gal). The process can destroy the majority of groundwater contaminants such as petroleum products; solvent-related organics, such as TCE, DCE, TCA, and vinyl chloride; and some aromatic compounds, such as toluene, benzene, and phenol with different degradation rates. Treatability of each individual pollutant, therefore, has to be considered when designing an effective system. Various sizes of UV oxidation systems are commercially available; however, contaminated water must have a concentration of heavy metal ions of less than 10 mg/L to allow proper transmission of UV light. Free-radical scavengers, including the excessive oxidizers, may inhibit contaminant destruction efficiency. Gates et al. (1995) conducted laboratory experiments to remove TCE from clay soil by adding hydrogen peroxide solutions. Bench-scale slurry studies showed that TCE reduction can be as much as 98% of the initial concentration.

3.3 Established Treatment Technologies and Their Markets

Chapter 2 discussed the distribution of hazardous pollutants in all major remediation programs and the contamination sites in the United States. The distribution of toxic pollutants in all major remediation programs and contamination sites in the United States corresponds to the following seven programs:

- National Priorities List (Superfund)
- Resource Conservation and Recovery Act (RCRA)
- Underground storage tanks (USTs)
- Department of Defense (DOD)
- Department of Energy (DOE)

- Other federal agencies
- States and private parties

Over the past 15 years, almost half a million sites with potential contamination have been reported to state or federal authorities. Over the next decade, federal, state, and local governments and private industries will continue their efforts in the remediation of contaminated sites. This continuing demand for site remediation services and for the technology to treat the different waste streams makes physicochemical treatment processes important for environmental remediation. Remediation technologies in use and factors that affect the demand for remediation technologies are presented here for each program.

3.3.1 National Priorities List Sites

3.3.1.1 Remedial Technology
Since Superfund was established, the approach to remediate contaminated sites has evolved from emphasizing the containment of waste to promoting waste treatment. Prior to 1987, the most common methods for remediating hazardous waste were to excavate the contaminated material and dispose of it at an off-site landfill. The preferred trend of permanent remedies for site cleanup caused the birth of the term *alternative treatment technologies*. The alternative treatment technologies are further divide into two groups: *established* and *innovative*. The established remediation technologies are those that have sufficient published costs and performance data to support their regular use for site remediation. Technologies without sufficient costs and performance data are referred to as innovative remediation technologies. In general, the number of innovative remedial technologies has increased at Superfund and other contaminated sites. During remediation of groundwater at 76% of the Superfund sites, a number of soil vapor extraction (SVE) and thermal desorption projects have been completed; however, the technology of pump and treatability is preferred.

Soil vapor extraction has become the preferred technology for both chlorinated and nonchlorinated VOCs in soil. Thermal desorption and bioremediation are also commonly used to treat VOCs. Bioremediation is usually applied to nonhalogenated VOCs, such as benzene.

Bioremediation and thermal desorption are the most frequently selected innovative technologies for NPL sites with SVOCs, which are the second most common contaminants found at NPL sites. Also, SVE has been selected for some of the most volatile SVOCs (e.g., phenols and naphthalenes). Current research efforts are focused on biodegradation of chlorinated aliphatic hydrocarbons, such as trichloroethylene (TCE) and vinyl chloride, which occur at many sites. Thermal desorption most effectively treats PAHs and PCBs, and it may be particularly useful to pretreat organics prior to metal treatment.

The most frequently selected technology for treating metals is solidification/stabilization. Out of all of the innovative technologies, soil washing is chosen most often to remediate soils that are contaminated with both metals and organics. No treatment technologies have yet been selected at NPL sites with low-level radioactive metals combined with other hazardous constituents (mixed waste). Usually, a combination of different technologies is used to treat media and waste that contain both metals and organics. The purpose of the combined use of several treatment technologies in a series is to:

- Reduce the volume of the material requiring subsequent treatment
- Prevent emission of volatile contaminant during excavation
- Address multiple contaminants within the same media

3.3.1.2 Remediation Cost

The USEPA has estimated that the market value for remediation of non-federal facilities is $6.7 billion (1996 dollars). This estimated cost does not include costs for site assessment and studies, design, operation and maintenance, long-term responses, site management, administrative cost, other agency support, oversight of potentially responsible parties (PRPs), and enforcement activities. The estimate was based on the assumption that PRPs will be responsible for at least 70% of the cost.

3.3.2 Resource Conservation and Recovery Act

3.3.2.1 Remedial Technologies

For facilities treating groundwater, the predominant technology chosen has been pump and treat. The innovative technologies selected include *in situ* bioremediation, *ex situ* bioremediation, thermal desorption, and chemical treatment. The facilities requiring soil treatment are selected based on established technologies, such as off-site disposal and incineration. The innovative technology most often selected has been SVE.

3.3.2.2 Remedial Cost

According to an estimate derived from the Regulatory Impact Analysis (RIA), the RCRA will cost $38.8 billion in 1996 dollars to implement. PRPs will incur most of the cost (about 89%), with the remaining 11% being incurred by federal facilities.

3.3.3 Underground Storage Tank Sites

3.3.3.1 Remedial Technology

Data on established or innovative technologies to remediate contaminated UST sites have not been centralized at the federal level. One study from the University of Massachusetts (1995), provided information based on a written survey collected from 49 states. Based on the responses, approximately 96,000 sites were undergoing remediation in these states. Natural attenuation and pump and treat were the most frequently selected groundwater technologies.

3.3.4 Department of Defense

3.3.4.1 Remedial Technology

The Department of Defense actively participates in technology innovations to meet its environmental restoration needs more effectively and efficiently. The following are examples of innovative technology applications at DOD sites:

- Bioremediation (VOCs and PAHs)
- *In situ* soil vapor extraction (VOCs, PAHs, gasoline, oil, and lubricants)
- Soil washing (PCBs and metals)
- *In situ* vapor extraction for VOCs in groundwater
- *Ex situ* vapor extraction
- *In situ* soil venting (lubricants and solvents)
- *In situ* bioventing (lubricants in soil and groundwater)
- *In situ* bioremediation (lubricants and solvents in soil and groundwater)
- *Ex situ* bioremediation (lubricants in soil and groundwater)
- *Ex situ* bioremediation (explosives and propellants in soil)
- Chemical detoxification of chlorinated aromatic compounds in soil
- *In situ* carbon regeneration
- Incineration of explosive-contaminated soils
- Infrared thermal destruction
- Low temperature thermal stripping of VOCs in soil
- Mobile rotary kiln incineration of soils
- Thermal destruction
- Radio frequency thermal soil decontamination (lubricants and solvents)
- Xanthale treatment (heavy metals in groundwater and wastewater)
- Stabilization/solidification (soil)

3.3.4.2 Remedial Cost

The DOD estimates that the cost of completing the remaining work at all DOD sites will be over $28.6 billion.

3.3.5   Department of Energy

3.3.5.1   *Remedial Technology*

The Department of Energy recognized that much of the remediation at its installations cannot be accomplished without new technological solutions, so the agency has developed the following innovative technologies for their sites:

- *Soil remediation* — Electrokinetics, innovative soil washing, and *in situ* vitrification
- *Groundwater* — Dynamic underground stripping, *in situ* bioremediation, biosorption of uranium, recirculating wells, and microbial filters
- *Facilities* — Gas-phase decontamination
- *Buried waste* — Cooperative telerobotic retrieval, automated waste, and conveyance system
- *Mixed low-level waste treatment* — Plasma hearth system
- *Characterization* — Expedited site characterization
- *High-level waste* — Efficient separation and robotic system

3.3.5.2   *Remedial Cost*

Characterization and assessment (C/A) activities are in progress at most installations and other locations. Although the DOE estimates that it will take 75 years (1996 to 2070) to complete the remediation, it expects to remediate nearly 80% of its currently known sites by 2021. The total estimated cost for the remediation of all installations is $227 billion.

3.3.6   Waste Sites Managed by Other Federal Agencies

3.3.6.1   *Remedial Technology*

The information available regarding the remediation work at the Civilian Federal Agency (CFA) is very limited. Most probably, the CFA will implement the innovative technologies used at the DOD and DOE sites.

3.3.6.2   *Remedial Cost*

Although the precise estimate of the potential costs for remediation at these facilities it not available, it has been estimated at $15 billion.

3.3.7   Sites Managed by States and Private Parties

3.3.7.1   *Remedial Technology*

Based on state actions listed on the CERCLA Information System database, states have selected either containment (on-site or off-site) or treatment (in-site and off-site), similar to DOE or DOD as the predominant technologies

3.3.7.2   *Remedial Cost*

The EPA estimates that it will take an average of 30 years for states to complete remediation at known sites. The total costs for both states and private parties is estimated to be $12.2 billion, based on information available from 45 states. In addition, the agency estimates that the 130,000 to 450,000 brownfield sites will cost over $650 billion to remediate.

3.4   How to Select Treatment Technology

The nature of the pollutants, the concentration of pollutants, and the media contaminated are the three most important determining factors when selecting the appropriate technology for treating a specific type of hazardous waste.

3.4.1   Nature of Pollutants

It is well documented that the structure of organic pollutants and speciation of inorganic pollutants determine the effectiveness of treatment by different technologies. For example, AOPs are used to destroy organic pollutants based on the molecular structures of the pollutants. In terms of reactivity of organic compounds, the following order may be expected:

PCBs > PAHs > chlorinated aliphatic alkanes > chlorinated aromatics > unsaturated chlorinated aliphatic compounds

Within a given class of compounds, the higher the degree of chlorination, the more difficult it becomes to oxidize the compound. On the other hand, it will be easier to reduce it.

The speciation of heavy metals determines the treatability of heavy metals. Furthermore, the size of heavy metals plays a critical role in terms of effectiveness of a physicochemical treatment technology. The following list summarizes the general treatability information provided in Table 3.2:

- When particle size is greater than 1000 μm, it is filterable. Metal precipitates complexed with organic particles will be filterable.

TABLE 3.2

Metal Speciation and Treatability

	← filterable →			
	← membrane filterable →			
	← dialysable →			
	← in true solution →			

Free Metal Ions	Inorganic Ion Pairs, Inorganic Complexes	Organic Complexes, Chelates	Metal Species Bound to High-Molecular-Weight Organic Material	Metal Species in the Form of Highly Dispersed Colloids	Metal Species Sorbed on Colloids	Precipitates, Organic Particle Remains of Inorganic Organisms
Diameter range		$10\ \text{Å}$		$100\ \text{Å}$	$1000\ \text{Å}$	
Cu^{3-} aq	$Cu_3(OH)_3^{2-}Pb$	Me-SR	Me-lipids	FeOOH	$Me_2(OH)_3$	Metal hydroxide
Fe^{3+} aq	$(CO_2)^3$	Me-OOCR	Me-humic-acid polymers	$Fe(OH)_3$	$MeCO_3$, MeS, etc. on clays	Metal oxide precipitates
Pb^{3+} aq	$CuCO_3$		"Lakes"	Mn(IV) oxides	FeOOH or Mn(IV) on oxides	
	AgSH		"Gelhstoffe"	$Mn_2O_{13} \cdot 5H_2O$		
	$CdOH^-$		Me-polysaccharides	$Na_2Mn_{14}O_{31}Ag_3S$		
	$CoOH^-$					
	$Zn(OH)^-$					
	$Ag_3S_3H_3^{2-}$					

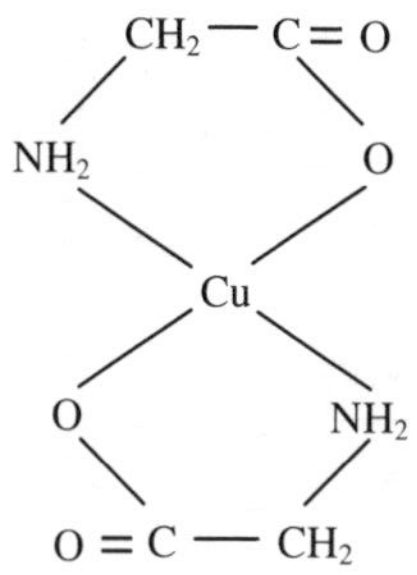

Structure (under Organic Complexes, Chelates):

$$
\begin{array}{ccc}
CH_2 - C = O & & NH_2 \\
& \diagdown\ Cu\ \diagup & \\
NH_2 & & CH_2 \\
& \diagup\quad \diagdown & \\
O & & O = C
\end{array}
$$

Source: Baes, C.F. Jr., and Mesmer, R.E. *The Hydrolysis of Cations,* Wiley-Interscience, New York, 1976. With permission.

- When particle size is between 100 and 1000 µm, the metal is membrane filterable. These particles include metal species bound to high-molecular-weight organic materials.

- When particle size is less than 100 µm, it can be dialyzed. Particles in this class include heavy metals complexed with organics and chelates, inorganic ion pairs, and free metal ions.

One important difference between inorganic and organic pollutants is that the former cannot be destroyed; therefore, inorganic pollutants such as heavy metals are either oxidized to form precipitates of higher valence or reduced to form elementary metals, which can be recovered.

Because bioremediation has *in situ* advantages, physicochemical treatment processes are often used to increase the biodegradability of organic pollutants. Factors affecting the biodegradability of organic pollutants include: (1) chemical structure, such as the presence of an aromatic ring; (2) pollutant concentration; (3) substituents of the target molecule (halo-, nitro-, SO_3H–); (4) pH; (5) alkalinity; and (6) inhibitory compounds. Biodegradability is inversely proportional to the chlorine content of an organic pollutant, as shown in Figure 3.2.

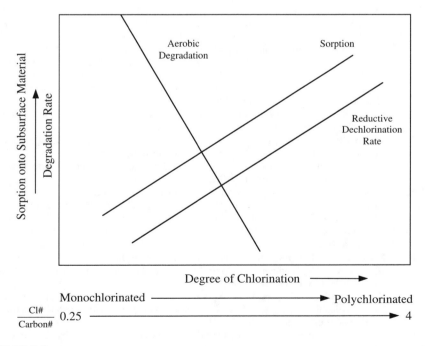

FIGURE 3.2
Effect of degree of chlorination on biological and physicochemical degradation. (From Norris, R.D. et al., *Handbook of Bioremediation*, Lewis Publishers, Boca Raton, FL, 1993. With permission.)

Extensive investigations have shown that molecular structure quantitatively affects the biodegradability of a compound. Table 3.3 summarizes quantitative structure–activity relationship (QSAR) studies in biodegradability.

Because hazardous organic pollutants are toxic to microorganisms, bioremediation has major limitations in terms of concentrations of organic pollutants. The concentration limit for biodegradation is based on the contaminants present and is shown in Table 3.4 (Jeffrey et al., 1995):

Although AOPs are economically unsuitable for wastewaters containing high concentrations of organic constituents, AOPs can be used to transform recalcitrant organic pollutants into biodegradable compounds rather than to

TABLE 3.3

Quantitative Structure and Activity Relationship (QSAR) Models and Biodegradability

Year	Event
1901	Meyer and Overton clearly showed that fat–water partition coefficients could be used to explain the narcotic action of simple organic compounds (Overton et al., 1901).
1930s	The first major step in QSAR analysis occurred when Louis Hammett proposed the Hammett sigma constant (σ_x) to assign numerical values for the electronic effects of substitution on an aromatic ring (Hammett et al., 1940).
1950s	Omerod and Hansen were the first to use σ_x in QSAR study, the former to explain hydrolysis and the latter to relate toxicity (Omerod et al. 1953; Hansen et al., 1962).
1950s	Taft made the next development by extending Hammett's idea to aliphatic and introducing a steric parameter (Taft, 1956).
1960s	Hansch and co-workers combined the concepts of Hammett and Taft to derive a hydrophobic parameter ($\log P$) for substituents that were more related to biological activities and also formulated a biological activity model known as the linear free energy relationship (LFER).
1960	Hall et al. (1960) reported on the relationship between electron density on the ring of substituted tryptophan and the rate of metabolism by *Escherichia coli*.
1970s	Another type of descriptor was derived from the electronic configuration such as steric parameters, van der Waals radii, and intermolecular forces.
1978	Dorn and Knackmuss et al. (1978) correlated the maximum reaction rate for ring scission of a series of substituted catechols with Hammett's σ value
1981	Geating's model combined discriminate analysis with pattern recognition techniques for the development of a series of predictive equations identifying biodegradable and nonbiodegradable compounds (Geating, 1981).
1981	The computer-based procedures drawing upon the USEPA's database cover the behavior of 25,000 chemicals (*USEPA QSAR System User Manual*, 1988).
1981	Mudder's (1981) work, which was a pioneer effort in the area of environmental QSARs, was based on real biodegradation data and used 25 molecular descriptors as described by Hansch and Leo.
1987	Vaishnav et al. (1987) used $\log P$ (K_{ow}) and Hammett's σ value and BOD data for several alcohols, ketones, and related aliphatic compounds to correlate structure with extent of oxygen use at 5 days.
1987	The structural features associated with degradable and persistent substances could be used as a basis for QSARs employing multivariate analysis (Niemi et al., 1987).
1990	Parsons and Govers et al. (1990) reviewed the recent work on QSARs and concluded that no general relationship exists between biodegradability and structure except with analogous groups or homologous series.

TABLE 3.4

Treatment Levels for Soil Contaminated with Restricted RCRA Hazardous Waste

Organics	Concen-tration (ppm)	Threshold Concentra-tion (ppm)	Percent Reduction Range	Technologies Achieved Recommended Effluent Concentration Guidance
Halogenated nonpolar aromatics	0.5–10	100	90–99.0	Biological treatment, low-temperature stripping, soil washing, thermal destruction
Dioxins	0.00001–0.05	0.5	90–99.9	Dechlorination, soil washing, thermal destruction
PCBs	0.1–10	100	90–99.9	Biological treatment, dechlorination, soil washing, thermal destruction
Halogenated phenols	0.5–40	400	90–99	Biological treatment, low-temperature stripping, soil washing, thermal destruction
Halogenated aliphatics	0.5–2	40	95–99.9	Biological treatment, low-temperature stripping, soil washing, thermal destruction
Halogenated cyclics	0.5–20	200	90–99.9	Thermal destruction
Nitrated aromatics	2.5–10.0	10,000	99–99.9	Biological treatment, soil washing, thermal destruction
Heterocyclics	0.5–20	200	90–99.9	Biological treatment, low-temperature stripping, soil washing, thermal destruction
Other polar organics	0.5–10	100	90–99.9	Biological treatment, low-temperature stripping, soil washing, thermal destruction

Source: USEPA, *Superfund LDR Guide No. 6A,* 2nd ed., U.S. Environmental Protection Agency, Washington, D.C., 1990.

mineralize them. For example, Shin and Lim (1996) attempted to improve the biodegradability of naphthalene sulfonic acid (NSA) by preozonation. Wastewater generated from refining the crude naphthalene with acid chemicals contains high concentrations of NSA. In conventional activated sludge plants, naphthalenes carrying the SO_3H group as substitutes resist biodegradation or are not degraded completely. Figure 3.3 shows that the biodegradability (BOD_5/COD) depends on the amount of ozone consumption. It illustrates that biodegradability increases as reaction times increase and decreases as initial NSA concentrations increase; therefore, preozonation can significantly increase COD removal and improve the biodegradability of NSA.

Adams et al. (1996) discussed the biodegradation of nonionic surfactants and the effects of oxidative pretreatment. Nonionic surfactants account for more than 30% of the surfactants used in the world. Ethylene oxide and propylene oxides are bio-recalcitrant, low-foaming surfactants used in many applications such as dishwashing and bottle washing. AOT with H_2O_2 and ozone has proven to be highly effective and enhances the biodegradability of the chemicals. The results of Adams et al. show an initial decrease in the rate of biodegradation of EO/PO, but this decrease was followed by an

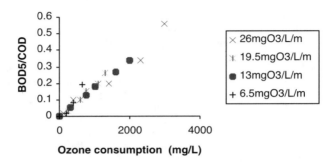

FIGURE 3.3
BOD$_5$/COD vs. ozone dosage. (From Shin, H.S. and Lim, J.L., *Environ. Sci. Health*, 31(5), 1009–1024, 1996. With permission.)

increase in the rate and extent of biomineralization. Advanced oxidation pretreatment using H_2O_2/O_3 at a 0.5 molar ratio has proven to be effective at enhancing the biodegradation of both alkyl phenol ethoxylates (APEs) and EO/PO copolymer.

Carberry and Benzing (1991) worked on peroxide preoxidation of recalcitrant toxic waste to enhance biodegradation. Land disposal is required for industrial chemicals that are not readily biodegraded. Such compounds can have adverse effects on the environment if they escape containment. Hydrogen peroxide is a potentially economical method of preoxidation utilized to enhance the biodegradation of persistent and recalcitrant organics in contaminated soil systems. This pre-oxidation technology was examined in a laboratory respirometer using three model toxic organic chemicals: toluene, trichloroethylene, and pentachlorophenol. Results indicated that prooxidation enhanced the biodegradation of trichloroethylene and pentachlorophenol. Toluene, in contrast, was not significantly oxidized by pretreatment with hydrogen peroxide, and the oxidation pretreatment process did not enhance its biodegradation rate.

3.4.2   Concentration of Pollutants

Figure 3.4 indicates that biological processes are effective when organic concentration is less than 1%. Physicochemical processes are effective for concentrations ranging from 1 to 5%, while thermal processes are the best for concentrations greater than 5%. Figure 3.5 shows that concentration plays a critical role in determining which physicochemical process is the best for a particular situation. For example, adsorption, ion exchange, and membrane filtration may only be suitable when the concentration of heavy metals is less than 0.1% by weight. Precipitation, encapsulation, and electrolysis may be applicable to concentrations less than 1, 5, and 10%, respectively.

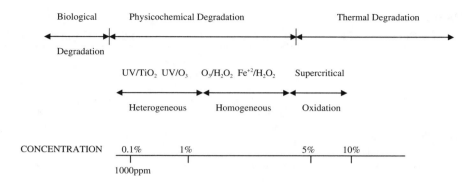

FIGURE 3.4
Treatment technologies and their suitable concentrations.

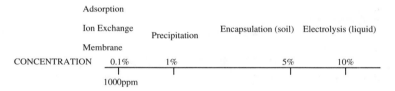

FIGURE 3.5
Treatment technologies of heavy metals and their suitable concentrations.

3.4.3   Contaminated Media

Contaminated media are classified as (1) groundwater, (2) soil, (3) air, and (4) sludge and sediments. The USEPA has developed general guidelines for selecting appropriate treatment technologies (see list below). These guidelines summarize the 48 technologies available, including *in situ* and *ex situ* biological, thermal, and physicochemical processes, and Table 3.5 highlights how eight factors are ranked according to quantitative rating criteria. The following steps, then, should be followed in order to select a suitable technology:

- Identify the contaminated media as soil, sediment, or sludge; as groundwater; or as air emission.
- Identify the class of contaminants as halogenated volatile organic compounds (HVOCs), halogenated semivolatile organic compounds (HSVOCs), non-HVOCs, non-HSVOCs, fuel hydrocarbons, pesticides, or inorganics.
- Select each technology according to the ranking of the nine factors shown in Table 3.6.
- Select a combination of treatment technologies to achieve the remediation goals.

TABLE 3.5

USEPA Definition of Rating Levels for Remediation Technologies of Groundwater

Factor	Definition	Worse	Average	Better
Overall cost	Per 1000 gallons	>$10	$3 – $10	<$3
Commercial availability	Number of vendors	<2	2 – 4	>4
Minimum contaminant concentration achievable	µg/liter	>100	5 – 100	<5
Time to complete cleanup	Years to clean up 10^6 gallons	>10	3 – 10	<3
System reliability and maintainability	Reliability ®	Low R	Average R	High R
	Maintainability (M)	High M	Average M	Low M
Awareness of remediation consulting community	Level of awareness	Unknown	Moderately known	Generally known
Regulatory/permitting acceptability	Level of acceptability	Below average	Average	Above average
Community acceptability	Level of acceptability	Below average	Average	Above average

Source: Means, J.L. et al., *The Application of Solidification/Stabilization to Waste Materials*, CRC Press, Boca Raton, Fl, 1995. With permission.

TABLE 3.6

Remediation Technology Matrix for VOCs and SVOCs

Technology	VOCs	Nonhalogenated VOCs (non-HVOCs)	SVOCs	Nonhalogenated SVOCs (non-HSVOCs)
Soil vapor extraction	g	g	y	y
Thermal desorption	g	g	g	g
Incineration	g	g	g	g
Liquid phase carbon adsorption	g	g	g	g
Dehalogenation	y	y	y	y
Soil washing	y	y	g	g
AOP oxidation	g	g	g	g
Air stripping	g	g	y	y
Vitrification	g	g	g	g

Note: g = potential effectiveness; y = average effectiveness.
Source: USEPA, *Remediation Technologies Screening Matrix and Reference Guide*, EPA 542-B-93-005, U.S. Environmental Protection Agency, Washington, D.C., 1993.

It is very important to note that the screening matrix developed by the USEPA can only be used as a general guide. The critical step that ensures the success of a treatment process is gaining an understanding of the reaction mechanisms and kinetics of an indicator chemical compound within its class.

In other words, treatability of a specific class of organic pollutants has to be evaluated before the process can be selected. This can be achieved through experimental study of an indicator compound. Alternatively, the degradation rate constants or removal efficiencies of the other organic pollutants of the same class may be predicted through QSAR analysis. Similar to the QSAR studies in biological treatment processes, QSAR analysis can also be used as a very powerful tool to assess and predict the treatability of organic pollutants by physicochemical processes, especially by advanced oxidation processes.

3.5 Summary

There are several different ways to classify physicochemical treatment technologies of hazardous wastes. There are huge markets for these technologies to remediate contaminated sites listed by governmental agencies such as the EPA, DOE, and DOD, etc. To select the best remediation technology, the nature of pollutants, the concentration of pollutants, and the contaminated media are the three determining factors. Once a treatment technology is selected, the treatability of the target pollutants should be evaluated so that an effective process can be designed.

References

Adams, C.D., Spitzer, S., and Cowan, R.M., Biodegradation of nonionic surfactants and effects of oxidative pretreatment, *J. Environ. Eng.*, 122(6), 477–483, 1996.

Aki, S. and Abraham, M., Catalytic supercritical water oxidation of pyridine: comparison of catalysts, *Indust. Eng. Chem. Res.*, 38(2), 358–367, 1999.

Akmehmet, I. and Arslan, I., Application of photocatalytic oxidation treatment to pretreated and raw effluents from the Kraft bleaching process and textile industrial, *Environ. Pollut.*, 103(2–3), 261–268, 1998.

Anitescu, G., Zhang, Z., and Tavlarides, L., Kinetic study of methanol oxidation in supercritical water, *Indust. Eng. Chem. Res.*, 38(6), 2231–2237, 1999.

Augugliaro, V. and Coluccia, S., VOCs abatement: photocatalytic oxidation of toluene in vapour phase on anatase TiO_2 catalyst, 1997.

Baes, C.F., Jr. and Mesmer, R.E., *The Hydrolysis of Cations*, Wiley-Interscience, New York, 1976.

Bagnasco, G., Peluso, G., and Russo, G., Ammonia oxidation over CuO/TiO_2: selectivity and mechanistic study, paper presented at the 3rd World Congress on Oxidation Catalysis, 1997, pp. 643–652.

Beziat, J., Besson, M., and Gallezot, P., Catalytic wet air oxidation of wastewater, 3rd World Congress on Oxidation Catalysis, 1997, pp. 615–621.

Beziat, J., Besson, M., Gallezot, P., and Durecu, S., Catalytic wet air oxidation on a Ru/TiO$_2$ catalyst in a trickle-bed reactor, *Indust. Eng. Chem. Res.*, 1310–1315, 38(4), 1999.

Biswas, P. and Zachariah, M.R., *In situ* immobilization of lead species in combustion environments by injection of gas phase silica sorbent precursors, *J. Environ. Sci. Technol.*, 31(9), 2455–2463, 1997.

Carberry, J.B. and Benzing, T.M., Peroxide pre-oxidation of recalcitrant toxic waste to enhance biodegradation, *Water Sci. Technol.*, 23, 367–376, 1991.

Cha, D.K., Song, J.S., and Sarr, D., Treatment technologies, *Water Environ. Res.*, 69, 676–689, 1997a.

Cha, D.K., Song, J.S., Sarr, D., and Kim, B.J., Hazardous waste treatment technologies, *Water Environ. Res.*, 68(4), 575–586, 1997b.

Chang, M.L., Wu, S.C., and Chen, C.Y., Diffusion of volatile organic compounds in pressed humid acid disks, *Environ. Sci. Technol.*, 31(8), 2307–2312, 1997.

Chaouki, J. et al., Incineration of PCB-contaminated soils: effect on soil properties, *Proc. Int. Conf. Fluidized Bed Combustion*, ASCE, New York, 2, 1171, 1995.

Charles, A.W., Ed., *Hazardous Waste Management*, McGraw-Hill, New York, 1989.

Cheremisinoff, P.C., Ed., *Encyclopedia of Environmental Control Technology*, Vol. 1: *Thermal Treatment of Hazardous Waste*, Gulf Publishing, Houston, TX, 1989.

Cheremisinoff, P.C., Ed., *Encyclopedia of Environmental Control Technology*, Vol. 4: *Hazardous Waste Containment and Treatment*, Gulf Publishing, Houston, TX, 1990a.

Cheremisinoff, P.C., Ed., *Encyclopedia of Environmental Control Technology*, Vol. 6: *Pollution Reduction and Contaminant Control*, Gulf Publishing, Houston, TX, 1990b.

Chu, H., *Preparation, Characterization and Evaluation of Heterogeneous Catalysts for Wastewater Treatment*, Hong Kong, 1998.

Crittenden, J., Suri, R., Perran, D., and Hand, D., Decontamination of water using adsorption and photocatalysis, *J. Water Res.*, 31(3), 411–418, 1997.

Dorn, E. and Knackmuss, H.J., Chemical structure and biodegradability of halogenated aromatic compounds: substituted effects on 1,2-dioxygenation of catechol, *Biochem. J.*, 174, 85, 1978.

Elliott, D., Neuenschwander, G., Phels, M., Hart, T., and Zacher, A., Chemical processing in high-pressure aqueous environments. 6. Demonstration of catalytic gasification for chemical manufacturing wastewater cleanup in industrial plans, *Indust. Eng. Chem. Res.*, 879–883, 1999.

Gates, D.D. and Siegrist, R.L., *In-situ* chemical oxidation of tricholorethylene using hydrogen peroxide, *J. Environ. Eng.*, 121, 639, 1995.

Geating, J., Literature study of the biodegradability of chemicals in water, Report EP-600/2-81-175, U.S. EPA, Cincinnati, OH, 1981.

Hall, A.N. et al., The degradation of some benzene substituted tryptophans by *Escherichia coli* tryptophanase, *Biochem. J.*, 74, 209, 1960.

Hammett, L.P., *Physical Organic Chemistry*, McGraw-Hill, New York, 1940.

Hansen, O.R., *Acta Chem. Scand.*, 16, 1593–1600, 1962.

Hayes, R. and Kolackowski, S., *Introduction to Catalytic Combustion*, Gordon and Breach Science, London, 1997.

Helland, B.R. et al., Reductive dechlorination of carbon tetrachloride with elementary iron, *J. Hazard. Mater.* 41, 205, 1995.

Hess, G., Froitzheim, H., and Baumgartner, C., The adsorption and catalytic decomposition of CO$_2$ on Fe(111) surfaces studied with high resolution EELSS, *J. Surface Sci.*, 331–333(A), 138–142, 1995.

Hewitt, A.D., Comparison of sample preparation methods for the analysis of volatile organic compounds in soil samples: solvent extraction vs. vapor partitioning, *J. Environ. Sci. Technol.*, 32(1), 143–149, 1998.

Horst, G. and Dieter, H., Analysis of vapor extraction data from applications in Europe, paper presented at the 6th National Conference on Hazardous Waste and Hazardous Materials, New Orleans, LA, April 12–14, 1989, pp. 542–547.

Huang, W. and Walter, W.J., Jr., A distributed reactivity model for sorption by soils and sediments. 10. Relationships between desorption, hysteresis, and the chemical characteristics of organic domains, *Environ. Sci. Technol.*, 31(9), 2562–2569, 1997.

Hudlicky, M., *Reduction in Organic Chemistry*, 2nd ed., ACS Monographs, American Chemical Society, Washington, D.C., 1996.

Imamura, S., Catalytic and not catalytic wet oxidation, *Indust. Eng. Chem. Res.*, 38(5), 1743–1753, 1999.

Johnson, P.C., Assessment of the contribution of volatilization and biodegradation to *in situ* air sparging performance, *Environ. Sci. Technol.*, 32(2), 276–281, 1998.

Kim, B.J. and Qi, S., Hazardous waste treatment technologies, *Water Environ. Res.*, 67, 560–570, 1995.

Kim, B.J., Qi, S., and Shanley, R.S., Hazardous waste treatment technologies, *Water Environ. Res.*, 66, 440–455, 1994.

Lin, S. and Gurol, M., Catalytic decomposition of hydrogen peroxide on iron oxide: kinetics, mechanism, and implications, *Environ. Sci. Technol.*, 32(10), 1417–1423, 1998.

MacNeil, J., Berseth, P., and Westwood, G., Aqueous catalytic disproportionation and oxidation of nitric oxide, *Environ. Sci. Technol.*, 32, 876–881, 1998.

Martino, C. and Savege, P., Oxidation and thermolysis of methoxy- and hydroxy-substituted phenols in supercritical water, *Indust. Eng. Chem. Res.*, 38(5), 1784–1791, 1999.

Matsumoto, M.R., Reed, B.E., and Jensen, J.N., Physicochemical processes, *Water Environ. Res.*, 67, 419–431, 1995.

Max, D., Ed., *How to Select Hazardous Waste Treatment Technologies for Soils and Sludges*, Noyes Data Corporation, 1989.

Maxymillian Technologies, www.maxymillian.com.

McCabe, M.M., Evaluation of a thermal technology for the treatment of contaminated soils and sludges, paper presented at the 6th National Conference on Hazardous Waste and Hazardous Materials, New Orleans, LA, April 12–14, 1989, pp. 339–342.

Means, J.L., Smith, L.A., Nehring, K.W. et al., *The Application of Solidification/Stabilization to Waste Materials*, CRC Press, Boca Raton, FL, 1995.

Mudder, T.I., Development of empirical structure–biodegradability relationships and testing protocol for slightly soluble and volatile priority pollutants, Ph.D. dissertation, University of Iowa, Iowa City, 1981.

Niemi, G.J. et al., Structure features associated with degradable and persistent chemicals, *Environ. Tox. Chem.*, 6, 515, 1987.

Nogueira, R., Alberici, R., Mendez, M., Jardin, W., and Eberlin, M., Photocatalytic degradation of phenol and trichloroethylene: online and real-time monitoring via membrane introduction mass spectroscopy, *Indust. Eng. Chem. Res.*, 38(5), 1754–1758, 1999.

Norris, R.D. et al., *Handbook of Bioremediation*, Lewis Publishers, Boca Raton, FL, 1993.

Omerod, W.E., *Biochem. J.*, 54, 701–704, 1953.

Overton, E., *Studien über die Narkose*, Gustav Fisher, Jena, Germany, 1901.

Parsons, J.R. and Govers, H.A.J., Quantitative structure–activity relationships for biodegradation, *Ecotoxicol. Environ. Safety*, 19, 212, 1990.

Peng, J. et al., Volatilization of selected organic compounds from quiescent water, *J. Environ. Eng.*, 120, 662, 1994.

Petrier, C., Jiang, Y., and Francois, M., Ultrasound and environment: sonochemical destruction of chloroaromatic derivatives, *Environ. Sci. Technol.*, 32(9), 1316–1318, 1998.

Pintar, A., Bercic, G., and Batista, J., Catalytic liquid-phase phenol oxidation over metal oxides and molecular sieve, paper presented at the 3rd World Congress on Oxidation Catalysis, 1997, pp. 633–642.

Reed, B.E., Lin, W., Matsumoto, M.R., and Jensen, J.N., Physicochemical processes, *Water Environ. Res.*, 69, 444–484, 1997.

Remediation Technologies Network, L.L.C., Categories of technologies, www.remedial.com, 2003.

Rivas, F., Kolaczkowski, S., Beltran, F., and McLurgh, D., Hydrogen peroxide promoted wet air oxidation of phenol: influence of operating conditions and homogeneous metal catalysts, *J. Chem. Technol. Biotechnol.*, 74(5), 390–398, 1999.

Rowe, D. and Lloyd, W., The catalytic purification of CO-rich air, *J. AWMA*, 49(3), 308–309, 1999.

Sasaoka, E., Sada, N., Hara, K., and Sakata, Y., Catalytic activity of lime for N_2O decomposition under coal combustion conditions, *Indust. Eng. Res.*, 38(4), 1335–1340, 1999.

Shin, H.S. and Lim, J.L., *Environ. Sci. Health*, 31(5), 1009–1024, 1996.

Stefan, M. and Bolton, J., Mechanism of the degradation of 1,4-dioxene in dilute aqueous solution using the UV/hydrogen peroxide process, *Environ. Sci. Technol.*, 32(11), 1588–1595, 1998.

Stone, G. and Brooks, D., Carbon clean, *Water Environ. Technol.*, 8(2), 40–43, 1996.

Straitz, J.F., Use incineration to destroy gases safely. *Environ. Eng. World*, 1, (4), 18, 1995.

Taft, R.W., Separation of polar, steric, and resonance effects in reactivity, in *Steric Effects in Organic Chemistry*, Newman, M.S., Ed., John Wiley & Sons, New York, 1956.

Thomas, R.G., Volatilization from soil, in *Handbook of Chemical Property Estimation Methods*, Lyman, W. J., Reed, W.F., and Rosenblatt, D.H., Eds., McGraw-Hill, New York, 1982.

Treatment Technologies Screening Matrix, www.frtr.gov/matrix2/section3/matrix.html, 2003.

USEPA, QSAR System User Manual: A Structure–Activity Based Chemical Modeling and Information System, U.S. Environmental Protection Agency, Environ. Res. Lab., Duluth, MN, and Montana State University, Bozeman, 1988.

USEPA, Terra Vac, *In-situ* vacuum extraction system, Applications Analysis Report, Cincinnati, OH, EPA/540/A5-89/003, 1989.

USEPA, Superfund LDR Guide No. 6A, 2nd ed., U.S. Environmental Protection Agency, Washington, D.C., 1990.

USEPA, Remediation Technologies Screening Matrix and Reference Guide, EPA 542-B-93-005, U.S. Environmental Protection Agency, Washington, D.C., 1993.

USEPA, Fifth Forum on Innovative Hazardous Waste Treatment Technologies: Domestic and International, Chicago, IL, USEPA, Office of Solid Waste and Emergency Response, Washington, D.C., 1994.

USEPA, Innovative Treatment Technologies: Annual Status Report Database (ITT Database), EPA-542–C-96–002, U.S. Environmental Protection Agency, Office of Solid Waste and Emergency Response, Technology Innovation Office, Washington, D.C., January, 1997.

USEPA, Hazardous waste clean-up information, U.S. Environmental Protection Agency, Ground-Water Remediation Technologies Analysis Center, Remediation Technologies, www.gwrtac.org/htm/tech_topic.htm, 2003.

Vaishnav, D.D. et al., Quantitative structure–biodegradability relationships for alcohols, ketones and alicylic compounds, *Chemosphere*, 16, 695, 1987.

Weir, B.A. and McLane, C.R., Design of an UV oxidation system for treatment of TCE-contaminated groundwater, *Environ. Progr.*, 15(3), 179–186, 1996.

Willard, E.M. and Ruby, M.M., Eds., *Environmental Hazards: Toxic Waste and Hazardous Material*. ABC-CLIO, Inc., 1991.

Wu, Y., Taylor, K., Biswas, N., and Bewtra, J., Kinetic model-aided reactor design for peroxidase-catalyzed removal of phenol in the presence of polyethylene glycol, *J. Chem. Technol. Biotechnol.*, 74(6), 519–526, 1999.

4

Advanced Oxidation Processes

4.1 Introduction

Any oxidation process in which hydroxyl radical is the dominant species is defined as an advanced oxidation process (AOP). For any oxidation reaction, two factors determine the rate of reaction. First, if a reaction has a high free energy or high electrical potential, the reaction is very likely to occur and it is considered to be thermodynamically favorable. The oxidation potentials for common oxidants suitable for environmental applications are listed in Table 4.1.

As can be seen in the table, the hydroxyl radical has an oxidation potential of 2.80 V. The hydroxyl radical is a short-lived and extremely potent oxidizing agent, according to its potential as shown in the table. Because they are extremely potent oxidizing agents, hydroxyl radicals react with organic compounds by three mechanisms: hydrogen abstraction, electron transfer, and hydroxylation (Huang et al., 1993). From a thermodynamic point of view, the higher the oxidation potential is, the stronger the oxidant species will be.

Another factor is how fast the reaction is. The fundamental theory underlining the mechanisms involved in AOPs is the transition state theory (TST), which provides theoretical guidance for the search of the most efficient AOP. According to the TST, hydroxyl radicals may accelerate the oxidation rates of an organic compound by several orders of magnitude compared with oxidation rates for common oxidants. This is because the radical reaction will have a much lower activation energy barrier than regular reactions do; therefore, oxidants such as oxygen, hydrogen peroxide, and ozone are combined with catalysts such as transition metals, metal oxides, photons, and ultrasound to generate hydroxyl radicals.

For each AOP, the degradation rate is investigated to search for the most efficient process. We begin this chapter with basic chemical kinetics followed by discussion on the TST, oxidants, and catalysts used in AOPs.

TABLE 4.1

Oxidation–Reduction Potentials of Chemical Reagents
for Water and Wastewater Treatment

Reactions	Potential in Volts (E°) at 25°C
$F_2 + 2e = 2F^-$	2.87
$OH\cdot + H^+ + e^- = H_2O$	2.33
$O_3 + 2H^+ + 2e = O_2 + H_2O$	2.07
$H_2O_2 + 2H^+ + 2e = 2H_2O_2$ (acid)	1.76
$MnO_4^- + 4H^+ + 3e = MnO_2 + 2H_2O$	1.68
$HClO_2 + 3H^+ + 4e = Cl^- + 2H_2O$	1.57
$MnO_4^- + 8H^+ + 5e = Mn^{2+} + 4H_2O$	1.49
$HOCl + H^+ + 2e = Cl^- + H_2O$	1.49
$Cl_2 + 2e = 2 Cl^-$	1.36
$HOBr + H^+ + 2e = Br^- + H_2O$	1.33
$O_3 + H_2O + 2e = O_2 + 2 OH^-$	1.24
ClO_2 (gas) $+ e = ClO_2^-$	1.15
$Br_2 + 2e = 2Br^-$	1.07
$HOI + H^+ + 2e = I^- + H_2O$	0.99
ClO_2 (aq.) $+ e = ClO_2^-$	0.95
$ClO^- + 2H_2O + 2e = Cl^- + 2OH^-$	0.9
$H_2O_2 + 2H_3O + 2e = 4H_2O$ (basic)	0.87
$ClO_2^- + 2H_2O + 4e = Cl^- + 4OH^-$	0.78
$OBr^- + H_2O + 2e = Br^- + 4OH^-$	0.70
$I_2 + 2e = 2 I^-$	0.54
$I_3 + 2e = 3 I^-$	0.53
$OI^- + H_2O + 2e = I^- + 2OH^-$	0.49
$O_2 + 2H_2O + 4e = 4OH^-$	0.40

Source: Lide, D.R. et al., *CRC Handbook of Chemistry and Physics*,
73rd ed., CRC Press, Boca Raton, FL, 1992. With permission.

4.2   Chemical Kinetics

Chemical kinetics focuses on the rate of a reaction through studying the
concentration profile with time. Based on the number of reactants involved
in the chemical reaction, the reaction can be classified as zero, first, or second
order. Third-order reactions are rare because the probability of three reactants
colliding and reacting is low. The following are simplified mathematic
descriptions of the chemical kinetics of the various orders.

4.2.1   Zero-Order Reactions

The rate law for a reaction that is zero order can be expressed as:

$$-\frac{dA}{dt} = k'\left[A_0\right]^0 = k \tag{4.1}$$

The above equation can be integrated as:

$$\int_{A_0}^{A} dA = -k\int_0^t dt \tag{4.2}$$

Therefore, the time required to reduce the concentration of reactant A to half is:

$$t_{1/2} = \frac{1/2A_0}{k} = \frac{A_0}{2k} \tag{4.3}$$

As a result, the rate constant can be found as:

$$k = \frac{A_0}{2t_{1/2}} \tag{4.4}$$

Because the general form of the units of rate constants is (time)$^{-1}$(concentration)$^{1-n}$, the unit of the rate constant of a zero-order reaction is (time)$^{-1}$(concentration)^1. The rate of zero-order reaction is independent of the concentration of the reactant, which is often encountered in heterogeneous reactions on the surface such as activated carbon adsorption.

4.2.2   First-Order Reactions

The following is a typical first-order reaction:

$$A \xrightarrow{k} B \tag{4.5}$$

The rate law for the first-order reaction is:

$$\frac{dA}{dt} = -k[A] \tag{4.6}$$

Integrating this equation, the time dependence of concentration A becomes:

$$-\ln[A] = -kt + \text{constant} \tag{4.7}$$

The concentration profile of reactant A is:

$$\frac{[A]_t}{[A]_0} = -kt \qquad (4.8)$$

Therefore, the $t_{1/2}$ can be found as follows:

$$k = \frac{\ln 2}{t_{1/2}} \qquad (4.9)$$

where the rate constant k is $(time)^{-1}$.

4.2.3   Second-Order Reactions

The second-order reaction has the following general form:

$$A + B \xrightarrow{\ k\ } C \qquad (4.10)$$

The rate law for a second-order reaction can be expressed as:

$$\frac{dA}{dt} = -k[A][B] \qquad (4.11)$$

Assume the x moles of reactants A and B have been reacted and x moles of C have been produced, then the production rate of C should be:

$$\frac{dx}{dt} = k(A_0 - x)^1 (B_0 - x)^1 \qquad (4.12)$$

$$\frac{dx}{k(A_0 - x)(B_0 - x)} = dt \qquad (4.13)$$

$$\int_0^x \frac{dx}{k(A_0 - x)(B_0 - x)} = \int_0^t dt \qquad (4.14)$$

$$\frac{a}{A_0 - x} + \frac{b}{B_0 - x} = \frac{1}{(A_0 - x)(B_0 - x)} \qquad (4.15)$$

$$k = \frac{1}{t} \cdot \frac{1}{B_0 - A_0} \ln \frac{B_0 - x}{A_0 - x} + I \qquad (4.16)$$

$$kt = \frac{1}{B_0 - A_0} \ln \frac{A_0(B_0 - x)}{B_0(A_0 - x)} \tag{4.17}$$

$$A = \frac{A_0}{1 + kA_0 t} \tag{4.18}$$

$$k = \frac{1}{t_{1/2} A_0} \tag{4.19}$$

where the unit of rate constant k should be $(time)^{-1}(concentration)^{-1}$.

4.2.4   *n*th Order Reactions

For the nth order reaction with respect to one reactant, the general solution for rate expression is:

$$r = \frac{dA}{dt} = -k[A]^n \tag{4.20}$$

A simple integration of this equation results in:

$$\frac{1}{n-1}\left(\frac{1}{[A]_t^{n-1}} - \frac{1}{[A]_0^{n-1}} \right) = kt \tag{4.21}$$

which can be rewritten as:

$$\left(\frac{1}{[A]_t^{n-1}} - \frac{1}{[A]_0^{n-1}} \right) = (n-1)kt \tag{4.22}$$

As mentioned before, the unit of the rate constant is $(time)^{-1}(concentration)^{1-n}$.

4.3   Transition State Theory

Chemical reactions are studied in terms of elementary reactions involving only one step for bond breaking, bond formation, or electron transfer. A characteristic of elementary reactions is the molecularity. In other words, if a reaction takes place in a single irreducible act at a molecular level without any detectable intermediates, the reaction is called as an *elementary reaction*:

$$A + B \xrightarrow{\ kp\ } C \tag{4.23}$$

However, when reactants A and B are approaching each other, most of the motions in a reacting molecular system are ordinary vibrations, rotations, and translations. Only one normal mode corresponding to the reaction coordinates is involved in breaking or forming a chemical bond to form a new molecule. The new chemical bond results in the rearrangement of atoms. The collection of these atoms is defined as the *reactive center*.

As the distance between the two atoms becomes shorter and shorter, the electron clouds interact with each other due to the rapid motion of electrons. As a result, a multidimensional and continuous potential surface is developed. The potential field becomes stronger and stronger as the two molecules approach each other. On this multidimensional surface exists a most economic energy path for reactants A and B to interact with each other. The reaction path is referred as the *reaction coordinate*. Along this coordinate, the highest energy along the most economic reaction path defines an *activated complex* (AC), which is expressed as $[AB]^{\ddagger}$. The contour plot of such potential energy can be seen in Figure 4.1. The energy between reactants A and B and $[AB]^{\ddagger}$ is defined as the *activation energy barrier*, as shown in Figure 4.2.

According to the TST, the following reaction will represent the formation of the activated complex and subsequent production of C:

$$A + B \overset{K^{\ddagger}}{=\!=\!=} [AB]^{\ddagger} \xrightarrow{\ kp\ } C \tag{4.24}$$

The AC usually has several characteristics. The molecules going over the barrier are in equilibrium with all the other reactant molecules, and the AC can be treated as a normal molecular species except that one of its vibrational modes is missing and must be replaced by translation along the reaction coordinate. Also, the rate of formation of product C through the AC is the universal frequency $v = \dfrac{k_B T}{h}$; therefore, the reaction rate k can be expressed as follows:

$$k = \frac{k_B T}{h} \cdot [AB]^{\ddagger} \tag{4.25}$$

because:

$$K^{\ddagger} = \frac{[AB]^{\ddagger}}{[A][B]} \tag{4.26}$$

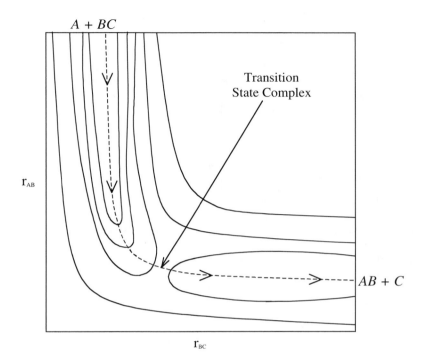

FIGURE 4.1
Potential energy surface with a late transition state. (From Boudart, M., *Kinetics of Chemical Processes*, Prentice Hall, Englewood Cliffs, NJ, 1968. With permission.)

Therefore,

$$k = \frac{k_B T}{h} K^{\ddagger} [A] [B] \tag{4.27}$$

Concentrations A and B in solution should be replaced by the activities of reactants A and B. Then, the rate expression becomes:

$$k = \frac{k_B T}{h} \cdot \frac{v_A v_B}{v^{\ddagger}} \cdot K^{\ddagger} [A] [B] \tag{4.28}$$

On the other hand, thermodynamics offers the following relationship:

$$K^{\ddagger} = e^{-\Delta G^{\ddagger}/RT} \tag{4.29}$$

Therefore, Equation (4.29) becomes:

$$k = \frac{k_B T}{h} \cdot \frac{v_A v_B}{v^{\ddagger}} \cdot e^{-\Delta G^{\ddagger}/RT} \tag{4.30}$$

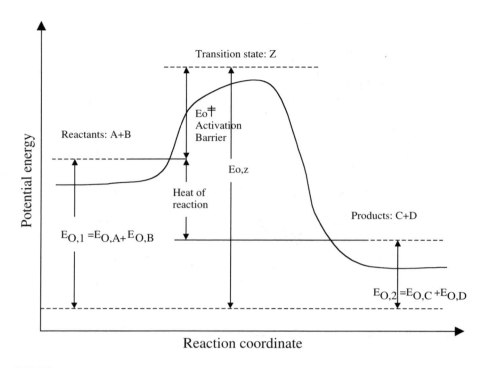

FIGURE 4.2
Potential energy profile for elementary reaction. (From Boudart, M., *Kinetics of Chemical Processes*, Prentice Hall, Englewood Cliffs, NJ, 1968. With permission.)

Furthermore,

$$\Delta G^{\ddagger} = \Delta H^{\ddagger} - T\Delta S^{\ddagger} \tag{4.31}$$

The equation takes the final form as follows:

$$k = \frac{k_B T}{h} \cdot \frac{v_A v_B}{v^{\ddagger}} \cdot e^{\Delta S^{\ddagger}/R} \cdot e^{-\Delta H^{\ddagger}/RT} \tag{4.32}$$

Experimentally, the Arrhenius equation describes the relationship between the rate constant k and the temperature as follows:

$$k = A \cdot e^{-E_a/RT} \tag{4.33}$$

By taking the natural logarithms of both sides, the equation becomes:

$$\ln k = -\frac{E_a}{R} \cdot \frac{1}{T} + \ln A \tag{4.34}$$

The above equation suggests that the activation energy barrier can be experimentally determined. In addition, if the transition state equation is compared with the Arrhenius equation, the following will also be true:

$$A = \frac{k_B T}{h} \cdot \frac{v_A v_B}{v^\ddagger} \cdot e^{\Delta S^\ddagger / R} \qquad (4.35)$$

Because E_a is the potential energy of activation, the difference between $\Delta H^\ddagger$ and E_a is the kinetic energy of activation, which is usually small compared to E_a. For a bimolecular reaction, this difference is RT. As a result, the activation energy will be the same as the change of activation enthalpy:

$$E_a = \Delta H^\ddagger \qquad (4.36)$$

It is important to point out that for exothermic reaction, the activation energy $\Delta E^\ddagger$ equals the activation energy:

$$\Delta E^\ddagger = E_a \qquad (4.37)$$

However, for endothermic reaction, the activation energy equals the activation barrier plus the heat of reaction ΔH:

$$\Delta E^\ddagger = E_a + \Delta H \qquad (4.38)$$

The transition state theory indicates that the rate of a reaction is not a matter of energy alone, but also requires a favorable configuration by a change of entropy. In addition, the rate of a reaction can be speeded up through the following methods. These methods are the guiding principles in the search for the most efficient AOPs:

- By increasing the temperature to increase the universal collision frequency $v = \frac{k_B T}{h}$, where k_B is the Boltzmann constant ($1.38*10^{-23}$ J•K^{-1}) and h is the Planck constant ($6.63*10^{-24}$ J•s); supercritical water oxidation uses this method.
- By increasing the ground energy of reactants using ultraviolet photons and ultrasound to reduce the activation energy barrier.
- By increasing pressure to increase the positive entropy of activation, $e^{\Delta S^\ddagger / R}$. In supercritical water oxidation, either air or oxygen is used as the oxidant. Although the activation energy barrier is extremely high, ΔS is large due to extremely high pressure. As a result, the reaction is still fast enough to oxidize concentrated organic waste.
- By decreasing the enthalpy of the reaction, $\Delta H^\ddagger$.

The last approach is the fundamental approach used in AOPs. In terms of thermodynamics, ordinary oxidants such as oxygen, ozone, and hydrogen peroxide will form activated complexes with organic pollutants with large enthalpy, $\Delta H^{\ddagger}$, and the reactions will be thermodynamically less favorable than reactions between hydroxyl radicals and organic compounds; therefore, one way to increase reaction rates is to convert these oxidants to hydroxyl radicals first. As a result, the enthalpy change when the hydroxyl radical attacks an organic molecule is several orders smaller than when a common oxidant attacks an organic molecule; thus AOPs can be found by any combination of oxidants such as oxygen, ozone, and hydrogen peroxide with catalysts such as UV photons, transition metals, and ultrasound. Based on these guiding principles in searching for AOPs, Table 4.2 provides possible AOPs with different combinations of oxidants and catalysts.

Because AOPs take advantage of the high reactivity of hydroxyl radicals, initial, propagation, promotion, recombination, and reversible reactions are commonly involved in the degradation of organic pollutants. Table 4.3 lists these major elementary reactions.

4.4   Oxidants

4.4.1   Oxygen

Because 20% of air is oxygen, it is not surprising that oxygen is the most common oxidant. The redox reaction of oxygen is:

$$O_2 + 2H_2O + 4e = 4OH^- \tag{4.39}$$

TABLE 4.2

Possible AOPs with Different Combinations of Oxidants and Catalysts

Catalyst	Metals and Ions			Metal Oxides		Oxidants			Photon	Ultra-sound	Electron
Oxidant	Fe^{2+}	Fe	Pt	TiO_2	Fe_2O_3	OH^-	O_3	H_2O_2	UV	US	e^- (reductant)
O_3	X	X	X		X	X		X	X	X	
H_2O_2	X	X	X	X	X		X		X	X	
O_2	X	X	X	X						X	
H_2O				X						X	X
TiO_2									X		

Note: X represents a combination that can generate hydroxyl radicals.

TABLE 4.3

Reactions in the Degradation of Organic Pollutants by Various AOPs

Initiation Reactions

(1) $O_3 + OH^- \longrightarrow \bullet O_2^- + \bullet HO_2$ $k_1 = 70\ M^{-1}\ s^{-1}$

(2) $O_3 + HO_2^- \longrightarrow \bullet HO_2 + \bullet O_3^-$ $k_2 = 5.5 \times 10^6\ M^{-1}\ s^{-1}$

(3) $O_3 + H_2O \xrightarrow{hv} O_2 + H_2O_2$

Propagation Reactions

(4) $O_3 + \bullet O_2^- \longrightarrow \bullet O_3^- + O_2$ $k_4 = 1.6 \times 10^9\ M^{-1}\ s^{-1}$

(5) $\bullet O_3^- + H^- \longrightarrow \bullet HO_3$ $k_5 = 5.2 \times 10^{10}\ M^{-1}\ s^{-1}$

(6) $\bullet HO_3 \longrightarrow \bullet OH + O_2$ $k_6 = 1.1 \times 10^5\ s^{-1}$

(7) $H_2O_2 \xrightarrow{hv} 2 \bullet [\bullet OH]$ —

Promotion Reactions

(8) $O_3 + \bullet OH \longrightarrow \bullet HO_2 + O_2$ $k_8 = 1.1 \times 10^8\ M^{-1}\ s^{-1}$

(9) $H_2O_2 + \bullet OH \longrightarrow \bullet HO_2 + H_2O$ $k_9 = 2.7 \times 10^7\ M^{-1}\ s^{-1}$

(10) $HO_2^- + \bullet OH \longrightarrow \bullet O_2^- H_2O$ $k_{10} = 7.5 \times 10^9\ M^{-1}\ s^{-1}$

(11) $\bullet HO_2 + \bullet OH \longrightarrow H_2O + O_2$ $k_{11} = 1.0 \times 10^{10}\ M^{-1}\ s^{-1}$

Recombination Reactions

(12) $\bullet HO_2 + \bullet HO_2 \longrightarrow H_2O_2 + O_2$ $k_{12} = 8.3 \times 10^5\ M^{-1}\ s^{-1}$

(13) $H_2O + \bullet HO_2 + \bullet O_2^- \longrightarrow H_2O_2 + O_2 + OH^-$ $k_{13} = 9.7 \times 10^7\ M^{-1}\ s^{-1}$

Reversible Reactions

(14) $\bullet HO_2 \longleftrightarrow H^- + \bullet O_2^-$ $pK_1 = 4.8$

(15) $H_2O_2 \longleftrightarrow HO_2^- + H^+$ $pK_2 = 11.8$

Although the redox potential of oxygen is only 0.4 V, it can form super-oxygen anion ($\bullet O_2^-$) in water by a rapid self-redox reaction to O_2 and H_2O_2, which is a two-electron reduction. The third electron will further convert H_2O_2 to $\bullet OH$, which can be achieved by adding Fe^{2+} salts. The fourth electron and one proton will produce water as the terminal reductive product of the above reaction. The pH effect on the thermodynamics of the reduction of oxygen is summarized in Figure 4.3.

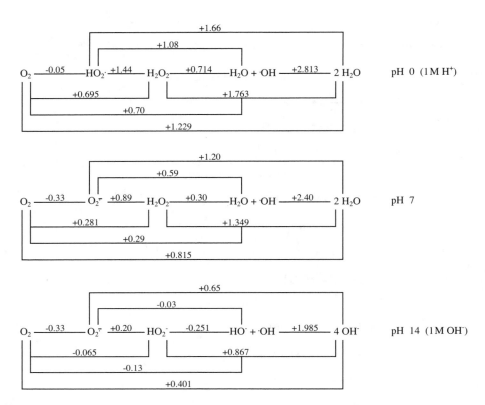

FIGURE 4.3
Standard potentials for redox reactions involving oxygen at 25°C. (From Sawyer, D.T. and Nanni, E.J., Jr., Redox chemistry of dioxygen species and their chemical reactivity, in *Oxygen and Oxy-Radicals in Chemistry and Biology*, Rodgers, M.A.J. and Powers, E.L., Eds., Academic Press, New York, 1981, 15–44. With permission.)

4.4.2   Hydrogen Peroxide

Thenard identified hydrogen peroxide, H_2O_2, as a chemical compound in 1818 (Schumb et al., 1995). It has been commercially available since the middle of 19th century, but the scale of manufacture and use has increased rapidly since around 1925, when electrolytic processes for its manufacture were introduced to the United States, and its use in industrial bleaching became increasingly important. Commercially, hydrogen peroxide is handled as an aqueous solution in a wide concentration range. Hydrogen peroxide has a covalent oxygen–oxygen bond and two covalent hydrogen–oxygen bonds. All oxidative properties of H_2O_2 originate from these bonds. Hydrogen peroxide is a liquid at normal temperatures and has a melting point of –0.43°C, and a boiling point of 15.2°C at 1 atm. Its commercial usage is in the form of an aqueous solution; therefore, the rate constant in aqueous solutions is more important than that of the pure compound.

4.4.2.1   Molecular Structure

The four atoms in a hydrogen peroxide molecule are structurally joined by simple covalent bonds, H–O–O–H, in a nonpolar structure. The structure can be defined by four parameters: the O–O distance, the O–H distance, the O–O–H angle, and the angle between two planes, each of which is defined by two oxygen atoms and one hydrogen atom. The best values for these parameters in the solid state are 1.453 ± 0.007 Å for the O–O distance; 0.988 ± 0.005 Å for O–H distance; 102.7 ± 0.3 Å for O–O–H angle; and 90.2 ± 0.6 Å for the dihedral angle between the O–O–H planes. The dihedral angle appears sensitive to its environment and thus may be different in the vapor phase or in other crystals containing H_2O_2 (Othmer, 1991).

4.4.2.2   Speciation of Hydrogen Peroxide

Hydrogen peroxide is 100% soluble in water. The freezing point of the mixture depends on the percentage of hydrogen peroxide by weight. Table 4.4 lists the freezing point changes with the composition of the mixture.

The eutectics exist at 45.2 and 61.2 wt% H_2O_2, and the compound $H_2O_2 \bullet 2H_2O$ (48.6 wt% H_2O_2) exists in the solid state. The evidence indicates that solid solutions are not formed in this system, although it is extremely difficult to obtain water-free solid H_2O_2. The recommended value for the heat of fusion of H_2O_2 is 87.84 cal/g. For the liquid/vapor phase relationship for aqueous hydrogen peroxide, partial pressures of the vapors over the liquid are lower than the calculated value for ideal solutions. Table 4.5 shows

TABLE 4.4

Freezing Point of Hydrogen Peroxide

H_2O_2 Concentration (wt%)	Freezing Point (°C)	H_2O_2 Concentration (wt%)	Freezing Point (°C)
0	0	50	−52.2
10	-6.4	60	−55.5
20	-14.6	61	−56.1[a]
30	-25.7	65	−49
35	33	70	−40.3
40	41.4	80	−24.8
45	51.7	90	−11.5
45.2	52.2[a]	100	−0.43
48.6	52[b]	—	—

[a] Eutectic.
[b] Compound, $H_2O_2 \bullet 2H_2O$45 (1966) 4256.
Source: Othmer, D.F., Hydrogen peroxide, in *Encyclopedia of Chemical Technology,* 4th ed., Howe-Grant, M. Ed., John Wiley & Sons, New York, 1991. With permission.

TABLE 4.5

Atmospheric Boiling Points of Hydrogen Peroxide Solutions

Liquid Composition (wt% H_2O_2)	Boiling Point (°C)	Vapor Composition (wt% H_2O_2)
0	100	0
10	101.7	0.9
20	103.6	2.1
30	106.2	4.2
35	107.9	5.8
40	109.6	7.6
50	113.8	13.0
60	119	20.8
70	125.5	33.4
80	132.9	51.5
90	141.3	75.0
100	150.2	100

Source: Othmer, D.F., Hydrogen peroxide, in *Encyclopedia of Chemical Technology,* 4th ed., Howe-Grant, M. Ed., John Wiley & Sons, New York, 1991. With permission.

atmospheric boiling points and related liquid and vapor compositions. The heats of vaporization for aqueous hydrogen peroxide at 25 and 60°C are provided in Table 4.6.

4.4.2.3   Thermodynamics of Hydrogen Peroxide

The values for a number of thermodynamic properties of H_2O_2 can be obtained from chemical engineering data handbook. The average heat capacity from 25 to 60°C for 100% H_2O_2 is 0.628 cal/g/°C). The mixing heat of H_2O_2 and 100% water ranges from –590 cal/mol H_2O_2 at 0°C to –1110 cal/mol H_2O_2 at 75°C. Table 4.6 presents the heats of vaporization for aqueous hydrogen peroxide (Othmer, 1991).

The formation heat of H_2O_2 from the reaction 4.40 at 25°C has been calculated at –32.52 kcal/mol:

$$H_2\ _{(gas)} + O_2\ _{(gas)} \rightarrow H_2O_2\ _{(gas)} \tag{4.40}$$

The free energy of formation of anhydrous H_2O_2 liquid at 25°C is calculated as –28.78 kcal/mol. The decomposition heat of pure liquid H_2O_2 to water and oxygen at 25°C is –23.44 kcal/mol.

4.4.2.4   Reaction Mechanism

Depending on its usage, hydrogen peroxide is a versatile and effective oxidizing agent as a source of active oxygen, compared with molecular oxygen

TABLE 4.6

Total Heat of Vaporization of Aqueous H_2O_2

H_2O_2 Concentration (wt%)	Heat of Vaporization (cal/g solution)	
	25°C	60°C
0	582.1	563.2
20	534.5	526.4
40	503.1	487.4
60	460.4	446.0
80	414.1	401.3
100	362.7	351.3

Source: Othmer, D.F., Hydrogen peroxide, in *Encyclopedia of Chemical Technology,* 4th ed., Howe-Grant, M. Ed., John Wiley & Sons, New York, 1991. With permission.

resulting from simple decomposition. The various reactions can be simplified to five general types, as follows:

1) Decomposition

$$2H_2O_2 \rightarrow 2H_2O + O_2 \tag{4.41}$$

2) Molecular addition

$$H_2O_2 + Y \rightarrow Y \bullet H_2O_2 \tag{4.42}$$

3) Substitution

$$H_2O_2 + RX \rightarrow ROOH + HX \tag{4.43}$$

$$H_2O_2 + 2RX \rightarrow ROOR + 2HX \tag{4.44}$$

4) H_2O_2 as a reducing agent

$$H_2O_2 + Oxidant \rightarrow Oxidant–H_2 + O_2 \tag{4.45}$$

5) H_2O_2 as an oxidizing agent

$$H_2O_2 + Reductant \rightarrow Reductant–O + H_2O \tag{4.46}$$

While undergoing these reactions, hydrogen peroxide may react as a molecule, or it may first ionize or be dissociated into free radicals. The mechanism is very complex in many cases and may depend on the types of catalyst and reaction conditions.

4.4.2.5   Ionization

Hydrogen peroxide exhibits a weak acidic character, having a dissociation constant of about 1.5×10^{-12}; thus, pure aqueous solutions of hydrogen peroxide have pH values below 7. The dissociation constant of hydrogen peroxide is 10^{-6}.

$$\text{H–O–O–H} \rightleftharpoons \text{H}^+ + \text{H–O–O}^- \qquad pKa = 10^{-6}$$

$$(4.47)$$

Depending on the pH of the medium, ionic species will have either nucleophilic character (pH > 7) or electrophilic character (pH < 7). In an alkaline medium, hydrogen peroxide reacts with hydroxide anions to give perhydroxyl anions HOO^- according to the equilibrium.

$$\text{H–O–O–H} + OH^- \rightleftharpoons HOO^- + H_2O$$

$$(4.48)$$

This equilibrium is shifted to the right when the basicity of the medium increases. This shows the nucleophilic character of the peroxidic linkage O–O due to formation of the HOO^- anion. This perhydroxyl anion is considered to be a supernucleophile, as its reactivity is about 200 times higher than that of the OH^- anion. The pH effect on the kinetic rate constant for the dismutation of superoxide is shown in Figure 4.4.

4.4.2.6   Free-Radical Formation

Hydrogen peroxide can dissociate into free radicals by breaking either an H–O bond or the O–O bond.

$$\text{H–O–O–H} \Big\langle \begin{array}{l} \text{H–O–O}^- + \text{H}^+ \\ \text{H–O–O}^{\bullet} + \text{H}^{\bullet} \end{array}$$

$$(4.49)$$

$$\text{H–O–O–H} \Big\langle \begin{array}{l} HO^+ + OH^- \\ HO^{\bullet} + OH^{\bullet} \end{array}$$

$$(4.50)$$

Depending on whether the electron pair of the broken bond is shared or not by the two new entities, the reaction sequence will involve either an ionic or a free-radical pathway, as shown in Equation (4.49) and Equation (4.50).

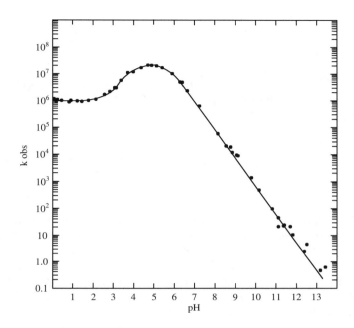

FIGURE 4.4
Effect of pH on the second-order rate constant for the dismutation of superoxide. (From Bielski, B.H.J., Reevaluation of the spectral and kinetic properties of $\cdot HO_2$ and of $\cdot O_2^-$ free radicals, *Photobiol.*, 28, 645–649, 1978. With permission.)

The heat for Reaction (4.49) is 88 kcal/mol, and for Reaction (4.50) it is about 50 kcal/mol. The latter reaction predominates in uncatalyzed vapor-phase decomposition and photochemically initiated reactions.

4.4.2.7 Decomposition

It is necessary to minimize decomposition hydrogen peroxide, as the release of oxygen and heat causes safety problems and results in the poor efficiency of utilization. In the absence of any reagent, the HOO^- anion is able to oxidize another molecule of hydrogen peroxide by an unstable transition complex with the release of molecular oxygen.

$$O_2 + H_2O + HO^- \tag{4.51}$$

This decomposition reaction can be completely overridden in the presence of an electrophilic substrate. In an acidic medium, hydrogen peroxide is more stable as the equilibrium is completely shifted to the left.

$$H-O-O-H \rightleftharpoons HOO^- + H^+ \qquad (4.52)$$

Due to the solvation of protons by hydrogen peroxide, an oxonium structure is observed, as in Equation (4.53); this structure changes the nature of the peroxide bond and thus increases its stability.

$$H-O-O-H + H^+ \rightleftharpoons H + H-O^+-O-H \qquad (4.53)$$

When a reaction medium contains a more powerful nucleophile such as an acid, alcohol, ketone or oxometal groups, these functional groups will protonate to generate a new electrophilic intermediate by reacting with hydrogen peroxide. This is a very important pathway in which the peroxidic linkage O–O is conserved and transferred into a new dissymmetric molecule referred to as a *peroxycompound*. These reactions are carried out on a large scale by industries that manufacture performic acid or peracetic acid, either of which leads to various kinds of epoxides, such as soybean oil and linseed oil.

4.4.2.8 H_2O_2 as an Oxidizing Agent

Hydrogen peroxide is a strong oxidant with standard oxidation potentials of 1.8 and 0.87 V at pH 0 and 14, respectively. Hydrogen peroxide oxidizes a large number of organic and inorganic compounds; however, the reaction mechanism varies greatly with reductants, the types of catalysts used, and other reaction conditions. Oxidized compounds range from iodide ions to organic cellulose fibers. Hydrogen peroxide has been applied to oxidize sulfides, hypochlorites, nitrites, cyanides, and chlorine in wastewater.

4.4.2.9 H_2O_2 as a Reducing Agent

Hydrogen peroxide can also reduce strong oxidizing agents such as chlorine, hypochlorites, potassium permanganate, and ceric sulfate, as follows:

$$Cl_2 + H_2O_2 \rightarrow 2HCl + O_2 \qquad (4.54)$$

$$2KMnO_4 + 5H_2O_2 + 3H_2SO_4 \rightarrow K_2SO_4 + 2MnSO_4 + 8H_2O + 5O_2 \qquad (4.55)$$

4.4.2.10   OH•H_2O_2 Complex

A new hydrogen bond in OH•H_2O_2 has been explored using theoretical calculations. The OH•H_2O_2 radical complex is formed through the addition of a hydroxyl radical to H_2O_2. The additional reaction exhibits anomalous temperature coefficients. A strong upturn of the rate constant occurs near 800 K. This property can be interpreted reasonably in terms of a complex-forming mechanism. *Ab initio* molecular orbital methods were used to search for hydrogen bonding between the OH radical and the H_2O_2 molecule. The structure, vibration spectrum, and binding energy of the $OH \bullet H_2O_2$ complex have been determined by Lee et al. (1988).

4.4.2.11   Geometries

The fully optimized geometries for the $OH \bullet H_2O_2$ complex and the monomers are depicted in Figure 4.5, which shows that the $OH \bullet H_2O_2$ complex involves two types of hydrogen bonds. One is the hydrogen bond between the H″ atom in the OH radical, as well as one of the oxygen atoms in H_2O_2. (H″···O′). The other hydrogen bond occurs between the oxygen atom in the hydroxyl radical and one of the hydrogen atoms in the hydrogen peroxide (O″···H). Therefore, $OH \bullet H_2O_2$ has a floppy five-member ring structure with the other hydrogen atom and in the H_2O_2 out of the plane.

Table 4.7 shows the rotational constants for the complex and the monomers. The distances between the hydrogen-bonded heavy atoms are presented in Table 4.8. It is evident that the $OH \bullet H_2O_2$ complex is an asymmetric rotor. Because this hydrogen bond has a permanent dipole moment that is somewhat larger than those of the monomers, it should be active in the microwave region of the spectrum.

4.4.2.12   Energetics

The total energies and the binding energies for the $OH \bullet H_2O_2$ complex are calculated by the MP2, CCSD(T), and B3LYP methods, respectively. The results predict almost the same binding energies for the $OH \bullet H_2O_2$ complex, regardless of whether or not the MP2 or B3LYP optimized geometries are used in the single-point energy calculations. The change in geometry leads to an energy change of only approximately 0.1 kcal/mol. This may be caused by the flatness of the potential energy surface for such a weakly bound complex. It is interesting to compare the stability of $OH \bullet H_2O_2$ with several analogous hydrogen-bonded complexes on the basis of the reported binding energies. The $OH \bullet H_2O_2$ complex has a dissociation energy similar to those for the $H_2O \bullet H_2O$ and $OH \bullet H_2O_2$ complexes; however, it is more tightly bound than the $OH \bullet H_2S$ complex by 2.4 kcal/mol and less stable than the $H_2O \bullet H_2O$ complex by approximately 2.8 kcal/mol.

4.4.2.13   Frequencies

Table 4.9 illustrates the harmonic vibrational frequencies and the corresponding infrared intensities for the $OH \bullet H_2O_2$ complex. In comparison with the

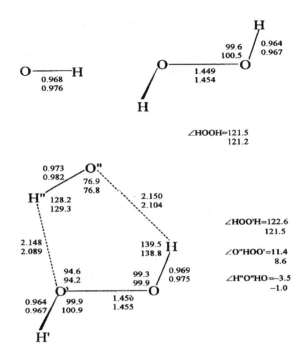

FIGURE 4.5
OH•H₂O₂ complex having two types of hydrogen bonds. (From Wang, B. et al., *Chem. Phys. Lett.*, 309, 274–278, 1999. With permission.)

TABLE 4.7

Molecular Characteristics of Different Species in H₂O₂

Species	Distances		Angles		Rotational Constants			U (μ)
	OO″	O′O″	HOO″	H″O″O′	B$_A$	B$_B$	B$_C$	
OH•H₂	2.955	2.854	28.2	36.2	2.5803	5.692	4.711	1.982
O₂	2.909	2.816	28.5	35.1	25.517	5.882	4.832	1.912
OH		—			569	569 083	—	1.908
					083	559 756		1.832
					559			
					756			
H₂O₂		—			303	26 736	25	1.812
					037	26 471	586	1.732
					303		25	
					382		355	

Note: Distances are in Å, angles are in degree, rotational constants are in Mhz, and the dipole moments are in U (in Debye) for the OH•H₂O₂ complex; the monomers are calculated at the UMP2 (full)/6-311 + + G(d,p) and UB#LYP/6-311 + + G(d,p).

Source: From Wang, B. et al., *Chem. Phys. Lett.*, 309, 274–278, 1999. With permission.

TABLE 4.8

Total Energies and Binding Energies (D_e and D_o, in kcal mol^{-1}) for the OH•H$_2$O$_2$ Complex

Methods	OH·H$_2$O$_2$	OH	H$_2$O$_2$	D_e	D_o
MP2//MP2	−226.88993	−75.59737	−151.28332	5.80	4.06
CCSD(T)//MP2	−226.93082	−75.61491	−151.30656	5.87	4.13
B3LYP//B3LYP	−227.37381	−75.76241	−151.60217	5.79	3.97
MP2//B3LYP	−226.88982	−75.59729	−151.28326	5.82	4.00
CCSD(T)//B3LYP	−226.93086	−75.61489	−151.30659	5.89	4.07

Source: From Wang, B. et al., *Chem. Phys. Lett.*, 309, 274–278, 1999. With permission.

vibrational frequencies of the isolated OH and H$_2$O$_2$, the frequency shifts of the complex relative to OH + H$_2$O$_2$ are also listed.

Although the UMP2 and UB3LYP methods yield similar geometries and energies for the OH•H$_2$O$_2$ complex, they seemingly predict different vibration spectra. Because no experimental studies of the OH•H$_2$O$_2$ complex have been conducted, it is difficult to determine which predictions should be more reliable; however, several important conclusions can still be drawn from the high-level theoretical calculations. Both the MP2 and B3LYP methods predict the presence of four strong bands in the infrared spectrum of the OH•H$_2$O$_2$ complex. They involve three intramolecular vibrational modes — HOO'H' torsion (v_8), OH symmetric stretching (v_2), and OH asymmetric bending (v_5) — and one intermolecular vibrational mode (v_7). It is interesting to note that these three intramolecular modes belong to the H$_2$O$_2$ moiety.

The MP2 and B3LYP methods predict very similar vibrational fundamental frequencies and infrared intensities for the five intermolecular modes (v_7, v_9, v_{10}, v_{11}, and v_{12}). Moreover, the v_7 mode is predicted to be one of the most intense bands in the infrared spectrum of the complex. The OH•H$_2$O$_2$ radical complex is supported by the observation of these vibrational modes in the laboratory.

Using both the UMP2 (full) and UB3LYP methods with the 6-311 ++ G(d,p) basis set, the equilibrium structure of the OH•H$_2$O$_2$ complex is considered to be a floppy, five-membered ring involving two distorted O$\cdots$H$\cdots$O hydrogen bonds with O–O distances of approximately 2.9 and 2.8 Å. The infrared spectrum of the complex is predicted. The OH•H$_2$O$_2$ complex may play an important role in oxidizing organic pollutants by different AOPs.

TABLE 4.9

Harmonic Vibrational Frequencies (cm^{-1}) and Intensities (km mol^{-1}) for the OH•H$_2$O$_2$ Complex at the UMP2 (full)/6-311 + + G(d,p) and UB3LYP/6-311 + + G(d,p) Levels

No.	Mode Description	Frequencies[a]		Intensities[b]	
		UMP2 (full)	UB3LYP	UMP2 (full)	UB3LYP
1	OH asymmetrical stretching	3852.2 (+0.1)	3780.2 (+3.2)	51.7 (3.2)	50.8 (3.8)
2	OH asymmetrical stretching	3781.6 (−69.5)	3650.1 (−125.9)	205.6 (2.9)	239.1 (3.5)
3	O′H″ stretching	3755.5 (−87.2)	3601.4 (−105.5)	13.5 (0.7)	21.2 (1.6)
4	OH symmetrical bending	1491.4 (+32.6)	1495.7 (+42.0)	8.6	13.2
5	OH asymmetrical stretching	1319.7 (+17.9)	1316.9 (+20.4)	109.8 (1.1)	98.6 (1.1)
6	OO′ stretching	922.5 (+0.1)	931.5 (-3.1)	1.5 (1.4)	1.9 (1.7)
7	Intermolecular	517.0	559.5	267.3	250.1
8	HOO′H′ torsion	407.5 (+12.7)	415.2 (+48.3)	253.4 (1.1)	276.6 (1.2)
9	Intermolecular	260.8	281.7	31.5	51.6
10	Intermolecular	237.1	225.1	85.6	72.7
11	Intermolecular	165.3	182.3	3.7	2.6
12	Intermolecular	129.5	143.5	45.2	36.0

[a] Numbers in the parentheses are frequency shifts relative to the monomers.

[b] Numbers in the parentheses are ratios of intensities for the complex relative to the monomers.

Source: From Wang, B. et al., *Chem. Phys. Lett.*, 309, 274–278, 1999. With permission.

4.4.2.14 *Environmental Applications of H$_2$O$_2$*

Hydrogen peroxide has several industrial applications and plays a significant role in protecting the environment (Venkatadri and Peters, 1993). It can be used in the surface-polishing industry for cleaning, decorating, protecting, and etching metals. H$_2$O$_2$ offers nontoxic alternatives for surface treatments, compared to hexavalent chromium. It has been used for the detoxification of wastewater that contains cyanide through oxidation of the cyanide to cyanate and in groundwater containing iron and manganese through converting the metal ions to insoluble hydroxides/oxides, which can be treated. It can supply oxygen, by disassociation into oxygen and water, to microorganisms in biological treatment facilities and can be used in the bioremediation of contaminated sites. As a disinfecting agent, it is used in the control of undesirable biofilm growth. For the treatment of photochemical effluents containing sulfites and silver, H$_2$O$_2$ converts them to sulfate and silver complex precipitates, respectively. For the degradation of formaldehyde in air emissions hydrogen peroxide can be used. It prevents odors by the conversion of sulfides, thioethers, disulfides, sulfites, and thiosulfates to elemental sulfur or sulfate in the washing solution. The treatment of sulfur and nitrogen oxides present in air emissions from coking, sulfuric acid production, pickling, and other chemical plant processes is carried out using hydrogen peroxide for conversion of these oxides to their corresponding acids. H_2O_2 can reduce excess chlorite and hypochlorite in wastewater, air, or bleached textiles to chloride salts.

Treatment with hydrogen peroxide can improve the biodegradability of organics that are inhibitory to biological treatment and reduce toxicity. Several organics that can be treated with hydrogen peroxide include nitrobenzene, aniline, cresols, monochlorophenols, dichlorophenols, and trichlorophenols. It is also used in the removal of organic contaminants such as phenols, fats, oils, grease, and suspended solids in wastewater. Generally, inorganics react faster than organics with hydrogen peroxide, and the reaction of trace organics is much slower, due to mass transfer limitations (Venkatadri and Peters, 1993). Factors affecting hydrogen peroxide treatment include pH, temperature, contact time, application rate, and reactivity of compounds.

Oxidation by the sole use of hydrogen peroxide is not effective for high concentrations of certain refractory contaminants. For example, highly chlorinated aromatic compounds and inorganic compounds such as cyanides have very low reaction rates at reasonable H_2O_2 concentrations. However, hydrogen peroxide is a very useful oxidation reagent and a potential source of oxygen. It is also easy to handle and is safer to use than many other chemicals. It seems to be a versatile reagent in treating contaminants in municipal drinking water, wastewater, and gaseous effluents. Transition metal salts, such as iron salts and UV light, can activate H_2O_2 to form hydroxyl radicals, which is the reason why hydrogen peroxide finds its wide applications in different AOPs.

4.4.3 Ozone

In an aqueous solution, ozone may act on various compounds according to the following two pathways (Hoigne and Bader, 1977): (1) direct reaction with the molecular ozone, and (2) indirect reaction with the radical species that are formed when ozone decomposes in water. These two basic reactions of ozone in water are illustrated in Figure 4.6.

4.4.3.1 Molecular Ozone Reactivity

The extreme forms of resonance structures in ozone molecules can be represented as shown in Figure 4.7. This structure illustrates that the ozone molecule has two types of dipoles, which can serve as electrophilic and nucleophilic agents, respectively. In organic solvents, the following three reactions can be observed:

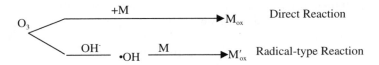

FIGURE 4.6
Reaction pathways of ozone. (From Langlais, B. et al., *Ozone in Water Treatment: Application and Engineering*, Lewis Publishers, Boca Raton, FL, 1991. With permission.)

FIGURE 4.7
Extreme forms of resonance structures in an ozone molecule. (From Langlais, B. et al., *Ozone in Water Treatment: Application and Engineering*, Lewis Publishers, Boca Raton, FL, 1991. With permission.)

- *Cyclo-addition (Criegee mechanism)* — As a result of its dipolar structure, an ozone molecule may lead to three dipolar cyclo-additions on unsaturated bonds, with the formation of primary ozonide (I) corresponding to the reaction shown in Figure 4.8. In a protonic solvent such as water, this primary ozonide decomposes into a carbonyl compound (aldehyde or ketone) and a zwitterion (II) that quickly leads to a hydroxy–hyperoxide (III) stage that, in turn, decomposes into a carbonyl compound and hydrogen peroxide (see Figure 4.9).

- *Electrophilic reaction* — The electrophilic reaction is restricted to molecular sites with a strong electronic density and, in particular, certain aromatic compounds. Aromatics substituted with electron donor groups (OH, NH_2) show high electronic densities on carbons located in the *ortho* and *para* positions; however, the aromatics substituted with electron-withdrawing groups (–COOH, –NO_2) are weakly ozone reactive. In this case, the initial attack of the ozone molecules takes place mainly on the least deactivated *meta* position. This reactivity results in the aromatic compounds bearing the electron donor groups *D* (for example, phenol and aniline), which react quickly with the ozone. This reaction is shown schematically in Figure 4.10. The initial attack of the ozone molecule leads to the formation of *ortho*- and *para*-hydroxylated by-products. These hydroxylated compounds are highly susceptible to further ozonation. The compounds lead to the formation of quinoid. Consequently, aliphatic products with carbonyl and carboxyl functions are a result of the opening of aromatic rings.

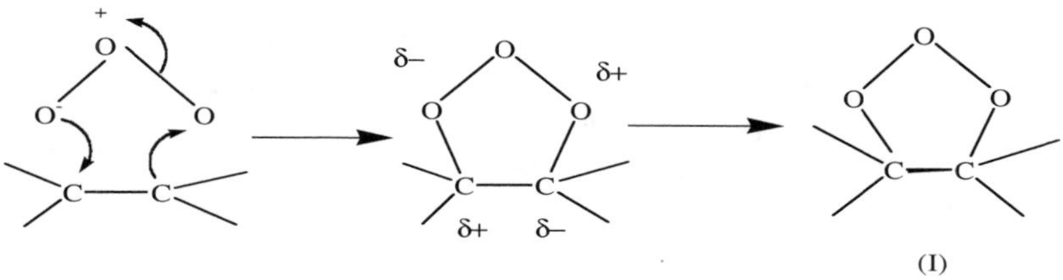

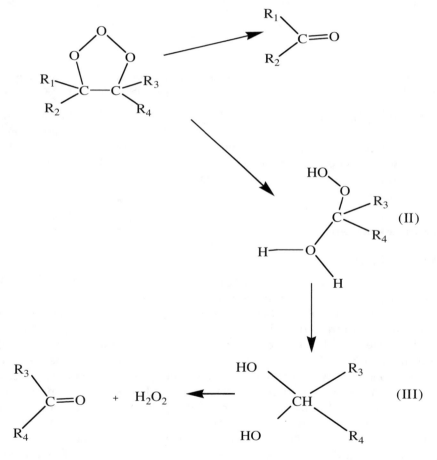

FIGURE 4.8
Dipolar cyclo-addition on unsaturated bonds. (From Langlais, B. et al., *Ozone in Water Treatment: Application and Engineering*, Lewis Publishers, Boca Raton, FL, 1991. With permission.)

FIGURE 4.9
Generation of zwitterion and hydroxy–hydroperoxide. (From Langlais, B. et al., *Ozone in Water Treatment: Application and Engineering*, Lewis Publishers, Boca Raton, FL, 1991. With permission.)

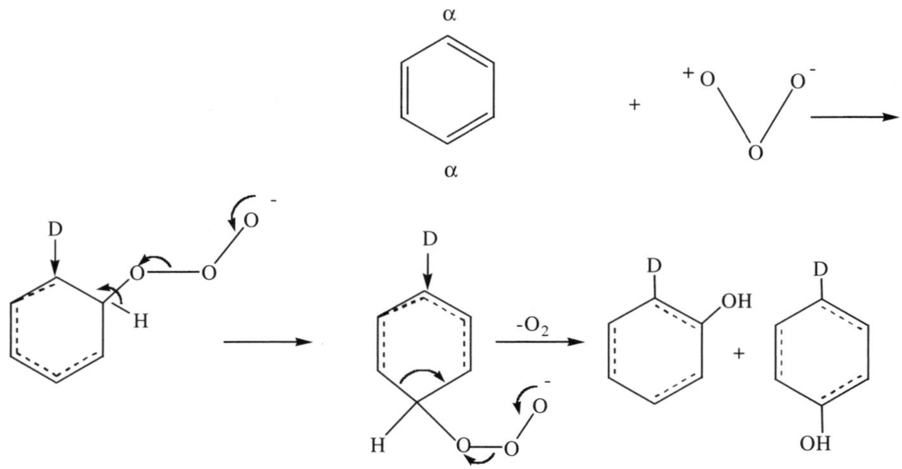

FIGURE 4.10
Phenol and aniline quick reaction with ozone. (From Langlais, B. et al., *Ozone in Water Treatment: Application and Engineering*, Lewis Publishers, Boca Raton, FL, 1991. With permission.)

- *Nucleophilic reaction* — The nucleophilic reaction is found locally on molecular sites showing electronic deficits and, more frequently, on carbons carrying electron-withdrawing groups. The molecular ozone reactions are extremely selective and limited to unsaturated aromatic and aliphatic compounds, as well as to specific functional groups.

4.5 Catalysts

In order to effectively generate hydroxyl radicals, catalysts such as metals, metal oxides, photons, and ultrasound are required, in addition to the presence of oxidants. Catalysis is a chemical reaction in which a substance, called a *catalyst*, is introduced in the reaction with the intent to increase the production rate of a desired product or the destruction rate of an undesired substrate. Another function of catalysts is to facilitate the contact between the two reactants by reducing the distance between them. On the surface of a catalyst, an organic compound reacts with another compound more easily than in solution. Catalysts that participate in a chemical reaction do not ultimately change; therefore, the catalysts may only lose their activity over time. Catalysts can be divided into two categories: (1) homogeneous catalysts, for which the catalysts and the reactants are in the same phase; and (2) heterogeneous catalysts, for which the catalysts are in different phases than the reactants. It is frequently desirable in heterogeneous catalysis to have a large surface area for a reaction. One way to achieve this is to spread out active catalysts on the surface of a support material.

4.5.1 Ultrasound

The propagation of an ultrasonic wave in water generates cavitation bubbles, which behave like a microreactor at the center of a high-energy phenomenon leading to the destruction of organic pollutants. The center of the sonochemical activity is the cavitation bubbles during pulsation and/or collapse, both high-energy phenomena. The sonochemical treatment is particularly efficient for the destruction of volatile chloroaromatic hydrocarbons, such as 1,2-dichlorobenzene, 1,3-dichlorobenzene, 1,4-dichlorobenzene, 1,3,5-trichlorobenzene, and 1-chloronaphthalene. Compounds with a high vapor pressure can enter the bubble, where they undergo thermolytic reactions. In addition, the •OH radical is generated according to the following equation:

$$H_2O \rightarrow H\bullet + \bullet OH \tag{4.56}$$

4.5.2 Photon

Photons, especially in the ultraviolet wavelength, are important for photocatalytic degradation processes such as UV/H_2O_2, UV/O_3, $UV/H_2O_2/O_3$, UV/TiO_2, and $UV/ferrioxalate/H_2O_2$. Photons can accelerate a chemical reaction mostly through generation of hydroxyl radicals. For example, in UV/H_2O_2 system, hydrogen peroxide is directly cleaved into two hydroxyl radicals. When combined with UV irradiation, the degradation efficiency of Fenton's reagent is enhanced, as follows:

$$Fe^{2+} + H_2O_2 \rightarrow Fe^{3+} + OH^- + \bullet OH \tag{4.57}$$

In $Fe(II)/H_2O_2$ acid media, oxidation occurs via hydroxyl radical. The use of ultraviolet in both cases results in the generation of HO radicals, formed by the reaction of Fe(II) with H_2O_2. In ferrioxalate systems, this reaction occurs after photolysis by photons with wavelengths less than 450 nm; therefore, it plays a significant role in the degradation of organic pollutants in natural water, due to solar irradiation.

As a result of photoexcitation of TiO_2 by UV photons, electron–hole (e–h) pairs are separated into conduction bands and valence bands, respectively:

$$TiO_2 + hv \rightarrow TiO_2 + e^- + h^+ \tag{4.58}$$

Once separated, this pair can induce chemical redox transformation with the species adsorbed on the surface. It is assumed that surface hydroxide ions act as hole traps by producing OH• radicals:

$$OH^- + h^+ \rightarrow \bullet OH \tag{4.59}$$

4.5.3 Transition Metals

Transition metals have the ability to form a complex with an electron donor. In this complex, the outermost set of five stable d electron orbitals of the transition metal is partially filled. When transition metal salts, such as ferrous sulfate, are combined with hydrogen peroxide, the resulting oxidation power is known as Fenton's reagent. Instead of forming a complex with hydrogen peroxide, the ferrous ion loses one electron to the peroxide bond of hydrogen peroxide. As a result, a hydroxyl radical and a hydroxide ion are formed.

4.6 Catalyst Support

Catalyst-supporting materials are used to immobilize catalysts and to eliminate separation processes. The reasons to use a catalyst support include: (1) to increase the surface area of the catalyst so the reactant can contact the active species easily due to a higher per unit mass of active ingredients; (2) to stabilize the catalyst against agglomeration and coalescence (fuse or unite), usually referred to as a *thermal stabilization*; (3) to decrease the density of the catalyst; and (4) to eliminate the separation of catalysts from products. Catalyst-supporting materials are frequently porous, which means that most of the active catalysts are located inside the physical boundary of the catalyst particles. These materials include granular, powder, colloidal, coprecipitated, extruded, pelleted, and spherical materials. Three solids widely used as catalyst supports are activated carbon, silica gel, and alumina:

- *Activated carbon* is often used as a catalyst support because of its high surface area and porosity. For example, Uchida et al. (1993) utilized activated carbon adsorbent as support for TiO_2 photocatalysts. It is reported that the adsorptive action of the activated carbon support enables the organic pollutants to concentrate around the loaded TiO_2, resulting in a high photocatalytic degradation rate.

- *Alumina* is a porous, high-surface-area form of aluminum oxide. The surface has more polar characteristics than silica gel does; therefore, it has both acidic and basic characteristics, reflecting the nature of the metal. Alumina has a high melting point, slightly over 2000°C, which is also a desirable property for a support due to its thermal stability. Alumina is composed of aluminum trihydroxides, $Al(OH)_3$; aluminum oxyhydroxides, $AlO(OH)$; and aluminum oxide, $Al_2O_3n(H_2O)$.

- *Silica gel* is an amorphous, highly porous, partially hydrated form of silica, which is a substance made from the two most abundant elements in the Earth's crust: silicon and oxygen. Silica gel powder is widely used in industries for catalyst support; as adsorbents, dehydrating agents, and the buffing media in toothpaste; and in chromatography. The principal structural unit of silica gel consists of Si^{4+} cations tetrahedrally coordinated to O_2 anions. The tetrahedral subunits are corners linked to form chains, rings, sheets, or three-dimensional assemblages. Unshared tetrahedral apices generally form hydroxyl groups.

4.7   Influence of Temperature and Pressure

According to the transition state theory, the appropriate operating temperature and pressure are important variables in the degradation of hazardous organic pollutants. Supercritical water and wet air oxidation are the systems that require high pressure and temperature to decompose organic pollutants. Increasing the temperature and pressure has been shown to increase the oxidation efficiency and decrease reaction time. Wet air oxidation can be further improved by adding different catalysts. Homogeneous transition metals (Fe^{2+}, Cu^{2+}) may be suitable catalysts; the dissolved ions need to be separated at the end of the process. Precious metals (Pt, Pd, Ru) deposited on supporting materials have been reported as active for supercritical water oxidation (CWAO). Beziat et al. (1997) have shown that a platinum catalyst supported on carbon could decompose formic oxalic and maleic acids very easily at atmospheric pressures and a temperature <100°C. Higher temperatures did not result in improvement of the performance of the platinum catalysts. Ruthenium was the most active catalyst among the precious metals examined in the oxidation of polyethyleneglycol at 200°C.

4.8   Summary

Hydroxyl radical is the dominant species in advanced oxidation process. According to transition state theory, catalysts such as ultrasounds, photons, transition metals, and their ions are all able to decompose oxidants, such as oxygen, hydrogen peroxide, and ozone, into hydroxyl radicals. As a result, UV/H_2O_2, UV/O_3, $UV/H_2O_2/O_3$, Fe^{2+}/H_2O_2, and UV/TiO_2 are the typical AOPs. Under supercritical temperature and pressure, supercritical water oxidation also produces reactive radicals such as hydroxyl radicals. Depending upon the nature of pollutants, treatability of each individual pollutant may vary greatly due to its molecular structure. Therefore, quantitative structure and activity relationship (QSAR) may play an important role in assessing and predicting the treatability of a specific pollutant.

References

Beziat, J., Besson, M., and Gallezot, P., Catalytic wet air oxidation of wastewater, paper presented at the 3rd World Congress on Oxidation Catalysis, 1997, pp. 615–621.

Bielski, B.H.J., Reevaluation of the spectral and kinetic properties of $\cdot HO_2$ and of $\cdot O_{2-}$ free radicals, *Photobiol.*, 28, 645–649, 1978.

Boudart, M., *Kinetics of Chemical Processes*, Prentice Hall., Englewood, Cliffs, NJ, 1968.

Charles, A.W., Ed., *Hazardous Waste Management*, McGraw-Hill, New York, 1989.

Hoigne, J. and Bader, H., *Proceeding of the International Ozone Association Symposium*, International Ozone Association, Toronto, 1977, p. 16.

Hoigne, J., Mechanisms, rates, and selectivities of oxidations of organic compounds initiated by ozonation in water, in *Handbook of Ozone Technology and Applications*, Rice, R.G. and Netzer, A., Eds., Ann Arbor Science Publishers, Ann Arbor, MI, 1982.

Huang, W. and Walter, W.J., Jr., A distributed reactivity model for sorption by soils and sediments. 10. Relationships between desorption, hysteresis, and the chemical characteristics of organic domains, *Environ. Sci. Technol.*, 31(9), 2562–2569, 1997.

Huang, C.P., Dong, C., and Tang, W.Z., Advanced chemical oxidation: its present role and potential future in hazardous waste treatment, *Waste Manage.*, 13, 361–377, 1993.

Langlais, B., Reckhow, D.A., and Brink, D.R., Eds., *Ozone in Water Treatment: Application and Engineering*, Lewis Publishers, Boca Raton, FL, 1991.

Lee, C., Yang, W., and Parr, R.G., *Phys. Rev. B*, 41, 785, 1988.

Lide, D.R. et al., *CRC Handbook of Chemistry and Physics*, 73rd ed., CRC Press, Boca Raton, FL, 1992.

Othmer, D.F., Hydrogen peroxide, in *Encyclopedia of Chemical Technology*, 4th ed., Howe-Grant, M. Ed., John Wiley & Sons, New York, 1991.

Remediation Technologies, Ground-Water Remediation Technologies Analysis Center, www.gwrtac.org/htm/tech_topic.htm, 2003a.

Remediation Technologies Network, LLC, Categories of technologies, www.remedial.com, 2003b.

Sawyer, D.T. and Nanni, E.J., Jr., Redox chemistry of dioxygen species and their chemical reactivity, in *Oxygen and Oxy-Radicals in Chemistry and Biology*, Rodgers, M.A.J. and Powers, E.L., Eds., Academic Press, New York, 1981, 15–44.

Schumb, W.C., Satterfield, C.W., and Wentworth, R.L., *Hydrogen Peroxide*, ACS Monograph Series, Reinhold Publishing, New York, 1995.

Treatment Technologies Screening Matrix, www.frtr.gov/matrix2/section3/matrix.html, 2003USEPA, Hazardous waste clean-up information, www.clu-in.com, 2003.

Uchida, H., Itoh, S., and Yoneyama, H., *Chem. Lett.*, 1993.

Venkatadri, R. and Peters, R.W., Chemical oxidation technologies: ultraviolet light/hydrogen peroxide, Fenton's reagent, and titanium dioxide-assisted photocatalysis, *Hazardous Waste Hazardous Mater.*, 10, 107–149, 1993.

Wang, B., Hou, H., and Gu, Y., Existence of hydrogen bonding between the hydroxyl radical and hydrogen peroxide: $OH \cdot H_2O_2$, *Chem. Phys. Lett.*, 309, 274–278, 1999.

5

Quantitative Structure–Activity Relationships

5.1 Introduction

The concept of Quantitative Structure–Activity Relationship (QSAR) analysis is to mathematically quantify the correlations between reactivity and molecular descriptors using chemical intuition and experience. Once a correlation between structure and activity is found, any number of compounds, including those not yet synthesized, can be readily assessed using a computer to determine the treatability of a given organic compound. Treatability of organic pollutants by a specific Advanced Oxidation Process (AOP) is usually carried out through laboratory studies; however, the combination of dozens of AOPs and thousands of toxic organic pollutants makes it impossible to evaluate the treatability of each pollutant by each AOP. In addition, experimental studies are usually very costly and time consuming. Fortunately, a class of organic pollutants reacts similarly to a given AOP because similar changes in structure produce similar changes in reactivity according to the QSAR theory; therefore, QSAR analysis can be used as an important tool to assess the treatability of a variety of organic pollutants by an AOP.

5.2 Fundamental Theory of QSAR

Quantitative structure–activity relationship studies are of great importance in modern chemistry. From their origin in the study of organic chemistry dating back to the 19th century, these studies have relied on some empirical and qualitative rules about the reactivity similarities of compounds with similar structures. The most significant development in QSARs occurred with the work of Louis Hammett (1894–1987), who correlated some electronic properties of organic acids and bases with their equilibrium constants and reactivity (Johnson, 1973). Hammett postulated that the effect of substituents on the ionization of benzoic acids could be used as a model system to estimate the electronic effect of substituents on similar reaction systems. The treatment is reasonably successful whether the substrates are attacked by electrophilic, nucleophilic, or free radical reagents. The most important concept developed is that the mechanism is the same as that in a given reaction series. Although the Hammett methodology has been criticized by theoreticians because of its empirical foundation, it is astonishing that substituent sigma (σ) constants, obtained simply from the ionization of organic acids in solution, can frequently and successfully predict equilibrium and rate constants for a range of organic reactions in solution (Hansch and Leo, 1995).

Chemical reactivity and biological activity can be related to molecular structure and physicochemical properties. QSAR models can be established among hydrophobic–lipophilic, electronic, and steric properties, between quantum-mechanics-related parameters and toxicity; and between environmental fate parameters such as sorption and tendency for bioaccumulation. The main objective of a QSAR study is to develop quantitative relationships between given properties of a set of chemicals and their molecular descriptors. To develop a valid QSAR model, the following steps are essential:

- Assume feasible reaction mechanisms.
- Select training set activity data for these reaction mechanisms.
- Select suitable structure descriptors influencing the reactions.
- Conduct statistical correlation analysis of these two sets of data.
- Validate the QSAR model using another set of data outside the training data.
- Define the application criteria.
- Modify the assumption and calibrate the QSAR model.

The general procedure for modeling of QSAR is illustrated in Figure 5.1.

The search for relationships among the dynamic and equilibrium properties of related series of compounds has been a paradigm of chemists for many years. The discovery of such unifying principles and predictive relationships has practical benefits. Numerous relationships exist among the structural characteristics, physicochemical properties, and/or biological qualities of classes of related compounds. Perhaps the best-known attribute relationships are the correlations between reaction rate constants and equilibrium constants for related reactions commonly known as linear free-energy relationships (LFERs). The LFER concept led to the broader concepts of QSARs, which seek to predict the environmental fate of related compounds based on correlations between their bioactivity or physicochemical properties and structural features. For example, therapeutic response, environmental fate, and toxicity of organic compounds have been correlated with

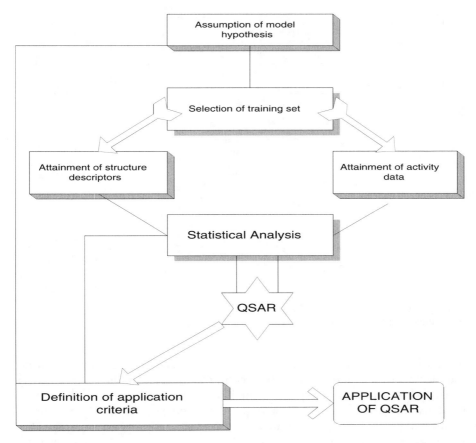

FIGURE 5.1
Principal elements involved in the development of a QSAR model.

various classes of structural descriptors such as electronic, topological, hydrophobic, and steric constants, as shown in Figure 5.2.

In this book, QSAR models are presented systematically for each AOP so that the treatability of organic compounds can be assessed once an AOP process is selected.

5.2.1 Effects of Molecular Structure on Reactivity

Most compounds within a given functional group do have more or less the same reaction mechanisms when this functional group is the reactive center; therefore, a large number of reactions can be classified together and can serve as an aid to understanding the reaction mechanism. For example, Benigni et al. (1995) compiled electrophilic reactivity data for 142 compounds and used the data to establish the contribution of different functional groups and molecular determinants for various properties. An equation containing ~20

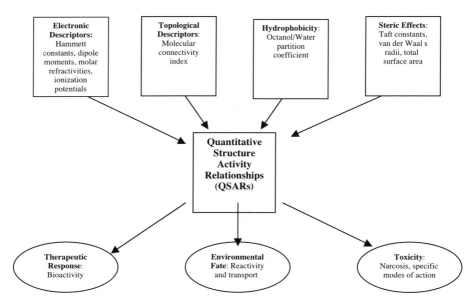

FIGURE 5.2
Illustration of environmental QSAR.

molecular determinants was derived which provided a system for estimating the electrophilic reactivity for other compounds. The mathematical model consisted of a pattern recognition method specifically designed for use in nonlinear situations. The addition of the estimated electrophilicity parameter increased the overall performance of the system by approximately 90%; however, a specific functional group does not always react the same way, and the activity may be significantly influenced by the rest structure of the molecule. The greatest variations may be expected when additional functional groups are present. The effects of structure on reactivity can be divided into three major types: field-inductive (polar), resonance (or mesomeric), and steric.

Substituents influence reactivity by altering electron density at the reaction site and by steric effects. Electronic effects are the easiest to be evaluated for aromatic compounds with substituents in *meta* and *para* positions, where steric influences (which are more difficult to predict) are minimized. Reactions favored by high electron density are accelerated by electron-donor substituents, while electron-withdrawing groups will accelerate reactions favored by low electron density. Electronic effects can be further divided as being field-inductive or resonance effects. The former is transmitted along a chain of atoms without reorganizing the chemical bonds and decreases rapidly with distance from a substituent group. The latter occurs in compounds with conjugated double bonds and produces changes in electron densities at certain positions in conjugated systems.

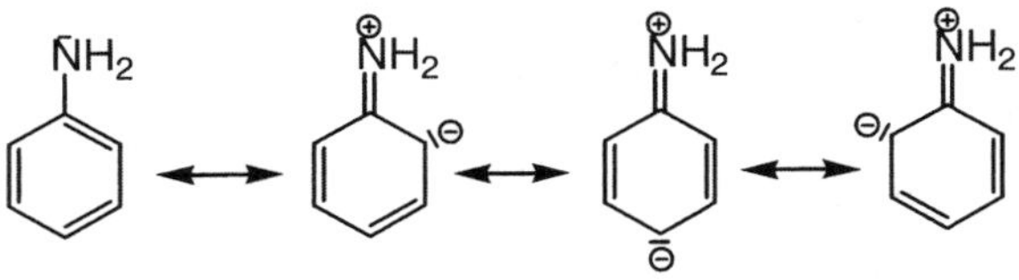

FIGURE 5.3
Resonance effect of aniline. (From Hine, J., *Structural Effects on Equilibria in Organic Chemistry*, John Wiley & Sons, New York, 1975. With permission.).

5.2.2 Electronic Effects

Resonance always results in a different distribution of electron density. For example, if the first term on the left in Figure 5.3 were the actual structure of aniline, the two unshared electrons of nitrogen would reside entirely on that atom. Because the real structure is actually a hybrid that includes contributions from the other canonical forms shown, the electron density of the unshared pair is spread over the ring. This decrease in electron density at one position and the corresponding increase elsewhere is called the *resonance effect*. When steric inhibition is present, resonance may be reduced or prevented because the atoms are sterically forced out of planarity. The resonance effect of a group operates only when the group is directly connected to an unsaturated system.

In ethane, the C–C bond is completely nonpolar because it connects two equivalent atoms; nevertheless, the C–C bond in chloroethane is polarized by the presence of the electronegative chlorine atom. This polarization is actually the sum of two effects. The first carbon atom (C-1), having been deprived of some of its electron density by the greater electronegativity of Cl, is partially compensated by drawing the C–C electrons closer to itself, resulting in a polarization of this bond and a slightly positive charge on the second carbon (C-2) atom. This polarization of one bond caused by the polarization of an adjacent bond is called the *inductive effect*. The effect is the greatest for adjacent bonds but may also be felt farther away; thus, the polarization of the C–C bond causes a slight polarization of the three methyl C–H bonds:

$$\overset{\delta\delta+}{\underset{2}{H_3C}} \rightarrow \overset{\delta+}{\underset{1}{CH_2}} \rightarrow\rightarrow \overset{\delta-}{Cl}$$

The other effect operates not through bonds but directly through space or solvent molecules and is called the *field effect*. Generally, the field effect depends on the molecule geometry, but the inductive effect depends only on the nature of the bonds. In most cases, the two types of effects are considered together; more recently, evidence suggests that field effects are a more important mode of transmission than resonance effects (Hansch and

Leo, 1995). The field-inductive effect operates through space, solvent molecules, or the σ bonds of a system, while the resonance effect operates through π electrons only.

5.2.3 Steric Effects

It occasionally happens that a reaction proceeds much faster or slower than expected on the basis of electronic effects alone. Very often in these cases steric effects are influencing chemical reaction rates. Sheer physical blockage of the attack of the reagent may be attributed to steric hindrance. Steric effects were originally thought to be a spatial effect caused by mechanical interference between groups (Finar, 1973). However, when a molecule undergoes chemical reaction via a transition state, the geometry of both the initial and transition states must be taken into consideration. Thus steric factors may affect the speed and/or mechanism of a reaction. It is often very difficult to distinguish steric and polar factors. Nevertheless, the effects of the steric factor can often be assessed in a certain qualitative manner. When steric effects decrease the rate of a reaction, that reaction is considered to be subject to steric hindrance.

Not all steric effects decrease reaction rates. When steric effects speed up a reaction, that reaction is said to be subject to steric acceleration or steric strain. Steric strain exists in a molecule when bonds are forced to make abnormal angles. In general, two kinds of structural features result in sterically caused abnormal bond angles. One of these is called *small-angle strain* and is found in small-ring compounds, where the angles must be less than those resulting from normal orbital overlap. The other is called *nonbonded interactions*, which arise when nonbonded atoms are forced into close proximity by the molecule geometry.

If two different three-dimensional arrangements of the atoms in a molecule are interconvertible merely by free rotation about bonds, they are *conformations*; otherwise, they are *configurations*. Configurations represent isomers that can be separated; however, conformers cannot be separated. Conformational effects on reactivity may be considered under the steric effects.

5.2.4 Molecular Descriptors

Quantitative structure–activity relationships represent an attempt to correlate activities with structural descriptors of compounds. These physicochemical descriptors, which include hydrophobicity, topology, electronic properties, and steric effects, are determined empirically or, more recently, by computational methods. The success of a QSAR method depends on two factors: the training dataset obtained by testing a group of chemicals and the descriptors obtained from some easily measurable or calculable property of the chemicals.

A preliminary and essential step in a QSAR study is to evaluate the database to identify any outliers and hidden patterns, trends, and major groupings. Outliers refer to certain members of the database exhibiting mechanistic behaviors so different that the outlier cannot belong to the bulk of the data. Selecting suitable molecular descriptors, whether they are theoretical or empirical or are derived from readily available experimental characteristics of the structures, is an important step in the development of sound QSAR models. Many descriptors reflect simple molecular properties and thus can provide insight into the physicochemical nature of the activity or property under consideration.

Most commonly, correlations are made to chemical constants that are defined by the effect of substituents on a reference reaction. They are usually designated σ and are applied to QSARs in the form of Hammett's equation or its various extensions. Alternatively, a descriptor can be a property of a substrate molecule that is available, is readily measurable, and/or can be calculated by independent means, such as octanol/water partition coefficients. Despite their numerous successful applications in QSAR studies, experimentally determined σ constants also have some disadvantages. They are available only for a limited set of substituents and are not of very good quality for uncommon functional groups. As an alternative, the use of quantum-chemical parameters may be a solution.

Recent progress in computational hardware and the development of efficient algorithms have assisted the routine development of molecular quantum-mechanical calculations. New semiempirical methods calculate realistic quantum-chemical molecular quantities in a relatively short computational time frame. Quantum-chemical calculations are thus an attractive source for molecular descriptors that can express all of the electronic and geometric properties of molecules and their interactions. Quantum-chemical methods can be applied to QSARs by direct derivation of electronic descriptors from the molecular wave function.

Descriptors derived via quantum chemistry are fundamentally different from experimentally measured quantities although there is some natural overlap. Unlike experimental measurements, quantum-chemical calculations have no statistical error. An inherent error is associated with the assumptions required to facilitate the calculations, but the computational error is considered to be approximately constant throughout the series when using quantum-chemistry-based descriptors with a series of related compounds. In most studies, the direction but not the magnitude of the error is known. The major weakness of quantum-chemical descriptors is the failure to directly address bulk effects of a molecule.

5.2.5 Linear Free-Energy Relationships

It is well known that no general relationship exists between the reaction thermodynamics and the rates at which they proceed toward equilibrium.

The absence of a general relationship is understandable in light of the transition-state theory: the reaction rates are controlled by the energy difference between reactants and a transition-state complex rather than by the energy difference between reactants and products. In spite of the lack of a universal relationship between reaction rates and their equilibria, correlations do exist between the rates and the energy of reactions for sets of related compounds. Linear free-activity relationships (LFERs) are useful in kinetics because they enable us to predict the reaction rates from more easily measured (or more readily available) equilibrium properties, and they are equally valuable in improving the understanding of reaction mechanisms and the rate controlling steps.

The term *linear free-activity relationships* is used for such correlations because they are usually linear correlations between the logs of rate constants and the logs of equilibrium constants for the reactions of the compounds. According to transition-state theory, rate constants are exponentially related to the free energy of activation ($\Delta G^{\neq}$), and equilibrium constants are similarly related to ΔG^0. The principle is that the reactivity of molecules can be divided into a reacting group (X) and a nonreacting residue (R) (Johnson, 1973). In the absence of strong specialized interactions between R and X, the following assumptions will hold:

- The changes in the value of $\Delta G^{\neq}$ for any reaction involving X which are produced by a series of changes in R are linearly related to the changes in ΔG^0 for the same reaction.

- The changes in the value of $\Delta G^{\neq}$ for one reaction involving a reacting group X_1 produced by a series of changes in R are linearly related to the changes in the corresponding values of another reaction involving X_1 and also to those for a reaction involving a different group, X_2.

An equation could be written for two reactions exhibiting a LFER as:

$$\ln k_2 - \ln k_1 = \alpha(\ln K_2 - \ln K_1) \tag{5.1}$$

or

$$\frac{-\Delta G_2^{\neq} + \Delta G_1^{\neq}}{RT} = \frac{\alpha(-\Delta G_2^0 + \Delta G_1^0)}{RT} \tag{5.2}$$

The free energy of activation is composed of an entropy activation ($\Delta S^{\neq}$) and an enthalpy of activation ($\Delta H^{\neq}$). The former is associated with the pre-exponential factor A of the Arrhenius equation and the latter with the experimental E_{act}, which defines the sensitivity of the reaction rate to temperature. The existence of an LFER for a set of reactants is equivalent to the statement that the free energy of activation is a constant fraction of the free energy of reaction for the reactants ($\Delta G^{\neq}/\Delta G^0 = \alpha$). If $\Delta S^{\neq}$ and ΔS^0 are constant for reactions in series, then the change in $\Delta G^{\neq}$ is proportional to the change in ΔG^0, and log k is proportional to log K. If the entropy terms are not constant, but the entropies and enthalpies are correlated (the compensation effect), then the LFER still holds.

Two major categories of LFERs that have been developed over the past 60 years are shown in Table 5.1. These relationships apply to a wide variety of classes of organic and inorganic compounds and a wide range of reactions.

5.2.6 Hammett LFER

5.2.6.1 *Sigma (σ) Constants*

Hammett's equation (and its extended forms) has been widely used for the study and interpretation of organic reaction mechanisms. The concept of the electron-donating and electron-withdrawing power of substituents, characterized by simple experimentally determined constants introduced by Hammett, revolutionized the entire physical organic chemistry. Hammett proposed the sigma (σ) constant to assign numerical values for the electronic effects of substitution on an aromatic ring (Johnson, 1973). In the dissociation of benzoic acid, Hammett observed that adding substituents to the aromatic ring of benzoic acid had an orderly and quantitative effect on the dissociation constant. Electron withdrawal by the nitro group increases dissociation, with the effect being less at the *meta* than at the *para* position. On the other hand, the electron-donating ethyl group decreases the equilibrium constant in the absence of steric effects.

The underlying assumption in this concept is the absence of steric effects. Using the rate and equilibrium constants for ionization of benzoic acid as references, Hammett found that equilibrium constants for a variety of reactions showed a linear relationship with σ and defined the LFER. Data for these equilibria are typically graphed as illustrated in Figure 5.4.

TABLE 5.1

Major Classes of LFERs Applicable to Reactions in Aquatic Systems

Relationship	Types of Reaction or Reactants	Basis of LFER
Hammett	Reactions of *para*- or *meta*-substituted aromatic compounds: hydrolysis, hydration of alkenes, substitution, oxidation, enzyme-catalyzed oxidations; some type II photooxidations	Electron withdrawal and/or donation from/to reaction site by substituents on aromatic rings via resonance effects
Taft	Hydrolysis and many other reactions of aliphatic organic compounds	Steric and polar effects of substituents

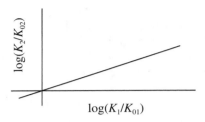

FIGURE 5.4
Linear free-energy relationship.

Because this relationship is linear, the following equation can be written:

$$\log(\frac{K_2}{K_{02}}) = \rho \log(\frac{K_1}{K_{01}}) = \rho\sigma \tag{5.3}$$

where K_{01} or K_{02} represents equilibrium constants for unsubstituted compounds and K_1 or K_2 represents substituted compounds. Hammett defined σ as a substituent parameter, which is a characteristic of a given substituent and its position on the ring. The effects of *meta* and *para* substituents are thought to be additive. The σ values are numbers that sum up the total electronic effects (resonance plus field induction) of a substituent when attached to a benzene ring. The magnitude of σ gives the relative strength of the electron-withdrawing or electron-donating properties of the substituents. The basic Hammett relationship does not apply to *ortho* substituents because they may exert steric as well as polar effects on the reaction centers.

The Hammett equation is an LFER that can be demonstrated as follows for the case of rate constants. For an unsubstituted reactant, as a reference reactant, the free energy of transition state complex (TSC) is:

$$\Delta G_0^{\neq} = -RT \ln k_0 \tag{5.4}$$

For a substituted reactant, as a reference reactant, the free energy of its TSC is as follows.

For each reaction, where X is any group:

$$\Delta G^{\neq} = -RT \ln k \tag{5.5}$$

The Hammett equation may be rewritten:

$$\text{Log } k - \log k_0 = \rho\sigma \tag{5.6}$$

so that

$$\frac{-\Delta G^{\neq}}{2.3RT} + \frac{\Delta G_0^{\neq}}{2.3RT} = \rho\sigma \qquad (5.7)$$

and

$$-\Delta G^{\neq} = 2.3\rho\sigma RT - \Delta G_0^{\neq} \qquad (5.8)$$

For a given reaction under a given set of conditions, ρ, R, T, and $\Delta G_0^{\neq}$ are all constant; therefore σ is linear with ΔG^-. A typical Hammett plot is shown in a plot of log k/k_0 for a series of substituted ethyl benzoates vs. the Hammett sigma constant of substituents in Figure 5.5.

5.2.6.2   Hammett's Reaction Constant ρ

The slope of line ρ (called the *reaction constant*) is a measurement of the sensitivity of such a reaction to the electronic effect of the substituents; nevertheless, it can be explained as a proportionality constant pertaining to a given equilibrium (i.e., it depends on the reaction and the solvent). It relates the effect of substituents on that equilibrium to the effect of those substituents on the benzoic acid equilibrium. That is, if the effect of substituents is proportionally greater than that on the benzoic acid equilibrium, then $\rho > 1$; if the effect is less than that on the benzoic acid equilibrium, then $\rho < 1$. By definition, for benzoic acid the parameter ρ is equal to 1 in water at 25°C. Many organic reactions proceed via a series of steps, each of which might have a different reaction constant ρ; hence, to achieve a successful correlation, ρ must be at least roughly additive.

The sensitivity of ρ to different reaction mechanisms is demonstrated very well in the extended selectivity relationship for electrophilic substitution in biphenyls, as shown in Figure 5.6 (Stock and Brown, 1963). The logarithms of partial rate factor f, which is defined as k/k_H, is nonlinear with the sensitivity factor ρ; however, solvolysis of substituted biphenyl holds the linear relationship. Twisting of one of the phenyl rings out of the plane of the other contributes to this variable resonance effect. For example, in the reactions of the least negative ρ, the angle of twist will be large. As ρ increases more negatively, the transition-state complex has a greater positive charge on the ring. This will increase the stabilizing conjugative interaction due to the planar conformation of the ring and the phenyl substituent.

Hammett's equation has certain limitations. To obtain good results based upon a model system, the reaction mechanism must be similar. Although σ is derived from ionization constants in water or in highly polar solvent systems such as 50:50 ethanol/water, the correlation of reactions in nonpolar solvents may not be good. This finding is especially true for substituents that have strong hydrogen-bonding capacity. Another problem with Hammett's equation concerns the additivity of σ. Although additivity seems to be true for two substituents that do not interact with each other, it is

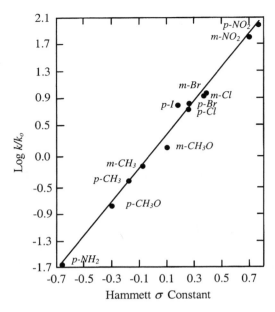

FIGURE 5.5
Log k/k_0 for acid dissociation of a series of substituted ethyl benzoates vs. the Hammett σ constant of the substituent groups. (From Tinsley, I.J., *Chemical Concepts in Pollutant Behavior*, John Wiley & Sons, New York, 1979. With permission.)

unknown how far it can be expected to hold. Still another problem with σ constants is the variable behavior of charged substituents such as $N^+(CH_3)_3$ or COO– for which the σ constants are heavily dependent on the ionic strength of the reaction medium but whose neutral substituents are not. Finally, large substituents and especially those with conformational freedom may not be well behaved.

As originally stated, Hammett's equation was expected to hold for aromatic systems similar to benzoic acid. In fact, it works best when the reaction center is insulated from resonance interaction with the aromatic ring (e.g., X-$C_6H_4CH_2Q$, where Q is the reaction center). Hammett's LFER applies to hydrolysis reactions, to substitution and oxidation reactions of aromatic compounds, and to enzyme-catalyzed reactions such as the oxidation of phenols and aromatic amines by peroxidase. The fact that σ applies to such a variety of reactions implies that it measures a fundamental property such as electron density at the reactive site. The substituent parameter σ decreases with decreasing K_a, and the extent of ionization decreases when electron density increases at the O–H bond in a carboxylic acid group. Thus, σ and electron density must be inversely related. Electron-withdrawing groups such as –Cl and NO_2 have positive σ values; electron donors such as –CH_3 and –NH_2 have negative σ values. Similarly, ρ measures a reaction's sensitivity to electron density. Whereas ρ is positive for nucleophilic reactions that are hindered by high electron density, electrophilic reactions that are accelerated by high

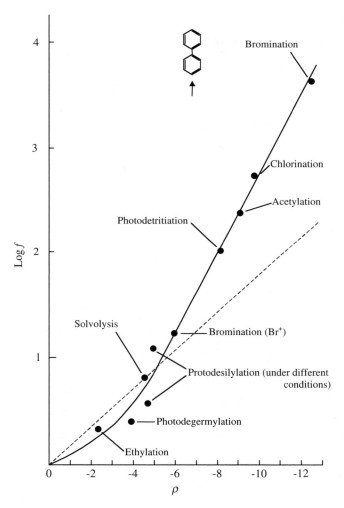

FIGURE 5.6
The extended selectivity relationship for biphenyl. (From Stock, L.M. and Brown, H.C., in *Advances in Physical Organic Chemistry*, Vol. 1, Gold, V., Ed., Academic Press, London. With permission.)

electron density have negative values. Hammett's σ constants for a large number of substituents are listed in Table 5.2. The constants are obtained from the ionization constants of the appropriately substituted benzoic acid in water at 25°C. By definition, σ for hydrogen is zero.

5.2.6.3 Sigma Minus (σ⁻) and Sigma Plus (σ⁺) Constants

A large class of important organic reactions that do not correlate well by Hammett's equation includes compounds connected with electrophilic

TABLE 5.2

Hammett Substituent Constants

Substituent	σ Meta	σ Para	Substituent	σ Meta	σ Para
Me	−0.07	−0.17	NHCO$_2$Et	0.09	−0.13
Et	−0.07	−0.15	NHAc	0.21	0.00
i-Pr	−0.07	−0.15	NHCHO	0.22	0.05
t-Bu	−0.10	−0.20	NHCOCF$_3$	0.35	0.14
CH$_2$Ph	−0.18	−0.11	NHSO$_2$Me	0.21	0.05
C≡CH	0.2	0.23	OH	0.12	−0.37
Ph	0.06	−0.01	Ome	0.12	−0.27
Picryl	0.27	0.31	Oet	0.10	−0.24
CH$_2$SiMe$_3$	−0.16	−0.21	Oph	0.25	−0.32
CH$_2$OMe	0.02	0.03	OCF$_3$	0.40	0.35
CH$_2$OPh	0.03	0.07	Oac	0.39	0.31
CHO	0.36	0.44	OSO$_2$Me	0.39	0.37
Ac	0.38	0.50	F	0.34	0.06
COPh	0.36	0.44	SiMe$_3$	−0.04	−0.07
CO$_2^-$	−0.1	0.00	PO$_3$H$^-$	0.2	0.26
CO$_2$Et	0.37	0.45	SH	0.25	0.15
CN	0.56	0.66	Sme	0.15	0.00
CH$_2$CN	0.15	0.17	SCF$_3$	0.40	0.50
CH$_2$I	0.07	0.09	Sac	0.39	0.44
CH$_2$Br	0.11	0.12	SOMe	0.52	0.49
CH$_2$Cl	0.09	0.12	SO$_2$Me	0.60	0.72
CF$_3$	0.43	0.54	SO$_2$CF$_3$	0.79	0.93
CF(CF$_3$)$_2$	0.37	0.53	SO$_2$NH$_2$	0.46	0.57
NH$_2$	−0.16	−0.66	SO$_3^-$	0.05	0.09
NMe$_2$	−0.15	−0.83	SME$_2^+$	1.00	0.90
N(CF$_3$)$_2$	0.4	0.53	SF$_5$	0.61	0.68
NMe$_3^+$	0.88	0.82	Cl	0.37	0.23
N$_3$	0.37	0.08	Br	0.39	0.23
N=NPh	0.3	0.35	I	0.35	0.28
N$_2^+$	1.76	1.91	IO$_2$	0.70	0.76
NO$_2$	0.71	0.78			

Source: Hine, J., *Structural Effects on Equilibria in Organic Chemistry,* John Wiley & Sons, New York, 1975. With permission.

substituents in aromatic systems. When the reaction site comes into direct resonance with the substituent, the σ constants of the substituents do not correlate with equilibrium or rate constants. For these cases, two new sets of σ values have been devised: σ⁻ for cases in which an electron-withdrawing group interacts with a developing negative charge in the transition state and σ⁺ values, where an electron-donating group interacts with a developing positive charge.

Substituents such as NO$_2$, C≡N, SO$_2$CH$_3$, CONH$_2$, and COOR can donate an electron pair to reaction centers such as OH, NH$_2$, OR, and SH. In the presence of a strong-resonance electron-withdrawing group and a strong-resonance electron-donating group *para* to each other on an aromatic ring, the deviations from Hammett's equation were all in one direction. Such resonance leads to a species that was more stable than would be expected from Hammett's equation. The strong electron-withdrawing substituents that can accept an electron pair have σ^- values significantly different from σ. This observation led to the formulation of σ^- parameters that have been derived from phenols or anilines for *para* substituents (σ_p^-). For the σ^- constants, substituents in the *meta* position cannot participate effectively in delocalizing a positive charge by resonance. The regular σ_m values can be used for σ^-.

For a reaction center that is capable of strongly withdrawing electrons from the ring by a resonance effect, a set of substituent constants for *para* substituents (σ^+) capable of resonance electron donation has been devised. For strong electron-attracting groups such as NO_2 and CN, σ^+ values are essentially the same as σ. The dual sigmas σ and σ^- or σ and σ^+ have been widely and successfully used. Table 5.3 lists a number of σ^- and σ^+ constants.

Although Hammett's equation has been modified and extended, σ constants still remain the most general means for estimating the electronic effect of substituents on reaction centers. The power of the simple Hammett σ values derives from the fact that they often take into account solvent effects on substituents, such as hydrogen bonding, dipole interactions, etc. Hammett's idea of using a simple experimental model system to correlate electronic effects with structural change has been extended to the development of steric and hydrophobic parameters. These extensions of Hammett's thinking have been used to tackle all kinds of structure–activity relationships (SARs) of chemical reactions (Hansch and Leo, 1995).

5.2.7 Taft's LFER

Hammett's success in treating the electronic effect of substituents on the equilibria rates of organic reactions led Taft to apply the same principles to steric, inductive, and resonance effects. The Hammett σ constants appear to be made up primarily of two electronic vectors: field-inductive effect and resonance effect. For substituents on saturated systems, such as aliphatic compounds, the resonance effect is rarely a factor, so the σ form the benzoic acid systems is not applicable. Taft extended Hammett's idea to aliphatics by introducing a steric parameter (E_s). He assumed that for the hydrolysis of esters, steric and resonance effects will be the same whether the hydrolysis is catalyzed by acid or base. Rate differences would be caused only by the field-inductive effects of R and R' in esters of the general formula ($XCOOR$), where X is the substituent being evaluated and R is held constant. Field effects of substituents X could be determined by measuring the rates of acid and base catalysis of a series $XCOOR$. From these rate constants, a value σ^* could be determined by Equation (5.9):

TABLE 5.3

σ⁻ and σ⁺ Constants

Substituent	σ_p Minus	σ_p Plus	Substituent	σ_p Minus	σ_p Plus
Br	0.25	0.15	$SOCH_3$	0.73	—
Cl	0.19	0.11	SO_2CH_3	1.13	—
F	−0.03	−0.07	SCH_3	0.06	−0.60
SF_3	—	—	$C\equiv CH$	0.53	0.18
SF_5	0.86	—	$CH=CH_2$	—	−0.16
I	0.27	0.14	$COCH_3$	0.84	—
NO_2	1.27	0.79	$COOCH_3$	0.64	0.49
N_3	0.11	—	$NHCOCH_3$	−0.46	−0.60
H	0.00	0.00	$N(CH_3)_2$	−0.12	−1.70
OH	−0.37	−0.92	Cyclopropyl	−0.09	−0.41
NH_2	−0.15	−1.30	$CH(CH_3)_2$	−0.16	−0.28
SO_2NH_2	0.94	—	$CH=C(CN)_2$	1.20	0.82
CF_3	0.65	0.61	$C(CH_3)_3$	−0.13	−0.26
OCF_3	—	—	C_6H_5	0.02	−0.18
SO_2CF_3	1.63	—	$N=NC_6H_5$	0.45	−0.19
CN	1.00	0.66	OC_6H_5	−0.10	−0.50
CHO	1.03	0.73	NHC_6H_5	−0.29	−1.40
$CONH_2$	0.61	—	COC_6H_5	0.83	0.51
CH_3	−0.17	−0.31	$CH_2C_6H_5$	−0.09	−0.28
$NHCONH_2$	—	—	$C\equiv CC_6H_5$	0.30	−0.03
OCH_3	−0.26	−0.78	$CH=CHC_6H_5$	0.13	−1.00

Source: Hansch, C. and Leo, A., *Exploring QSAR: Fundamentals and Applications in Chemistry and Biology,* American Chemical Society, Washington, D.C., 1995. With permission.

$$\sigma^* = \frac{1}{2.48}\left[\log\left(k/k_0\right)_B - \log\left(k/k_0\right)_A\right] \tag{5.9}$$

where the subscripts A and B denote acid- and base-catalyzed hydrolysis, respectively; the factor 2.48 puts σ^* on the same scale as Hammett's σ; the k value is the rate constant for the hydrolysis of substituted acetates; and the k_0 values are rate constants for acid and base hydrolysis of acetic acid esters. The methyl group was chosen as the reference substituent. In this case $\sigma^*_{CH_3}$ is zero by definition. Usually, R is an ethyl or a methyl group, so small R groups would not affect σ^* significantly, but in many cases the rate constants do not depend on the nature of R. Since σ^* is the substituent constant for a group of X substituents at a saturated carbon, only additive field-inductive effects will be measured by σ^*. Once a set of σ^* values has been obtained, the following equation will hold:

$$\log(k/k_0) = \rho^* \sigma^* \tag{5.10}$$

TABLE 5.4

Taft Substituent Constants

Substituent	σ^*	Substituent	σ^*
CCl_3	2.65	$CH=CHPh$	0.41
CHF_2	2.05	$CHPh_2$	0.4
CO_2Me	2.00	CH_2CH_2Cl	0.38
$CHCl_2$	1.94	$CH=CHMe$	0.36
Ac	1.65	$CH_2CH_2CF_3$	0.32
$C{\equiv}CPh$	1.35	CH_2Ph	0.22
CH_2SO_2Me	1.32	$CH_2CH=CHMe$	0.13
CH_2CN	1.30	$(CH_2)_3CF_3$	0.12
CH_2F	1.10	CH_2CH_2Ph	0.08
CH_2Cl	1.05	Me	0.00
CH_2Br	1.00	Et	−0.10
CH_2CF_3	0.92	n–Pr	−0.12
CH_2I	0.85	n–Bu	−0.13
CH_2OPh	0.85	Cyclohexyl	−0.15
CH_2OMe	0.64	CH_2Bu–t	−0.16
CH_2Ac	0.60	i–Pr	−0.19
Ph	0.60	Cyclopentyl	−0.20
CH_2OH	0.56	$CHEt_2$	−0.22
$CH_2CH_2NO_2$	0.50	CH_2SiMe_3	−0.26
H	0.49	t–Bu	−0.30

Source: Hine, J., *Structural Effects on Equilibria in Organic Chemistry,* Wiley, New York, 1975. With permission.

where ρ^* is analogous to Hammett's ρ. This equation can then be used as the secondary source for the calculation of substituent constants for groups if the corresponding ester hydrolysis kinetic data are not available. The value of σ^* for different substituents is listed in Table 5.4. The parameter σ^* is important because it allows one to evaluate substituent effects on aliphatic reaction rates by a formula analogous to Hammett's equation.

Because the electronic nature of substituents has little effect on the rate of acid-catalyzed hydrolysis of *meta-* or *para-*substituted benzoates (e.g., ρ for the acid hydrolysis of XC_6H_4COOR esters is close to zero), Taft suggested that the electronic nature of substituents will also have little effect on acid-catalyzed hydrolysis of aliphatic esters (Lowry and Richardson, 1987). Nevertheless, a strong electronic effect occurs in basic hydrolysis, as can be examined from the large ρ values for base-catalyzed hydrolysis of *meta-* or *para-*substituented benzoates. Hence, the effect of X on acid hydrolysis is purely steric but is a combination of steric and electronic effects in basic hydrolysis. Taft defined E_s, a steric substituent constant, by Equation (5.11):

$$\log(k / k_0)_A = E_s \tag{5.11}$$

E_s is defined in reference to hydrogen as zero. The reaction rates involving substituents with small polar effects can be correlated with E_s. For very large groups, the relationship may fail because of the presence of steric effects that are not constant. The equation also fails when X is resonant with the reaction center to a varying extent during the initial and final (or transition) states. Equation (5.12) states that the free energy of activation of a reaction with a substituted compound relative to that with an unsubstituted compound depends on independent contributions from polar and steric effects:

$$\text{Log } (k/k_0) = \rho^*\sigma^* + \delta E_s \qquad (5.12)$$

where δ is a measure of the reaction sensitivity to steric effects, as measured by E_s. E_s has proven to be a very effective molecular descriptor for improving correlations for which σ is not adequate.

In recent years, molecular descriptors such as the energy of the highest occupied molecular orbital (E_{HOMO}) and the energy of the lowest unoccupied molecular orbital (E_{LUMO}) have gained in popularity for QSAR analysis, as these descriptors are readily calculated from PC-based software such as SPARTAN. Before we discuss E_{HOMO} and E_{LUMO} further, a brief discussion of quantum chemistry is necessary.

5.2.8 Quantum-Chemical Calculations

Quantum chemistry is the foundation of molecular chemistry dealing with structure, properties, and interaction of molecules. The basic principles are offered by quantum mechanics. Quantum-chemical calculations are able to supply information needed for molecular descriptors for QSAR analyses. The use of quantum-chemical calculations is becoming common to establish molecular equilibrium geometries and conformations and to supply quantitative thermochemical and kinetic data.

In principle, quantum-chemical theory should be able to provide precise quantitative descriptions of molecular structures and their chemical properties; however, due to mathematical and computational complexities this seems unlikely to be realized in the foreseeable future. Thus, researchers need to rely on approximate methods that have now become routine and have found wide applications. In many cases, errors due to the approximate nature of quantum-chemical calculations and the neglect of the solvation effects are largely transferable within structurally related series (Karelson and Lobanov, 1996). Thus, relative values of calculated descriptors can be meaningful even though their absolute values are not directly applicable.

The theories underlying calculations have now evolved to a stage where a variety of important quantities, such as molecular equilibrium geometry and reaction energetics, may be obtained with sufficient accuracy. Closely related are the spectacular advances in computer hardware and software over the past decade that allow the processing of large numbers of costly calculations. Therefore, calculations are being used not only to interpret experimental data, but also to supplement limited data or even replace it entirely. Due to the increased computing power of PCs, graphics-based programs such as PC SPARTAN Pro (Wavefunction, Inc.) have become increasingly popular.

5.2.9 Principle of Quantum Mechanics

Quantum mechanics describes molecules in terms of interactions between nuclei and electrons and molecular geometry in terms of minimum energy arrangements of nuclei. All quantum-mechanical methods ultimately trace back to Schrödinger's (time-independent) equation, which may be solved exactly for the hydrogen atom. For a multinuclear and multielectron system, the Schrödinger equation may be defined as:

$$H\Psi = E\Psi \tag{5.13}$$

where H is termed the Hamiltonian operator, which describes both the kinetic energies of the particles (i.e., nuclei and electrons) which make up the molecule and the electrostatic interactions felt between individual particles; Ψ is a function (many-electron wave function) of the positions and momenta of all the particles; and E is the energy of the system. Unfortunately, the multiple-electron Schrödinger equation cannot be solved exactly. Based on the physical considerations of constructing simplified models of target structures, approximations have to be made.

Several different and well-documented computational models are supported by PC SPARTAN Pro. No one method of calculation is ideal for all applications. The most sophisticated quantum-chemical models may yield excellent results but may be too expensive for routine treatment, and it will usually be necessary to contend with lesser treatments. Practical models are not likely to offer the best possible treatment formulation. Compromise is almost always an essential component of model selection; therefore, when selecting models compromises must be made between the best formulation and approximation.

Most *ab initio* calculations are based on the orbital approximation (Hartree–Fock method). Hartree–Fock molecular orbital models that remain a mainstream of quantum-chemical techniques are generally satisfactory for a wide variety of thermochemical and kinetic comparisons. The one-electron solutions for many-electron molecules will closely resemble the one-electron solutions for the hydrogen atom. In general, this method provides better results for the larger number of atomic orbitals employed. Hartree–Fock models are particularly attractive for the structure determinations of medium-size organic and main-group inorganic molecules. The calculations are usually not practical for molecules with more than 100 atoms because it is impossible to estimate precisely how much computer time a particular calculation will require. Cost comparisons should be based on representative cases to help distinguish applications that are practical from those that are impossible.

5.2.10 Procedure for Quantum-Mechanical Calculations

The majority of organic molecules are made up of a relatively few elements and obey conventional valence rules. In PC SPARTAN Pro, organic molecules are constructed from atomic fragments, functional groups, and rings by model entry kits, which are builders of molecular structure. Figure 5.7 shows the built structure of carbofuran in SPARTAN. The initial molecular structure resulting from building may not be suitable, and it will generally be advantageous to perform preliminary structure refinement using empirical molecular mechanics techniques.

A molecular mechanics minimizer based on the MMFF94 force field is available. The quantum calculations are based on a Hartree–Fock molecular orbital model using $3-21G^{(*)}$ basis sets to calculate the single-point energy shown in Figure 5.8. The previous setup of quantum mechanics calculations is submitted to perform the required calculations. This procedure provides access to PC SPARTAN Pro's computation capabilities and also controls access for specifying and displaying molecular properties. The calculation time depends upon the complexity of the chemical structures and computer capacity. Finally, the molecule properties such as total energy, HOMO, LUMO, dipole moment, surface area and volume of a space-filling model, and symmetry point group (shown in Figure 5.9) are displayed by the program. Dipole moment, HOMO, and LUMO are molecular descriptors closely related to oxidation-reduction processes. These descriptors will be used extensively in assessing treatability of organic pollutants.

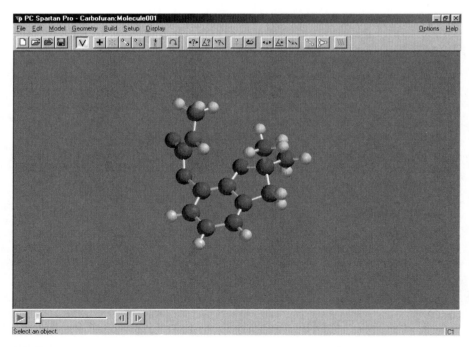

FIGURE 5.7
Molecule structure of carbofuran constructed in SPARTAN.

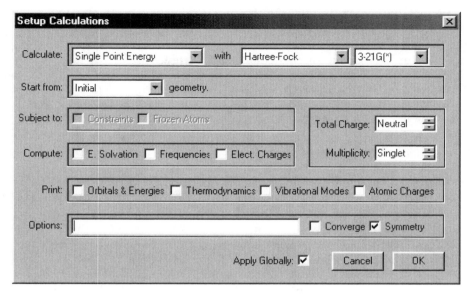

FIGURE 5.8
The setup for quantum calculations of compounds in SPARTAN.

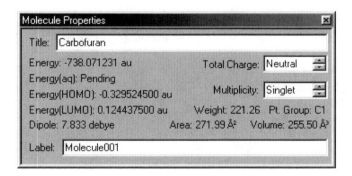

FIGURE 5.9
The calculated molecule properties of carbofuran in SPARTAN.

5.2.11 Dipole Moment

The most obvious and most often used quantity to describe polarity is the dipole moment of a molecule. Dipole moments result from electronegativity (i.e., attraction for electrons) differences of the atoms within a molecule. A molecule becomes dipolar if the electrons forming the bond accumulate toward more electronegative atoms, leaving fewer electronegative atoms with a slight positive charge. The dipole moment, μ, can be defined as the vector in the direction of negative to positive charge of magnitude qd, where q is the electronic charge and d is the distance between the charges (positive and negative centers), and is given by $\mu = q \times d$ (Finar, 1973). The unit of the dipole moment is the Debye (D). Typically, dipole moments of organics fall in the range of 0 to 5 D.

Two kinds of dipole moments are the induced and the permanent. The induced dipole is created by an electric field. Its value is temperature independent, because the molecule will reorient itself in the direction of the field after it is perturbed by a thermally agitated molecule. The permanent dipole is caused by the electric negativity differences of the atoms within the molecule and is temperature dependent. At higher temperatures, the random movement of molecules opposes their tendency to become oriented in the direction of the electric field (Lyman et al., 1982).

The dipole moments of molecules are often treated as being equal to the vector sum of the bond dipoles of the various bonds in the molecules. It is almost impossible to measure the dipole moment of an individual bond within a molecule. For example, molecules such as methane, carbon tetrachloride, and *p*-dichlorobenzene have no dipole moments, whereas molecules such as methylene chloride and *m*-dichlorobenzene do. The vector sum treatment could be made to agree quantitatively with all known dipole moments if the bond moments were treated as variables that depend on the nature of the particular molecule in which the bonds were located.

Dipole moments reflect overall charge distributions in molecules; therefore, they can be employed to judge the relative merits of one level of calculation over another in describing overall charge distribution. The dipole moment is a fairly significant parameter used to aid in the determination of molecular structure, bond angles, and resonance. Finally, dipole moment is useful in ascertaining the potential for a molecule to interact with its surrounding medium via hydrogen bonding, van der Waals forces, and dipole–dipole attractions. Van der Waals force differs from dipole–dipole attraction in that the former is an attraction between small, transient charges in molecules which are normally considered nonpolar, whereas the latter involves permanent molecular dipoles.

5.2.12   Energies of HOMO and LUMO

As reacting molecules approach closely enough, their orbitals will overlay each other. The interaction constitutes a perturbation that will mix the orbitals. The perturbation leads to the bonding and antibonding interactions when two separate orbitals are brought together to form a bond (Fleming, 1976). For example, the highest occupied orbital of the molecule on the left and the lowest occupied orbital of the molecule on the right will form a bond. The new molecular orbitals, shown in the center of Figure 5.10, will be an approximation of the transition state of the two orbitals.

The formation of the bonding orbital is exothermic (E_1); however, the formation of the antibonding orbital is endothermic (E_2). The interaction between two filled orbitals will cause little change in the total energy because

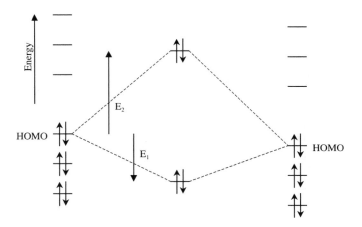

FIGURE 5.10
The interaction of the homo of one molecule with the homo of another molecule. (From Fleming, I., *Frontier Orbitals and Organic Chemical Reactions*, John Wiley & Sons, New York, 1976. With permission.)

one orbital moves down nearly as much as the other moves up. Combinations of unfilled orbitals with the other unfilled orbitals will have no effect on the energy of the system, because without electrons energy cannot be lost or gained. The significant interactions are therefore between the filled orbitals of one molecule and the empty orbitals of the other. Furthermore, because the strongest interactions will be between those orbitals that are close to each other in terms of energy levels, the most important interactions are between the highest occupied molecular orbital (HOMO) of one molecule and the lowest unoccupied molecular orbital (LUMO) of the other, as shown in Figure 5.11. It shows that the energy of the bonding combination decreases and the energy of the antibonding combination increases; therefore, there is no net effect on the actual energy of the system, because no electrons occupy that orbital. These orbitals are sometimes referred to as the *frontier orbitals*. If HOMO and LUMO interaction cannot occur, this stabilizing interaction is absent. The small energy increase arising from the filled level interaction will dominate, which prevents the reaction from taking place.

According to the frontier molecular orbital theory (FMO) of chemical reactivity, the formation of a transition state is due to an interaction between the frontier orbitals, such as HOMO and LUMO of reacting species. In general, the important frontier orbitals for a nucleophile reacting with an electrophile are HOMO (nucleophile) and LUMO (electrophile).

The energy of the HOMO (E_{HOMO}) is directly related to the ionization potential and characterizes the susceptibility of the molecule to attack by electrophiles. On the other hand, E_{HOMO} is directly related to the electron affinity and characterizes the susceptibility of the molecule toward attack by nucleophiles. Both the E_{HOMO} and E_{LUMO} energies are important in radical reactions. The concept of hard and soft nucleophiles and electrophiles has

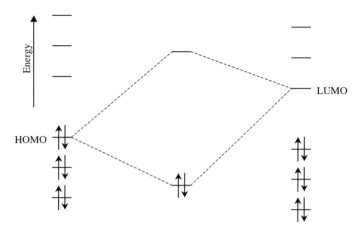

FIGURE 5.11
The interaction of the HOMO of one molecule with the LUMO of another molecule. (From Fleming, I., *Frontier Orbitals and Organic Chemical Reactions*, John Wiley & Sons, New York, 1976. With permission.)

been also directly related to the relative energy of the HOMO and LUMO orbitals. Following are the rules:

- Hard nucleophiles have a low E_{HOMO}.
- Soft nucleophiles have a high E_{HOMO}.
- Hard electrophiles have a high E_{LUMO}.
- Soft electrophiles have a low E_{LUMO}.

5.2.13 Octanol/Water Partition Coefficient

The octanol/water partition coefficient (K_{ow} or P) is defined as the ratio of the concentration of a chemical in the octanol phase to its concentration in the aqueous phase of an octanol/water system at equilibrium:

$$K_{ow} = \frac{\text{Concentration in Octanol Phase}}{\text{Concentration in Aqueous Phase}} \qquad (5.14)$$

The parameter is measured using low-solute concentrations less than 0.01 mol/L, where K_{ow} is not affected by solute concentration. Values of K_{ow} are usually measured at room temperature (20 or 25°C). Numerous studies have shown that it is useful to correlate K_{ow} with biological, biochemical, or toxic properties. In recent years, the octanol/water partition coefficient has become a key parameter in studies of the environmental fate of organics, because of its increasing use in the estimation of other properties or activities, such as the soil sorption coefficient.

Because K_{ow} represents the tendency of a chemical to partition itself between an organic phase and an aqueous phase, values of K_{ow} have some meaning in themselves. Chemicals with low K_{ow} values (for example, less than 10) may be considered relatively hydrophilic; conversely, chemicals with high K_{ow} values (for example, greater than 10^4) are extremely hydrophobic. Several methods are available for the estimation of octanol/water partition coefficients for organic chemicals. Leo's fragment constant method uses empirically derived atomic or group fragments and structural factors and is the most popular method for calculating K_{ow}.

Many software packages have been developed for calculation of molecular descriptors. Table 5.5 shows some software employed for molecular modeling, quantum-chemical calculations, molecular dynamics, and QSARs.

TABLE 5.5

Computer Software for QSAR Analysis

Name	Type
SYBYL	Modeling software
ALCHEMY	Molecular modeling
CHEM-X	Modeling software
FRODOS	Protein display and modeling
MIDAS	Molecular display modeling
CONCORD	Generation of 3D structures
WIZARD	Generation of 3D structures
MM2P	Force field calculations
MOPAC	Semiempirical methods
QCPE	Quantum-chemistry programs
CHEMLAB	Quantum mechanics, molecular display
AMBER	Molecular mechanics, dynamics
DISCOVER	Molecular dynamics
INSIGHT	Graphics interface to Discover
GRID	Potential contour maps
ADAPT	Statistics package for QSAR
SIMCA	Statistics package for QSAR
MEDCHEM	Log P calculations, database program

5.3 Chlorine Effect on Molecular Descriptors for QSAR Analysis

It has been frequently observed that the greater number of chlorine atoms a chlorinated compound contains, the more recalcitrant the chemical will be; therefore, it is important to quantitatively describe the effect so that the treatability of chlorinated compounds can be estimated in terms of the number of chlorine atoms contained by a chlorinated molecule. Table 5.6 is a compilation of QSAR models developed by Tang and Pierotti (2000).

TABLE 5.6

Different QSAR Models for Specific Number of Carbon Atoms

Chemical Class/Subclass	# of Carbons	Log P	HOMO	LUMO	ΔE	Dipole Moment
Alcohol	All	0.6000	-0.9721*	-0.9925*	0.9809*	-0.2946
All two-carbon	2	0.9915*	-0.9902*	-0.9992*	0.9943*	-0.0938
Without diol	2	—	-0.9974*	-0.9991*	0.9956*	—
Aldehyde	All	0.7530	-0.6208	-0.9513*	0.8862	-0.6468
Aromatic	All	0.8960*	-0.3061	-0.1356	0.1965	-0.0693
All 6 carbon	6	0.9193*	-0.5522	-0.9521*	0.4285	-0.0346
Benzene	6	0.9971*	-0.7175	-0.9910*	0.6930*	-0.0346
Phenol	6	0.9689*	-0.8390	-0.9937*	0.9085*	-0.4292
All 7 carbon	7	0.0574	—	—	—	—
Toluene	7	—	-0.6107	-0.9979*	0.8904	-0.0917
Benzoic acid	7	0.0100	—	—	—	—
Carboxylic acid	all	0.6272	-0.0283	-0.6222	0.3734	-0.3779
All two-carbon	2	0.9935*	-0.9961*	-0.9714	0.9467	-0.9525
Halogenated Aliphatic	all	0.2693	-0.0265	-0.7592	0.5560	-0.4347
All one-carbon	1	0.9745*	-0.9995*	-0.9948*	0.9884*	-0.9268*
All two-carbon	2	0.9473*	-0.3707	-0.3473	0.0224	-0.3917
Alkane	2	0.9671*	-0.3251	-0.5104	0.5769	-0.4490
Alkene	2	0.9434*	-0.5912	-0.9948*	0.9834	-0.6026
Ketone	All	0.7745	-0.4750	-0.1949	0.5754	-0.5742
Nitrile	All	0.9139*	—	—	—	—
All two-carbon	2	0.9922*	-0.9677	-0.9933*	0.9995*	-0.9998*

Note: (*) indicates significant correlation ($p < 0.10$).

5.3.1 Dipole Moment

Dipole moment is a measure of the polarity due to the difference in the partial positive (δ^+) and negative (δ^-) charge associated with a covalent bond between two atoms of different electronegativities. The presence of a single halogen atom, such as chlorine, can increase dipole moment; however, compounds with identical opposing atoms or functional groups and compounds with larger structures can cancel or reduce dipole moment. This may explain the reduction of dipole moment as the number of halogens increases and the poor correlation of several classes and larger molecules. Although dipole moment correlations may be poor for most chemical classes, the dipole moment should still be considered as a parameter in modeling small molecules for which dipole moments are important.

The dipole moment (DM) showed a poor correlation ($r < 0.49$) for alcohols, all aromatics, and all halogenated aliphatics; however, for small molecules (two carbons), dipole moment correlations were significant for halogenated aliphatics and nitriles, as shown in Equation (5.15).

$$DM = -0.6786 \text{ (chlorine no.)} + 3.0397, r^2 = 0.8589, n = 3,$$
$$r = -0.9268, 0.10 < p < 0.05 \tag{5.15}$$

For one-carbon halogenated aliphatics, the dipole moment decreases as the number of chlorines increases. The dataset consists of chloromethane, dichloromethane, chloroform, and carbon tetrachloride. The dipole moment represented 85.89% of the variance in the linear regression equation; therefore, the probability of getting a correlation of -0.9268 for a sample size of three is between 5 and 10%:

$$DM = -0.7201 \text{ (chlorine no.)} + 4.0252, r^2 = -0.9997, n = 3,$$
$$r = -0.9998, 0.02 < p < 0.01 \tag{5.16}$$

For two-carbon nitriles, the dipole moment decreases as the number of chlorines increases. The dataset contained chloroacetonitrile, dichloroacetonitrile, and trichloroacetonitrile. The dipole moment represented 99.97% of the variance in the linear regression equation; therefore, the probability of getting a correlation of -0.9998 for a sample size of three is between 1 and 2%.

E_{LUMO} is a measure of the ability of a compound to accept electrons (i.e., to act as an electrophile or undergo reduction). The above correlations show a decrease in E_{LUMO} as the number of chlorines increases. As E_{LUMO} decreases, the ability of a compound to behave as an electrophile increases; however, properties that increase stability increase LUMO energy and decrease reactivity. For example, between two-carbon alkanes and alkenes, $r = -0.5104$ and $r = -0.9948$, respectively. These data agree with Richard and Hunter (1996) in regard to the stability of alkanes over alkenes.

Overall, LUMO correlates well for several alcohol, aromatic, halogenated aliphatic, and nitrile groups; however, the LUMOs of two-carbon alkanes and ketones are not as well correlated with the number of chlorine. In this case, the correlation coefficients are only $r = 0.51$ and $r = -0.1949$, respectively, for alkanes and ketones.

$$E_{\text{LUMO}} = -0.038(\text{chlorine no.}) + 0.2382, \; r^2 = 0.9985, \; n = 4, \; r = 0.9992, \; p < 0.01$$

$$(5.17)$$

For two-carbon alcohols, LUMO decreases as the number of chlorines increases. As LUMO decreases, the ability of a compound to undergo reduction increases; therefore, an increase in chlorine increases the reactivity of the molecule. The dataset contains 2,2,2-trichloro-1,1-ethanediol; 2,2,2-trichloroethanol; 2,2-dichloroethanol; and 2-chloroethanol. LUMO represents 99.85% of the variance in the linear regression equation. The probability of getting a correlation of −0.9992 for a sample size of four is less than 1%.

$$E_{\text{LUMO}} = -0.0127(\text{chlorine no.}) + 0.1381, \; r^2 = 0.9064, \; n = 47,$$
$$r = -0.9521, \; p < 0.01 \qquad (5.18)$$

Figure 5.12 shows QSAR analysis for six-carbon aromatic compounds (see also Table 5.7).

LUMO decreases as the number of chlorines increases. LUMO represents 90.64% of the variance in the linear regression equation. The probability of getting a correlation of −0.9521 for a sample size of 47 is less than 1%.

$$E_{\text{LUMO}} = -0.0406(\text{chlorine no.}) + 0.2524, \; R^2 = 0.9896, \; n = 4,$$
$$R = 0.9948, \; p < 0.01 \qquad (5.19)$$

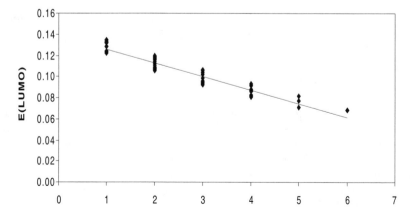

FIGURE 5.12
E_{LUMO} vs. number of chlorines for six-carbon aromatics.

TABLE 5.7

Compounds Used in the Correlation of LUMO vs.
Number of Chlorines for Six-Carbon Aromatics

1,2,3,4-Tetrachlorobenzene	2,4-Dichlorophenol
1,2,3,5-Tetrachlorobenzene	2,5-Dichloroaniline
1,2,3-Trichlorobenzene	2,5-Dichlorophenol
1,2,4,5-Tetrachlorobenzene	2,6-Dichloroaniline
1,2,4-Trichlorobenzene	2,6-Dichlorophenol
1,2-Dichlorobenzene	2-Chloroaniline
1,3,5-Trichlorobenzene	2-Chlorophenol
1,3-Dichlorobenzene	3,4,5-Trichloroaniline
1,4-Dichlorobenzene	3,4,5-Trichlorophenol
2,3,4,5-Tetrachloroaniline	3,4-Dichloroaniline
2,3,4,5-Tetrachlorophenol	3,4-Dichlorophenol
2,3,4,6-Tetrachlorophenol	3,5-Dichloroaniline
2,3,4-Trichloroaniline	3,5-Dichlorophenol
2,3,4-Trichlorophenol	3-Chloroaniline
2,3,5,6-Tetrachloroaniline	3-Chlorophenol
2,3,5,6-Tetrachlorophenol	4,6-Dichloro-1,3-benzenediol
2,3,5-Trichlorophenol	4-Chloroaniline
2,3,6-Trichlorophenol	4-Chlorophenol
2,3-Dichloroaniline	Chlorobenzene
2,3-Dichlorophenol	Hexachlorobenzene
2,4,5-Trichloroaniline	Pentachloroaniline
2,4,5-Trichlorophenol	Pentachlorobenzene
2,4,6-Trichlorophenol	Pentachlorophenol
2,4-Dichloroaniline	

For one-carbon halogenated aliphatics, LUMO decreases as the number of chlorines increases. The dataset is chloromethane, dichloromethane, chloroform, and carbon tetrachloride. LUMO represents 98.96% of the variance in the linear regression equation. The probability of getting a correlation of –0.9948 for a sample size of four is less than 1%.

For two-carbon nitriles, LUMO decreases as the number of chlorines increases. The dataset contains chloroacetonitrile, dichloroacetonitrile, and trichloroacetonitrile. LUMO represents 99.85% of the variance in the linear regression equation. The probability of getting a correlation of –0.9933 for a sample size of three is between 5 and 10%.

$$E_{LUMO} = 0.0203(\text{chlorine no.}) + 0.1582, \ r^2 = 0.9967, \ n = 3,$$
$$r = 0.9933, \ 0.10 < p < 0.05 \tag{5.20}$$

HOMO only showed significant correlations for halogenated aliphatic and for small molecules within alcohol and carboxylic acid groups. Aromatic compounds and larger molecules did not correlate well.

$$E_{HOMO} = -0.0117(\text{chlorine no.}) - 0.4197, \, r^2 = 0.9990, \, n = 4,$$
$$r = -0.9995, \, p < 0.02 \tag{5.21}$$

For one-carbon halogenated aliphatics, HOMO decreases as the number of chlorines increases. HOMO is a measure of the ability of a compound to donate electrons (i.e., to act as a nucleophile or undergo oxidation); therefore, an increase in chlorine numbers increases the reactivity of the molecule. The dataset is chloromethane, dichloromethane, chloroform, and carbon tetrachloride. HOMO represents 99.90% of the variance in the linear regression equation. The probability of getting a correlation of –0.9995 for a sample size of four is less than 2%.

5.3.2   ΔE

The difference between E_{HOMO} and E_{LUMO} energies is ΔE, which measures the ability of an electron to move to another energy level and may occur during a reaction or be involved in electron transfer steps (Benigni and Richard, 1996). In many QSAR models, HOMO can be incorporated into ΔE. ΔE correlates better than HOMO, as shown in Figure 5.13, for alkenes ($r = 0.9834$ and -0.5912, respectively) and phenols ($r = 0.9085$ and -0.8390, respectively); however, LUMO correlations were slightly better than ΔE for benzenes ($r = -0.9910$ and 0.6930, respectively) and toluenes ($r = -0.9979$ and 0.8904, respectively).

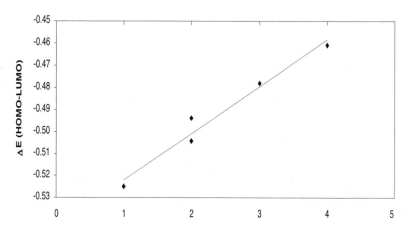

FIGURE 5.13
ΔE vs. number of chlorines for two-carbon alkenes.

$$\Delta E = 0.0211(\text{chlorine no.}) - 0.5430, R^2 = 0.9670, n = 5,$$
$$R = 0.9834, p < 0.01 \qquad (5.22)$$

For two-carbon halogenated alkenes, ΔE increases as the number of chlorines increases. The dataset is 1,1-dichloroethylene; 1,2-dichloroethylene; chloroethylene; tetrachloroethylene; and trichloroethylene. ΔE represents 96.70% of the variance in the linear regression equation. The probability of getting a correlation of 0.9834 for a sample size of five is less than 1% (see Figure 5.14).

$$\Delta E = 0.007(\text{chlorine no.}) - 0.4526, r^2 = 0.8254, n = 19,$$
$$r = 0.9085, p < 0.01 \qquad (5.23)$$

For phenols, ΔE increases as the number of chlorines increases. The dataset is found in Table 5.8. ΔE represents 82.54% of the variance in the linear regression equation. The probability of getting a correlation of 0.9085 for a sample size of 19 is less than 1%.

5.3.3   Octanol/Water Partition Coefficient (Log *P*)

Log *P* correlated well for six-carbon aromatics and small alcohols and carboxylic acids (in regard to number of carbons). Correlations for two-carbon alkanes were also better than the other descriptors (see Figure 5.15).

$$\text{Log } P = 0.7132(\text{chlorine no.}) + 1.4953, r^2 = 0.8451, n = 48,$$
$$r = 0.9193, p < 0.01 \qquad (5.24)$$

For six-carbon aromatics, log *P* increases as the number of chlorines increases. The dataset is shown in Table 5.9.

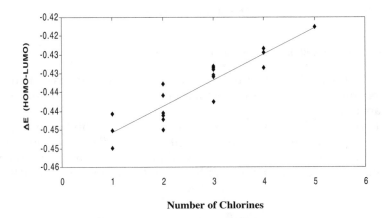

Number of Chlorines

FIGURE 5.14
ΔE vs. number of chlorines for phenols.

TABLE 5.8

Compounds Used in the Correlation of ΔE vs.
Number of Chlorines for Phenols

2,3,4,5-Tetrachlorophenol	2,5-Dichlorophenol
2,3,4,6-Tetrachlorophenol	2,6-Dichlorophenol
2,3,4-Trichlorophenol	2-Chlorophenol
2,3,5,6-Tetrachlorophenol	3,4,5-Trichlorophenol
2,3,5-Trichlorophenol	3,4-Dichlorophenol
2,3,6-Trichlorophenol	3,5-Dichlorophenol
2,3-Dichlorophenol	3-Chlorophenol
2,4,5-Trichlorophenol	4-Chlorophenol
2,4,6-Trichlorophenol	Pentachlorophenol
2,4-Dichlorophenol	

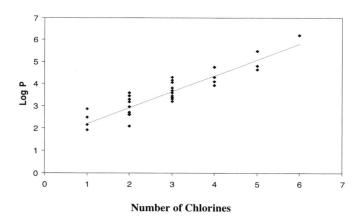

Number of Chlorines

FIGURE 5.15
Log P vs. number of chlorines for six-carbon aromatics.

Log P represents 84.51% of the variance in the linear regression equation.
The probability of getting a correlation of 0.9193 for a sample size of 48 is
less than 1%.

For two-carbon alcohols, log P increases as the number of chlorines
increases. The dataset is 2,2,2-trichloroethanol; 2,2-dichloroethanol; and 2-
chloroethanol. Log P represents 98.30% of the variance in the linear regression equation; therefore, the probability of getting a correlation of 0.9915 for
a sample size of three is between 5 and 10%.

$$\text{Log } P = 0.855(\text{chlorine no.}) - 1.09, \; R^2 = 0.983, \; n = 3,$$
$$R = 0.9915, \; 0.10 < p < 0.05 \tag{5.25}$$

TABLE 5.9

Compounds Used in the Correlation of Log P vs.
Number of Chlorines for Six-Carbon Aromatics

1,2,3,4-Tetrachlorobenzene	2,4-Dichloroaniline
1,2,3,5-Tetrachlorobenzene	2,4-Dichlorophenol
1,2,3-Trichlorobenzene	2,5-Dichloroaniline
1,2,4,5-Tetrachlorobenzene	2,5-Dichlorophenol
1,2,4-Trichlorobenzene	2,6-Dichloroaniline
1,2-Dichlorobenzene	2,6-Dichlorophenol
1,3,5-Trichlorobenzene	2-Chloroaniline
1,3-Dichlorobenzene	2-Chlorophenol
1,4-Dichlorobenzene	3,4,5-Trichloroaniline
2,3,4,5-Tetrachloroaniline	3,4,5-Trichlorophenol
2,3,4,5-Tetrachlorophenol	3,4-Dichloroaniline
2,3,4,6-Tetrachlorophenol	3,4-Dichlorophenol
2,3,4-Trichloroaniline	3,5-Dichloroaniline
2,3,4-Trichlorophenol	3,5-Dichlorophenol
2,3,5,6-Tetrachloroaniline	3-Chloroaniline
2,3,5,6-Tetrachlorophenol	3-Chlorophenol
2,3,5-Trichlorophenol	4,6-Dichloro-1,3-benzenediol
2,3,6-Trichlorophenol	4-Chloroaniline
2,3-Dichloroaniline	4-Chlorophenol
2,3-Dichlorophenol	Chlorobenzene
2,4,5-Trichloroaniline	Hexachlorobenzene
2,4,5-Trichlorophenol	Pentachloroaniline
2,4,6-Trichloroaniline	Pentachlorobenzene
2,4,6-Trichlorophenol	Pentachlorophenol

$$\text{Log } P = 0.885(\text{chlorine no.}) - 0.9067, \ r^2 = 0.9871, \ n = 3,$$
$$r = 0.9935, \ 0.10 < p \ 0.05 \tag{5.26}$$

For two-carbon carboxylic acids, log P increases as the number of chlorines increases. The dataset is chloroacetic acid; dichloroacetic acid; and trichloroacetic acid. Log P represents 98.71% of the variance in the linear regression equation; therefore, the probability of getting a correlation of 0.9935 for the sample size of three is between 5 and 10% (see Figure 5.16).

$$\text{Log } P = 0.6432(\text{chlorine no.}) + 0.4283, \ r^2 = 0.9352, \ n = 9, \ r = 0.9671, \ p < 0.01 \tag{5.27}$$

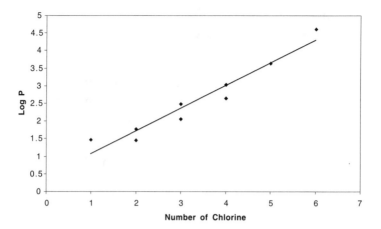

FIGURE 5.16

Log P vs. number of chlorines for two-carbon alkanes.

For two-carbon alkanes, log P increases as the number of chlorines increases. The dataset consists of 1,1,1,2-tetrachloroethane; 1,1,1-trichloroethane; 1,1,2,2-tetrachloroethane; 1,1,2-trichloroethane; 1,1-dichloroethane; 1,2-dichloroethane, chloroethane; hexachloroethane; and pentachloroethane. Log P represents 93.52% of the variance in the linear regression equation. The probability of getting a correlation of 0.9671 for a sample size of nine is less than 1%.

Log P is used to represent the permeability of a membrane. Because the transport of chemicals across a membrane depends on overall permeability of the membrane, hydrophobic or lipophilic compounds pass more easily through a membrane than do hydrophilic compounds. Because log P provides an indication of lipophilicity and the ability for a compound to be transported through a lipid membrane, it is considered rare not to have log P present in a QSAR for heterogeneous AOP, such as UV/TiO$_2$. The mechanistic role of log P involves a common nonspecific mode of action that affects transportation processes, such as movement across a membrane or adsorption, before it reaches the target molecule (Nendza and Russom, 1991).

Several significant correlations between the numbers of chlorines and the molecular descriptors with different number of carbons have been developed. Alcohols show a significant correlation with HOMO, LUMO, and ΔE; aldehydes show an inverse correlation with LUMO; and aromatic compounds show a direct correlation with log P. Other classes and descriptors did not show significant correlations (see Figure 5.17).

$$E_{HOMO} = -0.0135(\text{chlorine no.}) - 0.4107, R^2 = 0.9440, n = 8, R = -0.9721, p < 0.01$$

$$(5.28)$$

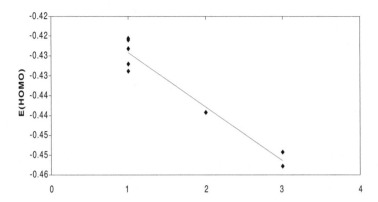

FIGURE 5.17
E_{HOMO} vs. number of chlorines for alcohols with different number of carbons.

For alcohols, E_{HOMO} decreases as the number of chlorines increases. The dataset is 1-chloro-2-propanol; 2,2,2-trichloro-1,1-ethanediol; 2,2,2-trichloroethanol; 2,2-dichloroethanol; 2-chloroethanol; 3-chloro-1,2-propanediol; 3-chloropropanol; and 4-chloro-1-butanol. HOMO represents 94.4% of the variance in the linear regression equation; therefore, the probability of getting a correlation of –0.9721 for a sample size of eight is less than 1% (see Figure 5.17).

$$E_{HOMO} = -0.04(\text{chlorine no.}) + 0.2439, \; R^2 = 0.9851, \; n = 8, \; R = 0.9925, \; p < 0.01 \tag{5.29}$$

Figure 5.18 shows that LUMO decreases as the number of chlorines increases for alcohols. The dataset is 1-chloro-2-propanol; 2,2,2-trichloro-1,1-ethanediol; 2,2,2-trichloroethanol; 2,2-dichloroethanol; 2-chloroethanol; 3-chloro-1,2-propanediol; 3-chloropropanol; and 4-chloro-1-butanol. LUMO represents 98.51% of the variance in the linear regression equation; therefore, the probability of getting a correlation of –0.9925 for a sample size of eight is less than 1%.

$$E_{LUMO} = 0.0265(\text{chlorine no.}) - 0.6547, \; r^2 = 0.9621, \; n = 8, \; r = 0.9809, \; p < 0.01 \tag{5.30}$$

For alcohols, ΔE increases as the number of chlorines increases. The dataset is 1-chloro-2-propanol; 2,2,2-trichloro-1,1-ethanediol; 2,2,2-trichloroethanol; 2,2-dichloroethanol; 2-chloroethanol; 3-chloro-1,2-propanediol; 3-chloropropanol; and 4-chloro-1-butanol. ΔE represents 96.21% of the variance in the linear regression equation; therefore, the probability of getting a correlation of 0.9834 for a sample size of five is less than 1% (see Figure 5.19).

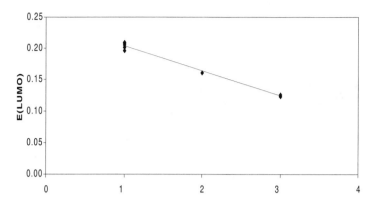

FIGURE 5.18
E_{LUMO} vs. number of chlorines for alcohols with different number of carbons.

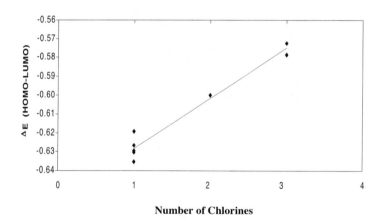

Number of Chlorines

FIGURE 5.19
ΔE vs. number of chlorines for alcohols with different number of carbons.

$$\Delta E = -0.0185(\text{chlorine no.}) + 0.1449, R^2 = 0.9049, n = 4,$$
$$R = 0.9513, 0.05 < p < 0.02 \tag{5.31}$$

For aldehyde, LUMO decreases as the number of chlorines increases. The dataset is 2,2,3-trichlorobutanal; 2-chloropropanal; chloroacetaldehyde; and dichloroacetaldehyde. LUMO represents 90.49% of the variance in the linear regression equation; therefore, the probability of getting a correlation of −0.9513 for a sample size of four is between 2 and 5% (see Figure 5.20).

$$E_{LUMO} = 0.6883(\text{chlorine no.}) + 1.6196, r^2 = 0.8029, n = 53, r = 0.8960, p < 0.01 \tag{5.32}$$

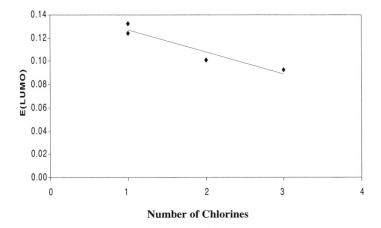

FIGURE 5.20
E_{LUMO} vs. number of chlorines for aldehydes with different number of carbons.

TABLE 5.10

QSAR Models for Alcohol and Aldehydes

Chemical Class	Descriptor	n	r	p
Alcohol	HOMO	8	–0.9721	$p < 0.01$
	LUMO	8	–0.9925	$p < 0.01$
	DE	8	0.9809	$p < 0.01$
Aldehyde	LUMO	4	–0.9513	$0.05 < p < 0.02$
Aromatic	Log P	53	0.8960	$p < 0.01$

Table 5.10 summarizes the QSAR models between different molecular discriptors and the number of chlorine atoms contained in a molecule of alcohols, aldehydes, and aromatic compounds.

5.4 QSAR in Elementary Hydroxyl Radical Reactions

Advanced oxidation processes are characterized by the formation of hydroxyl radicals. QSAR models predicting degradation rate constants in hydroxyl radical reactions will help us understand AOPs. The bond dissociation energy of hydroxyl radicals is 428 (kJ/mol), as determined by the radiolysis method. This value is significantly high when compared with free radicals such as $^{\bullet}HO_2$ and $^{\bullet}H_2$. The most reactive radicals will have the greatest bond dissociation energies. When combined with hydrogen, the hydroxyl radical is considered to be the most reactive radical, next only to the element fluorine. Hydroxyl radicals also have the highest ionization potential among several radicals, including $\cdot HO_2$ and $\cdot H_2$ (Field and Franklin, 1957).

The reactions of $\cdot OH$ with organic compounds can be classified into three major types: (1) hydroxyl addition, (2) hydrogen abstraction, and (3) electron transfer. Which type of reaction takes place depends upon the nature of the substrates present in the medium. Organic compounds containing aromatic ring or carbon–carbon multiple bonds undergo hydroxylation due to the rich π-electron cloud on the aromatic ring. For example:

$$\cdot OH + C_6H_6 \rightarrow \cdot C_6H_6OH \tag{5.33}$$

while hydrogen abstraction is the reaction with unsaturated organic compounds:

$$\cdot OH + CH_3CH_2OH \rightarrow \cdot CH_2CH_2OH + H_2O \tag{5.34}$$

Electron transfer is usually found in reactions between hydroxyl radicals and inorganic ions:

$$\cdot OH + Fe^{2+} \rightarrow OH^- + Fe^{3+} \tag{5.35}$$

The degradation rates of organics by AOPs are greatly enhanced when compared with conventional oxidation processes because of the high reactivity of hydroxyl radicals. In contrast to most other oxidants, reactions of hydroxyl radicals with organics containing C–H or C–C multiple bonds generally proceed with rate constants approaching the diffusion-controlled limit (10^{10} M^{-1} s^{-1}); therefore, oxidation rates are usually limited by the rates of $HO\cdot$ generation and competition by other $HO\cdot$ scavengers in solution rather than by inherent reactivity of the radical. Some compounds, particularly those containing multiple halogens or oxygen atoms and few hydrogen atoms, react with $HO\cdot$ relatively slowly (10^7–10^8 M^{-1} s^{-1}) or, in the case of *per*-halogenated compounds, not at all. In such cases, oxidation rates and efficiencies can be low in waters containing natural scavengers due to significant competition by other scavengers. Therefore, knowledge of $HO\cdot$ reaction rate constants is essential in predicting oxidation rates and efficiencies in a variety of systems.

Because hydroxyl radicals have indiscriminate reactivity, they can react with almost all types of organic and inorganic compounds. Most aromatic compounds undergo radical attack on the aromatic ring in a manner similar to that of benzene systems. The products and the rate constants for hydroxyl radical attack on aromatic compounds are listed in Table 5.11. The data were obtained from the pulse radiolysis studies (Buxton et al., 1988).

The rate constants between hydroxyl radicals and aromatic compounds are from 3×10^9 M^{-1} s^{-1} to 14×10^9 M^{-1} s^{-1} and are diffusion limited. It is known

TABLE 5.11

Rate Constants for Hydroxyl Radical Attack on Aromatic
Compounds

Aromatic Compounds	Rate Constants ($10^7 \ M^{-1} \ s^{-1}$)
Phenol (pH 6 to 9)	1400
Fluorobenzene	1000
Biphenol	950
Phenylmethanol	840
Benzene	780
Styrene (ethenylbenzene)	600
Chlorobenzene	550
Iodobenzene	500
Benzoic acid (pH = 3)	430
Nitrobenzene	390
Toluene	300

Source: Data from Buxton, G.V. et al., *J. Phys. Chem. Ref. Data*, 17,
513, 1988.

that electron-donating substituents such as OH increase the hydroxyl radical
addition rate, while withdrawing groups such as NO_2, I, and Cl transform the
substituted benzene ring to being less reactive compared with nonsubstituted
benzene. The substitution effect on the oxidation rate constants has the following order:

$$NO_2 < I < Cl < CH_2CH_3 < H < F < OH$$

Hart (1952, 1954) studied the oxidation of formic acid by the radiolysis
method. In the presence of oxygen, hydroxyl radicals abstract hydrogen from
HCO_2H. Both the carboxyl radicals and formyl radicals are formed. These
radicals undergo oxygen addition and subsequently dissociate:

$$^{\bullet}OH + HCO_2H \rightarrow H_2O + {}^{\bullet}CO_2H \text{, where } k = 1.3 \times 10^8 \ M^{-1} \ s^{-1} \quad (5.36)$$

$$^{\bullet}CO_2H + O_2 \rightarrow CO_2 + {}^{\bullet}HO_2 \text{, where } k = 2.4 \times 10^9 \ M^{-1} \ s^{-1} \quad (5.37)$$

Tang and Hendrix (1998) developed QSAR models for different classes of
organic compounds reacting with hydroxyl radicals as follows.

5.4.1 Substituted Alcohols

For substituted alcohols reacting with hydroxyl radicals, the least substituted
alcohol, methanol, was used as a reference compound for Hammett correlation analysis. The σ^* values were taken from Hansch et al. (1995) and rate
constants were taken from the U.S. Department of Commerce (USDOC,
1977). These values are believed to be the most accurate rate constants in

the literature. To describe the substitution effect on hydrogen abstraction by hydroxyl radicals from the substituted alcohols, the molecular descriptor σ^* was used to formulate the correlations. A strong correlation between the rate constant and the number of carbons in the chain was observed. The rate constants increased in value with an increase in the number of carbons in the alcohol chain, as shown in Figure 5.21. The correlation coefficient is 0.9175 for data size of 7 points.

The above Hammett correlation shows a direct correlation between the number of carbon atoms and the rate constant for alcohols. The kinetic rate constant is observed to increase with the number of carbon atoms on the alcohol chain. The reaction rate between alcohols and hydroxyl radicals can be described as follows: methanol < ethanol < propanol < butanol < pentanol < hexanol < heptanol. Heptanol has the fastest reaction constant among all the alcohols and is about 10 times faster than the reference compound of methanol. The reaction pathway for substituted alcohols is shown in Figure 5.21.

Hydrogen abstraction by the OH radical may result in the formation of a positive transition state complex in nature because of the negative slope of the Hammett correlation for substituted alcohols, as shown in Figure 5.22.

5.4.2   Chlorinated Alkanes

A correlation for chlorinated alkanes was also established using σ^* as a descriptor. As discussed earlier, σ^* can be used to describe the behavior of aliphatic compounds. Figure 5.23 demonstrates the Hammett correlation for chlorinated alkanes.

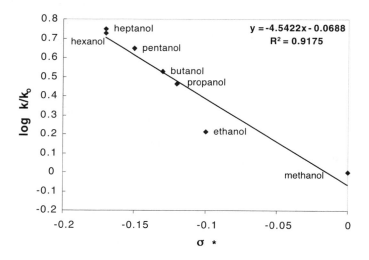

FIGURE 5.21
Hammett's equation of rate constants of substituted alcohols reacting with hydroxyl radical. Experimental conditions: k values obtained at pH ~9, k values defined as $(1/m^*sec^*10^8)$.

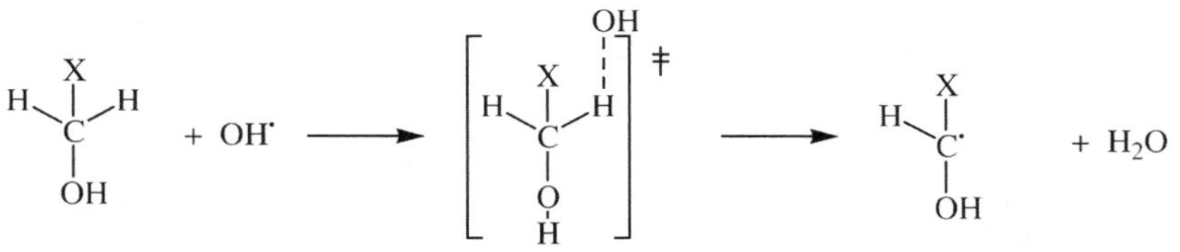

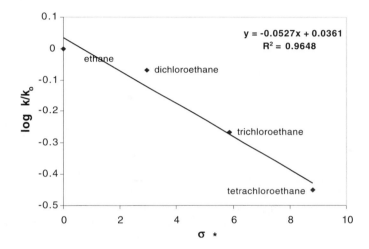

FIGURE 5.22
Reaction pathway for substituted alcohols reacting with hydroxyl radicals.

$$y = -0.0527x + 0.0361$$
$$R^2 = 0.9648$$

FIGURE 5.23
Hammett's correlation for rate constants of chlorinated alkanes reacting with hydroxyl radical. Experimental conditions: elementary hydroxyl radical kinetic data, pH 9.

5.4.3   Substituted Phenols

Hammett's equation was also established for substituted phenols from the elementary hydroxyl radical rate constants. The Hammett resonance constant was used to derive a QSAR model for substituted phenols. The simple Hammett equation has been shown to fail in the presence of electron-withdrawing or electron-donating substituents, such as an –OH group (Hansch and Leo, 1995). For this reason, the derived resonance constants such as σ^0, σ^-, and σ^+ were tested in different cases. In the case of multiple substituents, the resonance constants were summed. Figure 5.24 demonstrates a Hammett correlation for substituted phenols. The least-substituted compound, phenol, was used as a reference compound. Figure 5.24 shows the effects of different substituents on the degradation rates of phenols. Nitrophenol reacted the fastest, while methoxyphenol and hydroxyphenol reacted at a slower rate. This Hammett correlation can be used to predict degradation rate constants for compounds similar in structure.

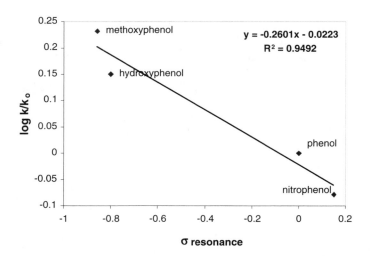

FIGURE 5.24
Hammett's equation for rate constants of substituted phenols reacting with hydroxyl radical
Experimental conditions: elementary hydroxyl radical rate constants $(1/m*sec*10^8)$, pH 9.

The QSAR models for phenols using resonance constants are apparently more accurate than QSAR models using Hammett constants. The R^2 values for QSAR models using resonance constants, $\sigma_{resonance}$, and Hammett constants, σ, were 0.9492 and 0.5473, respectively. The R^2 value demonstrates that the QSAR model using resonance constants has a better fit. Figure 5.25 shows the reaction pathway for hydroxyl radical attack on phenols.

The NO_2 substituent of nitrobenzene is a strongly electron-withdrawing group and decreases the electron cloud around the benzene ring, making $OH^•$ radical attack more difficult. The CH_3 and OH are electron-donating groups and increase the electron density on the benzene ring, making hydroxyl radical attack easier. The electron donating groups activate the ring toward electrophilic substitution, while the electron-withdrawing groups deactivate the ring.

5.4.4 Substituted Carboxylic Acids

The Hammett correlation for substituted carboxylic acids is demonstrated in Figure 5.26. The Hammett constant σ_m fails for aliphatic compounds, and the derived constant σ^* must be used to predict accurate Hammett correlations. The least-substituted carboxylic acid, formic acid, was used as the reference compound. The Hammett correlation for substituted carboxylic acids (CAs) demonstrates that the CAs substituted by electron-withdrawing substituents, such as Cl, oxidize the fastest. CAs substituted by electron-donating groups, such as CH_3 and NH_2, oxidize more slowly than those substituted by electron-withdrawing substituents. The reaction pathway for substituted carboxylic acids is shown in Figure 5.27. These trends are different for phenols and alkanes, because the reaction site is at the electron pair located at the oxygen atom.

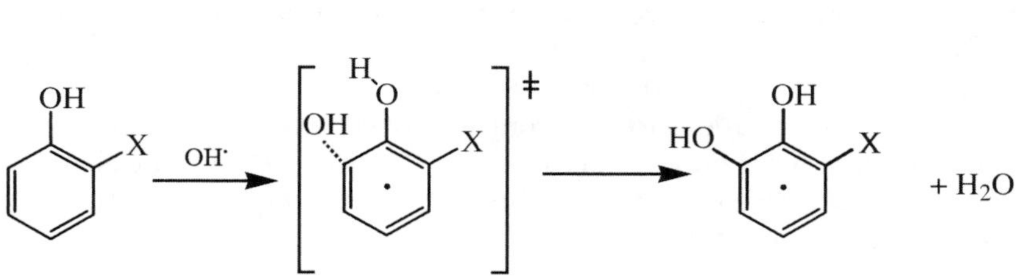

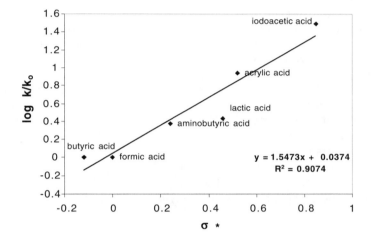

FIGURE 5.25
Reaction pathway for substituted phenols reacting with hydroxyl radical.

FIGURE 5.26
Hammett's correlation for rate constants of substituted carboxylic acids reacting with hydroxyl radical. Experimental conditions: elementary hydroxyl radical rate constants $(1/m*sec*10^8)$, pH 9.

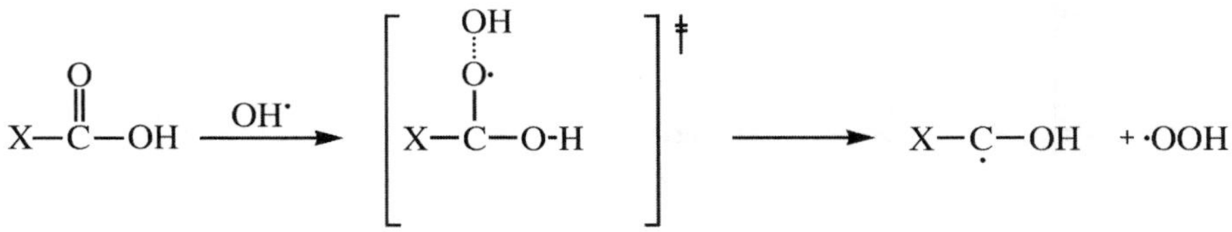

FIGURE 5.27
Reaction pathway for carboxylic acids reacting with hydroxyl radical.

5.4.5 Substituted Benzenes

The Hammett correlation was also established for substituted benzenes and halogenated benzenes. A good correlation could not be established using halogenated and substituted benzenes in the same equation. Halogenated species were found to behave differently than alkyl- or hydroxy-substituted species. Figure 5.28 and Figure 5.29 demonstrate Hammett correlations for substituted and halogenated benzenes.

For Hammett correlations for rate constants of substituted benzenes reacting with hydroxyl radicals under experimental conditions of pH 9, the elementary hydroxyl radical rate constants ($1/m*sec*10^8$) containing toluene, dimethoxy-benzene, benzene, and nitrobenzene correlate with $\sigma_{resonance}$ as follows:

$$\text{Log } (k/k_0) = -1.4006\sigma_{resonance} + 0.0994, \ n = 4, \ r^2 = 0.8152 \qquad (5.38)$$

The negative slope suggests that the TCS is negative due to the alkaline pH of 9 (see Figure 5.28).

A poor correlation between substituted benzenes and σ is observed with a coefficient of 0.81; therefore, the reaction mechanism for the degradation of substituted benzenes cannot be accurately suggested from the correlation. The correlation for halogenated benzenes was more accurate with a goodness of fit of 0.9739 than for substituted benzenes, which had a goodness of fit of 0.8152. The better fit for halogenated compounds may be explained by sim-ilar behavior of the substituents. The most electron-withdrawing substituent, fluorine, oxidizes the fastest. Chlorine and iodine are less electron-withdraw-ing substituents than F and oxidize more slowly. The reaction pathway for halogenated benzenes is shown in Figure 5.29..

The reaction mechanism for the reaction of substituted benzenes with hydroxyl radical appears to be hydroxylation. Hydroxyl radical attack on

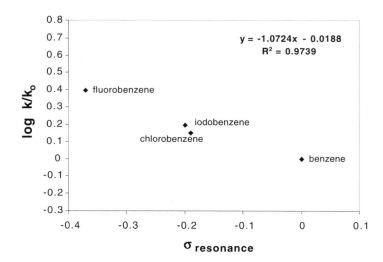

FIGURE 5.28
Hammett's correlation for rate constants of halogenated benzenes reacting with hydroxyl rad-ical. Experimental conditions: elementary hydroxyl radical rate constants ($1/m*sec*10^8$).

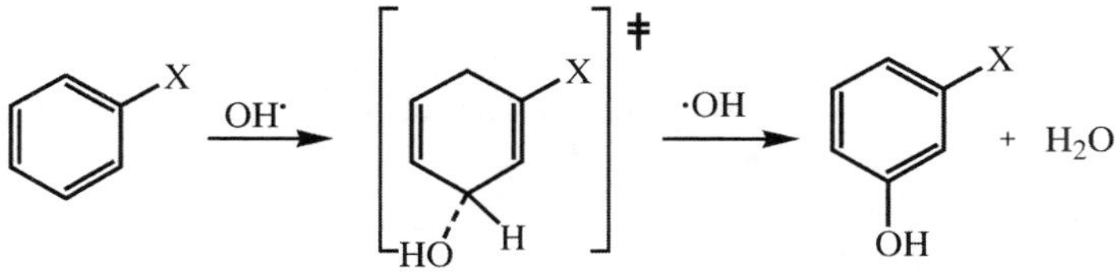

FIGURE 5.29
Reaction pathway for halogenated benzenes reacting with hydroxyl radical.

the benzene ring cleaves one of the C=C bonds and allows for the OH$^{\bullet}$ radical to bond to the benzene ring. Once the OH$^{\bullet}$ radical has bonded the ring, the ring is stabilized

5.4.6 Substituted Alkenes

A Hammett correlation was derived using σ^* for substituted alkenes, as shown in Figure 5.30. As mentioned earlier, aliphatic compounds can be described using the Taft–Hammett constant σ^*. Figure 5.30 demonstrates that the degradation of substituted alkenes will proceed faster than for cyclo-hepatriene, due to the ring structure. The reaction mechanism for the hydrolysis of substituted alkenes is shown in Figure 5.30.

The reaction pathway for substituted alkenes proceeds by hydrolysis. The OH radical attacks the C=C bond to form a positive transition state complex. This positive transition state complex corresponds to the negative slope of the Hammett correlation in Figure 5.30. The reaction proceeds until both carbons of the C=C bond have been hydroxylated.

For Hammett's equation for rate constants of substituted alkenes reacting with hydroxyl radicals under experimental conditions of pH 9, the elementary hydroxyl radical rate constants $(1/m^*sec^*10^8)$ containing butene, propylene, butadiene, and cyclohepatriene correlate with σ^* as follows:

$$\text{Log } (k/k_0) = -0.1016\sigma^* + 0.0049, \ n = 4, \ r^2 = 0.9276 \qquad (5.39)$$

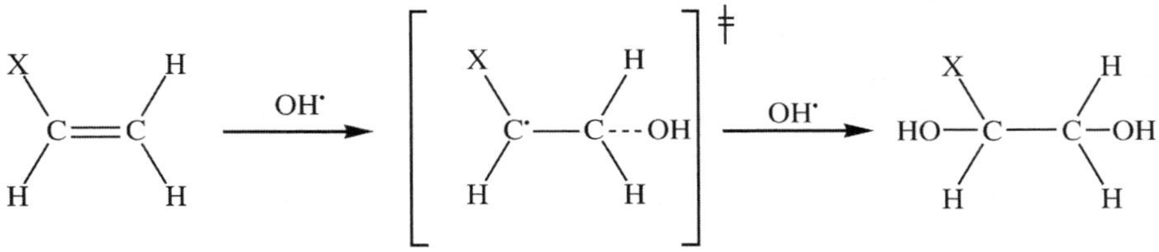

FIGURE 5.30
Reaction pathway for substituted alkenes reacting with hydroxyl radical.

A Hammett correlation for substituted ketones was also made using the derived Hammett constant σ^*. The least-substituted ketone, acetone, was used as the reference compound in the correlation. Only three ketones were used for the correlation; therefore, the model may not be as robust as the other QSAR models.

In Hammett correlations, the descriptors, such as $\sigma_{resonance}$ (or σ_{res}) and σ^*, can be used to derive equations for aromatic and aliphatic compounds, respectively. For aromatic compounds, the σ_{res} descriptor formulated better Hammett correlations than the σ_m descriptor. Given the value of a molecular descriptor, a Hammett correlation for a particular chemical class may be used to predict kinetic rate constants for compounds with similar chemical structure. The QSAR models for each class of compounds studied by elementary hydroxyl radicals are summarized in Table 5.12.

TABLE 5.12

QSAR Models for Elementary Hydroxyl Radical Reactions

Chemical Class	QSAR Model	R^2	k_0 (1/ m*sec*10^8)
Alcohols	Log $k/k_0 = -4.5422\sigma^* - 0.0688$	0.9175	11
Indoles	Log $k/k_0 = 0.3729\sigma_m + 0.0241$	0.9071	137
Phenols	Log $k/k_0 = -0.855\sigma_{res} + 0.0116$	0.9766	1.6
Carboxyclic acids	Log $k/k_0 = 1.5473\sigma^* + 0.0374$	0.9074	1.6
Benzenes	Log $k/k_0 = -1.4006\sigma_{res} + 0.0994$	0.8152	32
Halogenated benzenes	Log $k/k_0 = -1.0724\sigma_{res} - 0.0188$	0.9739	32
Chlorinated alkanes	Log $k/k_0 = -0.0527\sigma^* + 0.0361$	0.9648	83
Ketones	Log $k/k_0 = -7.027\sigma^* + 0.0956$	0.9359	0.9

5.5   QSAR Models between $K_{HO}\cdot$ in Water and $K_{HO}\cdot$ in Air with Molecular Descriptor (MD)

Predictive equations based on literature values were determined by correlating sets of aqueous-phase data with either gas-phase data or σ constants for the same compounds (Haag and Yao, 1992). A correlation of hydroxyl radical H-atom abstraction rate constants for substituented alkanes in water vs. the gas phase was developed. The 19 compounds were predominantly (82%) straight chained and contained four or fewer carbon atoms; 18% were C_5-C_8; and a few were cyclic or branched hydrocarbons. Some chemicals deviated noticeably from the best-fit line and were then omitted from the correlation. Most of the rate constants lie within a factor of three of the regression line given by:

$$\text{Log } k_{HO\cdot}(\text{water}) = 1.682 + 0.805 \log k_{HO\cdot}(\text{air}) \tag{5.40}$$

where both rate constants are in units of mol^{-1} s^{-1}. Experimental data for alkanes gave good agreement with the values calculated from the gas-phase data; therefore, the correlation could be used with reasonable confidence for other alkanes as well. Aromatics and alkenes do not fall on the same correlation line as alkanes, apparently because they can react rapidly by addition to double bonds rather than H-atom abstraction; therefore, they correlate with rate constants for aromatics in water with Hammett σ constants, as shown in Figure 5.31.

$$\text{Log } k_{HO\cdot}(\text{water}) = 9.829 - 0.318\Sigma\sigma \tag{5.41}$$

This correlation includes seven methy- and four halo-substituted benzenes, three heterocyclic compounds, and two phenols, along with benzene, naphthalene, and $-NH_2$, $-OCH_3$, $-CN$, $-NO_2$, $-CHO$, vinyl, and phenyl-substituted benzene. For aromatics, the range of values was only a factor of 5, and all data are within a factor of 1.6 of the correlation line; therefore, simple calculations were likely to give data of about the same accuracy as experimental measurements (see Figure 5.31).

Tratnyek and Hoigne (1994) investigated 25 substituted phenoxide anions for QSARs that can be used to predict rate constants for the reaction of additional phenolic compounds oxidized by chlorine dioxide (OClO). Correlating oxidation rates of phenols in aqueous solution is complicated by the dissociation of the phenolic hydroxyl group. The undissociated phenol and the phenoxide anion react as independent species and exhibit very different properties. The correlation analysis should be performed on the two sets of rate constants separately.

The σ^- scale was developed to incorporate the effect of through-resonance. Both types of correlations included *ortho-* as well as *meta-* and *para-*substituted phenoxides, and the only outliers are compounds that exhibit strong intramolecular hydrogen bonding because this effect is not incorporated in most molecular descriptors. The rate constants for oxidation of phenoxide anions give good Hammett correlations to σ^- constants, as shown in the following equation:

$$\text{Log } k_{ArO^-} = 8.2(\pm0.2) - 3.2(\pm0.4)\Sigma\sigma^-_{0,m,p} \tag{5.42}$$

$$n = 23, \; s = 0.39, \; r = 0.97$$

This regression line is illustrated in Figure 5.32. The second-order rate constants for oxidation of the undissociated forms of substituted phenols are about six orders of magnitude smaller than the corresponding values for phenoxide anions, indicating that only the reaction of phenoxide anions will

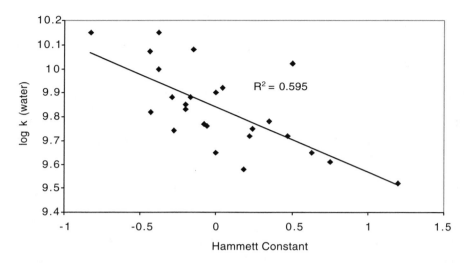

FIGURE 5.31
Correlation of hydroxyl radical rate constants for aromatics in water vs. Hammett's σ constants. (From Haag, W.R. and Yao, C.C.D., _Environ. Sci. Technol._, 26(5), 1005–1013, 1992. With permission.)

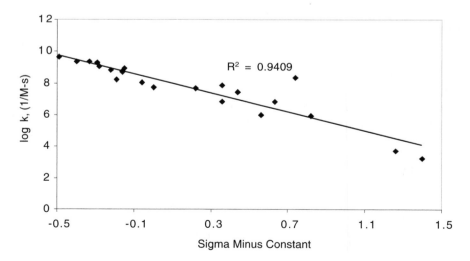

FIGURE 5.32
Rate constant vs. σ^- for substituted phenoxide anions. (From Tratnyek, P.G. and Hoigne, J., _Water Res._, 28(1), 57–66, 1994. With permission.)

be significant under the conditions of water treatment with chlorine dioxide. The rate constants for undissociated phenols also have much greater relative uncertainties than those for the phenoxide anions and give much less satisfactory correlations.

References

Benigni, R. and Richard, A.M., QSARs of mutagens and carcinogens: two case studies illustrating problems in the construction of models for noncongeneric chemicals, *Mutation Research,* 37(1), 29-46, 1996.

Benigni, R., Andreoli, C., Cotta-Ramusino, M., Giogi, F., and Gallo, G., The electronic properties of carcinogens and their role in SAR studies of noncongeneric chemicals, *Toxicity Modeling,* 1, 157–167, 1995.

Buxton, G.V., Greenstock, C.L., Helman, W.P., and Ross A.B., Critical review of rate constants for reactions of hydrated electrons, hydrogen atoms and hydroxyl radicals ($\bullet$OH/$\bullet$O$^-$) in aqueous solution, *J. Phys. Chem. Ref. Data,* 17, 513, 1988.

Field, F.H. and Franklin, J.L., *Electron Impact Phenomena,* Academic Press, New York, 1957.

Finar, I.L., *Organic Chemistry,* Longman, London, 1973.

Fleming, I., *Frontier Orbitals and Organic Chemical Reactions,* John Wiley & Sons, New York, 1976.

Haag, W.R. and Yao, C.C.D., Rate constants for reaction of hydroxyl radicals with several drinking water contaminants, in *Environ. Sci. Technol.,* 26(5), 1005–1013, 1992.

Hansch, C. and Leo, A., *Exploring QSAR: Fundamentals and Applications in Chemistry and Biology,* American Chemical Society, Washington, D.C., 1995.

Hansch, C., Leo, A., and Hoekman, D., *Exploring QSAR: Hydrophobic, Electronic, and Steric Constants,* American Chemical Society, Washington, D.C., 1995.

Hansch, C., Hoekman, D., and Gao, H., Comparative QSAR: toward a deeper understanding of chemicobiological interactions, *Chem. Rev.,* 96, 1045, 1996.

Hart, E.J., The radical pair yield of ionizing radiation in aqueous solutions of formic acids, *J. Am. Chem. Soc.,* 56, 594–599, 1952.

Hart, E.J., γ-Ray-induced oxidation of aqueous formic acid–oxygen solutions: effect of pH, *J. Am. Chem. Soc.,* 76, 4198–4201, 1954.

Hine, J., *Structural Effects on Equilibria in Organic Chemistry,* John Wiley & Sons, New York, 1975.

Lowry, T.H. and Richardson, K.S., *Mechanism and Theory in Organic Chemistry,* 3rd ed., Harper-Collins Publishers, New York, 1987.

Lyman, W.J., Roehl, W.F., and Rosenblatt, D.H., *Handbook of Chemical Property Estimation Methods,* McGraw-Hill, New York, 1982.

Karelson, M. and Lobanov, V., Quantum chemical descriptors in QSAR/QSPR studies, *Chem. Rev.,* 96, 1027, 1996.

Nendza, M. and Russom, C.L., QSAR modeling of the ERL-D fathead minnow acute toxicity database, *Xenobiotica,* 21, 147–170, 1991.

Richard, A.M. and Hunter, E.S. Quantitative structure–activity relationships for the developmental toxicity of haloacetic acids in mammalian whole embryo culture. *Teratology,* 53, 352–360, 1996.

Stock, L.M. and Brown, H.C., in *Advances in Physical Organic Chemistry*, Vol. 1, Gold, V., Ed., Academic Press, London, 1963.

Taft, R., *Steric Effect in Organic Chemistry*, John Wiley & Sons, New York, 1956.

Tang, W.Z. and Hendrix, T., Development of QSAR models to predict kinetic rate constants for elementary hydroxyl radical reactions, in *Water and Wastewater Industry and Sustainable Development: In Honor of Professor Xu Baojiu's 80th Anniversary*, Tsinghua University Press, Beijing, China, 1998, pp. 440–451.

Tang, W.Z. and Pierotti, A.J., WWW database of disinfection by-product properties and related QSAR information, in *Handbook on Quantitative Structure Activity Relationships (QSARs) for Pollution Prevention, Toxicity Screening, Risk Assessment and Web Applications*, Walker, J.D., Ed, SETAC Press, Pensacola, FL, 2000.

Tinsley, I.J., *Chemical Concepts in Pollutant Behavior*, John Wiley & Sons, New York, 1979.

Tratnyek, P.G. and Hoigne, J., Chlorine dioxide (OClO) in Water. II. Quantitative structure–activity relationships for phenolic compounds, *Water Res.*, 28(1), 57–66, 1994.

USDOC, *Selected Specific Rates of Reactions of Transients from Water in Aqueous Solution*, U.S. Department of Commerce, Washington, D.C., 1977.

6

Fenton's Reagent

6.1 Introduction

In 1881, Fenton published a brief description of the powerful oxidizing properties of a mixture of hydrogen peroxide and ferrous salts. This mixture became known as *Fenton's reagent,* and the reaction has become known as the *Fenton's reaction.* Initially, Fenton applied this reaction to oxidize organic acids such as formic, glycolic, lactic, tartronic, malic, saccharic, mucic, glyceric, benzoic, picric, dihydroxytartaric, dihydroxymaleic, and acetylenedicarboxylic (Fenton, 1900). In the absence of ferrous salt, the degradation of hydrogen peroxide proceeds at very slow rates, with little or no oxidation of the organic acids (Fenton, 1899, 1900). In addition, Cross et al. (1900) further confirmed that ferrous salts significantly enhance the kinetics of hydrogen peroxide decomposition. Goldhammer (1927) investigated the effect of Fenton's reagent on phenols and found that for each equivalent of Fe^{2+} three equivalents of H_2O_2 were decomposed. They also noted that in concentrated hydrogen peroxide solutions each mole of Fe^{2+} decomposed 24 equivalents of hydrogen peroxide.

Haber and Weiss (1934) were the first to propose that free radicals existed as intermediates during the chemical reactions in solution. The next year, Haber and Weiss further investigated the Fenton chemistry and concluded that Fenton's reaction can be expressed as a series of chain reactions with reaction pathways dependent on the concentration of the reactants. The study disproved the original theory of Fenton's reaction, which suggested that the interaction between an intermediate, six-valent, iron–oxygen complex and hydrogen peroxide was the most significant reaction step. In 1934, Haber and Weiss proposed that breaking rate of chain length was increased at lower pH so the propagation cycle was extended before termination. The concentration of free hydroxyl radicals was determined to be directly proportional to the concentration of hydrogen peroxide.

Baxendale and Wilson (1957) reported that in an oxygen-free environment Fenton's reagent initiates very rapid polymerization of methyl acrylate, methacrylic acid, methyl methacrylate, acrylonitrile, and styrene, and the reaction is a function of the concentration of hydroxyl radicals. In the presence of oxygen, no polymerization occurs. Barb et al. (1951) conducted an extensive investigation of Fenton's reagent chemistry. When $[H_2O_2]/[Fe^{2+}]$ ratios are low, the reaction rate is second order and stoichiometry is $2[Fe^{2+}] \cong [H_2O_2]$; however, in the presence of polymerizable vinyl compound the reaction remains second order but the stoichiometry changes to $[Fe^{2+}] \cong [H_2O_2]$. Thus, they concluded that polymerization of vinyl compounds occurs and results in a polymer with terminal hydroxyl groups. An inhibition effect of hydroxyl radicals due to the higher concentration of hydrogen peroxide was also suggested. To explain this mechanism, it was proposed that hydroxyl radicals react with hydrogen peroxide to form hydrogen dioxide. This process decreases the hydroxyl radicals generated by the reaction between ferrous iron and hydrogen peroxide. In addition, Barb et al. (1951) suggested that hydrogen dioxide is not a strong oxidizing agent capable of breaking the bonds of vinyl compound or oxidizing other organics.

Merz and Waters (1949) showed that oxidation of organic compounds by Fenton's reagent could proceed by chain as well as non-chain mechanisms, which was later confirmed by Ingles (1972). Kremer (1962) studied the effect of ferric ions on hydrogen peroxide decomposition for Fenton's reagent. It was confirmed that once ferric ions are produced the ferric–ferric system is catalytic in nature, which accounts for relatively constant concentration of ferrous ion in solutions.

In the late 1970s, two major theories were considered: the free radical mechanism by Walling and Cleary (1977) and complex formation by Kremer and Stein (1977). Walling proposed that Fenton's oxidation predominantly takes place by the free-radical mechanism. On the other hand, Kremer proposed that complexation between the iron and the organic molecules has a significant role and thus concluded that both mechanisms occur simultaneously. In the late 1980s a simultaneous effort was made to apply Fenton's reagent to the field of environmental science. Various contaminants were studied in the laboratory to determine the optimum conditions. Practical applications of Fenton's reagent to treat contaminants have also been examined by pilot-plant and continuous treatment systems in textile wastewater, etc.; for example, Bigda (1996) applied Fenton's reaction to the design of a reactor for treatment of organic contaminants.

6.2   Kinetic Models

Although Fenton (1894) studied the violet color in caustic alkali during oxidation of tartaric and racemic acids by ferrous salt and hydrogen peroxide, no reaction kinetic model was offered. Fenton reported that the color disappeared when acid was added. Also, it has been observed that fresh external air is more active than room air. Fenton performed different experiments using various amounts of ferrous and hydrogen peroxides and proposed that iron catalyzed this reaction. For example, a small amount of iron is sufficient to determine oxidation of an unlimited amount of tartaric acid. In tartaric acid, two atoms of hydrogen are removed from a molecule of acid, resulting in the production of dihydroxymaleic acid. Among common oxidants such as chlorine, potassium permanganate, atmospheric oxygen, and electrolysis, the most effective oxidizing agent is hydrogen peroxide. Fenton's work was extended to alcohols (Fenton, 1899) and other organic acids (1900). Attempts to identify the intermediates and products of several organic acids and alcohols were made without success.

6.2.1 Chain Reaction Mechanism by Merz and Waters

Goldschmidt and Pauncz (1933) suggested that Fenton's reaction is a chain reaction involving the same reactive intermediates occurring during catalytic decomposition of H_2O_2 rather than via formation of peroxides of iron:

$$2H_2O_2 = 2H_2O + O_2 \qquad (6.1)$$

It was also shown that the ratio of oxidized alcohol to oxidized Fe^{2+} could be greater then one. Baxendale and Wilson (1957) showed that hydroxyl radical initiating the chain polymerization of olefins by hydrogen peroxide was the same process as the rapid oxidation of glycolic acid. Merz and Waters (1947) confirmed that simple water-soluble alcohols are oxidized rapidly by Fenton's reagent. The primary alcohols are oxidized to aldehydes, which are further oxidized at comparable rates by exactly the same mechanism. Merz and Waters proposed a mechanism of chain oxidation of alcohols and aldehydes by sodium persulfate, hydrogen peroxide, and an excess of ferrous salt as follows:

1. Chain initiation:

$$Fe^{2+} + H_2O_2 = Fe^{3+} + OH^- + OH^\bullet \qquad (6.2)$$

2. Chain propagation:

$$RCH_2OH + OH^\bullet = R–CHOH^\bullet + H–OH \text{ (reversible)} \qquad (6.3)$$

$$R–CHOH^\bullet + HO–OH = R–CHO + OH^\bullet + H_2O \qquad (6.4)$$

3. Chain ending at low substrate concentration:

$$Fe^{2+} + OH^\bullet = Fe^{3+} + OH^- \qquad (6.5)$$

4. Chain ending at high substrate (alcohol) concentration:

$$2R–CHOH\bullet = R–CHO + R–CH_2OH \text{ (disproportionation)} \qquad (6.6)$$

In 1949, Merz and Waters determined the values for the ratio of rate constants k_2/k_3 that indicated which particular radical reduced hydrogen peroxide. Based on the reaction pathways, they classified the reacting compounds into two groups. The first group of substrates reacts by chain process. Only a small amount of reducing agent is required. The second group is comprised of substrates that react by non-chain processes — in this case, the oxidation is caused by the hydroxyl radical, and considerable loss of hydroxyl radical occurs. For the first group, the reaction rate can be expressed by Equation (6.7):

$$d[H_2O_2]/d[RH] = 1 + k_2[Fe^{2+}]/k_3[RH] \qquad (6.7)$$

For non-chain reactions, the kinetic rates are described by Equation (6.8):

$$d[H_2O_2]/d[RH] = 2 + k_2[Fe^{2+}]/k_3[RH] \qquad (6.8)$$

The values for the ratio of rate constants k_2/k_3 can be determined from the intercept of their graphs. The results will suggest which particular radical reduced hydrogen peroxide.

6.2.2   Redox Formulation by Barb et al.

Barb et al. (1951) gave a redox formulation that involves the following reaction sequence:

$$Fe^{3+} + H_2O_2 \overset{k_1}{=} Fe^{2+} + HO_2^{\bullet} + H^+ \qquad (6.9)$$

$$Fe^{2+} + H_2O_2 \overset{k_2}{=} Fe^{3+} + OH^- + OH^{\bullet} \qquad (6.10)$$

$$HO^{\bullet} + H_2O_2 \overset{k_3}{=} H_2O + HO_2^{\bullet} \qquad (6.11)$$

$$HO_2^{\bullet} + Fe^{3+} \overset{k_4}{=} Fe^{2+} + H^+ + O_2 \qquad (6.12)$$

$$HO_2^{\bullet} + Fe^{2+} \overset{k_5}{=} Fe^{3+} + OH_2^- \qquad (6.13)$$

$$OH^{\bullet} + Fe^{2+} \overset{k_6}{=} Fe^{3+} + OH^- \qquad (6.14)$$

where k_1 and k_2 showed inverse [H$^+$] dependence.

6.2.3 Complex Mechanism by Kremer and Stein

The following scheme was presented by Kremer and Stein (1959), and further elaborated by Kremer (1963):

$$\text{Fe}^{3+} + \text{H}_2\text{O}_2 \underset{k_b}{\overset{k_a}{\rightleftharpoons}} \text{FeOOH}^{2+} + \text{H}^+ \qquad (6.15)$$

$$\text{FeOOH}^{2+} \overset{k_c}{=} \text{HO}^- + \text{FeO}^{3+} \qquad (6.16)$$

$$\text{FeO}^{3+} + \text{H}_2\text{O}_2 = \text{Fe}^{3+} + \text{H}_2\text{O} + \text{O}_2 \qquad (6.17)$$

Let $C_1 = $ [H$^+$] and $C_2 = $ [FeO^{3+}], k_a and k_d showed inverse [H$^+$] dependence and $k_b \gg k_a \gg k_c$, C_1 could be taken as a low concentration intermediate to a good approximation

$$[C_1] = K[\text{H}_2\text{O}_2][\text{Fe}^{3+}], \ K = k_a/k_b \qquad (6.18)$$

$$[\text{Fe}^{3+}]_t = [C_2] + [\text{Fe}^{2+}] \qquad (6.19)$$

$$-d[\text{H}_2\text{O}_2]/dt = k_c K[\text{Fe}^{3+}]_t[\text{H}_2\text{O}_2] + (k_d - k_c K)[C_2][\text{H}_2\text{O}_2] \qquad (6.20)$$

$$d[\text{O}_2]/dt = k_d[C_2][\text{H}_2\text{O}_2] \qquad (6.21)$$

$$d[C_2]/dt = k_c K[\text{Fe}^{3+}]_t[\text{H}_2\text{O}_2] - (k_d + k_c K)[C_2][\text{H}_2\text{O}_2] \qquad (6.22)$$

[C_2] rises continually during the reaction, approaching a saturation value of $k_c K[\text{Fe}^{3+}]_t/(k_c K + k_d)$, and $-d[\text{H}_2\text{O}_2]/dt$ is always greater than twice $d[\text{O}_2]/dt$. At the end of the reaction, some hydrogen peroxide will be stored as C_2, and less than 0.5 mol of O$_2$ will be liberated per mole of H$_2$O$_2$ decomposed.

6.2.4 Walling's Modified Kinetic Model

Walling and Kato (1971) modified the reaction mechanism proposed by Merz and Waters as follows:

$$\text{Fe}^{2+} + \text{H}_2\text{O}_2 = \text{Fe}^{3+} + \text{OH}^- + \text{OH}^{\bullet}, \ k_1 = 76 \qquad (6.23)$$

$$OH^\bullet + Fe^{2+} = Fe^{3+} + OH^-, \, k_2 = 3 \times 10^8 \qquad (6.24)$$

$$OH^\bullet + R_iH \overset{k_{3i}}{=} H_2O + R_i^\bullet \qquad (6.25)$$

$$OH^\bullet + R_jH \overset{k_{3j}}{=} H_2O + R_j^\bullet \qquad (6.26)$$

$$OH^\bullet + R_kH \overset{k_{3k}}{=} H_2O + R^\bullet, \, k_3 = 10^7 - 10^{10} \qquad (6.27)$$

$$R_i\bullet + Fe^{3+} \overset{k_4}{=} Fe^{2+} + product \qquad (6.28)$$

$$2R_j^\bullet \overset{k_5}{=} product \, (dimer) \qquad (6.29)$$

$$R_k\bullet + Fe^{2+} \overset{k_6}{\underset{H^+}{=}} Fe^{3+} + R_kH \qquad (6.30)$$

where k is in L/mol/s, taken from the literature. The reaction conditions were chosen to minimize the competing processes as follows:

$$HO^\bullet + H_2O_2 = H_2O + HO_2, \, k = (1.2\text{--}4.5) \times 10^7 \qquad (6.31)$$

$$2HO^\bullet = H_2O_2, \, k = 5.3 \times 10^9 \qquad (6.32)$$

Thus, the stoichiometry is:

$$R = 2ar \, (1 - R) + b \qquad (6.33)$$

where $R = \Delta[Fe^{2+}]/2\Delta[H_2O_2]$, $a = k_2/\Sigma k_3$, $r = [Fe^{2+}]/2[RH]$, and $b = (k_{3j} + 2k_{3k})/2\Sigma k_3$. This mechanism is referred to as the *free-radical mechanism*.

6.2.5   Ingles' Approach

In 1972, Ingles reported his studies of Fenton's reagent using redox titration. He found evidence in support of Kremer's complex mechanism theory and concluded that, when suitable complexes are formed, substrates are not oxidized by free radical; rather, electron transfer processes might be involved. Fenton's reaction scheme was modified by Ingles for the case when substrate is present in large amounts in the form of substrate/iron-peroxide complexes. Ingles suggested that electron transfer occurs within this complex.

$$R^I - Fe^{II} - O - OH \longrightarrow R^I - Fe^{III} = O + \cdot OH \quad (6.34)$$

All substrates were considered to compete as ligands in iron complexes and to modify the reaction characteristics of each other and of the complex. Reaction 6.34 yields hydroxyl radicals, so the free-radical mechanism proposed by Walling appeared to be possible; however, Equation (6.35) to Equation (6.38) involve electron transfer and do not lead to formation of hydroxyl radicals. Equation (6.37) and Equation (6.38) involve ionic mechanisms:

$$R^I - Fe^{II} - O - OH \longrightarrow R^I - Fe^{III} = O + OH^- + \cdot R^2 \quad (6.35)$$

$$R^I - Fe^{II} - O - OH \longrightarrow \cdot R^I + Fe^{III} = O + OH^- \quad (6.36)$$

$$R^I - Fe^{II} - O - OH \longrightarrow {}^+R^I + Fe^{II} = O + OH^- \quad (6.37)$$

$$R^I - Fe^{II} - O - OH \longrightarrow R^I - Fe^{III} = O + OH^- + {}^+R^2 \quad (6.38)$$

6.2.6 Transition State Approach by Tang and Huang

6.2.6.1 Competitive Method

When modeling oxidation kinetics of chlorophenols by Fenton's reagent, elementary rate constants are critical to obtain quantitative stoichiometric data in terms of optimal dosages for H_2O_2 and Fe^{2+} to achieve a given removal efficiency. If an elementary reaction rate constant for a given compound is not available in the literature, another method can be used to determine it experimentally. For example, the rate constants of 2,4-dichlorophenol (DCP) and 2,4,6-trichlorophenol (TCP) were determined by an alternative method by Tang and Huang (1996a). The equation used to calculate the rate constants is as follows:

$$k_{HO\bullet,S} = \frac{\ln([S]/[S_0])}{\ln([R]/[R_0])} k_{HO\bullet,R}$$ (6.39)

where:

$k_{HO^\bullet,S}$ = rate constant between any organic compound and hydroxyl radical.

$k_{HO^\bullet,R}$ = rate constant between reference compound and hydroxyl radical.

[S] = concentration of the substrate at any time.

[S_0] = initial concentration of the substrate.

[R] = concentration of the reference compound at any time.

[R_0] = initial concentration of the reference compound (2-chlorophenol).

In their work, the reference compound is 2-chlorophenol, with a rate constant of 8.2×10^9 M^{-1} s^{-1}. Either 2,4-DCP or 2,4,6-TCP was mixed with 2-chlorophenol in a reactor, separately. Then, H_2O_2 was mixed with the organic compound and the pH was adjusted to 3.5. The organic concentrations were measured by gas chromatography (GC) before and after Fe^{2+} was added. The results are shown in Figure 6.1. According to the slopes of the straight line, the rate constants between hydroxyl radicals and 2,4-DCP and 2,4,6-TCP can be determined as 7.22×10^9 M^{-1} s^{-1} and 6.27×10^9 M^{-1} s^{-1}, respectively. The hydroxylation rate constants for 2,4-DCP and 2,4,6-TCP are clearly smaller than that for 2-chlorophenol; therefore, increasing chlorine content on the aromatic ring decreases the reactivity of the chlorinated phenols toward hydroxyl radical attack.

6.2.6.2   Dechlorination Kinetic Model

6.2.6.2.1   Pseudo First-Order Kinetic Model

When an excess of H_2O_2 and Fe^{2+} is added at constant concentrations to the system, a steady-state concentration of hydroxyl radical can be assumed. The concentration of both H_2O_2 and Fe^{2+} can be considered as constant; therefore, the pseudo first-order kinetic can be developed as follows:

$$\text{Chlorinated phenols} + {}^\bullet OH \overset{k_1}{=}$$
$$\text{intermediates (chlorinated aliphatic compounds)}$$ (6.40)

where: k_1 is the pseudo first-order rate constant of oxidation.

$$\text{Intermediates} + {}^\bullet OH \overset{k_2}{=} \text{chloride ion} + CO_2 + \text{other products}$$ (6.41)

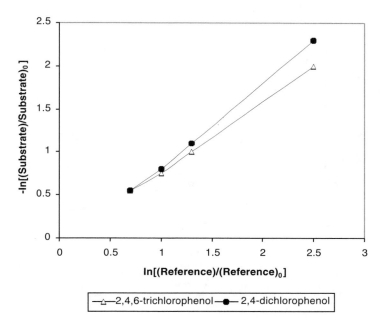

FIGURE 6.1
The competitive oxidation kinetics of different chlorinated phenols at optimal ratio of H_2O_2 to Fe^{2+} of 25 and optimal pH of 3.5. Experimental conditions: (2-CP) = (2,4-DCP) = (2,4,6-TCP) = 5×10^{-4} M; $H_2O_2 = 5 \times 10^{-3}$ M; $Fe(ClO4)_2 = 2 \times 10^{-4}$ M; pH = 3; ionic strength = 0.05 M as Na_2SO_4.

where k_2 is the pseudo first-order rate constant of dechlorination. The degradation kinetics can be modeled as the following:

$$d(CP)/dt = -k_1(CP) \tag{6.42}$$

$$d(I)/dt = k_1(CP) - k_2(I) \tag{6.43}$$

$$d(Cl^-)/dt = k_2(I) \tag{6.44}$$

where CP is the concentration of chlorinated phenols at any time t; I is the concentration of intermediates formed at any time t; and Cl^- is the concentration of chloride ion. The integrated form of the above equation is:

$$(CP)/(CP)_0 = \exp(-k_1 t) \tag{6.45}$$

$$(Cl)/(CP)_0 = 1 + [k_1\exp(-k_2 t) - k_2\exp(-k_1 t)]/(k_2 - k_1) \tag{6.46}$$

where $(CP)_0$ is the initial concentration. Figure 6.2 shows both the experimental data and the concentration profile predicted by the kinetic model for the oxidation and dechlorination of 2,4,6-TCP.

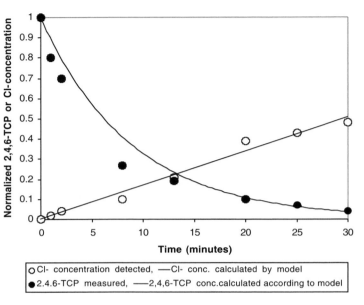

FIGURE 6.2
Kinetic modeling of 2,4,6-trichlorophenol oxidation and chloride ion dissociation (in the mathematical model, $k_1 = 0.150$ 1/min and $k_2 = 0.032$ 1/min). Experimental conditions: 2,4,6-TCP = 5×10^{-4} M; $H_2O_2 = 5 \times 10^{-3}$ M, $Fe(ClO4)_2 = 2 \times 10^{-4}$ M; pH = 3.5; ionic strength = 0.05 M as N_2SO_4.

It is important to note that both H_2O_2 and Fe^{2+} have to be overdosed to maintain a steady-state concentration of hydroxyl radical and to obtain a satisfactory approximation of the mathematical model with the experimental data. When H_2O_2 and Fe^{2+} concentrations are 5×10^{-3} M and 2×10^{-4} M, respectively, the relative rate constants of 2-chlorophenol (2-CP) and 2,4,6-TCP with respect to 2,4-DCP can be calculated. The oxidation and dechlorination constants of 2,4-DCP were found to be 0.995 1/min (k_1) and 0.092 1/min (k_2), as reported in a previous study (Tang and Huang, 1996). For comparison, Table 6.1 summarizes all the kinetic constants as determined in this study and in the related literature.

TABLE 6.1

Kinetic Rate Constants of Chlorinated Phenols by Fenton's Reagent

	2-CP	2,4-DCP	2,4,6-TCP
Elementary rate constants (M^{-1} s^{-1})	8.2×10^9	7.2×10^9	6.3×10^9
Measured oxidation constants (1/min)	1.666	0.995	0.15
Measured dechlorination constants (1/min)	1.386	0.092	0.032

TABLE 6.2

Relative Ratios of Kinetic Constants Using 2,4-DCP as the Reference Compound
($k/k_{2,4-DCP}$)

	2-CP	2,4-DCP	2,4,6-TCP
Elementary rate constants	1.14	1	0.88
Observed oxidation constants	1.67	1	0.15
Observed dechlorination constants	15.07	1	0.35

To evaluate the effect of the number of chlorines on the degradation rate
constants of different chlorophenols, Table 6.2 shows the rate constants of
elementary, oxidation, and dechlorination for the ratios of $k_{2-CP}/k_{2,4-DCP}$ and
$k_{2,4,6-TCP}/k_{2,4-DCP}$. The relative rate constants are plotted against the number of
sites unoccupied by chlorine atoms on the chlorinated phenols in Figure
6.3. A linear correlation between the rate constants and the number of sites
available is found with a standard deviation of 0.132. Clearly, the more
chlorine atoms the aromatic rings contain, the fewer sites are available for
hydroxyl radical attack; however, the correlation should not be used for

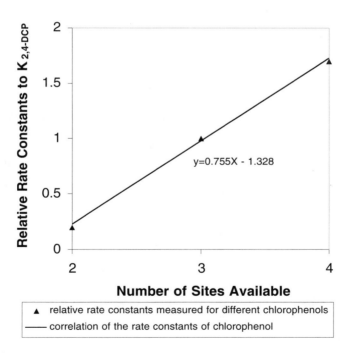

FIGURE 6.3
The correlation between oxidation constants and sites available on aromatic ring for hydroxyl
radical attack. Experimental conditions: 2-CP = 2,4-DCP = 2,4,6-TCP = 5×10^{-4} M; $H_2O_2 = 5 \times 10^{-3}$ M; $Fe(ClO4)_2 = 2 \times 10^{-4}$ M; pH = 3.5; ionic strength = 0.05 M as N_2SO_4.

predicting the rate constants of *tetra*-chlorophenol and *penta*-chlorophenol due to steric hindrance. Figure 6.3 indicates that the oxidation rate decreases with the increasing degree of chlorine content in the following order (in terms of both elementary rate constants and the observed pseudo first-order rate constants): 2-CP > 2,4-DCP > 2,4,6-TCP.

6.2.6.2.2   Dechlorination Kinetic Model Using Transition State Theory

When both H_2O_2 and Fe^{2+} are not overdosed, the concentrations of both H_2O_2 and Fe^{2+} will change. The pseudo first-order kinetic model developed above does not apply. In order to quantitatively model the effect of H_2O_2 and Fe^{2+} on the dechlorination kinetics by Fenton's reagent, the dechlorination kinetic model is developed as follows. First, hydroxyl radicals are generated by H_2O_2 decomposition by Fe^{2+}:

$$Fe^{2+} + H_2O_2 \rightarrow Fe^{3+} + {}^{\bullet}OH + OH^- \quad k_i = 51 \; M^{-1} s^{-1} \qquad (6.47)$$

where k_i is the initiation rate constant of hydroxyl radical generation. Then, the following termination reactions occur simultaneously in the reactor:

$$2{}^{\bullet}OH \rightarrow H_2O_2 \; k_{t1} = 5.3 \times 10^9 \; M^{-1} s^{-1} \qquad (6.48)$$

$$Fe^{2+} + {}^{\bullet}OH \rightarrow Fe^{3+} + OH^- \; k_{t2} = 3 \times 10^8 \; M^{-1} s^{-1} \qquad (6.49)$$

$${}^{\bullet}HO + H_2O_2 \rightarrow H_2O + HO_2 \; k_{t3} = 2.7 \times 10^7 \; M^{-1} s^{-1} \qquad (6.50)$$

Tang and Huang (1996a) reached two conclusions. First, hydroxyl radicals will attack unoccupied sites of the aromatic ring; second, chlorine atoms will be released from the chlorinated aliphatic intermediates instead of the aromatic ring. When chlorinated phenols are present in the system, the following reaction mechanisms can be assumed:

$$CP + {}^{\bullet}OH \underset{k_b}{\overset{k_f}{\rightleftharpoons}} (CP - OH)^{*} \qquad (6.51)$$

$$(CP - OH)^{*} \overset{k_m}{\rightarrow} \text{chlorinated aliphatic intermediates} \qquad (6.52)$$

$$\text{Chlorinated aliphatic compounds} + {}^{\bullet}OH \overset{k_p}{\rightarrow} CO_2 + Cl^- + H_2O \qquad (6.53)$$

where k_f is the rate constant for formation of activated complex; k_b is the rate constant for decomposition of activated complex after hydroxyl radical attack on chlorinated phenols; k_m is the rate constant for the formation of intermediates; and k_p is the rate constant for the formation of products.

Because of the high reactivity of hydroxyl radicals, activated complex, and chlorinated intermediates, their concentrations are extremely low at the steady state; therefore, a pseudo first-order steady state can be assumed for the kinetic modeling. As a result, the steady-state concentration of the activated complex can be obtained by setting the change of its concentration to zero:

$$\frac{d(C)^*}{dt} = k_f(^\bullet OH)^*(CP) - k_b(C)^* - k_m(C)^* = 0 \tag{6.54}$$

where (C^*) is the concentration of the activated complex.

Thus, the steady-state concentration of the activated complex should be:

$$(C)^* = \frac{k_f(^\bullet OH)^*(CP)}{k_b + k_m} \tag{6.55}$$

For chlorinated aliphatic intermediates, the steady-state concentration can be derived by the same principle:

$$\frac{d(I)}{dt} = k_m(C^*) - k_p(I)(^\bullet OH) = 0 \tag{6.56}$$

Then, the concentration can be expressed as follows:

$$(I) = \frac{k_m(C^*)}{k_p(^\bullet OH)} \tag{6.57}$$

Similarly, the change of hydroxyl radical concentration should also equal zero:

$$\frac{d(^\bullet OH)}{dt} = k_i(H_2O_2)(Fe^{2+}) - k_f(^\bullet OH)(CP) + (C^*)(k_b - k_m)$$

$$- k_{t1}(^\bullet OH)^2 - k_{t2}(^\bullet OH)(Fe^{2+}) - k_{t3}(^\bullet OH)(H_2O_2) = 0 \tag{6.58}$$

Substituting Equation (6.55) into Equation (6.58), we obtain:

$$k_i(H_2O_2)(Fe^{2+}) - k_f(^\bullet OH)(CP) + k_f(^\bullet OH)(CP)\frac{(k_b - k_m)}{(k_b + k_m)}$$

$$- k_{t1}(^\bullet OH)^2 - k_{t2}(^\bullet OH)(Fe^{2+}) - k_{t3}(^\bullet OH)(H_2O_2) = 0 \tag{6.59}$$

Simplifying this equation, we get:

$$k_i(H_2O_2)(Fe^{2+}) - 2k_f(\text{·OH})(CP)\frac{(k_m)}{(k_b + k_m)}$$

$$-k_{t2}(\text{·OH})(Fe^{2+}) - k_{t3}(\text{·OH})(H_2O_2) = 0 \qquad (6.60)$$

For simplicity, we assume that the following expression containing rate constants is also a constant k:

$$k = \frac{2k_f k_m}{(k_b + k_m)} \qquad (6.61)$$

Then, the final form of hydroxyl radical concentration can be expressed as follows:

$$(\text{·OH}) = \frac{k_i(H_2O_2)(Fe^{2+})}{k(CP) + k_{t2}(Fe^{2+}) + k_{t3}(H_2O_2)} \qquad (6.62)$$

The rate equation for dechlorination of chlorinated phenols by Fenton's reagent can be expressed as follows:

$$r_{Cl^-} = \frac{d(Cl^-)}{dt} = k_p(I)(\text{·OH}) = k_m\frac{k_f(CP)(\text{·OH})}{k_b + k_m} = \frac{k}{2}(CP)(\text{·OH}) \qquad (6.63)$$

so the final expression for dechlorination is:

$$\frac{d(Cl^-)}{dt} = \frac{k/2(CP)k_i(H_2O_2)(Fe^{2+})}{k(CP) + k_{t2}(Fe^{2+}) + k_{t3}(H_2O_2)} \qquad (6.64)$$

Because Fenton oxidation of chlorophenols should follow the same mechanism, the activated complex in the transition state should have a similar structure. Therefore, Equation (6.64) can be applied to mono-, di-, and trichlorophenols. It is not certain, however, that it can be applied to *tetra-* and *penta-* chlorophenols due to steric hindrance; therefore, when the above general equation is applied to chlorophenols, the equation becomes:

$$\frac{1}{r_0} = \frac{2[k(CP)_0 + k_{t3}(H_2O_2)_0]}{k\,k_i(H_2O_2)_0(CP)_0}\frac{1}{Fe^{2+}} + \frac{2k_{t2}}{k\,k_i(H_2O_2)_0(CP)_0} \qquad (6.65)$$

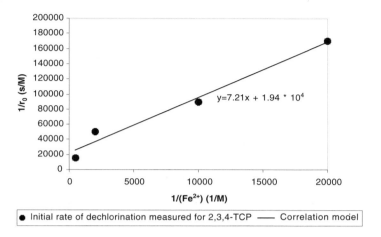

FIGURE 6.4
The plot of reciprocal initial dechlorination rate vs. reciprocal concentration of Fe^{2+} during 2,3,4-TCP oxidation. Experimental conditions: $H_2O_2 = 5 \times 10^{-3}$ M; 2,3,4-TCP = 5×10^{-4} M; pH = 3.5; ionic strength = 0.05 M as N_2SO_4.

The k can be obtained using the initial rate method by overdosing H_2O_2. For example, at the overdosed H_2O_2 concentration of 5×10^{-3} M and constant pH of 3.5, the k can be obtained by plotting the $1/r_0$ vs. $1/Fe^{2+}$ during the dechlorination of 2,3,4-trichlorophenol.

Figure 6.4 shows that the slope of the line is 7.21 (s). As a result, the k can be calculated as 7×10^5 (1/s).

When Fe^{2+} is overdosed, the k value can be obtained through the following equation:

$$\frac{1}{r_0} = \frac{2\left[k(TCP)_0 + k_{t2}(Fe^{2+})_0\right]}{k\,k_i(Fe^{2+})_0(TCP)_0}\frac{1}{H_2O_2} + \frac{2k_{t3}}{k\,k_i(Fe^{2+})_0(TCP)_0} \quad (6.66)$$

At the overdosed Fe^{2+} concentration of 5×10^{-3} M and constant pH of 3.5, the $1/r_0$ vs. $1/H_2O_2$ during the dechlorination of 2,3,4-TCP is shown in Figure 6.5. From the slope of 54.1 (s), the k can be calculated as 7.3×10^5 (1/s), which is about the same as when H_2O_2 was overdosed.

Using the developed model, the k values for 2-CP, 3-CP, and 4-CP are 1.12 $\times 10^7$, 1.004×10^9, and 1.005×10^8 (1/s), respectively; therefore, the dechlorination constants for monochlorophenols follow a decreasing order: 3-CP > 4-CP > 2-CP. Because chloride ion can be released only after the rupture of the aromatic ring, the faster the hydroxylation of the parent compounds, the faster the dechlorination process should be. Therefore, the above order can be understood in terms of the effect of the substituents on the reactivity of their parent compounds. It is known that both OH and Cl are *ortho* and *para* directors. Under the influence of these directors, the following preference of hydroxyl radical attack is expected:

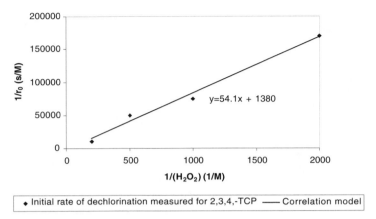

FIGURE 6.5

The plot of reciprocal initial dechlorination rate vs. reciprocal concentration of H_2O_2 during 2,3,4-TCP oxidation. Experimental conditions: $Fe^2 = 5 \times 10^{-3}$ M; 2,3,4-TCP $= 5 \times 10^{-4}$ M; pH = 3.5; ionic strength $= 0.05$ M as N_2SO_4.

$$(6.67)$$

In Relation (6.67), the solid arrow presents a stronger directory effect by the hydroxyl group than that by the chlorine group, which is presented by the dash arrow. It can be seen that 3-CP has three *ortho* and *para* positions enhanced by OH and Cl directors. For 4-CP and 2-CP, however, no position is enhanced by both OH and Cl directors. Because of these directors, intermediates with higher degrees of oxidation are expected to be produced in oxidation of 3-CP compared to the oxidation of 4-CP and 2-CP. Therefore, the dechlorination rate constants of 4-CP and 2-CP will be smaller than that of 3-CP. On the other hand, 2-CP will have some steric hindrance effect due to the OH and Cl groups on the aromatic ring being located closer than is the case for 4-DCP. As a result, 2-CP will be more difficult to oxidize than 4-CP. In other words, because the *ortho* position of chlorine is closer to the hydroxyl group than the *meta* and *para* positions, it will be subject to more steric strain than other congeners and have a greater change of free energy after dechlorination; therefore, hydroxylation at the *ortho* position will experience more steric strain than *meta* or *para* monochlorophenols.

Figure 6.6 presents the plots of dechlorination kinetic rate constants vs. $H_2O_2/(CP)$ ratio during the oxidation of 2-MCP, 2,4-DCP, and 2,4,6-TCP at a constant H_2O_2/Fe^{2+} ratio of 2.5 and optimal pH of 3.5. When the ratio of H_2O_2/DCP increases, the dechlorination rate constants increase. Furthermore, the difference between dechlorination rate constants becomes more

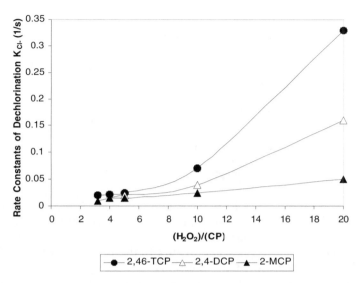

FIGURE 6.6
Dechlorination rate constants for 2-CP, 2,4-DCP, and 2,4,6-TCP vs. H_2O_2/organic concentrations for 2-CP, 2,4-DCP, and 2,4,6-TCP. Experimental conditions: $H_2O_2 = 5 \times 10^{-3}$ M; $Fe^{2+} = 2 \times 10^{-3}$ M; pH = 3.5; ionic strength = 0.05 M as N_2SO_4.

and more apparent when the H_2O_2/DCP ratio increases from 1 to 20. For a constant H_2O_2/CP, the H_2O_2/Fe^{2+} ratio is constant. Therefore, the steady-state concentration of hydroxyl radicals is constant. However, as the ratio of H_2O_2/DCP increases, more hydroxyl radicals are available to react with CP. Because chloride ion is released through numerous hydroxylation and hydrogen abstraction steps, more hydroxyl radicals are available for a DCP molecule, and the faster the dechlorination kinetics will be. The effect of the H_2O_2/CP ratio on the dechlorination constants can be clearly seen in Figure 6.6. The dechlorination rate follows this decreasing order: $k_{TCP} > k_{DCP} > k_{MCP}$. At the optimal H_2O_2/Fe^{2+} ratio, the amount of H_2O_2 available for 1 mol of chlorinated phenols affects dechlorination rate constants significantly, as shown in Figure 6.9. When the ratio of H_2O_2/CP is 20, the dechlorination constants for 2-CP, 2,4-DCP, and 2,4,6-TCP are 0.05, 0.16, and 0.33 (1/s), respectively. The relative dechlorination constants of these chemicals are 1:3:6. At a constant H_2O_2/CP ratio, the maximal capacity of Fenton's reagent is constant, because the number of available sites not occupied by chlorine atoms is the major factor responsible for dechlorination. If we assume that each site has an equal probability of being attacked by hydroxyl radicals, the efficiency of dechlorination by 1 mol of hydroxyl radicals should be 1/4, 2/3, and 3/2 for 2-CP, 2,4-DCP, and 2,4,6-TCP, respectively. This gives 1:2.6:6 as a relative dechlorination rate constant. The theoretical prediction agrees fairly well with the experimental result of 1:3:6. This is the reason why no linear correlation could be found between dechlorination constants and the nonchlorinated sites available. All the sites unoccupied by chlorine

TABLE 6.3

Dechlorination Rate Constants during
Oxidation of Dichlorophenols

Dichlorophenol	Rate Constants ($\times 10^7$ 1/s)
2,4-Dichlorophenol	2.28
2,3-Dichlorophenol	8.61
2,5-Dichlorophenol	11.90
2,6-Dichlorophenol	13.46
3,5-Dichlorophenol	13.82
3,4-Dichlorophenol	14.12

atoms on the aromatic ring of the chlorinated phenols have the same reactivity toward hydroxyl radicals when chlorine atoms occupy the 2, 4, and 6 positions. Table 6.3 shows that the effect of the chlorine position on dechlorination rate constants decreases according to the following order: 2,5-DCP > 3,5-DCP > 2,3-DCP > 2,6-DCP > 2,4-DCP.

The *ortho* and *para* director nature of Cl seems to play a less important role when the number of chlorine atoms increases from one to three; dechlorination has the following order in terms of decreasing rate constants: 2,4,6-TCP > 2,4,5-TCP > 2,3,4-TCP. This order suggests that the steric hindrance becomes a determining factor in preventing hydroxyl radical attack on the unoccupied sites of aromatic rings. For example, 2,4,6-TCP has much larger space than 2,4,5-TCP due to separation of the chlorine atoms on the aromatic ring. 2,4,5-TCP, in turn, has a larger space than does 2,3,4-TCP in which chlorine atoms can locate closely to one another. Therefore, 2,4,6-TCP is affected the least by steric hindrance; 2,4,5-TCP is subjected to an average steric hindrance effect; and 2,3,4-TCP is subjected to the greatest steric hindrance. The dechlorination rate constants seem to follow the same order suggested by steric hindrance. The nature of the *ortho* and *para* directors of Cl group seems to have diminished to a certain degree; in other words, if the directory effects of Cl group are predominant, then the order of decreasing dechlorination rate constants should be somewhat reversed.

From Equation (6.64), the following conditions have to be valid so that the initial dechlorination rate is independent of the organic concentration:

$$k(CP) \gg [k_{t2}(Fe^{2+}) + k_{t3}(H_2O_2)] \tag{6.68}$$

Under this condition, the numerical value of this expression is 5.1×10^{-4} (M^{-1} s^{-1}), which is one magnitude larger than the experimental average value of 5.1×10^{-5} (M^{-1} s^{-1}), as shown in Figure 6.7. Nevertheless, this implies that the limiting step in the oxidation of chlorinated phenols is the generation of hydroxyl radicals through Fenton's reagent.

$$r_0 = k_i(H_2O_2)_0(Fe^{2+})_0 \tag{6.69}$$

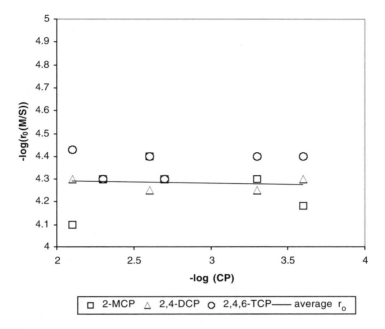

FIGURE 6.7
Dechlorination initial rate vs. different organic concentrations. Experimental conditions: H_2O_2 = 5×10^{-3} M; Fe^{2+} = 2×10^{-3} M; pH = 3.5; ionic strength = 0.05 M as N_2SO_4.

6.2.6.2.3 Oxidation Model of Unsaturated Aliphatic Compounds

The same transition complex approach and steady-state assumptions were used to develop the kinetic model of unsaturated chlorinated aliphatic compounds such as trichloroethylene (TCE). The model reflects the effects of H_2O_2, Fe^{2+}, and organic compounds on the oxidation kinetics as follows:

$$r_s = k_{observed} * \left\{ \frac{k_i(H_2O_2)(Fe^{2+})}{k_{t2}(Fe^{2+}) + k_{t3}(H_2O_2)} \right\}^n (S) \qquad (6.70)$$

where:

$$k_{observed} = \frac{k_p k_f}{k_b + k_p} \qquad (6.71)$$

The above equation agrees with the experimental observation that oxidation kinetics is the pseudo first-order reaction with respect to the concentration of substrates. The rate constant of $k_{observed}$ can be experimentally determined by keeping two variables constant and varying one variable among the concentrations of H_2O_2, Fe^{2+}, and substrate. When the logarithm of both sides of Equation (6.70) is taken, the following linear relationship can be obtained:

$$\log(r_0) = \log(k_{observed}) + n * \log\left\{\frac{k_i(H_2O_2)_0(Fe^{2+})_0}{k_{t2}(Fe^{2+})_0 + k_{t3}(H_2O_2)_0}\right\} + \log(S_0) \quad (6.72)$$

Let X be equal to the following:

$$X = \frac{k_i(H_2O_2)_0(Fe^{2+})_0}{k_{t2}(Fe^{2+})_0 + k_{t3}(H_2O_2)_0} \quad (6.73)$$

The equation can be written as:

$$\log(r_0) = n * \log X + \log(k_{observed}) + \log(S_0) \quad (6.74)$$

Using the experimentally determined constants such as k_i, k_{+2}, and k_{+3}, the kinetic models for dichloroethylene (DCE), TCE, and *tetra*-CE can be expressed as follows:

$$r_s = -\frac{d(DCE)}{dt} = 0.21 * \left\{\frac{k_i(H_2O_2)(Fe^{2+})}{k_{t2}(Fe^{2+}) + k_{t3}(H_2O_2)}\right\}^{0.439} * (DCE)$$

$$\quad (6.75)$$

$$r_s = -\frac{d(TCE)}{dt} = 19.95 * \left\{\frac{k_i(H_2O_2)(Fe^{2+})}{k_{t2}(Fe^{2+}) + k_{t3}(H_2O_2)}\right\}^{0.261} * (TCE)$$

$$\quad (6.76)$$

$$r_s = -\frac{d(Tetra\text{-}CE)}{dt} = 3.71 * \left\{\frac{k_i(H_2O_2)(Fe^{2+})}{k_{t2}(Fe^{2+}) + k_{t3}(H_2O_2)}\right\}^{0.333} * (Tetra\text{-}CE) \quad (6.77)$$

It is important to point out that the above treatment is not applicable to different oxidation mechanisms. For example, the kinetic data obtained from dichloroethane oxidation does not give a straight line between $\log(r_0)$ and $\log(X)$, because the oxidation pathway of dichloroethane is different from that in oxidizing chlorinated ethylenes. For example, dichloroethane is one of the saturated chlorinated aliphatic compounds, and the first oxidative step is hydrogen abstraction instead of hydroxylation (Walling, 1975). The unsaturated bond formed after hydrogen abstraction is then attacked by the addition of hydroxyl radicals. As a result, hydrogen abstraction must precede hydroxylation in the oxidation of dichloroethane.

In order to obtain the optimal ratio of H_2O_2 to Fe^{2+}, Equation 6.76 can be differentiated with respect to H_2O_2, assuming that Fe^{2+} is the optimal concentration of $(Fe^{2+})_{opt}$. We set:

$$\frac{dr_s}{d(H_2O_2)} = 0 \tag{6.78}$$

The optimal concentration of H_2O_2 can be expressed as follows by solving the above equation.

$$(H_2O_2)_{opt} = -\frac{k_{t2}(Fe^{2+})_{opt}(S)}{2k_{t3}} \tag{6.79}$$

The optimal concentration of Fe^{2+} can be derived from the same mathematical approach. We set the derivative of r_s with respect to Fe^{2+} concentration equal to zero:

$$\frac{dr_s}{d(Fe^{2+})} = 0 \tag{6.80}$$

Then, the optimal concentration of Fe^{2+} is:

$$(Fe^{2+})_{opt} = -\frac{k_{t3}(H_2O_2)_{opt}(S)}{2k_{t2}} \tag{6.81}$$

When Equation (6.79) is divided by Equation (6.81), the optimal ratio of H_2O_2/Fe^{2+} can be obtained:

$$\left\{ \frac{(H_2O_2)}{(Fe^{2+})} = \frac{(k_{t2})^2}{(k_{t3})^2} * \frac{(Fe^{2+})}{(H_2O_2)} \right\}_{opt} \tag{6.82}$$

Therefore, the final form of the optimal ratio between H_2O_2 and Fe^{2+} can be expressed as:

$$\left\{ \frac{(H_2O_2)}{(Fe^{2+})} \right\}_{opt} = \frac{k_{t2}}{k_{t3}} \tag{6.83}$$

Substituting the numerical values into the above equation, we obtain the optimal ratio of H_2O_2 to Fe^{2+} as the following:

$$\left\{\frac{(H_2O_2)}{(Fe^{2+})}\right\}_{opt} = \frac{k_{t2}}{k_{t3}} = \frac{3*10^8}{2.7*10^7} = 11 \qquad (6.84)$$

Experimental data indicate that the optimal ratio of H_2O_2 to Fe^{2+} is about 5 to 11. To investigate the effect of a low H_2O_2/Fe^{2+} ratio, the Fe^{2+} concentration was overdosed at a constant value of 10^{-3} M. At a high H_2O_2/Fe^{2+} ratio, where H_2O_2 was overdosed at 10^{-2} M, the extrapolated maximum initial rate of 1.8 mM/min also occurs at an H_2O_2/Fe^{2+} ratio of 11. This value reasonably agrees with the theoretical value of 11 as the optimal ratio of H_2O_2/Fe^{2+}. At a constant H_2O_2/Fe^{2+}, the H_2O_2 concentration required for iso-percentage release of chloride ion in dichloroethylene is shown in Figure 6.8.Tang and Huang (1997) concluded that the amount of H_2O_2 required for a specific percentage removal of the organic compounds depends upon the initial organic concentration to be oxidized. This is also true for the total percentage of chloride ion released at different initial organic concentrations. The typical percentage removal of organic compounds and percentage release of chloride ion have been studied at 100, 70, 50, 40, 30, 20, 10, and 1%. The amount of H_2O_2 required to achieve a certain percentage removal follows the order of TCE < *tetra*-CE < DCE << DCEA (dichloroethane) at a Fe^{2+} concentration of 10^{-3} M. However, the amount of chloride ion detected at an H_2O_2 concentration of 10^{-2} M follows the order of DCEA << DCE < TCE < *tetra*-CE. It is much more difficult to remove chloride atoms from saturated aliphatic compounds such as DCEA than from unsaturated aliphatic compounds.

6.3   Oxidation of Organic Compounds

6.3.1   Trihalomethanes

Trihalomethanes (THMs) are priority pollutants listed by the U.S. Environmental Protection Agency (EPA). They are recalcitrant in nature, thus their destruction is difficult. The most commonly encountered THMs in drinking water threatening human health are chloroform, bromodichloromethane, dibromochloromethane, and bromoform. Tang and Tassos (1997) studied the oxidation kinetics and mechanisms of these four THMs.

The effect of the ratio of H_2O_2 to Fe^{2+} on oxidation kinetics, the oxidation kinetics of THM mixtures, and the effect of the number of chlorine atoms in a THM on its oxidation were all investigated. Bromoform is the easiest to oxidize of the four THMs. Bromoform concentrations used in the study of Fenton's reagent ratio and oxidation kinetics were 49.2, 98.3, and 295 µg/L. As the ratio of H_2O_2 to Fe^{2+} increases, the removal efficiency increases with

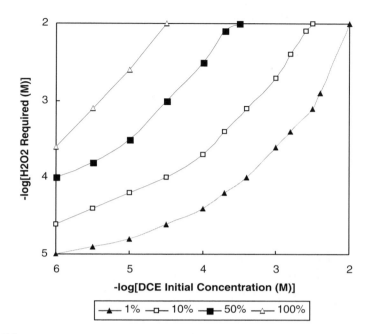

FIGURE 6.8
H_2O_2 concentration required for iso-percentage release of chloride ion in dichloroethylene (DCE) oxidation at optimal conditions. Experimental conditions: $H_2O_2/Fe^{2+} = 5$; pH = 3.5; ionic strength = 0.05 M as Na_2SO_4.

the initial concentration of bromoform. This may indicate that the hydroxyl radical has a preference toward organic compounds, resulting in proportionately less scavenging effect by $H_2O_2:Fe^{2+}$ and MeOH. For a higher $H_2O_2:Fe^{2+}$ ratio of 10:1, the amount of bromoform removal appeared to show dependence on initial bromoform concentration. As the initial organic substrate concentration increases, less scavenging of OH$^\bullet$ occurs. At an $H_2O_2:Fe^{2+}$ ratio of 100:1, bromoform removal was only 25%, while at a ratio of 5:1, 83% removal was observed. As the ratio decreased from 5:1 to 2:1, no increase in removal of bromoform was observed. Thus, the optimum ratio must be maintained to achieve maximum degradation.

Both H_2O_2 and Fe^{2+} are able to scavenge hydroxyl radicals generated through Fenton's reagent. If any one of them is not present at the optimum dosage, either H_2O_2 or Fe^{2+} will be able to scavenge hydroxyl radicals and reduce its availability to the substrate. Oxidative destruction of THMs was found to have slower kinetics, because THMs are saturated aliphatics and have only one C–H bond. Thus, oxidation of THMs is predominantly due to hydrogen abstraction, which has low kinetic rates. Because the second-order rate constant for hydrogen abstraction by hydroxyl radicals is 4×10^8 (M^{-1} s^{-1}), a higher concentration of hydroxyl radicals is required, which further implies that higher concentrations of Fe^{2+} and hydrogen peroxide are needed. Hydroxyl radicals are also scavenged by organic compound as well as chloride and bromide ions, which can be illustrated by the following reactions:

$$OH^\bullet + RH \rightarrow H_2O + R^\bullet \tag{6.85}$$

$$OH + R \rightarrow ROH^\bullet \tag{6.86}$$

$$CHBr_3 + OH^\bullet \rightarrow {}^\bullet CBr_3 + H_2O \tag{6.87}$$

$${}^\bullet CBr_3 + OH^\bullet \rightarrow HO\text{-}CBr_3 \tag{6.88}$$

$$HO = CRr_3 + OH^\bullet \rightarrow O = CBr_2 + H_2O + Br^- \tag{6.89}$$

$$Br^- + OH^\bullet \rightarrow BrOH^- \tag{6.90}$$

The inhibitory effect of chloride and bromide ions is not significant unless free ion concentration exceeds a range from 1×10^{-2} to 5×10^{-2} M. Dehalogenation is believed to be a slower process than oxidation of the parent compound, and the proposed general reaction is as follows:

$$THM + OH^\bullet \rightarrow \text{halogenated intermediates} \rightarrow X^- + CO_2 + H_2O + \text{end products} \tag{6.91}$$

Fluctuations in bromoform concentration were observed even after a triplicate run. This was explained by recombination of the bromoform radical as follows:

$$CHBr_3 + OH^\bullet \rightarrow {}^\bullet CBr_3 + H_2O \tag{6.92}$$

$$Fe^{2+} + {}^\bullet CBr_3 + H^+ \rightarrow Fe^{3+} + CHBr_3 \tag{6.93}$$

The above recombination reactions require the presence of either hydrogen radicals ($H^\bullet$) or both H^+ and e^-. As the experiments were carried out in acidic conditions (pH = 3.5), electron transfer was possible because Fenton's chemistry does not generate hydrogen radicals (Huang et al., 1993).

The oxidation rates for bromoform were slower than the oxidation rates of unsaturated chlorinated aliphatic compounds, including the TCE. Because the hydroxylation rate constant of TCE is 10^9 M^{-1} s^{-1} and the hydrogen abstraction of bromoform is 1.1×10^8 M^{-1} s^{-1}, aromatics and alkenes react more rapidly by hydroxyl addition to double bonds than does the more kinetically difficult hydrogen atom abstraction. No oxidative destruction of chloroform by Fenton's reagent was experimentally observed; an explanation for this is that both H_2O_2 and Fe^{2+} have rate constants about one magnitude higher with respect to hydroxyl radicals than chloroform.

Tang and Tassos (1997) reported that the oxidative degradation of THMs decreases as the number of chlorine atoms present in the substrate molecule increases. The relationship between removal rate and number of chlorine atoms was shown to be linear. This phenomenon is due to the fact that the bromine substituents are better leaving groups than chlorine substituents (Solomons, 1988). Another consideration is electronegativity and bond energy. Sharp (1990) derived a relation between the bond energy between atoms A and B and the electronegativity as:

$$D_{AB} = 0.5 \ (D_{AA} + D_{BB}) + 23 \ (X_A - X_B)^2 \tag{6.94}$$

where D_{AB}, D_{AA}, and D_{BB} are the bond energies between A and B, A and A, and B and B, respectively. Bond energy decreases as electronegativity decreases. Thus, ease of dehalogenation of an organic compound is directly proportional to the bond energy between the carbon and halogen atoms. Because the bond energies for C–Cl and C–Br bonds are 95 kcal/mol and 67 kcal/mol, respectively, brominated compounds are more easily oxidized than those containing proportionately more chlorine (Tang and Tassos, 1997).

6.3.2 Hydroxymethanesulfonic Acid

Hydroxymethanesulfonic acid (HMSA) is a complex formed from formaldehyde and S(IV). It has been detected in atmospheric liquids (i.e., rain and snow). The complex has high resistance to oxidation by oxygen as well as ferric ions and oxygen. Martin et al. (1989) first studied the oxidation of HMSA. Graedel et al. (1986) proposed that Fenton-type reactions are possible in atmospheric liquid water.

Martin et al. (1989) studied the oxidation of HMSA by Fenton's reagent and investigated the decomposition of both hydrogen peroxide and HMSA. They determined an estimate of the absolute rate of reaction between HMSA and hydroxyl radicals. The decomposition of hydrogen peroxide follows the first-order kinetics and can be described as follows:

$$-d(H_2O_2)/dt = k(Fe^{2+})(H_2O_2) \tag{6.95}$$

where k is 0.044 M^{-1} s^{-1} at pH 2 and temperature 25°C.

The actual rate of oxidation by free radicals was established by subtracting the rate of formation of SO_4^{-2} from the decomposition of HMSA. Experimental results showed good agreement with the first-order rate of decomposition of HMSA. Doubly ionized HMSA decomposes at a higher rate compared to singly ionized HMSA. The rate levels off until second ionization is complete, which would occur at high pH. Similar experiments were performed for acetaldehyde–bisulfite complex HESA. Acetaldehyde complex HESA was not as effective as HMSA in preventing S(IV) from being oxidized. Fenton's reagent studies were carried out at the pH levels of 1, 2, 3, and 4. Results showed that the oxidation is a first-order process with respect to different initial Fe^{2+} concentrations and different pH at 10^{-1} M H_2O_2 and 10^{-2} M HMSA. At higher concentrations of iron ion, a slight fall off was observed, which was attributed to less hydrogen peroxide and thus fewer hydroxyl radicals in the system. Similar experiments were carried out at various initial hydrogen peroxide concentrations, and the oxidation was seen to be of first order; however, at larger concentrations it deviated from first order due to the smaller amount of Fe^{2+}. The maximum rate has been observed at pH 3.5. An empirical rate approximating the oxidation kinetics of HMSA at pH 1 to 3 is shown below:

$$-d(HMSA)/(HMSA)dt = k(Fe^{2+})(H^+)^{-1}(H_2O_2)^{2/3} \qquad (6.96)$$

with $k = 1.4 \pm 0.2 \times 10^{-3}$ $(l/mol)^{2/3}$ s^{-1}; thus, if 10^{-5} M H_2O_2 and 10^{-6} M Fe are present at pH 3, the oxidation rate was 2.3×10^{-4} % h^{-1}, which represents very low kinetics.

A study of oxidation of HMSA was done relative to pinacol to estimate the absolute rate of oxidation of HMSA with OH radicals in solution. Pinacol was oxidized to acetone in Fenton's oxidation. Anbar and Neta (1967) reported reaction rates of OH radicals with pinacol and acetone of 3.2×10^8 M^{-1} s^{-1} and 4.3×10^7 M^{-1} s^{-1}, respectively. Table 6.1 presents the oxidation rates of pinacol (10^{-2} M) and pinacol–HMSA (10^{-2} M each). The concentrations of the reactants were $Fe^{2+} = 10^{-4} M$, $H_2O_2 = 0.1$ M, and pH 2. Oxidation rates for each molecule were different in separated and mixed reactions, as the steady-state concentration of free radicals depends on the chemistry of organic substrates in solution. HMSA is more reactive than pinacol by a factor of 3.9 ± 0.8. If the absolute rate of reaction of pinacol with OH radicals was calculated to be 3.2×10^8 M^{-1} s^{-1}, then:

$$HMSA + {}^{\bullet}OH = products \qquad (6.97)$$

and $k = 1.25 \pm 0.25 \times 10^9$ M^{-1} s^{-1}. This rate constant suggested that HMSA may be consumed fairly rapidly in tropospheric clouds, as OH from other sources should be abundant to give a large reaction rate.

6.3.3 Phenolic Waste

Phenolic wastes are one of the most prevalent forms of chemical pollutants in industry today. The major sources of phenolic waste are insulation fiberglass manufacturing, petroleum refineries, textile mills, steel making, plywood, hardboard production, manufacture of organic chemicals, paint stripping, and wood preservatives. Eisenhauer (1964) first studied oxidation of phenolic wastes with Fenton's reagent. It has been demonstrated that the oxidation of phenol involves the intermediate formation of catechol and hydroquinone (Merz and Waters, 1949; Stein and Weiss, 1951; Wieland and

Franke, 1928). Catechol can be oxidized in high yield to muconic acid by hydrogen peroxide and ferrous salt (Pospisil and Ettel, 1957). Thus, Eisenhauer showed that phenol oxidation proceeds according to Equation (6.98).

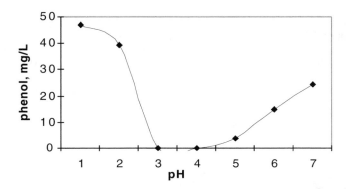

$$(6.98)$$

Series of experiments for phenol concentrations of 50 ppm, at pH 3 and ambient temperature of 10°C, have demonstrated that oxygen plays a major role in reaction by increasing the reaction rate and driving the reaction to completion. Optimum results were obtained when the reaction was carried out using 1 mol of ferrous salt and 3 mol of hydrogen peroxide per mole of phenol. Optimum results were obtained in a pH range of 3 to 4, as shown in Figure 6.9.

6.3.4 Substituted Phenols

The oxidation of a number of substituted phenolic wastes by Fenton's reagent was also studied by Eisenhauer (1964). It was observed that the greater the degree of substitution, the slower the rate of reaction, especially when substituents were *ortho* and *para* directing. When all available positions were blocked, as in the case of pentachlorophenol, no reaction occurred. Halophenols were rapidly oxidized and exhibited the following gradually decreasing order: Cl > Br > I, which is attributed to the change in electronegativity of the halogen groups. Phenols containing *meta*-directing groups such as carboxyl and nitro groups were very rapidly oxidized by Fenton's reagent, and methyl-substituted phenols (cresols) were more resistant to

FIGURE 6.9
Effect of pH on oxidation of phenol with Fenton's reagent. (From Eisenhauer, H.R., *J. Water Pollut. Contr. Fed.*, 36, 1116-1128, 1964. With permission.)

oxidation. Effluents from a refinery, steel plants, and insulation plants have been examined. A ratio of H_2O_2/Fe^{2+} of 9:1 has been able to completely oxidize the phenol. Steel plant effluent contained some cyanide, which interfered with the reaction; also, 16 mol of hydrogen peroxide per mole of phenol decreased the phenol level to 35%. For insulation plants, Fenton's reagent was also successful in reducing phenol and chemical oxygen demand (COD).

Barbeni et al. (1987) studied 18 different phenolic compounds. The reaction was started at pH 5 to 6, because pH decreased in the system due to generation of protons by Fenton's reagent. Table 6.4 indicates that all the mono-substituted phenols were readily oxidized. The effect of oxidation of di-substituted phenols was dependent on character and position of ring substituents. Dichlorophenols were more readily oxidized than electron-donating substituents of dimethyl phenols. Because Fenton's reagent is a free-radical mechanism, any substituent that increases the electron density of the ring will slow the reaction. So, methyl phenol, with an electron-donating group, will oxidize slowly.

It was recommended that catalyst concentration levels from 10 to 20 g/g iron ion should be used for phenolic compounds no greater than 2000 µg/l. Higher phenol concentrations require at least 100:1 phenol:iron ratios, and optimum results occur when phenol solution is initially at a pH between 5 and 6.

Barbeni et al. (1987) also compared the half-lives of different chlorophenols. They demonstrated that the appearance of chloride ion is independent of the disappearance of parent organic compounds during Fenton's oxidation

TABLE 6.4

Oxidation of Phenolic Compound, 3:1 Hydrogen Peroxide:Phenol Mole Ratio

Phenolic Compound	% Oxidized (1 hr, no Fe^{2+})	% Oxidized (1 hr, with Fe^{2+})
Phenol	0	100
2-Chlorophenol	9	100
3-Chlorophenol	50	100
4-Chlorophenol	20	100
2,4-Dichlorophenol	0	100
2,5-Dichlorophenol	2	74
Pentachlorophenol	6	100
o-Cresol	0	100
m-Cresol	0	100
p-Cresol	6	100
2,4-Dimethyl phenol	10	72
2,5-Dimethyl phenol	6	0
2-Nitrophenol	34	100
4-Nitrophenol	0	100
2,4-Dinitrophenol	2	30
2,5-Dinitrophenol	10	73
α-Naphthol	8.5	100
β-Naphthol	9.5	100

Source: Data from Barbeni et al., *Chemosphere*, 16, 2225–2237, 1987.

of 2-chlorophenol, and it took 30 minutes to oxidize 50 mg/L of aqueous chlorophenol (CP). Among the members of the groups of monochlorophenols, *meta* chloro substitution induces more rapid degradation than either *ortho* or *para* substitution. The formation of Cl⁻ was slower than the disappearance of 3,4-DCP, which suggested that chloroaliphatic intermediate(s) may be formed after the opening of the benzene ring, and the time lag between appearance of Cl⁻ ion and disappearance of phenol depends on the number of chloro substituents on the aromatic ring. Experiments utilizing various ferrous concentrations have shown that increasing ferrous concentration in solution for a fixed hydrogen peroxide at 5×10^{-3} M increases the rate of decomposition proportionally. Similar experiments were also carried out with ferric ions. It was observed that when equal molar amount of ferrous and ferric ions were used in the same reaction mixture, the reaction half time ($t/2$) was ~20 min, whereas it was ~50 min if only ferrous ions were used. Barbeni et al. (1987) indicated that the OH• formed as a result of reaction between Fe^{2+} and H_2O_2 are good electrophiles. This electrophilic interaction of OH• with the aromatic ring is favored by the presence of an electron-donating group (OH).

$$(OH)ClC_6H_3H + OH^{\bullet} = (OH)ClC_6^{\bullet}H_3 \begin{smallmatrix} H \\ \diagup \\ \diagdown \\ OH \end{smallmatrix} \qquad (6.99)$$

$$(OH)ClC_6H_3H(OH) = (OH)ClC_6H_3OH + H^+ \qquad (6.100)$$

Thus, the degradative oxidation of chlorophenols proceeds by a hydroxylated species (Equation 6.99 and Equation 6.100), followed by ring opening to yield aldehydes and ultimate degradation of CO_2 and Cl⁻. It was suggested that the first step is the formation of radical cation by acid-catalyzed dehydration of radicals formed due to the interaction of OH• with chlorophenols.

Oxygen acts as radical scavenger and forms $HO_2^{\bullet}$; this $HO_2^{\bullet}$ species successively reacts with either Fe^{2+} or Fe^{3+} to regenerate H_2O_2 and Fe^{2+}, as shown in Equation (6.101) to Equation (6.103).

$$\qquad (6.101)$$

$$Fe^{2+} + \rightarrow HO_2^{\bullet}\ Fe^{3+} + HO_2 - \rightarrow H_2O_2 \qquad (6.102)$$

$$Fe^{3+} + HO_2^{\bullet} \rightarrow Fe^{2+} + H^+ + O_2 \qquad (6.103)$$

Under batch and semibatch conditions, Potter and Roth (1993) examined the oxidation kinetics of three monochlorophenol isomers and five of the six dichlorophenol isomers. The pH was maintained at 3.5, and the ultimate oxidation to CO_2, H_2O, and HCl was not reached; however, the extent of oxidation could be estimated from the measured parameters. The mechanism of oxidation of these species is important in determining if the kinetics is feasible. The proposed reaction pathway suggested by Potter and Roth is two possible pathways for mineralization of *p*-chlorophenol, as shown in Figure 6.10.

While Metelitsa (1971) proposed mineralization by substitution of hydroxyl radicals, Pieken and Kozarich (1989) proposed mineralization by elimination (Figure 6.11). The experiments conducted by Potter and Roth (1993) showed agreement with previous work suggesting that monochlorophenols tend to mineralize to a greater extent than do dichlorophenols (Barbeni et al., 1987). They showed that mineralization may be favored as the initial concentration of organic species decreases, and mineralization occurs primarily through other intermediates.

As discussed earlier, the effects of the *meta, para*, and *ortho* positions of chlorine on the dechlorination kinetics of monochlorophenols, dichlorophenols, and trichlorophenols during Fenton oxidation were evaluated by comparing the rate constants of the kinetic model (Tang and Huang, 1995). This study proposed a pseudo first-order steady state with respect to organic concentration. The proposed reaction pathways considered that the hydroxyl radicals would attack unoccupied sites of the aromatic ring.

6.3.5 Pentachlorophenol

Pentachlorophenol, a widely used wood preservative, is considered to be moderately biorefractory with a biodegradation rate constant of 3×10^{-12} L/cell/hr, a log K_{ow} of 5.01, and a vapor pressure of 1.1×10^{-4} mmHg at 20°C. Watts et al. (1990) carried out completely mixed batch tests by treating *penta*-chlorophenol-contaminated soils with Fenton's reagent. Mineralization of pentachlorophenol (PCP) was studied in commercially available silica sand and two natural soils by removal of parent compound and total organic carbon with corresponding stoichiometric recovery of chloride. The soluble iron concentration decreased over the first 3 hr of treatment, and the concentration remained relatively constant thereafter. A possible mechanism for iron precipitation was proposed as follows:

$$Fe^{2+} + 1/2O_2 + 2OH^- \rightarrow \gamma\text{-FeOOH} + H_2O \qquad (6.104)$$

which has a reaction half-life of 25 min at neutral pH (Sung and Morgan, 1980). Watts et al. (1990) proposed zero-, first-, and second-order kinetics models but concluded that first-order was the best with $r^2 > 0.9$ for plots of the natural logarithm of concentration as a function of time. Although the reactions occurring in Fenton's system are complex, the empirical fit of experimental data to the first-order model has provided the most

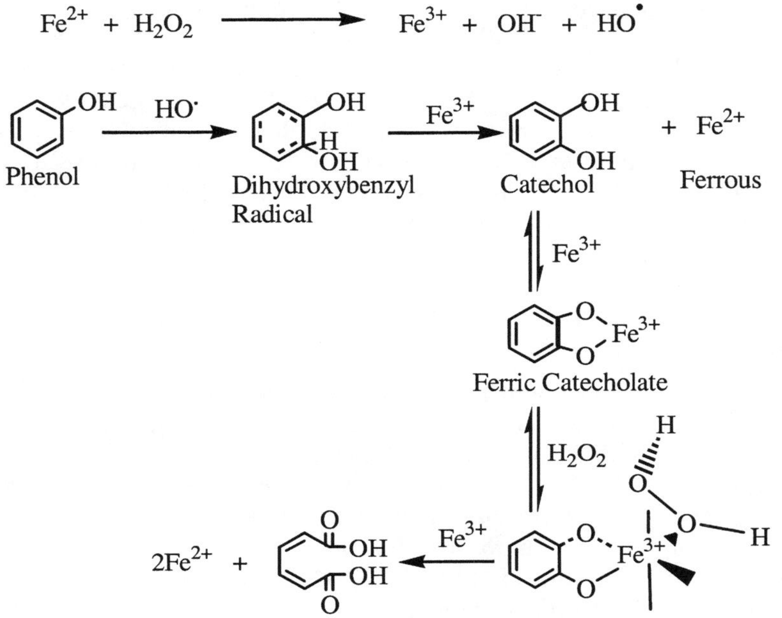

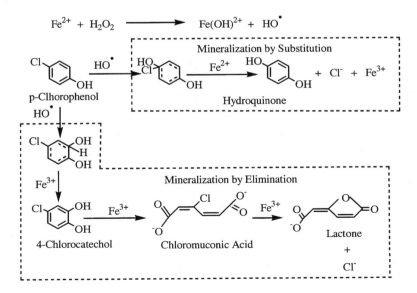

$$Fe^{2+} + H_2O_2 \longrightarrow Fe^{3+} + OH^- + HO^\bullet$$

FIGURE 6.10
Reaction pathways for oxidation of phenol. (From Potter, F.J. and Roth, J.A., *J. Hazardous Waste Hazardous Mater.*, 10(2), 151, 1993. With permission.)

$$Fe^{2+} + H_2O_2 \longrightarrow Fe(OH)^{2+} + HO^\bullet$$

Mineralization by Substitution

Hydroquinone

Mineralization by Elimination

FIGURE 6.11
Two possible pathways for degradation of p-cholorphenol. (From Potter, F.J. and Roth, J.A., *J. Hazardous Waste Hazardous Mater.*, 10(2), 151, 1993. With permission.)

accurate means of comparing different treatment conditions. These tests also showed that organic carbon was removed rapidly after PCP degradation, which illustrates that hydroxyl radical attack on the products is faster than on the parent compound. The addition of Fe^{2+} significantly increased the degradation rate constant at low pH. Table 6.5 shows the results of the kinetics at different pH levels with or without the addition of soluble iron. Based on the first-order rate constant for PCP degradation at pH 3, 5 h were required for 99% PCP degradation with Fe^{2+} addition and 39 h without Fe^{2+} degradation.

6.3.6 Nitro Phenols

Nitrophenols are persistent pollutants and commonly found in industrial wastewaters. Degradation of nitrophenols to less dangerous materials or mineralization is difficult by natural process involving biodegradation or oxidation. Nitrophenols are commonly found in degradation of pesticides such as parathion and nitrofen. Kiwi et al. (1994) showed efficient photo- and dark oxidation via Fenton-like reactions on 2 and 4-nitrophenols. Photolysis of acidic solutions of H_2O_2 give $OH^\bullet$ radicals as primary photoproducts (Baxendale and Wilson, 1957) as follows:

$$H_2O_2 \xrightarrow{h\nu} 2OH^\bullet \tag{6.105}$$

along with small amounts of HO_2 radicals:

$$H_2O_2 \xrightarrow{h\nu} H^\bullet + HO_2^\bullet \tag{6.106}$$

The irradiated 2-nitrophenol forms a triplet state that reacts with H_2O_2 (Kiwi et al., 1994).

$$HO - C_6H_4 - NO_2^* + H_2O_2 \rightarrow HO - C_6H_9 - NO_2H + HO_2 \tag{6.107}$$

TABLE 6.5

Fenton's Reagent Treatment Efficiencies

pH	2	3	4	k_{PCP}/k_{H2O2} 5	6	7	8
Silica + Fe	13.7	25	10.5	11.6	0.9	1.2	1.1
Soil 1 (no Fe)	12.5	4	6.7	0.67	0.63	0.18	0.12
Soil 1 + Fe	5	3.4	2	1.4	1	0.13	0.082
Soil 2 (no Fe)	8	17.6	2	11	4.5	1.1	0.36
Soil 2 + Fe	7.3	11.3	8.2	4.6	3.7	0.72	0.3

Source: Watts, R.J. et al., *Hazardous Waste Hazardous Mater.*, 7(4), 335, 1990. With permission.

Results of irradiation showed that photoproducts were formed as hydroquinones, benzoquinones, and nitrophenols. Under experimental conditions, Fe^{3+} ion is known to promote decomposition of H_2O_2 as follows:

$$Fe^{3+} + H_2O_2 \rightarrow HO_2^{\bullet} + H^+ + Fe^{2+} \tag{6.108}$$

or according to the mechanism:

$$Fe^{3+} + H_2O \xrightarrow{h\nu} HO^{\bullet} + H^+ + Fe^{2+} \tag{6.109}$$

with Fe^{2+} ion additionally producing $OH^{\bullet}$ radicals as follows:

$$Fe^{2+} + H_2O_2 \rightarrow HO^{\bullet} + OH^- + Fe^{3+} \tag{6.110}$$

Comparisons of the degradation of 2-nitrophenol under various conditions, such as direct photolysis, Fe^{3+} solution, H_2O_2/dark, and H_2O_2/light, were studied and the results are shown in Figure 6.12.

The effect of temperature was found to be beneficial on light-activated degradation. The time was reduced by 40% for dark processes and 50% for similar processes under light. 2-Nitrophenol degradation could be induced by sunlight when temperatures above 30°C were available in nature under favorable conditions. Light significantly improves the reaction kinetics of the system Fe^{2+}–H_2O_2 (see Equation 6.109). The degradation of 2-nitrophenol seems to favor the regeneration of Fe^{2+} ions in Equation (6.109). Fe^{2+} produces many $OH^{\bullet}$ radicals in contrast to the dark Fenton system, where Fe^{2+}:$OH^{\bullet}$ = 1:1.

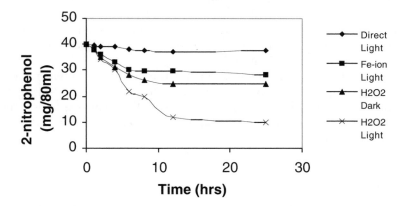

FIGURE 6.12
Photodegradation of 2-nitrophenol. (From Kiwi, J. et al., *Appl. Catalysis*, 3, 335–350, 1994. With permission.)

The overall oxidation of 2-nitrophenol by $OH^{\bullet}$ radicals could be expressed as follows:

$$O_2N\text{-}C_6H_4\text{-}OH + OH^{\bullet} \xrightarrow{-H} O_2N\text{-}C_6H_3\text{-}(OH)_2$$

$$\rightarrow \text{intermediates} \rightarrow CO_2 + H_2O + H^+ + NO_2^-/NO_3^- \qquad (6.111)$$

The overall reaction is:

$$C_6H_5NO_3 + 6.OH(3H_2O_2) + 1/2O_2 \rightarrow 6CO_2 + 5H_2O_2 + HNO_3 \quad (6.112)$$

Reaction (6.110) to Reaction (6.114) show that only H_2O_2 is consumed during the light degradation process, as seen below:

$$Fe^{2+} + H_2O_2 \rightarrow Fe^{3+} + OH^{\bullet} + OH^- \qquad (6.113)$$

$$\bullet H + Fe^{3-} + OH^{\bullet} \rightarrow \bullet OH + Fe^{2+} + H^+ \qquad (6.114)$$

$$\bullet H + H_2O_2 \rightarrow \bullet OH + H_2O \qquad (6.115)$$

Two processes are taking place simultaneously and leading to degradation of 2-nitrophenol. This process is represented by Kiwi et al. (1994) in Figure 6.13. The overall rate for total organic carbon (TOC) or dissolved organic carbon (DOC) degradation at 60°C in the first 2 hr is 10 times faster than the rate subsequently observed for this reaction between 2 and 5 hr. Under light, 4-nitrophenol degradation proceeded two times faster than 2-nitrophenol degradation. The rates of $OH^{\bullet}$ oxidation are about the same for both. The nitro group is an electron-withdrawing group. In a benzenic ring, the nitro-withdrawing effect is preferentially marked in the *ortho* and *para* positions, and in 2-nitrophenol hydroxylation of the 6 position leads to dihydroxy compound, as shown in Figure 6.13.

In the case of hydroxylated 4-nitrophenol, the two bonds between adjacent –OH groups might undergo easy oxidative scission, but in 2-nitrophenol only one bond of this type is found in hydroxylation. This explains why the photo-Fenton degradation of 2-nitrophenol proceeds at about half the rate of that for 4-nitrophenol.

6.3.7 Benzenes

Chlorobenzenes have been widely used as solvents and degreasing agents; in pesticides, including dichlorodiphenyltrichloroethane (DDT), and in dielectric fluids; as industrial precursors in the production of phenols; and as dyestuff intermediates. Chlorobenzenes represent a unique class of compounds because their hydrophobicity increases with chlorine substitution,

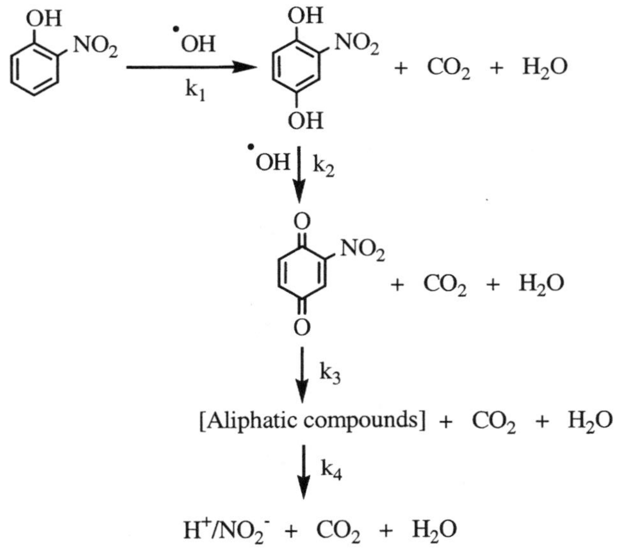

FIGURE 6.13
Degradation pathway of nitrophenol. (From Kiwi, J. et al., *Appl. Catalysis*, 3, 335–350, 1994. With permission.)

while the rates of hydroxyl radical attack on the ring are nearly equal regardless of the degree of chlorination (Watts et al., 1997). Sedlak and Andren (1991a) investigated the degradation of chlorinated aromatic organics (CAHs) using Fenton's reagent for bench-scale experiments. Twelve different intermediates were identified in the oxidation process, including 2-chlorophenol; chlorobenzoquinone; 3-chlorophenol; 4-chlorophenol; 2,2′-DCB (dichlorobenzene); 2,3′-DCB; 2,4′-DCB; 3,3′-DCB; 3,4′-DCB; 4,4′-DCB; MHClBps (monochlorobiphenyls); and DHClBps (dichlorobiphenyls).

Chlorobenzene and all intermediate products disappeared in the first 2 hr of reaction. The presence of oxygen at pH 3 and approximately 5 mol of H_2O_2 per mol of chlorobenzene were required to completely remove the aromatic intermediates from solution. Reactions in the absence of oxygen yielded much higher concentrations of dichlorobiphenyls when compared to those in air. The DOC analysis indicated that the aromatic intermediates undergo ring cleavage prior to mineralization. The reaction almost ceases after approximately 4 hr, as observed by the stabilization of DOC concentration, pH, and Fe^{2+}. The reaction mechanism is shown in Figure 6.14.

The first step in this sequence is the hydroxyl radical attack on chlorobenzene (reaction 2), which likely results in the formation of chlorohydroxycyclohexadienyl (ClHCD) radical I (Dorfman et al., 1962). This may initiate one of several possible further reactions. In the absence of strong oxidants, two predominant reactions are dimerization, to produce dichlorobiphenyls (reaction 3), and bimolecular disproportionation, to produce chlorophenol and chlorobenzene (reaction 4). Both reactions showed the stoichiometry of 2 mol

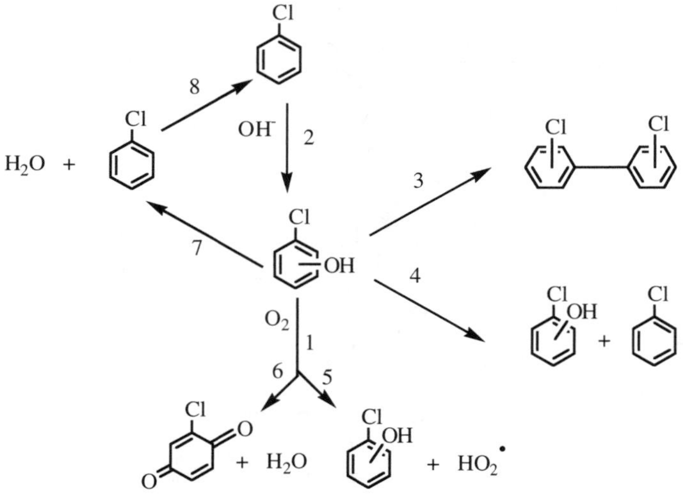

FIGURE 6.14
Proposed reaction pathway for oxidation of chlorobenzene with Fenton's reagent. (From Sedlak, D.L. and Andren, A.W., *Environ. Sci. Technol.*, 25, 777–782, 1991a. With permission.)

of H_2O_2 per mol of chlorobenzene oxidized. In the presence of oxygen or strong oxidants several additional reactions contribute to product formation. Reactions of oxidant with ClCHCD radical (reactions 5 and 6) predominate because they are first order with respect to the ClCHCD radical, whereas reactions 3 and 4 are second order with respect to the radical. The presence of oxygen or other strong oxidants favors the direct oxidation pathway and follows a stoichiometry in which less hydrogen peroxide is required to degrade the CAHs. The oxidation of chlorobenzene and chlorophenol isomers is most likely attributable to the formation of phenolic polymers. The disappearance of broadband absorption and dimers as the reaction progressed suggests that these polymers and dimers are ultimately amenable to oxidation.

Watts et al. (1997) investigated the effects of hydrogen peroxide concentration and desorption rates on the oxidation of four chlorobenzenes sorbed on hematite, which is a naturally occurring iron oxide found in soils and subsurface systems. Hydrogen peroxide concentrations from 0.1 to 5% were used, and the oxidation rates were compared to the rates of desorption. The desorption experiments showed that 0.1 mmol/kg 1,3,5-trichlorobenzene, 1,2,3,4-tetrachlorobenzene, pentachlorobenzene, and hexachlorobenzene partitioned on hematite. Desorption followed first-order kinetics with respect to each chlorobenzene, and rate constants were inversely proportional to chlorine substitution and the octanol/water partition coefficient. The degradation of 1,3,5-trichlorobenzene with initial concentrations of 0.1, 1, 2, and 5% H_2O_2, trichlorobenzene desorption, and H_2O_2 consumption were studied using deionized water as control. The data suggested that oxidation occurred at least in the sorbed state with hydrogen peroxide concentrations $\geq 2\%$. An optimum oxidation rate is reached at 2% H_2O_2, because 2% is the minimum H_2O_2 concentration capable of oxidizing sorbed contaminants. Although the rate constants for hydroxyl radical attack on chlorobenzenes were nearly equal, the rates of degradation decreased from lower to higher chlorine substitution. Chlorobenzene oxidation by hematite-catalyzed, Fenton-like reactions was highly dependent on its desorption rate. For example, hexachlorobenzene was not desorbed or oxidized, which shows that hydroxyl radicals do not degrade even reactive substrates while sorbed on the surface of hematite.

Watts et al. (1994) studied sorption and degradation of hexachlorobenzene on geothite as the iron catalyst and concluded that mineral-catalyzed, Fenton-like reactions are controlled by desorption, H_2O_2 concentration, and contaminant structure.

Equation (6.117) shows the heterogeneous decomposition of H_2O_2 on mineral surface:

$$S + H_2O_2 \rightarrow S^+ + OH^\bullet + OH^- \tag{6.116}$$

where S is the mineral surface and S^+ is the oxidized region of the surface. Hydroxyl radicals generated by the above reaction oxidize all four chlorobenzenes in the aqueous phase (not sorbed):

$$PhH_nCl_{6-n(aqueous)} + OH^\bullet \rightarrow products \tag{6.117}$$

$$PhH_nCl_{6-n(sorbed)} + OH^\bullet \rightarrow no\ reaction \tag{6.118}$$

Higher concentrations of H_2O_2 increase the concentration of hydroxyl radicals. In other words, the higher system redox potential may increase the oxidation state on mineral surfaces (S^{n+}), which might lead to oxidation of sorbed 1,3,5-trichlorobenzene, 1,2,3,4-tetrachlorobenzene, and pentachlorobenzene:

$$S^+ + nH_2O_2 + S^{n+} \rightarrow nOH^\bullet + OH^- \, [H_2O_2] = 2\ \% \tag{6.119}$$

$$S^+ + nOH^\bullet + nH^+ \rightarrow S^{n+} + nH_2O \, [H_2O_2] = 2\% \tag{6.120}$$

where S^{n+} is the more oxidized mineral surface relative to S^+ and is capable of degrading chlorobenzenes with at least one C–H bond but not perhalogenated hexachlorobenzene:

$$PhH_nCl_{6-n(sorbed)} + S^{n+} \rightarrow products\ (n = 1) \tag{6.121}$$

$$PhCl_{6(sorbed)} + S^{n+} \rightarrow no\ reaction \tag{6.122}$$

6.3.8   Toluenes

2,4-Dinitrotoluene (DNT) is generated by ammunition factories and other chemical industries. It is used to make 2,4-diaminotoluene for isocyanate production and for the production of dyes and explosives, for organic synthesis, and as a propellant additive. Mohanty and Wei (1993) carried out an experiment to study the effects of H_2O_2:DNT ratio, temperature, and sequential addition on the oxidation rates of DNT and its removal from the waste stream. The effect of the H_2O_2:DNT ratio was studied using ratios from 5:1 to 80:1. The ratio of Fe^{2+}:H_2O_2 was 0.024, and the initial pH varied from 4 to 5. The concentration of by-products decreased with increasing H_2O_2. The consumption of H_2O_2 increased with the amount of DNT applied. At higher levels of H_2O_2, self-decomposition increased as the $OH^{\bullet}$ generated reacted with H_2O_2, as shown in Equation (6.123):

$$H_2O_2 + OH^{\bullet} \rightarrow OH_2^{\bullet} + H_2O \qquad (6.123)$$

The rate of removal of DNT was seen to follow a pseudo first-order reaction with a rate coefficient of 0.035/min at a molar ratio of 15:1:1.8 of H_2O_2:DNT:Fe^{2+}. Three different experiments were performed to study the effect of sequential addition. During the course of the reaction, the resulting excess of Fe^{2+} decreases the kinetic significance of the initiation reaction. Sequential dosing of H_2O_2 produced lower concentrations of reaction by-products when compared to experiments performed without sequential dosing. The retention time of these by-products was higher then for their parent compounds. One of these by-products might be a dimer. If an excess of H_2O_2 is available, Equation (6.123) and Equation (6.124) become significantly more important:

$$Fe^{2+} + HO_2^{\bullet} \rightarrow Fe^{3+} + HO_2^{-} \qquad (6.124)$$

The minimum removal of TOC indicates the losses of H_2O_2 and shows the importance of Equation (6.124). No removal of TOC was observed during the first 30 min due to lower Fe^{2+} concentrations and thus limited hydroxyl radical is available for oxidation of organics. In the presence of oxygen, the rate of degradation and extent of oxidation increased dramatically, but no major difference was observed in the removal of TOC and COD. The reaction was also run at three different temperatures (21, 30, and 40°C), with all other conditions constant. Figure 6.15 shows that temperature affects the reaction rate considerably.

Hydrogen peroxide was depleted in 5, 4, and 2.5 hr for reactions at temperatures of 21, 30, and 40°C, respectively. TOC removal was also greater at

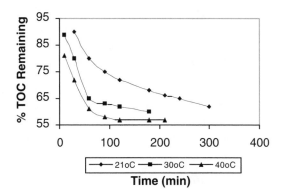

FIGURE 6.15
Effect of temperature on removal of TOC. (Mohanty, N.R. and Wei, I.W., *J. Hazardous Waste Hazardous Mater.*, 10(2), 171, 1993. With permission.)

higher temperatures. The initial presence of Fe^{3+} and Cu^{2+} enhanced the evolution of oxygen in the absence of substrates. The removal rate of DNT was pseudo first order, and the decomposition rate was a straight line with regression coefficient of 0.98.

6.3.9   BTX

Benzene, toluene, and xylene (BTX) are highly soluble in water and are found in oil-contaminated groundwater. Garroguino et al. (1992) reported that the toxicity of contaminated groundwater is primarily due to BTX. BTX present in groundwater in dilute concentrations presents a challenge to drinking-water supply resources. The most common pollution sources are underground oil storage tanks, and the pollution patterns are described by increased concentrations within close proximities to gas stations, factories, and oil-refinery plants. Due to the high concentrations of these tanks in densely populated areas, the potential adverse effects on human health are significant risk factors.

Lou and Lee (1995) oxidized BTX in a batch reactor using Fenton's reagent for simulating hydrocarbon-contaminated groundwater from a spill site. Changes in BTX concentration over time; the effects of pH, H_2O_2, and Fe^{2+} concentration on BTX destruction; the optimum ratio of H_2O_2:BTX:Fe^{2+}; and the rate expression of BTX have been reported (Lon and Lee, 1995). Results were obtained in terms of destruction efficiency (DRE) of BTX, which is expressed as follows:

$$DRE\ (\%) = \frac{C_0 - C_t}{C_t} \times 100 \qquad (6.125)$$

where C_0 is the initial concentration of BTX (mg/L) and C_t is the concentration of BTX after t min (mg/L). At an H_2O_2:BTX:Fe^{2+} ratio of 2.4:1:12 (mg/L), the BTX is significantly oxidized by Fenton's reagent, and benzene and toluene removal was shown to be about 86%; similarly, xylene removal was 80%. During the first 10 min, however, the concentration of BTX remained almost unchanged, followed by 90 to 94% removal within 60 min.

The DRE of BTX decreased with increasing pH value at pH greater than 4. For pH less than 4, the DREs of benzene, toluene, and xylene are about 87, 88, and 83%, respectively. At fixed pH and Fe^{2+}, the DRE of BTX increased almost linearly with increasing H_2O_2 concentration until 60 mg/L; with an increase in concentration of H_2O_2, the DRE of BTX remained at 80 to 90%, revealing a zero-order reaction.

As Haber and Weiss (1934) suggested, at lower H_2O_2 concentrations and fixed Fe^{2+} the oxidation reaction approaches second order; however, when the ratio of H_2O_2:Fe^{2+} increases, the reaction kinetic approaches zero order and the reaction process depends on the competition between hydroxyl radicals and superoxide radicals. If an excess of hydrogen peroxide is present, then the reactions as shown in Equation (6.123) and Equation (6.124) for 2,4-dinitrotoluene are dominant. The amount of H_2O_2 was used up quickly in this study, indicating the importance of Equation (6.123). At concentrations of Fe^{2+} greater than 600 mg/L, the DRE of BTX reached a maximum value at approximately 82% for benzene and toluene and 73% for xylene.

Lou and Lee (1995) found that the DRE of BTX caused by Fenton's reagent cannot be further raised due to lack of available H_2O_2; the H_2O_2 was consumed rapidly, and the DRE of BTX was minimal. To study the optimal ratio of H_2O_2:BTX:Fe^{2+}, the concentration of BTX was kept at 50 mg/L and the pH varied from 4 to 5. The range of the H_2O_2 to Fe^{2+} ratio studied was from 0 to 20. The DRE increased with an increase in ratio from 0 to 5. The optimal ratio was found to be 0.2; however, further increases in the ratio actually decreased the DRE of BTX. The effect of H_2O_2:BTX ratio on BTX removal was also studied by varying this ratio from 2.4 to 12, at the H_2O_2:Fe^{2+} ratio of 0.2. Examining the destruction of BTX vs. the H_2O_2:BTX ratio reveals that the optimal DRE of BTX occurred around 98 to 100% for xylene, benzene, and toluene when the H_2O_2:BTX ratio was 12:1. After combining these results, Lou and Lee concluded that BTX can be effectively oxidized at an H_2O_2:BTX:Fe^{2+} ratio of 12:1:60 (mg/L) in less than 10 min. The results of chromatograms indicated that the concentration of by-product was insignificant and very low. The plot of reaction time vs. BTX concentration shows that the kinetics follows a pseudo first-order reaction at 20°C. At an applied concentration ratio of 2.4:1:12 (H_2O_2:BTX:Fe^{2+}), the rate coefficient of BTX was calculated to be 4.03(hr^{-1}).

The removal of BTX was found to be pseudo first order. The coefficients of linear regression were found to be 0.991 for xylene and 0.994 for benzene and toluene.

The rate expression for benzene and toluene was given as follows:

$$\ln C = 3.92 - 4.03(t) \qquad (6.126)$$

and for xylene as:

$$\ln C = 3.89 - 2.91(t) \qquad (6.127)$$

where C is the concentration of benzene, toluene, or xylene or t hours (in ppm), and t is the reaction time (in hours).

6.3.10   Polychlorinated Biphenyls

In most wastes and wastewater, polychlorinated biphenyls (PCBs) and particulate matter are found in the aqueous phase. The fraction of PCBs associated with each phase depends on the hydrophobicity. The congeners containing more chlorine substituents have a stronger tendency to associate with particulate. PCBs sorbed to surfaces such as diatomaceous earth are not oxidized by aqueous $OH^\bullet$ at an appreciable rate relative to the reaction rate of $OH^\bullet$ with solution-phase PCBs. Sedlak and Andren (1994) performed a quantitative evaluation of the effect of sorption to particulate matter on the rate of PCB oxidation by $OH^\bullet$. The transformations of three PCB congeners — 2-monochlorobiphenyl (MClBp); 2,2′,5-trichlorobiphenyl (TrClBp); and 2,2′,4,5,5′-pentachlorobiphenyl (PeClBp) — were studied at an initial concentration of 1 μM of PCB solution. Data from the experiments were compared with predictions from quantitative kinetic models that used independently determined data on reaction rates and $OH^\bullet$ concentrations.

Hydroxyl radical was generated through the photo-Fenton reaction. The first step in the reaction was photoreduction of ferric hydroxy complex by radiation from a black light:

$$FeOH^{2+} + h\nu \rightarrow Fe^{2+} + OH^\bullet \qquad (6.128)$$

Ferrous iron produced is then oxidized by H_2O_2 in Fenton's reaction:

$$Fe^{2+} + H_2O_2 \rightarrow FeOH^{2+} + OH^\bullet + OH^- \qquad (6.129)$$

The rate of H_2O_2 consumption and the $OH^\bullet$ production were directly related to total iron concentration. The concentrations of hydroxyl radical produced were controlled by the rate of reaction with dissolved constituents. Rate constants for adsorption (k_a) and desorption (k_d) of PCBs from particles were calculated by regression of data from 1.5 to 5 hr. Adsorption rate constants were estimated from Equation (6.130) assuming that the partitioning rate constants between 2 and 5 hr without $OH^\bullet$ could be used for calculation of equilibrium partition coefficients (K_p)

$$K_p = k_a / k_d \qquad (6.130)$$

K_p were calculated by averaging the partitioning data collected after 2 hr in experiments conducted in the absence of OH•. Equation (6.130) may be employed to model partitioning kinetics over short time intervals, but applying the particles used in these experiments as equilibrium models may not adequately predict partitioning over extended time intervals for many types of particles. Partitioning data and reaction rate constants calculated for 2,2′,5-trichlorobiphenyl and 2,2′,4,4′,5-PeClBp by Sedlak and Andren (1994) are shown in Table 6.6.

Sedlak and Andren (1991b) modeled hydroxyl radical reaction kinetics in the presence of particulate. They found that the reaction kinetics for PCB oxidation in the presence of particulate resulted from the complex interplay between solution-phase OH• reactions and reversible adsorption–desorption reactions. A model predicting the reaction kinetics can be described by the following equation:

$$ \text{PCB} = \text{S} \underset{k_a}{\overset{k_d}{\rightleftarrows}} \text{PCB} + \text{S} \tag{6.131} $$

$$ \text{OH} \bullet + \text{PCB} \overset{k_{OH}}{\rightarrow} \text{PCB} - \text{OH} \tag{6.132} $$

where:

PCB = S is the particle-associated PCB concentration (mol/g).
PCB is the dissolved PCB concentration (M).
S is the particle concentration (g/L).
OH• is the steady-state hydroxyl radical concentration (M).
PCB-OH is the hydroxylated PCB radical concentration (M).

TABLE 6.6

Partitioning Data and Reaction Rate Constants Calculated for 2,2′,5-Trichlorobiphenyl and 2,2′,4,4′,5-Pentachlorobiphenyl

Variable	2,2′5-Trichlorobiphenyl	2,2′,4,5,5′-Pentachlorobiphenyl
k_d ($\times 10^4$ s^{-1})	2.04 (±0.10)	4.53 (±0.03)
k_a (1 g^{-1} s^{-1})	0.39 (±0.15)	1.41 (±0.34)
K_p (1 g^{-1})	1860.0 (±615)	31,600.0 (±5,400)
k_{OH} ($\times 10^9$ cm^3 M^{-1} s^{-1})	6.9	4.6

Source: Sedlak, D.L. and Andren, A.W., *Water Res.*, 28(5), 1207–1215, 1994. With permission.

Several options were possible for modeling Equation (6.131) and Equation (6.132), and the choice of an appropriate model depended on the relationship between the rate constants and the degree of accuracy desired. When the substrate was not strongly associated with the solid phase or when the reaction rate was much lower than the desorption rate, it was possible to model the transformation as a pseudo first-order process, based on the assumption that Reaction 6.131 was insignificant relative to the reaction shown in Equation (6.132) (i.e., $[PCB] = PCB_{tot}$). The transformation rate was then approximated by Equation (6.133).

$$-d[PCB]/dt = k_{OH} [OH^\bullet] [PCB] \qquad (6.133)$$

where k_{OH} is the second-order reaction rate constant (M^{-1} s^{-1}). Assuming a steady-state $OH^\bullet$ concentration, transformation of the solute could be modeled as a pseudo first-order process with an apparent first-order rate constant of $k_{OH}[OH^\bullet]$. This model was applied to results for MClBp, and good predictions were obtained for the transformation rates of PCB. The pseudo first-order model can be applied when only a small fraction of total PCB present was associated with the particulate phase. The main advantage of this model was that accurate predictions of reaction rates could be made without determination of adsorption–desorption rate constants. Because particle-associated PCB concentrations were low and desorption rates fast, the adsorption or desorption kinetics was not the rate-limiting step.

When a larger fraction of HOC was associated with the particulate phase, numerical solutions of Equation (6.132) and Equation (6.133) were required to predict PCB transformation rates. The instantaneous concentration of dissolved PCB was estimated by incorporating the terms S and $OH^\bullet$ into the rate constant expressions for Equation (6.133) and Equation (6.134).

$$[PCB] = \frac{k_d[PCB]_0}{\beta - \alpha} [e^{-\alpha t} - e^{-\beta t}] \qquad (6.134)$$

where $[PCB]_0$ is the initial total PCB concentration, and

$$\alpha\beta = k_d k_{OH}[OH^\bullet]$$

$$\alpha + \beta = k_d + k_a [S] + k_{OH} [OH^\bullet]$$

The rate of PCB transformation by $OH^\bullet$ was then calculated using Equation (6.134) at discreet time intervals, while the dissolved PCB concentrations were predicted by Equation (6.132). The predictions of the numerical solution model were compared with experimental data for TrClBp. The model showed good agreement with the data, and the experimental concentrations were within approximately ±10% of the predicted values. Transformation rates were much more sensitive to changes in k_d than OH$^\bullet$. The strong affinity of PeClBp for the particulate phase prevented the transformation of significant amounts of PeClBp over the time period.

These experiments provided valuable insight into the nature of HOC transformations in the presence of particles. When a substantial fraction of HOC was associated with the particulate phase, more complex models based on solution of kinetics expressions describing reversible sorption and OH$^\bullet$ transformation reactions were required to predict the transformation kinetics. HOC, almost completely associated with the particulate phase, underwent transformation reactions at very slow rates.

6.3.11   Pesticides

Atrazine (2-chloro-4-(ethylamino)-6-(isopropylamino)-s-triazine) is one of the most commonly used pesticides in the United States, and its usage accounts for 12% of the total volume of pesticides. Arnold et al. (1995) studied the effects of different reaction conditions on the efficacy of Fenton's reagent for degrading atrazine. Atrazine degradation and product formation rates were determined as a function of FeSO$_4$ and H$_2$O$_2$ concentrations, ratios, and solution pH. The ratios of FeSO$_4$:H$_2$O$_2$ used were 1:1, 1:200, and 2:1 at concentrations from 0.1 to 25 mM. The FeSO$_4$ and H$_2$O$_2$ ratio of 1:1 and its effect on atrazine degradation was also studied. During this study, seven major products identified were CDIT, CIAT, CDET, CEAT, ODIT, CDAT, and CAAT, as shown in Table 6.7.

At concentrations of 2.69 mM of hydrogen peroxide and ferrous iron, the reaction mixtures were depleted of CDIT, CIAT, CEAT, and ODIT, but CDAT, CAAT, and six minor unidentified atrazine derivatives persisted at Fenton's reagent concentrations up to 25 mM. Fenton's reagent treatment was ineffective in degrading CDAT and CAAT due to the low reactivity of oxidized products toward OH$^\bullet$. At low FeSO$_4$:H$_2$O$_2$ ratios, complete treatment was achieved with lower Fe^{2+} concentrations as compared to 2.69-mM treatment, but increasing the H$_2$O$_2$ concentration 100-fold lowered the efficiency of the reactions. An excess of hydrogen peroxide usage might have led to dealkylation, thus decreasing dechlorination. Increases in Fe^{2+} concentration lowered the reaction efficiency, as excess Fe^{2+} reacts with OH$^\bullet$ at a FeSO$_4$:H$_2$O$_2$ ratio of 2:1.

Atrazine degradation rates were studied at an atrazine concentration of 132 μM with Fenton's reagent (1:1) in the range of 1.06 to 2.69 mM. Almost 98% of the atrazine was removed in less than 30 s. After 24 hr, the degradation of remaining atrazine slowed down and 1% of the initial amount remained. Using Fenton's reagent with concentrations ≥1.42 mM, atrazine was degraded in <30 s to below detection limits. Fenton's reagent with concentrations of 2.69 mM showed that atrazine transformation products accounted for less than 5% each of initial atrazine concentration and were depleted within 3 hr. After 11.5 hr of treatment, the only products that remained were

TABLE 6.7

Chemicals and Abbreviations of Atrazine and Fenton's Reagent-Generated Degradation Products

Common Name	Chemical Name	Abbreviation
Atrazine	2-Chloro-4-(ethylamino)-6-(isopropylamino)-*s*-triazine	CIET
Atrazine amide	2-Acetamido-4-6-(isopropylamino)-*s*-triazine	CDIT
Deethylatrazine	2-Amino-4-chloro-6-(isopropylamino)-*s*-triazine	CIAT
Simazine amide	2-Acetamido-4-chloro-6-(ethylamino)-*s*-triazine	CDET
Deisopropylatrazine	2-Amino-4-chloro-6-(ethylamino)-*s*-triazine	CEAT
Hydroxyatrazine amide	2-Acetamido-4-hydroxy-6-(isopropylamino)-*s*-triazine	ODIT
Deisopropylatrazine amide	2-Acetamido-4-amino-6-chloro-*s*-triazine	CDAT
Chlorodiamino-*s*-triazine	2-Chloro-4,6-diamino-*s*-triazine	CAAT
Ammeline	2,4-Diamino-6-hydroxy-*s*-triazine	OAAT

Source: Arnold, S.M. et al., *Environ. Sci. Technol.*, 29, 2083, 1995. With permission.

CDAT and CAAT. Excess H_2O_2 resulted in slower degradation of atrazine and followed pseudo first-order kinetics. The half-life $(t_{1/2})$ of atrazine at a $FeSO_4$:H_2O_2 ratio of 1:100 was 1.9 hr. Table 6.8 lists the half-lives of atrazine according to various technologies.

Table 6.8 shows that the fastest depletion of atrazine occurs under Fenton's reagent. Arnold et al. (1995) demonstrated that $OH^\bullet$ may react with Cl^- at low pH to produce $HOCl^{\bullet-}$ and Cl_2, causing underestimation of Cl^- concentrations, but the results suggest that Cl^- scavenging by $OH^\bullet$ was minimal. Dechlorination and dealkylation occur simultaneously, and the batch treatment showed that dechlorination occurred more readily with alkylated *s*-triazines. Chlorinated products accounted for a large part of *s*-triazines present upon completion of Fenton's reagent.

Three different degradation mechanisms were proposed. In the first mechanism, the hydroxyl radical attacks atrazine by hydrogen abstraction from the secondary carbon of the ethylamino side chain, producing a free radical as shown in Equation (6.135).

$$HO^\bullet + RNHCH_2CH_3 = RNHC^\bullet HCH_3 + H_2O \qquad (6.135)$$

$$(RNHCH_2CH_3 = \text{atrazine}) \qquad (6.136)$$

$$O_2 + RNC^\bullet HCH_3 = RNHC(OO^\bullet)HCH_3 \qquad (6.137)$$

$$RNHC(OO^\bullet)HCH_3 + Fe^{2+} + H^+ = RNHC(OOH)HCH_3 + Fe^{3+} \qquad (6.138)$$

TABLE 6.8

Half-Lives of Atrazine Determined by Various Technologies

Technology	Atrazine (μM)	Half-Life ($t_{1/2}$) (m)	Ref.
Fenton's reagent(1.42 mM; 1:1)	132	0.5	Arnold et al., 1995
TiO$_2$/UV	116	20	Pelizzetti et al., 1990
H$_2$O$_2$/UV	96	1.8	Beltran et al., 1993
Ozone	465	16	Kearney et al., 1988
Fe(ClO$_4$)$_3$ (0.26 mM)/UV	5.2	1.4	Larson et al., 1991

Source: Arnold, S.M. et al., *Environ. Sci. Technol.*, 29, 2083, 1995. With permission.

Molecular oxygen reacts with the free-radical, giving a peroxy radical of atrazine as shown in Equation (6.137). This is reduced by Fe^{2+} to form a hydroperoxide. The rearrangement of the hydroperoxide forms amide through oxidation of secondary C with loss of a water molecule, as shown in Equation (6.139).

$$RNHC(OOH)HCH_3 = RNHC(O)CH_3 + H_2O \qquad (6.139)$$

The second mechanism proposed was dealkylation by hydrogen abstraction of secondary carbon followed by the introduction of an oxygen atom from either HO$^\bullet$ or a high-valent iron–oxo species forming an alcohol adjacent to electronegative nitrogen. The alcohol is unstable and decomposes to aldehyde and *N*-dealkylated *s*-triazine. The third mechanism, proposed by Potter and Roth (1993), is that dechlorination could occur by OH$^\bullet$ radical attack of the *s*-triazine ring at the position occupied by the chlorine group oxidizing the aromatic heterocyclic ring of atrazine. Reduction of the ring by Fe^{2+} results in dechlorination to give hydroxylated *s*-triazine, Fe^{3+}, and Cl$^-$.

The degradation pathway determined by a study by Arnold et al. (1995) accounted for the type and pattern of by-products (Figure 6.16). Reactions (a) to (e) show that dechlorination, alkyl side chain oxidation, and/or cleavage occur as parallel reactions, giving CDIT, CIAT, CDET, CEAT, or ODIT. Reactions (f) and (g) show that dealkylation gives CIAT and CDAT, which undergo side-chain oxidation and/or cleavage to form CAAT, as in reactions (j), (l), and (m). Similar reactions occur with CDET and CEAT to produce CDAT and CAAT as the main chloro-*s*-triazine compounds upon completion of Fenton's reagent treatment (reactions (h), (i), (k), (m), and (n)). Similar side-chain cleavage is expected to occur with dechlorinated products, as in reaction (p). It was suggested that dechlorination of alkylated *s*-triazine derivatives in reaction (q) is most likely due to sequential treatment of terminal products CDAT and CAAT resulting in major transformations to become dominated and/or dechlorinated products such as OAAT. The amide was unstable during the isolation procedure. CDIT, CDET, and CDAT formed small amounts of CIAT, CEAT, and CAAT, respectively, which indicated that the amide was easily cleaved.

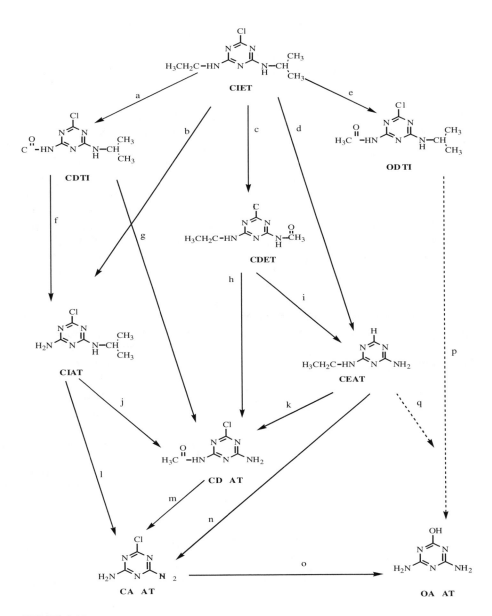

FIGURE 6.16
Degradation pathway for atrazine treated with Fenton's reagent. (From Arnold, S.M. et al., *Environ. Sci. Technol.*, 29, 2083, 1995. With permission.)

6.3.12   Herbicides

Pignatello (1994) studied the pesticide degradation and mineralization of herbicides in combination with visible light. 2,4-Dichlorophenoxy acetic acid (2,4-D) at a concentration of 0.1 mM was treated with [Fe^{2+}] and [H_2O_2] both greater than 1 mM. 2,4-D was transformed in less than 1 min. The reaction took nearly 2 min at a lower [Fe^{2+}] and [H_2O_2] concentration (0.25 mM) and gave 76 to 88% transformation. Transformation was independent of initial pH, but when Fe^{3+} (instead of Fe^{2+}) was used, the disappearance of the parent compound was more sensitive to pH. The optimum conditions for the transformation of 2,4-D by Fe^{3+}/H_2O_2 in both perchlorate and sulfate solution was at [Fe^{3+}] = 1 mM; [H_2O_2] = 2.5 to 10 mM; pH = 2.7 to 2.8; [2,4-D] = 0.1 mM. Transformation of 2,4,5-trichlorophenoxy acetic acid (2,4,5-T) at pH 2.8 was slower than 2,4-D. It was also observed that the transformation was inhibited by methanol or chloride due to scavenging of active oxidant and by sulfate due to complexation of Fe^{3+}. The intermediates were polychlorophenols (PCPs) and trichlorophenol (TCP). This reaction was sensitive to anion in solution. 2,4-D was rapidly converted to highly hydrophilic intermediates. Herbicide concentration (0.1 vs. 0.5 mM 2,4-D) had no effect on mineralization. Fe^{2+} was substituted instead of Fe^{3+}; however, a substantial effect on the extent or rate of mineralization of 2,4-D has not been observed.

The transformation and mineralization of herbicides by Fe^{3+}/H_2O_2 were promoted by irradiation with visible light with a small ultraviolet (UV) component. In a bright $Fe^{3+}/H_2O_2/h\nu$ system, transformation of 2,4,5-T was complete in less than 0.3 hr and mineralization was complete in less than 2 hr. In dark conditions, the transformations were observed to be 2.7 and 1.6 times faster for 2,4-D and 2,4,5-T, respectively, compared to the use of Fe^{3+}/H_2O_2. Fenton's reagent transformed the herbicides faster than Fe^{3+}/H_2O_2. It was more important than Fe^{3+}/H_2O_2 if removal of parent compound is required without mineralization. Retardation of 2,4-D at low pH was attributed to prior complexation of Fe^{3+} and H_2O_2, whereas retardation at pH above 3 was attributed to precipitation of Fe^{3+} to amorphous oxyhydroxides ($Fe_2O_3 \bullet nH_2O$). It was also observed that hydroxyl radicals could be scavenged by chloride and sulfate as follows:

$$OH^{\bullet} + Cl^- \overset{H^+}{\leftrightarrow} HOCl^{\bullet} \overset{H^+}{\leftrightarrow} Cl^{\bullet} + H_2O \overset{Cl^{\bullet}}{\leftrightarrow} Cl_2 \qquad (6.140)$$

$$OH^{\bullet} + HSO_4^- \rightarrow H_2O + SO_4^- \qquad (6.141)$$

$$OH^{\bullet} + 2,4\text{-D} \rightarrow \text{initial products} \qquad (6.142)$$

Reaction (6.142) was predicted to be slower than Reaction (6.140). Chloride strongly inhibited 2,4-D transformation which suggests that the inhibition of 2,4-D was mainly due to scavenging of $OH^{\bullet}$. Scavenging of $OH^{\bullet}$ by chloride at a pH of ~3 would begin to be significant when [Cl^-] was above a few millimoles. Dechlorination occurred rapidly after herbicide transformation. Phenoxy acetic acid decarboxylated to yield phenoxymethyl radical ($PhOCH_2^\bullet$). When the corresponding (aryloxy) methyl radical was produced from 2,4-D and 2,4,5-T, a pathway to phenol was oxidation, hydrolysis, and elimination as follows:

$$ArOCH_2^\bullet \rightarrow ArOCH_2^+ \rightarrow \left[ArOCH_2OH\right] \rightarrow ArOH \qquad (6.143)$$

Irradiation with weak-intensity light above 300 nm strongly accelerated degradation. The total mineralization according to Equation (6.141) was observed when the ratio of 2,4-D:Fe^{3+}:H_2O_2 was 1:4:5.

$$C_8H_6Cl_2O_3 + 7H_2O_2 + 4O_2 \rightarrow 2HCl + 8CO_2 + 9H_2O \qquad (6.144)$$

Pignatello and Baehr (1994) studied the herbicides such as 2,4-D and Metolachlor. The herbicides showed contrasting behavior in that Metolachlor was mostly sorbed and 2,4-D was mostly in solution. The behavior of herbicides, Fe(II) and Fe(III) complexes in soil, and herbicide degradation were studied. The following ligands: hydroxyethyleniminodiacetic acid (HEIDA), gallic acid (GAL), picolinic acid (PIC), and rhodizonic acid (RHO), as shown in Figure 6.17, were also investigated during herbicide degradation.

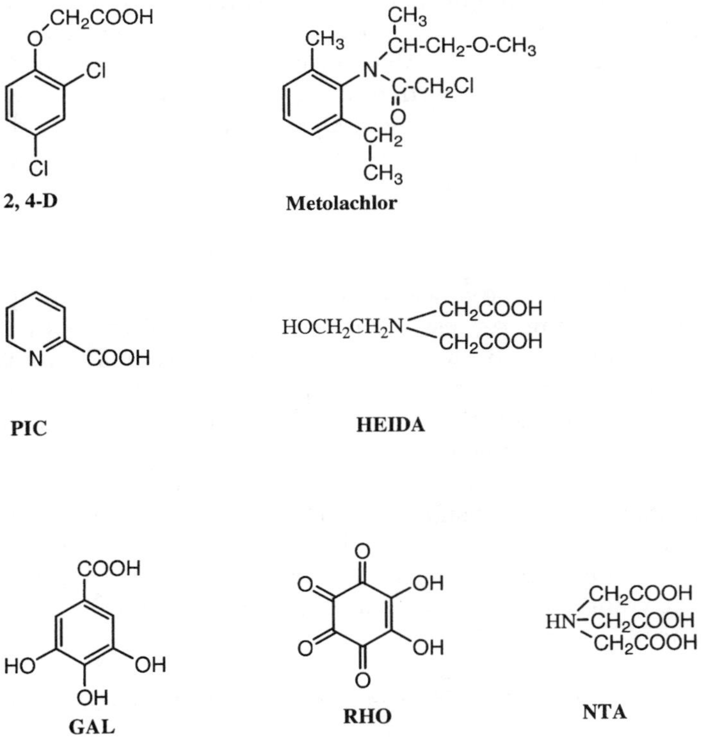

FIGURE 6.17
Molecular structure of herbicides in the Fenton's reaction system.

About 10 to 16% of total 2,4-D was adsorbed, whereas 89% of total Meto-lachlor was adsorbed. The complexes of Fe-GAL, -HEIDA, and -NTA have comparatively greater ability to stay in solution. Studies were carried out with 2000 mg/kg of 2,4-D equilibrated for 24 hr and treated with 0.1 mol/kg and 1×10^{-2} mol/kg Fe-L or corresponding free ligand. It was observed that an Fe-L system was superior to free ligand alone and that free ligand did not extract enough Fe from soil to achieve practical levels of degradation. Fe-GAL was inferior to Fe-HEIDA and Fe-NTA. The Fe-HEIDA and Fe-NTA complexes gave significantly less removal at 0.1 mol/kg hydrogen peroxide and nearly complete removal at 0.5 mol/kg H_2O_2. Metolachlor was treated using Fe-NTA and 0.5 mol/kg H_2O_2. About 92% was removed, and the results indicate that even compounds that initially are mostly sorbed can react. It is unclear whether desorption proceeds before sorbed molecules are attacked. A mass ratio of H_2O_2 to herbicide of 5.6 to 8.5 is required to remove the parent compound. The degree of ring and carboxy C mineralization increased in the order GAL > HEIDA > NTA. The mineralization was favored as H_2O_2 increased from 0.1 to 0.5 mol/kg but decreased with further increase to 1 mol/kg. As H_2O_2 increased, the extractable C decreased in the order GAL > HEIDA > NTA. The remaining radioactivity increased with H_2O_2 concentration and followed the order GAL > HEIDA > NTA.

In addition, the catalytic efficiencies of Fe(III)-L with Fe(II) were compared. Simple Fe^{2+} removed 61% of 2,4-D and 7% of the Metolachlor; Fe-NTA removed 99.3% 2,4-D and 87% Metolachlor. The combination of NTA with Fe^{2+} was found to be as effective as Fe-NTA in removing Metolachlor, indicating that Fe^{2+} is oxidized by H_2O_2 in milliseconds *in situ* due to the reactive Fe(III)-L complex.

6.3.13   Dyes

The major problem with dye wastewater is the color produced by residual dyes during the dyeing process. If the concentration of dyes in wastewater increases, then the color of the streams will grow darker. The transparency of streams will then be reduced, causing plants to perish and ecosystems to suffer. Several amino-substituted azo dyes are mutagenic and carcinogenic.

Kuo (1992) suggested that dye decolorization can result from both oxidation and coagulation processes. During the oxidation process, hydroxyl radicals may attack an organic substrate such as an unsaturated dye molecule. Thus, the chromophore of the dye molecule would be destroyed and decolorized. Ferric ions generated from Fenton's reaction might form ferric–hydroxo complexes with hydroxide ions, as shown in Equation (6.145) and Equation (6.146):

$$[Fe(H_2O)_6]^{2+} + H_2O \leftrightarrow [Fe(H_2O)_5OH]^{2+} + H_3O^+ \tag{6.145}$$

$$[Fe(H_2O)_5OH]^{2+} + H_2O \leftrightarrow [Fe(H_2O)_4(OH)_2]^{2+} \tag{6.146}$$

These complexes have a pronounced tendency to polymerize at pH 3.5 to 7, as shown in Equation (6.147) to Equation (6.149):

$$2[Fe(H_2O)_5OH]^{2+} \leftrightarrow [Fe_2(H_2O)_8(OH)_2]^{4+} + 2H_2O \tag{6.147}$$

$$[Fe_2(H_2O)_8(OH)_2]^{4+} + H_2O \leftrightarrow [Fe_2(H_2O)_7(OH)_3]^{3+} + H_3O^+ \tag{6.148}$$

$$[Fe_2(H_2O)_7(OH)_3]^{3+} + [Fe(H_2O)_5OH]^{2+} \leftrightarrow [Fe_3(H_2O)_5(OH)_4]^{5+} + 2H_2O \tag{6.149}$$

The experiments carried out showed that hydrogen peroxide and ferrous ions at pH less than 3.5 were more stable than when pH was greater than 3.5. In basic solutions, hydrogen peroxide was unstable and decomposed easily. Disperse dyes showed notable color removal above pH 9 which was attributed to coagulation generated by ferrous ions. The ferrous ions release electrons in basic solutions and form ferric ions; these ferric ions may form ferric–hydroxo complexes, which can coagulate the disperse dyes. The increased H_2O_2 dosage increases the COD removal efficiency. Disperse dyes have very low activity and are not readily reactive with hydrogen peroxide. Thus, H_2O_2 would not deplete quickly and would not be prompted to decompose further. Disperse dyes contain some dispersants that may react with the active substances, which might accelerate hydrogen peroxide decomposition.

Increased concentrations of ferrous sulfate lead to faster kinetic rates. When the dosage of ferrous sulfate is increased, it makes the redox reaction complete and causes coagulation, thus improving decolorization. Kuo (1992) found that low-temperature reactions require more time than do high-temperature reactions. Wastewaters from dyeing and finishing mills are usually equal to 50°C, which is considered a high temperature and favors decolorization of dye wastewater.

Kuo (1992) also studied five different types of dye wastewater randomly selected from mills, and the results were in good agreement with those from the laboratory studies. The average percent COD removal was about 90%, and average decolorization was above 97%. The effective pH was 3.5 and below. The effective dosage of H_2O_2 and ferrous sulfate was affected by the various types of dyes, and decolorization was affected by different dye structures, the auxiliary group on the dye molecule, and temperature. Spadaro et al. (1994) concluded that the hydroxyl radical attacks the azo linkage connecting an amine and a hydroxy-substituted ring during Fenton oxidation of azo dyes. The hydroxyl radical adduct breaks to produce phenyldiazene and a phenoxy radical. Hydroxyl radicals or molecular oxygen can readily oxidize phenyldiazene to yield the phenyldiazene radical, which is extremely unstable. The unstable phenyldiazene radical cleaves homolytically to generate a phenyl radical and molecular nitrogen. The phenyl radical might abstract hydrogen radical from •O₂H or dye degradation products to produce benzene. Phenyl radicals are not likely to be scavenged by molecular oxygen as they react very slowly with oxygen. The phenoxy radical might react with the hydroxyl radical and oxygen to result in aromatic ring degradation. Thus, the study showed that benzene generation occurs with phenylazo-substituted azo dyes. Because benzene is a priority pollutant, further treatment might be needed to reach discharge standards.

Lin and Peng (1995) investigated the treatment of textile wastewater from a large dyeing and finishing mill by a continuous process of combined chemical coagulation, Fenton's reagent, and activated sludge process. The textile wastewater reservoir held 300 L of screened raw textile wastewater. This wastewater was pumped into a chemical coagulation tank, the front section of which contained polyaluminum chloride at a concentration of 100 mg/L; a mixer operated at 150 rpm. Polymer was added to the back section of the tank at a concentration of 0.5 mg/L; the mixer speed here was 50 rpm. The retention time was 2 min in each section. The sedimentation tank followed by chemical coagulation had a retention time of 60 min. After this, the wastewater went to a Fenton's reactor (25 × 25 × 88 cm). The treated textile wastewater went from the Fenton's reactor to a second sedimentation tank for a 60-min retention time. The top textile wastewater was finally passed to an activated sludge tank, where it was mixed with equal volumes of activated sludge. A 1:1 ratio of $H_2O_2:Fe^{2+}$ was used. The decolorization of azo dyes, such as active yellow lightfast (AYL) 2KT, has been studied by Solozhenko et al. (1995). AYL elimination depended on the initial concentration of H_2O_2, and decolorization was accelerated by raising the temperature or under the influence of sunlight.

6.4   QSAR Models

Quantitative structure–activity relationships (QSARs) are important for predicting the oxidation potential of chemicals in Fenton's reaction system. To describe reactivity and physicochemical properties of the chemicals, five different molecular descriptors were applied. The dipole moment represents the polarity of a molecule and its effect on the reaction rates; E_{HOMO} and E_{LUMO} approximate the ionization potential and electron affinities, respectively; and the log P coefficient correlates the hydrophobicity, which can be an important factor relative to reactivity of substrates in aqueous media. Finally, the effect of the substituents on the reaction rates could be correlated with Hammett constants by Hammett's equation.

6.4.1   Dipole Moment

The polarity of the reacting substrate is an important molecular property that may affect the reaction rates. The polarity of a molecule can be measured by its dipole moment (μ), which is caused by different electronegativities within the atoms of a molecule. It describes the strength and orientation behavior of a molecule in an electrostatic field. The dipole moment can affect the hydrogen bonding potential of a molecule and thus the oxidation reaction rates in aqueous media. For halogenated aliphatics, the dipole moment has a significant correlation with the reaction rate constant. Larger dipole moments tend to decrease the reaction rates; Figure 6.18 shows that the dipole moment has a linear effect on the log value of the rate constant. This linear trend was found to hold for a range of H_2O_2:Fe^{2+} ratios from 2 to 10, as shown in Figure 6.18. Figure 6.19 shows the relation between the computed dipole moments of three halogenated alkanes — $CHBr_3$, $CHClBr_2$, and $CHCl_2Br$ — and experimentally determined reaction rates in Fenton's oxidation system. Chlorine atoms have considerably larger electronegativities compared to Br, and with the increase of the number chlorine atoms the dipole moment increases as well.

6.4.2 Highest Occupied Molecular Orbital Energies

The highest occupied molecular orbital (HOMO) is the highest energy level in the molecule that contains electrons. E_{HOMO} is an important factor governing molecular reactivity and properties. Molecules that react as a Lewis base, which are capable of donating an electron pair in bond formation, supply electrons from the highest occupied molecular orbital of reacting molecules. The magnitude of E_{HOMO} determines the thermodynamic conditions for donating the electron pair; therefore, the E_{HOMO} descriptor is often used to measure the nucleophilicity of a molecule. Thus, molecules with higher

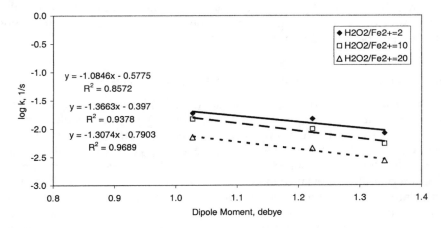

FIGURE 6.18
Rate constant vs. dipole moment for one-carbon halogenated alkanes. (Data adapted from Tang, W.Z. and Tassos, S., *Water Res.*, 31(5), 1117–1125, 1997.)

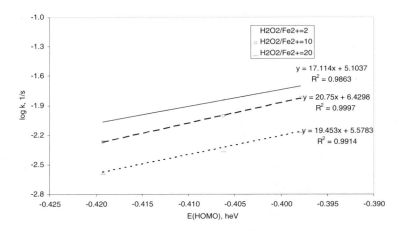

FIGURE 6.19

Rate constant vs. E_{HOMO} for halogenated alkanes. (Rate constants obtained from Tang, W.Z. and Tassos, S., *Water Res.*, 31(5), 1117–1125, 1997.)

E_{HOMO} values are expected to be relatively more reactive compared to molecules with lower E_{HOMO}.

Figure 6.19 shows the effect of computed E_{HOMO} values on the oxidation rates of three halogenated alkanes: dichlorobromomethane, chlorodibromomethane, and bromoform. As the energy of HOMO increases, the nucleophilic properties of the substrate increase. The hydroxyl radicals accept electrons and therefore act as the oxidant; they will have a greater tendency to oxidize substrates with higher E_{HOMO}. Consequently, the observed oxidation rates will be higher. The trend shown in Figure 6.19 demonstrates that E_{HOMO} can be applied to predict the oxidation properties of various substrates in Fenton's oxidation system. It is interesting to note that the trend is valid for a range of oxidant to catalyst ratios; the figure shows linearity for ratios of $H_2O_2:Fe^{2+}$ between 2 and 20.

6.4.3   Lowest Unoccupied Molecular Orbital Energies

The lowest unoccupied molecular orbital (LUMO) is the lowest energy level in the molecule that does not contain electrons. When a molecule acts as a Lewis acid in bond formation, incoming electron pairs are received in its LUMO. The value of the lowest energy level, E_{LUMO}, is a factor that determines the thermodynamic conditions under which a molecule accepts an electron pair. Its importance in governing molecular reactivity and its properties are similar to those for the HOMO energies. Molecules with lower E_{LUMO} will have a higher affinity to accept electrons than those with high E_{LUMO}. Thus, the molecular descriptor E_{LUMO} can be applied to quantify the electrophilicity of a molecule. Contrary to the effect of E_{HOMO}, higher E_{LUMO} values of substrate can be expected to have a negative effect on the reaction rates with hydroxyl radicals.

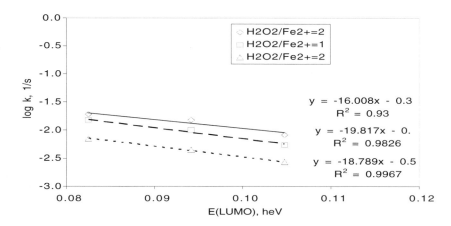

FIGURE 6.20

Rate constant vs. E_{LUMO} for one-carbon halogenated alkanes. (Rate constants obtained from Tang, W.Z. and Tassos, S., *Water Res.*, 31(5), 1117–1125, 1997.)

Figure 6.20 shows a plot of log (k) vs. E_{LUMO} for three halogenated alkanes: dichlorobromomethane, chlorodibromomethane, and bromoform. The increase of E_{LUMO} decreases the reaction rates and yields a linear relation. The relation is valid for three sets of data with variable oxidant to catalyst ratios ($H_2O_2:Fe^{2+}$) between 2 and 20.

Figure 6.21 indicates that the higher the E_{LUMO} is, the higher the rate constants during Fenton oxidation of halogenated phenols. The sign of slope is opposite to chlorinated aliphatic compounds. Because hydrogen abstraction is the dominant mechanism in degrading saturated aliphatic compounds, electrophilic addition of hydroxyl radical to the aromatic ring results in the opposite slope.

6.4.4 Octanol/Water Partition Coefficient

The octanol/water partition coefficient (log P) indicates the hydrophobicity of a substrate. Higher hydrophobicity of a chemical compound will decrease its aqueous solubility. Because hydroxyl radicals have reaction rates that are most likely diffusion controlled, the reaction rates of hydrophobic compounds in aqueous systems, such as Fenton's reaction, will be negatively affected. Figure 6.22 shows the effect of the octanol/partition coefficient on the oxidation rates of 2-chlorophenol, 2,4-dichlorophenol, 2,4,6-trichlorophenol, and pentachlorophenol. Pentachlorophenol has the lowest aqueous solubility and correspondingly its reaction rates are significantly affected.

6.4.5 Hammett's Constants

The effect of the substituents can also be expressed by Hammett's equation, which shows the value of log k/k_0, as a function of the σ values of the

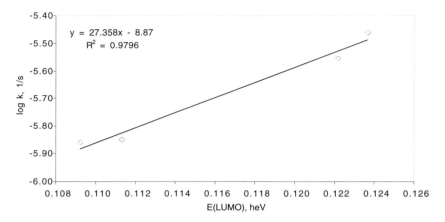

FIGURE 6.21
Rate constant vs. E_{LUMO} for halogenated phenols. (Rate constants obtained from Tang, W.Z. and Huang, C.P., *Waste Manage.*, 15(8), 615–622, 1995.)

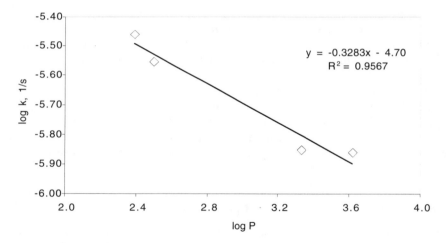

FIGURE 6.22
Rate constant vs. log *P* for halogenated phenols. (Rate constants obtained from Tang, W.Z. and Huang, C.P., *Waste Manage.*, 15(8), 615–622, 1995.)

substituent. A substituent may affect the reaction kinetics by altering the electron density at the reaction site and by steric effects. Electron–donor substituents accelerate reactions favored by high electron densities, and electron-withdrawing groups enhance reactions favored by low electron densities. Substituents can induce two categories of electronic effects: field-inductive and resonance. For aliphatic, a modified form of the σ values is used that is derived considering the inductive effects of the constituents; it is denoted by σ*.

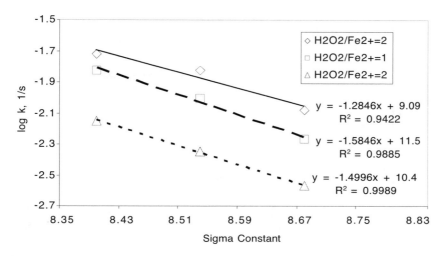

FIGURE 6.23

Rate constant vs. σ* for halogenated alkanes. (Rate constants obtained from Tang, Z.W. and Tassos, S., *Water Res.*, 31(5), 1117–1125, 1997.)

Figure 6.23 shows the effect of σ* on the reaction kinetics. Variations in the initial conditions result from varying oxidant to catalyst ratios, resulting in a shift of the linear trend with changing rate constants; however, the linear trend still holds for H_2O_2:Fe^{2+} ratios of 2 to 20.

Halogenated phenols correlate well with Hammett's constant (σ). Figure 6.24 shows the linear trend of the σ values of 3-chlorophenol, 4-chlorophenol, 3,4-dichlorophenol, and 3,5-dichlorophenol with respect to their first-order reaction rate constants. Aromatic compounds with substituents in the *meta* and *para* positions have pronounced steric differences and result in a well-defined trend. Table 6.9 summarizes all the QSAR models discussed above.

TABLE 6.9

QSAR Models for Fenton Oxidation

Chemical Class	Dataset	QSAR Model	R^2
Alkane	Dichlorobromomethane, chlorodibromomethane, bromoform	$Log\ k = -1.3074\mu - 0.7903$	0.9689
		$Log\ k = -18.789E_{LUMO} - 0.5922$	0.9967
		$Log\ k = 20.75E_{HOMO} + 6.4298$	0.9997
	Trichloromethane, 1,1,2-trichloroethane, 1,2-dichloropropane, and 1,2-dibromo-3-chloropropane	$Log\ k = -0.1645\sigma^* + 1.0086$	0.9989
		$Log\ k = -1.4996\sigma^* + 10.452$	0.9934
Phenol	3-Chlorophenol, 4-chlorophenol, 3,4-dichlorophenol, 3,5-dichlorophenol	$Log\ k = 27.358E_{LUMO} - 8.8711$	0.9796
		$Log\ k = -0.8668\sigma - 5.2609$	0.9411
		$Log\ k = 0.3283\ log\ P - 4.709$	0.9567

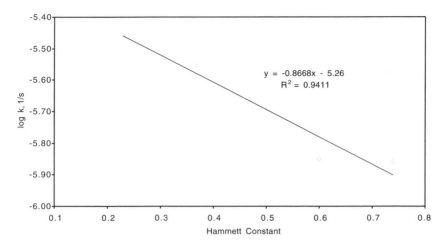

FIGURE 6.24
Rate constant vs. σ for halogenated phenols. (Rate constants obtained from Tang, W.Z. and Huang, C.P., *Waste Manage.*, 15(8), 615–622, 1995.)

6.5   Engineering Applications

Environmental applications require detoxification of hazardous substances to a level of parts per million (ppm) and even parts per billion (ppb). These purity levels, which were rarely considered in product synthesis, are now possible for wastewater due to Fenton's reagent. Fenton's oxidant is cost effective and relatively fast in destroying many toxics (Bigda, 1996). It attacks all reactive substrate concentrations under acidic conditions. Hydrogen peroxide is used to remove such contaminants as cyanide, sulfides, sulfites, chrome, and heavy metals by varying batch conditions. With an iron catalyst, the process often oxidizes organics, as well as reducing hexavalent chrome to trivalent precipitable form.

Batch reactors give great flexibility and can be designed to treat many different types of wastewater. Commercial Fenton reactors have been used for industrial wastewater treatment during the past 10 to 15 years (Bigda, 1996). An important application illustrating this is removal of phenol, formic acid, and other organics from paint stripper rinse water in the aerospace industry (Bigda, 1996). Large volumes of rinse water are generated when paint is removed from commercial and military aircraft; this rinse water contains paint flakes, dirt, pigments, solvents, oils, and chrome. Phenol levels of over 20,000 ppm in the rinse water have been reduced to less than 1 ppm using an automated Fenton system (Bigda, 1996). One commercial system has a computer with a programmable logic circuit which is programmed with 58 sequential commands to control the reactor and produce purified wastewater that meets all local and EPA regulations. The treatment process can be set for simple neutralization, oxidation of organics, removal of volatile organic compounds (VOCs), reduction of chrome VI, and precipitation of heavy metals.

Fenton's reagent can be used either as a single process or as one step in combination with other processes for wastewater treatment. For example, Fenton's reagent is used for treating specialty chemicals and process waters, in refinery and fuel terminals, and for engine and metal cleaning, groundwater detoxification, and site remediation. It can also be used to treat wastewater generated from various industries, including those manufacturing paint strippers containing phenol and formic acid, wood treating chemicals containing creosols and copper, and plastics and adhesives containing phenol formaldehyde. Wastewater from wood-preserving plants, which employ cresols, chlorophenols, arsenic, and copper compounds, can also be treated. Groundwater contaminated by BTX and phenols from refinery wastewater streams and tank bottoms at fuel terminals can also be remediated using Fenton's reaction.

Fenton's reagent has been used successfully to treat process waters generated by industries manufacturing explosives (e.g., TNT and RDX), insecticides, dyes and ink, photochemicals, pharmaceuticals, and other chemicals (Bigda, 1996). Most photochemicals can be economically treated with catalyzed peroxide. Munitions and demilitarization operations produce pink water containing explosives. Aqueous solutions of TNT, RDX, HMX, and similar substrates can also be oxidized using Fenton's reagent. Fenton's chemistry has been applied to remediate sites and groundwater contaminated with pesticides, chlorinated organics, explosives, and a variety of other toxics. Processes have also been cited to treat wastewater generated from soil washing.

Fenton's reagent is applicable in highly toxic environments where traditional technologies such as biological treatment processes are not effective. In addition, Fenton's reaction exhibits considerably faster kinetic patterns compared to aerobic or anaerobic biological processes. In decolorizing dye wastewaters, many commercial dyes are toxic to the microorganisms involved; Fenton's reagent can effectively decolorize textile wastewaters and substrates that may include high-molecular-weight polycyclic aromatic hydrocarbon (PAHs) and perchloroethylene (PCE). Fenton's reagent has an advantage over UV/H_2O_2 methods, where colored organic compounds interfere with the UV/H_2O_2 process due to UV light absorption. Fenton's reagent is a chemical destruction process; thus it destroys the contaminant rather than transferring pollutants from one phase to another. Capital equipment costs for application of Fenton's reagent can be expected to be very low compared to UV/H_2O_2 process. The primary factor contributing to chemical cost is the cost of hydrogen peroxide; it is important due to this reason to optimize the amount of peroxide required. Fenton's reagent has lower operation and maintenance costs compared to the ozonation methods, because the operating costs for Fenton's reaction depend primarily on the reagent concentrations and have very low energy requirements as opposed to the high energy requirements of ozonation.

The optimum pH range for effective treatment by Fenton's reagent is in the acidic range of pH higher than 2 but lower than 4. This implies that a posttreatment step is required for pH adjustment. This treatment technology may not be applicable to alkaline soils and sludges with strong buffering capacities due to scavengers such as HCO_3^- and CO_3^-. Proper sludge disposal is required after the treatment. Fenton's reaction has very low reaction rates for the following compounds: acetic acid, acetone, carbon tetrachloride, chloroform, maleic acid, malonic acid, methylene chloride, oxalic acid, *n*-paraffins, tetrachloroethane, and trichloroethane (Bigda, 1996). Fenton's reaction is highly exothermic. At temperatures lower than 65°F, Fenton's reaction may exhibit slower starting kinetics. Because initiation of the process may be sudden, especially if the peroxide concentration is high, care must be taken to add the peroxide slowly before temperature rises above 80°F.

References

Anbar, M. and Neta, P., A compilation of specific bimolecular rate constants for the reaction of hydrated electrons, hydrogen atoms and hydroxyl radicals with inorganic and organic compounds in aqueous solution, *Int. J. Appl. Radiat. Isotopes*, 18, 493–523, 1967.

Arnold, S.M., Hickey, W.J., and Harris, R.F., Degradation of atrazine by Fenton's reagent: condition optimization and product quantification, *Environ. Sci. Technol.*, 29, 2083, 1995.

Barb, W.G., Baxendale, J.H., George, P., and Hargrave, K.R., Reactions of ferrous and ferric ions with hydrogen peroxide, *Trans. Faraday Soc.*, 47, 591, 1951.

Barbeni, M., Minero, C., Pelizzetti, E., Borgarello, E., and Serpone, N., Chemical degradation of chlorophenols with Fenton's reagent, *Chemosphere*, 16, 2225–2237, 1987.

Baxendale, J.H. and Wilson, J.A., The photolysis of hydrogen peroxide at high light intensity, *Trans. Faraday Soc.*, 53, 344–356, 1957.

Bigda, J.R., Fenton's chemistry: an effective advanced oxidation process, *Natl. Environ. J.*, 6(3), 36–39, 1996.

Cross, C.F., Bevan, E.J., and Herberg, T., *Ber.*, 33, 2015, 1900.

Dorfman, C.M., Taumb, I.A., and Buhler, R.E., Pulse radiolysis studies. I. Transient spectra and reaction rate constant in irradiated aqueous solution of benzene, *J. Chem. Phys.*, 36, 3051–3061, 1962.

Eisenhauer, H.R., Oxidation of phenolic wastes, *J. Water Pollut. Contr. Fed.*, 36, 1116–1128, 1964.

Fenton, H.J.H., Oxidation of tartaric acid in the presence of iron, *J. Chem. Soc.*, 65, 899, 1894.

Fenton, H.J.H., Oxidation of certain organic acids in the presence of ferrous salts, *Proc. Chem. Soc.*, 15, 224, 1899.

Fenton, H.J.H., Oxidation of organic acids in presence of iron, *Chem. Soc. J.*, 77, 69–76, 1900.

Fenton, H.J.H. and Jackson, H., Oxidation of polyhydric alcohols in presence of iron, *J. Chem. Soc.*, 75, 1–11, 1899.

Garroguino, M.L. Gersberg, R.M., Dawsey, W.J., and Bradly, M.D., *Environ. Contam. Toxicol.*, 49, 241, 1992.

Goldhammer, *Biochemistry*, 85, 189, 1927.

Graedel, T.E., Mandich, M.L. and Weschler, C.J., Kinetic model studies of atmospheric droplet chemistry, *J. Geophys. Res.*, 91, 5205–5221, 1986.

Haber, F. and Weiss, J., The catalytic decomposition of hydrogen peroxide by iron salts, *J. Proc. Royal Soc. Assoc.*, 147, 332, 1934.

Huang, C.P., Dong, C., and Tang, Z., Advanced chemical oxidation: its present role and potential future in hazardous waste treatment, *Waste Manage.*, 13, 361–377, 1993.

Ingles, D.L., Studies of oxidations by Fenton's reagent using redox titration, *Australian J. Chem.*, 25, 87–95, 1972.

Kearney, P.C., Zeng, Q., and Ruth, J.C., *Treatment and Disposal of Pesticide Waste*, American Chemical Society, Washington, D.C., 1984, pp. 195–209.

Kearney, P.C., Muldoon, M.T., Somich, C.J., Ruth, J.M., and Voaden, D.J., *J. Agric. Food Chem.*, 36, 1301–1306, 1988.

Keating, E.J. et al., Phenolic problems solved with hydrogen peroxide and a ferrous salt reagent, in *Proc. of the 33rd Industrial Waste Conf.*, Purdue University, W. Lafayette, IN, 1979.

Kirk, R.E. and Othmer, E.F., Hydrogen peroxide, in *Encyclopedia of Chemical Technology*, 3rd ed., Vol. 5, Wiley, New York, 1979, pp. 745–753.

Kitajima, N., Fukuzumi, S., and Ono, Y., Formation of superoxide ion during the decomposition of hydrogen peroxide on supported metal oxides, *J. Phys. Chem.*, 82, 1505, 1978.

Kiwi, J., Pulgarin, C., and Peringer, P., Effect of Fenton and photo-Fenton reactions on the degradation and biodegradability of 2 and 4-nitrophenols in water treatment, *Appl. Catalysis*, 3, 335–350, 1994.

Kremer, M.L., Nature of intermediates in the catalytic decomposition of hydrogen peroxide by ferric ion, *Trans. Faraday Soc.*, 58, 702–706, 1962.

Kremer, M.L., *Trans. Faraday Soc.*, 59, 2535, 1963.

Kremer, M.L., Complex versus free radical mechanism for the catalytic decomposition of H_2O_2 by ferric ions, *Int. J. Chem. Kinetics*, 17, 1299–1314, 1975.

Kremer, M.L. and Stein, G., *Trans. Faraday Soc.*, 55, 959, 1959.

Kremer, M.L. and Stein, G., Kinetics of the Fe^{3+} ion–H_2O_2 reaction: steady-state and terminal-state analysis, *Int. J. Chem. Kinetics*, 9, 179–184, 1977.

Kuo, W.G., Decolorizing dye wastewater with Fenton's reagent, *Water Res.*, 26, 881, 1992.

Larson, R.A. and Waber, E.J., *Reaction Mechanisms in Environmental Organic Chemistry*, CRC Press, Boca Raton, FL, 130–132, 1994.

Lin, H.S. and Peng, F.C., A continuous Fenton's process for treatment of textile wastewater, *J. Environ. Technol.*, 16, 693–699, 1995.

Lou, C.J. and Lee, S.S., Chemical oxidation of BTX using Fenton's reagent, *J. Hazardous Waste Hazardous Mater.*, 12(2), 185–193, 1995.

Martin, L.R., Easton, M.P., Foster, J.W., and Hill, M.W., Oxidation of hydroxymethane-sulfonic acid by Fenton's reagent, *Atmospheric Environ.*, 23(3), 563, 1989.

Matsue, T., Fujihira, M., and Osa, T., Oxidation of alkylbenzenes by electrogenerated hydroxyl radical, *J. Electrochem. Soc.: Electrochem. Sci. Technol. (Japan)*, 128, 2565–2569, 1981.

Merz, J.H. and Waters, W.A., Mechanism of oxidation of alcohols with Fenton's reagent, *Disc. Faraday Soc.*, 2, 179–188, 1947.

Merz, J.H. and Waters, W.A., The oxidation of aromatic compounds by means of the free hydroxyl radical, *J. Chem. Soc.*, 15, 2427–2433, 1949.

Metelitsa, D.I., Mechanism of the hydroxylation of aromatic compounds, *Russian Chem. Rev.*, 40, 7, 1971.

Mohanty, N.R. and Wei, I.W., Oxidation of 2,4-dinitrotoluene using Fenton's reagent: reaction mechanisms and their practical applications, *J. Hazardous Waste Hazardous Mater.*, 10(2), 171, 1993.

Pelizzetti, E., Minero, C., Carlin, V., Vincenti, M., Parmauro, E., and Dolci, M., *Chemosphere*, 24, 891–910, 1971.

Pieken, W.A. and Kozarich, J.W., *J. Org. Chem.*, 54, 510, 1989.

Pignatello, J.J. and Baehr, K., Ferric complexes as catalysts for Fenton degradation of 2,4-D and metachlor in soil, *J. Environ. Qual.*, 23, 365–370, 1994.

Pospisil, J. and Ettel, V., Oxidation of pyrocatechol to muconic acid, *Chem. Prumysl (Czechoslovakia)*, 7, 244, 1957.

Potter, F.J. and Roth, J.A., Oxidation of chlorinated phenols using Fenton's reagent, *J. Hazardous Waste Hazardous Mater.*, 10(2), 151, 1993.

Sedlak, D.L. and Andren, A.W., Oxidation of chlorobenzene with Fenton's reagent, *Environ. Sci. Technol.*, 25, 777–782, 1991a.

Sedlak, D.L. and Andren, A.W., Aqueous-phase oxidation of polychlorinated biphenyls by hydroxyl radicals, *Environ. Sci. Technol.*, 25, 1419–1427, 1991b.

Sedlak, D.L. and Andren, A.W., The effect of sorption on oxidation of polychlorinated biphenyls (PCBs) by hydroxyl radicals, *Water Res.*, 28(5), 1207–1215, 1994.

Sharp, D.W.A., *The Penguin Dictionary of Chemistry*, 2nd ed., Penguin Books, London, 1990.

Solomons, T.W.G., *Organic Chemistry*, 4th ed., John Wiley & Sons, New York, 1988, pp. 137–138.

Solozhenko, E.G., Soboleva, N.M., and Goncharuk, V.V., Decolorization of azodye solution by Fenton's oxidation, *Water Res.*, 29, 2206, 1995.

Spadaro, J.J., Isabella, L., and Renganathan, V., Hydroxyl radical mediated degradation of azo dyes: evidence for benzene generation, *Environ. Sci. Technol.*, 28, 1389–1393, 1994.

Stein, G. and Weiss, J., Some free radical reaction of phenol: the action of hydrogen peroxide–ferrous salt reagent and of x-rays on aqueous solutions of phenols, *J. Chem. Soc. (British)*, 3265, 1951.

Sudoh, M., Kodera, T., Sakai, K., Zhang, J.Q., and Koide, K., Oxidative degradation of aqueous phenol effluent with electrogenerated Fenton's reagent, *J. Chem. Eng. Jpn.*, 19(6), 513–518, 1986.

Sundstrom, D.W., Klei, H.E., Nalette, T.A., Reidy, D.J., and Weir, B.A., Destruction of halogenated aliphatic by ultraviolet catalyzed oxidation with hydrogen peroxide, *J. Hazardous Waste Hazardous Mater.*, 3, 101, 1986.

Sung, W. and Morgan, J.J., Kinetics and product of ferrous iron oxygenation in aqueous systems, *Environ. Sci. Technol.*, 14, 561–568, 1980.

Tang, W.Z. and Chen, R., Decolorization kinetics and mechanisms of commercial dyes by H_2O_2/Fe powder system, *Chemosphere*, 32(5), 947–958, 1996.

Tang, W.Z. and Huang, C.P., Photocatalytic oxidation of organic compounds by CdS: improvement of the photostability of CdS, in *Environmental Engineering*, O'Melia, C.R., Ed., Proc. of the 1990 ASCE Specialty Conference, Arlington, VA, July 8–11, 1990, pp. 917–918.

Tang, W.Z. and Huang, C.P., pH effect on the oxidation pathways during photocatalytic oxidation of 2,4-dichlorophenols by CdS, paper presented at *Emerging Technologies in Hazardous Waste Management V*, Atlanta, GA, Sept. 27–29, 1992.

Tang, W.Z. and Huang, C.P., The effect of chlorine position of chlorinated phenols on their dechlorination kinetics by Fenton's reagent, *Waste Manage.*, 15(8), 615–622, 1995.

Tang, W.Z. and Huang, C.P., 2,4-Dichlorophenol oxidation kinetics by Fenton's reagent, *Environ. Technol.*, 17, 1371–1378, 1996a.

Tang, W.Z. and Huang, C.P., Effect of chlorine content of chlorinated phenols on their oxidation kinetics by Fenton's reagent, *Chemosphere*, 33(8), 1621–1635, 1996b.

Tang, W.Z. and Tassos, S., Oxidation kinetics and mechanisms of trihalomethanes by Fenton's reagent, *Water Res.*, 31(5), 1117–1125, 1997.

Venkatadri, R. and Peters, R.W., Chemical oxidation technologies: ultraviolet light/ hydrogen peroxide, Fenton's reagent, and titanium dioxide: assisted photocatalysis, *J. Hazardous Waste Hazardous Mater.*, 10, 107–149, 1993.

Walling, C., Fenton's reagent revisited, *Acc. Chem. Res.*, 8, 125–131, 1975.

Walling, C. and Cleary, M., Oxygen evolution as a critical test of mechanism in the ferric-ion catalyzed decomposition of hydrogen peroxide, *Int. J. Chem. Kinetics*, 9, 595–601, 1977.

Walling, C. and Goosen, A., Mechanism of the ferric ion catalyzed decomposition of hydrogen peroxide: effect of organic substrates, *J. Amer. Chem. Soc.*, 95(9), 2987–2991, 1973.

Walling, C. and Johnson, R.A., Fenton's reagent V: hydroxylation and side-chain cleavage of aromatics, *J. Am. Chem. Soc.*, 97(2), 363–367, 1975.

Walling, C. and Kato, S., The oxidation of alcohols by Fenton's reagent: the effect of copper ion, *J. Amer. Chem. Soc.*, 93, 4275, 1971.

Walling, C., El-Taliawi, G.M., and Johnson, R.A., Fenton's reagent IV: structure and reactivities in the reactions hydroxyl radicals and the redox reactions of radicals, *J. Am. Chem. Soc.*, 96, 133–139, 1974.

Walling, C., Camaioni, D.M., and Kim, S.S., Aromatic hydroxylation by peroxydisulfate, *J. Am. Chem. Soc.*, 100, 4814–4818, 1978.

Waters, W.A., *Mechanisms of Oxidation of Organic Compounds*, Spottiswoode Ballantyne, London, 1964, pp. 29–48.

Watts, R.J., Udell, M.D., and Rauch, P.A., Treatment of pentachlorophenol-contaminated soils using Fenton's reagent, *J. Hazardous Waste Hazardous Mater.*, 7, 4, 335–345, 1990.

Watts, R.J., Kong, S., Dippre, M., and Barnes, W.T., Oxidation of sorbed hexachlorobenzene using catalyzed hydrogen peroxide, *J. Hazard. Mater.*, 39, 33–37, 1994.

Watts, R.J., Jones, A.P., Chen, P., and Kenny, A., Mineral catalyzed Fenton-like oxidation of sorbed chlorobenzenes, *Water Res.*, 69(3), 269–275, 1997.

Wieland, H. and Franke, W., Mechanism of oxidation processes. XIV. Activation of oxygen by iron, *Ann. (Germany)*, 464, 101, 1928.

Yoshida, K., Shigeoka, T., and Yamauchi, F., Evaluation of aquatic environmental fate of 2,4,6-trichlorophenol with a mathematic model, *Chemosphere*, 15, 825–860, 1986.

7

Ultraviolet/Hydrogen Peroxide

7.1 Introduction

Ultraviolet/hydrogen peroxide process (UV/H_2O_2) can be applied to remove a wide range of toxic hazardous substances from groundwater, drinking water, and municipal and industrial wastewaters. After Koubek (1975) at the U.S. Naval Academy was awarded a patent for his work on "oxidation of refractory organics in aqueous waste streams by hydrogen peroxide and ultraviolet light," the first commercial application of UV/H_2O_2 processes emerged in 1977. Since then, the UV/H_2O_2 process has been successfully applied for groundwater remediation and wastewater treatment. This chapter discusses four major topics of the UV/H_2O_2 process: (1) fundamental principles involved in the process; (2) degradation kinetics and mechanisms of major classes of organic pollutants; (3) QSAR models for different classes of organic compounds and their treatability; and (4) engineering applications as impacted by water quality and various operation parameters. Table 7.1 briefly summarizes the major research areas of the past several decades.

7.2 Fundamental Theory

Ultraviolet light without hydrogen peroxide is not very effective for the degradation of organics. When UV light and H_2O_2 are combined, the overall oxidative reaction potential is greatly enhanced even under ambient temperature and pressure. The efficiency of direct photolysis depends on: (1) UV absorbance by substrates; (2) quantum yield of photolysis; (3) presence of other competitive UV absorbents; and (4) intensity of UV sources.

UV/H_2O_2 process may degrade organic contaminants either directly by photolysis or indirectly by hydroxyl radicals. If the photon wavelength is greater than 254 nm, hydroxyl radicals are largely responsible for initiating oxidation reactions. The hydroxyl radical is a short-lived, extremely potent

TABLE 7.1

Historical Aspects in UV/H_2O_2 Research

Period	Description of Study
1950s	Hydrogen peroxide photolysis was studied (Baxendale and Wilson, 1957).
1960s	Radical chain reaction was established in a hydrogen peroxide solution with UV radiation (Sehested et al., 1968).
1970s	Direct photoreaction for carbaryl, a carbamate insecticide, and trifluoroaniline, a dinitroaniline type herbicide, was studied at different wavelengths of sunlight.
1980s	Prengle (1983) reported quantitatively overall degradation of the parent species, oxidation of intermediates to form secondary intermediate and fragment, and further oxidation to form small and stable organic acid.
	Braun et al. (1988) developed a computer program named ACUCHEM for modeling complex reaction systems.
	Leifer (1988) reviewed fundamental theory and practice of the kinetics of aquatic reactions to express relevant direct and indirect photochemical reactions in natural water.
	Glaze and Kang (1989a,b) presented a model describing the photooxidation of water-soluble hazardous organic waste and also observed the rate increase with a decrease in pH of reaction medium.
	UV/H_2O_2 oxidation of aromatic hydrocarbon and phenolic compound and an empirical power-law rate expression were used for the destruction of such compounds (Sundstrom et al., 1989).
1990s	USEPA (1990) completed the applications analysis report on Ultrox International Ultaviolet Radiation/Oxidation Technology.
	A kinetic model for H_2O_2/UV process was developed using ACUCHEM software (Yao et al., 1992).
	Froelich (1992) described the UV/H_2O_2 commercial system, Perox-Pure™, according to operating data obtained from full-scale treatment systems.
	A kinetic model for the UV/H_2O_2 process based on UV/H_2O_2-induced OH radical oxidation of butylchloride in the presence of humic acid was developed (Liao and Gurol, 1995).
	OH radical reaction rate constants for phenol were estimated by De et al. (1999).
2000s	UV/H_2O_2 was compared with other AOPs according to kinetic rate constants by Beltran-Heredia et al. (2001).

oxidizing agent capable of oxidizing organic compounds, primarily by either hydroxylation or hydrogen abstraction. These reactions generate organic radicals, which continuously react with OH• to produce the final products, such as carbon dioxide, water, and inorganic salts. The rate and efficiency of generating hydroxyl radicals primarily depend on (1) the energy required to homolyze a given chemical bond, and, to a large extent, (2) the concentration of hydrogen peroxide.

7.2.1 H_2O_2 Photolysis

The photolysis of H_2O_2 is accomplished through the cleavage of the molecule to generate two hydroxyl radicals by each quantum of radiation absorbed as shown in Equation (7.1):

$$H_2O_2 + h\nu \rightarrow 2HO^\bullet \quad d[H_2O_2]/dt = \phi_{PER}I_a \tag{7.1}$$

The photolysis rate of aqueous H_2O_2 is strongly dependent on pH, while alkaline conditions will generally enhance the process. This dependence is most likely due to the higher molar absorption coefficient (ε_{PER}) of the peroxide anion at 253.7 nm. When photon wavelength is shorter than 242.0 nm, photolysis of water is energetically possible:

$$H_2O + h\nu \rightarrow HO^\bullet + H^\bullet \tag{7.2}$$

Under even shorter wavelengths (e.g., below 192 nm), hydrogen peroxide may form a hydroxyl radical by emitting an electron; however, considering that conventional mercury lamps are not capable of emitting such low wavelengths, it is unlikely that this reaction pathway bears significance in the overall oxidation sequence.

$$2H_2O_2 + h\nu \rightarrow e^- + HO^\bullet + H_3O^+ \tag{7.3}$$

The enhanced oxidizing ability of H_2O_2 under UV light may be attributed to the production of hydroxyl radicals ($HO^\bullet$), perhydroxyl radicals ($HO_2^\bullet$), and superoxide anion ($O_2^{\bullet-}$), as shown in the following reactions.

Propagation:

$$HO^\bullet + H_2O_2/HO_2^- \rightarrow HO_2^\bullet/O_2^- + H_2O, \text{ where}$$
$$k = 2.7 \times 10^7 + 7.5 \times 10^9 \times 10^{pH-11.8} \ M^{-1} \ s^{-1} \tag{7.4}$$

$$HO_2^\bullet + H_2O_2 \rightarrow HO^\bullet + H_2O + O_2, \ k = 3.7 \ M^{-1} \ s^{-1} \tag{7.5}$$

$$HO_2^\bullet + HO_2^- \rightarrow HO^\bullet + OH^- + O_2 \tag{7.6}$$

Termination:

$$HO_2^\bullet + HO_2^\bullet \rightarrow H_2O_2 + O_2, \text{ where } k_7 = 8.3 \times 10^5 \ M^{-1} \ s^{-1} \tag{7.7}$$

$$O_2^- + HO_2^\bullet \rightarrow HO_2^- + O_2 \tag{7.8}$$

$$HO^\bullet + HO_2^\bullet \rightarrow H_2O + O_2, \ k = 8 \times 10^9 \ M^{-1} \ s^{-1} \tag{7.9}$$

The perhydroxyl radical is a relatively weak and short-lived oxidizing agent. Together with their conjugate bases (i.e., $HO_2^{\bullet}/O_2^{\bullet-}$), these radicals disappear in aqueous solution in the absence of the other reactants by pH-dependent disproportionation. In addition, the disproportionation reaction of $HO_2^{\bullet}/O_2^{\bullet-}$ contributes to the regeneration of H_2O_2, as shown below:

$$HO_2^{\bullet} + O_2^{\bullet-} + H_2O \rightarrow H_2O_2 + O_2 + OH^-, \ k = 9.7 \times 10^7 \ M^{-1} \ s^{-1}$$

$$(7.10)$$

$$HO_2^{\bullet} \Leftrightarrow O_2^{\bullet-} + H^+, \ pK_a = 4.8 \qquad (7.11)$$

Termination reactions involving reactions of $HO^{\bullet}$ with $HO_2^{\bullet}$ and $O_2^{\bullet-}$ usually reduce the overall reaction rates. Because UV-irradiated aqueous H_2O_2 solutions contain a complex mixture of transient radicals, the observed reaction orders are usually not contributed by a single oxidation reaction of a given radical species (Beltran et al., 1996a–c). In addition, hydrogen peroxide can also dissociate by a dismutation reaction, as shown in Equation (7.12), with the maximum rate occurring at a pH equal to its pK_a value:

$$H_2O_2 + HO_2^- \rightarrow H_2O + O_2 + HO^{\bullet} \qquad (7.12)$$

For the convenience of discussion, all the elementary reactions involved in UV/H_2O_2 processes are listed in Table 7.2.

To assess the efficiency of the reactions listed in Table 7.2, Bolton et al. (1996) proposed a generally applicable standard for a given photochemical process. The proposed standard provides a direct link to the electrical efficiency. In this model, electrical energy per unit mass is calculated according to the quantum yield of the direct photolysis rate. Braun et al. (1997) calculated the quantum yield (Φ) according to Equation (7.13):

$$\Phi = \frac{\text{Number of converted molecules per unit time}}{\text{Number of absorbed photons per unit time}} = \frac{kC_0}{P_0(1-10^{-A(\lambda)})} \qquad (7.13)$$

where k is rate constant, C_0 is the concentration at the beginning of reaction ($t = 0$), P_0 is the emitted light intensity, and $A(\lambda)$ is the absorbance of the solution at the irradiation wavelength λ. Therefore, the quantum yield for the UV/H_2O_2 process, which can be defined as the number of moles of hydrogen peroxide decomposed per mole of light photon absorbed, has been estimated as 1.0 for the overall quantum yield ($\Phi_{PER,T}$) and as 0.5 for the primary quantum yield (Φ_{PER}). The primary quantum yield for peroxide decomposition is defined as the fraction of photon absorbed, resulting in a certain primary process. The primary quantum yield sometimes differs from the experimentally observed overall quantum yield, $\Phi_{PER,T}$, due to the free-radical chain reactions, as listed above. In hydrogen peroxide-rich solutions, the quantum yield for the decomposition of hydrogen peroxide should be unity in water containing hydroxyl radical scavengers, as well as pure water systems.

According to the definition of quantum yield and the Beer–Lambert law, the overall decomposition rate of H_2O_2 in pure water, where hydrogen peroxide is the only absorber, can be described as follows:

$$R_{H_2O_2} = \Phi_{PER,T} I_a = \Phi_{PER,T} I_0 \left[1 - \exp\left(-2.303 \varepsilon_{PER} b [H_2O_2] \right) \right]$$

where I_a is the light intensity absorbed by H_2O_2; I_0 is the incident light intensity; ε_{PER} is the absorptivity (or extinction coefficient) of H_2O_2, which is 17.9 M^{-1} cm^{-1} at 254 nm (Glaze et al., 1995); b is the effective light path length of the photoreactor that can be measured by dosing low concentrations of initial peroxide; and $[H_2O_2]$ is the concentration of H_2O_2. The rate of H_2O_2 decomposition can also be described as:

$$R_{H_2O_2} = \left(2.303 \Phi_{PER,T} \varepsilon_{PER} b I_0 f_{PER} \right) [H_2O_2] \tag{7.14}$$

where

$$f_{PER} = \frac{1 - 10^{-\varepsilon_{PER} b [H_2O_2]}}{2.303 \varepsilon_{PER} b [H_2O_2]} \tag{7.15}$$

The fraction of radiation absorbed by H_2O_2, f_{PER}, approaches 1 at relatively small concentrations of H_2O_2.

7.2.2 Degradation Mechanism of Organic Pollutants

In treating organic wastewaters, substantial differences exist between the reaction mechanisms of UV/H_2O_2 oxidation and UV irradiation or H_2O_2 oxidation alone.

7.2.2.1 *Photolysis*

When wastewater containing organic pollutants is subjected to UV irradiation, simultaneous photolysis of reactant and oxidant initiates the reaction by generating excited species of the substrate (M*) and free-radical oxidants (B*). Many mercury-vapor lamps used to produce photons are theoretically sufficient to disrupt π electrons in double bonds and break the single bonds of alkanes. When bombarding large molecules with photons, the molecules absorb photons and subsequently dissipate the energy through bond vibration. In other words, big molecules have a greater tendency to wobble away energy than to use it in reactions. In addition, bond cleavage in a molecule depends not only upon bond energies, but also on the stability of the resulting radical. In general, the process is subject to the effects of radical scavengers and competitive UV absorbers, which may limit its effectiveness.

The excited states of substrate species are very reactive. They react with available free-radical oxidants with subsequent generation of partially oxidized intermediates. Although organic compounds are assumed to react predominantly with $HO^\bullet$ and $HO_2^\bullet/O_2^{\bullet-}$, degradation of certain compounds can also take place directly by activation caused by UV, which improves the ability of the organics to be oxidized by H_2O_2 or hydroxyl radicals. Certain aromatics and olefins can also react with H_2O_2 directly. Phenol is an extensively studied example. These reactions may further enhance the oxidation of aromatics and olefins. The reaction intermediates (Int) are unstable and are further oxidized to mineral species (CO_2, H_2O, and HCl):

$$M_{(reactant)} + h\nu \rightarrow M^* \tag{7.16}$$

$$B_{(oxidants)} + h\nu \rightarrow B^*_{(free\ radical\ oxidants)} \tag{7.17}$$

$$M^* + B^* \rightarrow Int_{(partially\ oxidized\ intermediates)} \tag{7.18}$$

$$Int + B^* + h\nu \rightarrow CO_2 + H_2O + HCl \tag{7.19}$$

The rate of direct photolysis of a substrate is:

$$\left[-\frac{d[M]}{dt} \right]_{M,UV} = \Phi_M I_{a,M} = \Phi_M f_M I_0 [1 - \exp(-2.3 A_t)] \tag{7.20}$$

where Φ_M is the quantum yield of M for the direct photolysis, f_M is the fraction of radiation absorbed by M, and A_t is the total absorbency of the solution. Even though M is the only absorber present initially, if by-products of M act as strong UV absorbers, shielding M from photons, the net effect will be reduced.

The reaction of photolysis may proceed under two distinct conditions (Kang and Lee, 1997): (1) the concentration of by-product or the UV absorptivity is negligibly small, so f_M is close to 1; (2) the UV absorbance of the by-product is significantly high, shielding effective UV radiation for the photolysis of M. Under the first condition, Equation (7.20) can be integrated to yield Equation (7.21):

$$\ln \frac{[\exp(2.3\varepsilon_M b[M]) - 1]}{[\exp(2.3\varepsilon_M b[M]_0) - 1]} = -2.3\Phi_M I_0 \varepsilon_M bt \tag{7.21}$$

TABLE 7.2

Elementary Reactions in UV/H_2O_2 System

No.	Reactions	Rate Constant ($M^{-1}\,s^{-1}$)	Ref.
1	$H_2O_2 / HO_2^- + h\upsilon \rightarrow 2HO^\bullet$	k_1 is the measure for specific system (s). or $r_{UV,H2O2} = -r_{HOA}/2 = \phi_{H2O2}\,I_0$ $f_{H2O2}\,(1-e^{-A})$ $A = 2.303b(\varepsilon_{H2O2}C_{H2O2}+\varepsilon_{R1}C_{R1} + \varepsilon_{R2}C_{R2}+ \varepsilon_s C_s + \varepsilon_{HO2^-}C_{HO2^-})$. $f_{H2O2} = 2.303b(\varepsilon_{H2O2}\,C_{H2O2}+ \varepsilon_{HO2^-}C_{HO2^-})/A$. $\varepsilon_{H2O2} = 17.9-19.6\,M^{-1}\,cm^{-1}$, $\varepsilon_{HO2^-} = 228\,M^{-1}\,cm^{-1}$. $\phi_{H2O2} = \phi_{HO}{}^{2-} = 0.5$.	a
2	$H_2O_2 + HO^\bullet \rightarrow H_2O + HO_2^\bullet$	$k_2 = 2.7 \times 10^7$	Buxton et al. (1988)
3	$HO^\bullet + HO_2^- \rightarrow HO_2^\bullet + OH^-$	$k_3 = 5 \times 10^9$	Christensen et al. (1982)
4	$H_2O_2 + HO_2^- \rightarrow HO^\bullet + H_2O + O_2$	$k_4 = 3.0$	Koppenol et al. (1978)
5	$H_2O_2 + O_2^{\bullet-} \rightarrow HO^\bullet + O_2 + OH^-$	$k_5 = 0.13$	Weinslein et al. (1979)
6	$HO^\bullet + CO_3^{2-} \rightarrow CO_3^{\bullet-} + OH^-$	$k_6 = 3.9 \times 10^8$	Buxton et al. (1988)
7	$HO^\bullet + HCO_3^- \rightarrow CO_3^{\bullet-} + H_2O$	$k_7 = 8.5 \times 10^6$	Buxton et al. (1988)
8	$HO^\bullet + HPO_4^{2-} \rightarrow HPO_4^{\bullet-} + OH^-$	$k_8 = 1.5 \times 10^5$	Maruthamuthu et al. (1978)
9	$HO^\bullet + H_2PO_4^- \rightarrow HPO_4^{\bullet-} + H_2O$	$k_9 = 2.0 \times 10^4$	Maruthamuthu et al. (1978)
10	$H_2O_2 + CO_3^{\bullet-} \rightarrow HCO_3^- + HO_2^\bullet$	$k_{10} = 4.3 \times 10^5$	Draganic et al. (1991)
11	$HO_2^- + CO_3^{\bullet-} \rightarrow CO_3^{2-} + HO_2$	$k_{11} = 3.0 \times 10^7$	Draganic et al. (1991)
12	$H_2O_2 + HPO_4^{\bullet-} \rightarrow H_2PO_4^- + HO_2^\bullet$	$k_{12} = 2.7 \times 10^7$	Nakashima et al. (1970)
13	$HO^\bullet + HO^\bullet \rightarrow H_2O_2$	$k_{13} = 5.5 \times 10^9$	Buxton et al. (1988)
14	$HO^\bullet + HO^\bullet \rightarrow H_2O + O_2$	$k_{14} = 6.6 \times 10^9$	Sehested et al. (1968)
15	$HO_2^- + HO_2^\bullet \rightarrow H_2O_2 + O_2$	$k_{15} = 8.3 \times 10^5$	Bielski et al. (1985)
16	$HO_2^- + O_2^{\bullet-} \rightarrow HO_2^- + O_2$	$k_{16} = 9.7 \times 10^7$	Bielski et al. (1985)
17	$HO^\bullet + O_2^{\bullet-} \rightarrow O_2 + OH^-$	$k_{17} = 7.0 \times 10^9$	Beck (1969)
18	$HO^\bullet + CO_3^{\bullet-} \rightarrow ?$	$k_{18} = 3.0 \times 10^9$	Holcman et al. (1987)
19	$CO_3^{\bullet-} + O_2^{\bullet-} \rightarrow CO_3^{2-} + O_2$	$k_{19} = 6.0 \times 10^8$	Eriken et al. (1985)
20	$CO_3^{\bullet-} + CO_3^{\bullet-} \rightarrow ?$	$k_{20} = 3.0 \times 10^7$	Huie et al. (1990)
21	$HO^\bullet + R_1 \rightarrow ?$	k_{21}	b
22	$HO^\bullet + R_2 \rightarrow ?$	k_{22}	b

-- *continued*

TABLE 7.2 *(continued)*

Elementary Reactions in UV/H_2O_2 System

No.	Reactions	Rate Constant ($M^{-1}\,s^{-1}$)	Ref.
23	$HO^\bullet + R_1H \rightarrow ?$	K_{23}	[b]
24	$HO^\bullet + R_2H \rightarrow ?$	k_{24}	[b]
25	$HO^\bullet + S \rightarrow ?$	k_{25}	[b]
26	$CO_3^{\bullet-} + R_1 \rightarrow ?$	k_{26}	[b]
27	$CO_3^{\bullet-} + R_2 \rightarrow ?$	k_{27}	[b]
28	$HPO_4^{\bullet-} + R_1 \rightarrow ?$	k_{28}	[b]
29	$HPO_4^{\bullet-} + R_2 \rightarrow ?$	k_{29}	[b]
30	$O_2^{\bullet-} + R_1 \rightarrow ?$	K_{30}	[b]
31	$O_2^{\bullet-} + R_2 \rightarrow ?$	K_{31}	[b]
32	$HO_2^\bullet + R_1 \rightarrow ?$	k_{32}	[b]
33	$HO_2^\bullet + R_2 \rightarrow ?$	k_{33}	[b]
34	$R_1 + h\upsilon \rightarrow ?$	k_{34} is the measure for a specific system (s). $r_{UV,R1} = \phi_{R1}I_0 f_{R1}(1 - e^{-A})$. $f_{R1} = 2.303\, b\varepsilon_{R1}c_{R1}/A$.	[a]
35	$R_1 + h\upsilon \rightarrow ?$	k_{35} is the measure for a specific system (s). $r_{UV,R1} = \phi_{R2}I_0 f_{R2}(1 - e^{-A})$. $f_{R2} = 2.303b\varepsilon_{R2}c_{R2}/A$.	[a]
36	$S + h\upsilon \rightarrow ?$	$k_{36} = 0$ (measure for specific system (s). $r_{UV,R1} = \phi_{R2}I_0 f_S(1 - e^{-A}) = 0$.	[c]
37	$H_2CO_3^* \leftrightarrow H^+ + HCO_3^-$	$pK_{a,1} = 6.3$	Stumm et al. (1972)
38	$HCO_3^- \leftrightarrow H^+ + CO_3^{2-}$	$pK_{a,2} = 10.3$	Stumm et al. (1972)
39	$H_3PO_4 \leftrightarrow H^+ + H_2PO_4^-$	$pK_{a,3} = 2.1$	Stumm et al. (1972)
40	$H_2PO_4^- \leftrightarrow H^+ + HPO_4^{2-}$	$pK_{a,4} = 7.2$	Stumm et al. (1972)
41	$H_2O_2 \leftrightarrow HO_2^-$	$pK_{a,5} = 11.6$	Perry et al. (1981)
42	$HO_2A \leftrightarrow H^+ + O_2A^-$	$pK_{a,6} = 4.8$	Perry et al. (1981)
43	$R_1H \leftrightarrow H^+ + R_1$	$pK_{a,7}$	[b]
44	$R_2H \leftrightarrow H^+ + R_2$	$pK_{a,8}$	[b]

[a] Photolysis rate.
[b] Depend on target organic compound.
[c] Inorganic humic substance loss from photolysis.
Source: Crittenden, J.C. et al., *Water Res.*, 33(10), 2315–2328, 1999. With permission.

Under the second condition, that the by-product (BYP) formation is significant, the direct photolysis reaction of M with respect to time is given by Equation (7.22):

$$-\frac{d[M]}{dt} = \frac{\Phi_M \varepsilon_M b[M] I_0 [1 - \exp(-2.3A_r)]}{\varepsilon_M b[M] + \sum \varepsilon_{BYP} b[BYP_i]} \tag{7.22}$$

The high decomposition rate of pollutants is dependent on the high value of the molar extinction coefficient at 254 nm, the quantum yield of M, and the concentration of oxidation by-products.

The activation of the organic compounds can involve direct oxidation by UV, the formation of organic radicals, or other reactive intermediates. The oxidation mechanism of organic compound M in this system includes the reaction of M with $HO^\bullet$ and $HO_2^\bullet / O_2^{\bullet-}$ as follows:

$$M + HO^\bullet \rightarrow M_{oxi} \; k_M = 3 \times 10^9 \; M^{-1} \; s^{-1} \tag{7.23}$$

$$M + HO_2^\bullet / O_2^{\bullet-} \rightarrow M^*_{oxi} \tag{7.24}$$

Specifically, hydroxyl radicals can oxidize organics by hydroxylation, hydrogen abstraction, electrophilic addition, and electron transfer, depending upon the nature of organic compounds.

7.2.2.2 Electrophilic Addition

Electrophilic addition of hydroxyl radicals to organic π systems will lead to organic radicals, as shown in the following equation:

$$\tag{7.25}$$

Electrophilic addition is of particular interest for a mechanistic interpretation of the rapid dechlorination of chlorinated phenols and leads to generation of chloride ions. One possible pathway could be electrophilic addition of the radical to the π bond followed by a subsequent fragmentation of the intermediate chlorohydrol as shown in Equation (7.26):

$$\tag{7.26}$$

7.2.2.3   Hydrogen Abstraction

When a hydroxyl radical reacts with a saturated aliphatic compound, it will abstract a hydrogen atom from the compound as follows:

$$^{\bullet}OH + CH_3COH_3 \rightarrow {}^{\bullet}CH_2COCH_3 + H_2O \tag{7.27}$$

7.2.2.4   Electron-Transfer Reactions

When hydrogen abstraction or electrophilic addition reactions may be inhibited by multiple halogen substitutions or steric hindrance, a hydroxyl radical can be reduced to a hydroxide anion by an organic substrate shown in Equation (7.28):

$$^{\bullet}OH + RX \rightarrow OH^- + {}^{\bullet}RX^+ \tag{7.28}$$

7.2.2.5   Radical–Radical Reactions

Generated at high local concentrations, hydroxyl radicals may dimerize to H_2O_2:

$$2HO^{\bullet} \rightarrow H_2O_2 \tag{7.29}$$

Under conditions of excess H_2O_2, hydroxyl radicals will also produce hydroperoxyl radicals, as shown in Equation (7.30), which are much less reactive and do not contribute to the oxidative degradation of organic substrates. The concentration of $HO_2^{\bullet}$ is controlled by the pH of the reaction system, which in turn determines the efficiency of superoxide dismutation.

$$H_2O_2 + HO^{\bullet} \rightarrow H_2O + HO_2^{\bullet} \tag{7.30}$$

When direct photolysis by UV radiation is negligible, the oxidation rate (R_M) of model compound M can be expressed by the following equation:

$$R_M = \{k_M[HO^{\bullet}] + k_M^*([HO_2^{\bullet}] + [O_2^{\bullet}])\}[M] \tag{7.31}$$

$$R_M = -d[M]/dt = k_{M,OH}[M][OH]_{ss} \tag{7.32}$$

where $k_{M,OH}$ is the second-order rate constant for the reaction of M with hydroxyl radicals, and $[OH]_{ss}$ is the steady-state concentration of hydroxyl radicals formed in the UV/H_2O_2 process.

A typical reaction sequence occurring in the oxidation of organic substrates by UV/H_2O_2 is shown in Figure 7.1. H_2O_2 undergoes photolytically

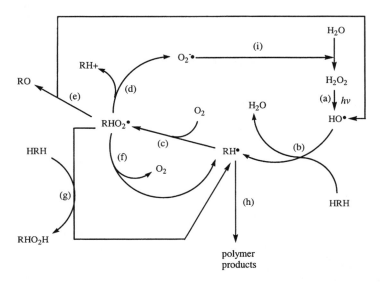

FIGURE 7.1
Reaction pathways for UV/H_2O_2 process. (From Legrini, O. et al., *Chem. Rev.*, 93, 671–698, 1993. With permission.)

induced dissociation, forming hydroxyl radicals (path *a* in Figure 7.1), according to Equation (7.1). The organic molecule, represented by the symbol HRH, is oxidized by the hydroxyl radical either by abstracting a hydrogen atom or by adding to a double bond, which produces an organic radical (RH•) (path *b*). If dissolved oxygen is present, the organic radical will quickly combine with it to form an organic peroxyl radical (RHO$_2$•) (path *c*), initiating subsequent oxidation reactions. When oxygen is not available, the organic radicals can combine to form organic molecules (path *h*) or they can react with H_2O_2. Legrini et al. (1993) proposed three different reaction paths to be followed by either peroxyl radicals or their tetraoxide dimers: (1) heterolysis and generation of organic cations, as well as superoxide anion (path *d*); (2) 1,3-hydrogen shift and homolysis into hydroxyl radicals and carbonyl compounds (path *e*); and (3) back reaction to RH• and O_2 (path *f*). The majority of peroxyl radicals decay following paths *d* and *e*, while far fewer follow path *f*.

Hydrogen abstraction by RHO$_2$• could also participate in the process of initiating a chain of thermal oxidation reactions (path *g*). In aqueous systems, cations will further react by solvolysis, and superoxide anion will readily disproportionate to yield H_2O_2 (path *i*). This is in contrast to the fate of superoxide anions in ozonation advanced oxidation processes (AOPs), where they react primarily with ozone to produce hydroxyl radical. This description of the chemical pathways of UV/H_2O_2 oxidation of organics illustrates that, when oxygen is present, the major paths directly or indirectly create more

H_2O_2, a source for hydroxyl radical production; therefore, the oxygen concentration in the contaminated water becomes an important parameter that determines whether the process will produce more H_2O_2 or the organic radicals will recombine to form by-products and deplete H_2O_2. This ability to regenerate oxidant decreases the required H_2O_2 dosage; therefore, when the process is optimized for field use, it could have a cost advantage over other AOPs.

7.3   Degradation Mechanisms of Organic Pollutants

Degrading organic compounds in aqueous systems, such as contaminated groundwater and industrial wastewaters by UV/H_2O_2 systems has been extensively investigated. Generally, halogenated alkanes such as trihalogenated ethanes are very difficult to oxidize, because carbon–chlorine (C–Cl) bonds are relatively inert with regard to potential radical substitution by hydroxyl radicals. Table 7.3 shows that the second-order rate constants for vinyl chloride and carbon tetrachloride with hydroxyl radicals are 7.1×10^9 and $5 \times 10^6 \ M^{-1} \ s^{-1}$, respectively; therefore, the kinetic rate constant decreases proportionally with an increasing number of chlorine atoms in the molecule. These compounds are amenable to direct photolysis, particularly at wavelengths below 250 nm; therefore, perchlorinated compounds such as chloroform exhibit comparatively slow degradation rates when hydroxyl radicals are the dominant oxidation species. In order to have a better understanding of the UV/H_2O_2 process, a discussion of the degradation kinetics and mechanisms of different classes of organic pollutants follows.

7.3.1   Alcohol

Pace et al. (1995) have attempted to correlate the conductivity of organic pollutants with their degradation rates. The photolysis reactions were carried out in a 1825-mL beaker and equipped with a 125-W high-pressure, mercury-vapor lamp (the peak of emission radiation was in the range of 280 to 380 nm), operating at constant temperature of $25.0 \pm 0.1°C$. The changes of the concentrations of organic substrate (2-chloroethanol), H_2O_2, and Cl^- were analyzed in order to determine a procedure that could be used to determine the correlation between conductivity and oxidation rate.

The interaction of 2-chloroethanol with hydroxyl radicals will yield primarily two different organic radicals that undergo fast conversion with the formation of Cl^- ions and several degradation products:

$$ClCH_2CH_2OH \xrightarrow[-H_2O]{HO\bullet} Cl\dot{C}\cdot HCH_2OH + ClCH_2\dot{C}HOH \qquad (7.33)$$

TABLE 7.3

Rate Constants for the Hydroxyl Radical

Compound (M)	$k_{M,OH} \times 10^{-9}$ (M^{-1} s^{-1})
Benzene	7.8
Toluene	3.0
1-Butanol	4.2
Vinyl chloride	7.1
Trichloroethylene	4.0
Tetrachloroethylene	2.3
Pyridine	3.8
Chlorobenzene	4.5
Nitrobenzene	3.9
Hydroperoxide ion, HO_2^-	7.5
Dichloromethane	0.058
Chloroform	0.005
Carbon tetrachloride	NR
Bicarbonate ion	0.0085
Carbonate ion	0.39

Source: Glaze, W.H. et al., *Water Pollut. Res. J. Can.*, 27(1), 23–42, 1992. With permission.

$$HO \cdot ClCH\ CH_2OH \xrightarrow{\text{fast}} \text{oxidation products} + Cl^- \qquad (7.34)$$

These results show an exponential decrease of organic substrate with respect to time and increasing Cl^- concentration. An inverse relationship exists between the disappearance of 2-chloroethanol and the formation of Cl^-; however, the sum of $[ClCH_2CH_2OH] + [Cl^-]$ remains constant throughout the irradiation. The relationship between increasing $[Cl^-]$ and increasing specific conductivity of the reaction solution suggests that the latter arises from Cl^- (as HCl), and the contributions of other conducting species (such as weak, undissociated organic acids) are most likely negligible. Acetic, glycolic, and formic acids, and acetaldehyde were formed in the reaction at low concentrations.

7.3.2 Alkane

De Laat et al. (1994) examined the effect of physicochemical parameters on the efficiency of the UV/H_2O_2 process for the degradation of chloroethanes in dilute aqueous solution. The kinetics of hydrogen peroxide photodecomposition in pure water and the rate of destruction of an organic solute by UV/H_2O_2 process were tested in cylindrical UV reactors equipped with low-pressure, mercury-vapor lamps. The following assumptions were made: (1) the degradation of M is due only to a direct reaction between M and hydroxyl radicals; (2) the rate of photolysis of hydrogen peroxide is not affected by organic and inorganic solutes; and (3) the steady-state approximation can be used for the concentration of hydroxyl radicals. The experiments showed that the four chloroethanes tested are not degraded by UV radiation alone or by hydrogen peroxide alone.

Glaze et al. (1995) proposed a kinetic model for the UV/H_2O_2 oxidation using radical and the direct photolysis of the organic compounds. The model utilized the literature values of the rate constants for radical formation and substrate oxidation. It was applied to interpret the data from the oxidation of 1,2-dibromo-3-chloropropane (1,2-DBCP) at low levels (less than 500 μg/L) in simulated and actual groundwater.

The effect of UV intensity, the initial concentration of hydrogen peroxide, and the various inorganic salts were investigated. Nitrate and bicarbonate/carbonate have a detrimental effect on the rate of oxidation of DBCP. Nitrate delays the reaction due to UV shielding, while bicarbonate/carbonate decrease oxidation efficiency due to HO• scavenging. The oxidation rate of DBCP is enhanced and the optimum peroxide level is lowered at low carbonate alkalinity, suggesting that presoftening of groundwater prior to oxidation of halogenated alkanes is necessary.

Liao and Gurol (1995) investigated a UV/H_2O_2 process in a continuous-flow, stirred-tank reactor under various initial and operating conditions. Synthetic solutions of a model organic compound, *n*-chlorobutane, were oxidized at various pH levels and in the presence of various amounts of humic materials and carbonate/bicarbonate ions to examine the effect of water quality on the process efficiency. Carbonate/bicarbonate ions were found to compete effectively with the model compound for the consumption of HO•, causing a dramatic reduction in the oxidation rate of model compound. However, carbonate ions did not affect the overall quantum yield for H_2O_2 decomposition, which was predicted to be 1.0 in the presence as well as the absence of carbonate ions. The oxidation efficiency for the target pollutant was determined by the rate of photolytic decomposition of H_2O_2 as well as the water quality. On the other hand, highly colored wastewaters containing larger concentrations of humic acids demonstrated significant recalcitrance for treatment with UV/H_2O_2. The reaction rates significantly deteriorate, making the process highly inefficient.

7.3.3 Alkene

Hirvonen et al. (1995) evaluated the feasibility of the UV/H_2O_2 process for the removal of trichloroethylene (TCE) and erythromycin (perchloroethylene [PCE]) in contaminated groundwater. The formation of chloroacetic acids (CAs) was used as an indication of partial degradation. The dominant by-product, dichloroacetic acid (DCA), accounted for the major part of the total yield of CAs. The observed concentrations of trichloroacetic acid (TCA) and DCA were relatively low compared with the total amount of TCE and PCE degraded. The effect of initial concentrations of the parent compounds, hydrogen peroxide, and bicarbonate on the yield of by-product was investigated and the reaction rates were analyzed for various combinations. In addition, the formation rates of TCA and DCA during UV and UV/H_2O_2 treatment proved that direct photolysis cannot be considered as a practical means to destroy TCE and PCE, due to its slow rate and the increased formation of partially degraded intermediates that may slow down the oxidation process.

7.3.4 Bentazone

The photodegradation of bentazone (3-isopropyl-1H-2,1,3-benzothiadiazine-4-3H-one 2,2 dioxide) has been conducted using direct polychromatic UV radiation and a UV/H_2O_2 process (Beltran-Heredia et al., 1996). The experiments were performed at various temperatures and pH values, and the initial hydrogen peroxide concentration in the combined process was also varied. The influence of the temperature and pH on the degradation was positive in both processes, although pH had a slighter effect. An irradiation influx rate model was used to evaluate the radiation flow rate absorbed by the reacting medium for each reaction time.

A kinetic study of direct photolytic decomposition of organics showed that in aqueous solution organic substrates can be decomposed by UV radiation emitted by a polychromatic source. In UV/H_2O_2 photolytic destruction, the reaction between the organic compound and the hydroxyl radicals generated in the H_2O_2 photodecomposition is dominant. In addition, the kinetics must consider the reaction of hydroxyl radicals with the radical scavengers if they are present. The Arrhenius correlation was proposed to express the rate constants of the reaction between hydroxyl radicals and bentazone as a function of the temperature:

$$k = 1.65 \times 10^{11} \exp(-1179/T) \text{ dm}^3 \text{ mol}^{-1} \text{ s}^{-1} \qquad (7.35)$$

7.3.5 Aromatic Hydrocarbons

Beltran et al. (1996a) investigated the advanced oxidation of the aqueous solution of three polynuclear aromatic hydrocarbons (PAHs): fluorene, phenanthrene, and acenaphthene. The oxidative mechanism of substrates by O_3 alone, O_3 combined with H_2O_2, UV radiation (254 nm) alone, or a combination of UV/H_2O_2 has been analyzed. In addition, the influence of different water sources, such as surface and organic-free waters, and type of occurring oxidation was also studied.

The application of the UV process alone showed low efficiency, which was attributed to the presence of UV-absorbing material. In addition, the photolysis of fluorene compared with the two other PAHs was significantly retarded when the fluorene was present in surface waters; however, the oxidation rates of the three PAHs were essentially independent of the water type. It was concluded that the elimination of PAHs is mainly due to HO$^\bullet$ generated from hydrogen peroxide photolysis. In order to simplify the kinetic treatment, the pseudo first-order kinetics is used to determine the reaction half-lives of PAHs. The application of UV radiation alone leads to the lowest degradation rates, with half-lives of more than 6 min. As a general rule, the oxidation rates in surface waters are slower than those in organic-free waters. Among the PAHs studied, fluorene has shown the lowest oxidation rates regardless of type of advanced oxidation methods applied.

Beltran et al. (1996a–c) studied oxidation of the same PAHs in water and applied UV radiation combined with hydrogen peroxide. The PAHs were found to be refractory to oxidation when hydrogen peroxide was applied alone. The decomposition rates of PAHs under the UV/H_2O_2 process were substantially higher compared with UV radiation alone, provided that appropriate concentrations of hydrogen peroxide and pH have been applied. The direct photolysis contribution decreases with increasing hydrogen peroxide concentration and is the main pathway in degradation under acid pH conditions. The bicarbonate ions did not retard the oxidation rate of PAHs, and the process rate remained unaltered compared to the blank experiment carried out in the absence of bicarbonate ions. Both the UV radiation and UV/H_2O_2 oxidation of PAHs produce a series of intermediate compounds. Most of these compounds, initially formed as complexes, were ultimately destroyed when the oxidation time was extended.

Wang et al. (1995) investigated the photodegradation of phenanthrene in water in the presence of photosensitizer fulvic acids (FAs) by UV/H_2O_2. Humic materials can act as photosensitizers or precursors for the production of reactive oxygen species such as singlet oxygen, hydroxyl radicals, peroxy radicals, hydrogen peroxide, and solvated electrons. A schematic representation was proposed as follows:

$$HS + light \rightarrow H_2O_2, ROO\bullet, O_2, e_{aq}, O_2 \bullet \quad \text{Reactive oxygen species}$$

$$\xrightarrow{\text{Environmental chemicals}} \text{Degradation products}$$

(7.36)

Because hydroxyl radicals can cause nucleophilic oxidation of aromatic rings when dissolved O_2 is present, the reaction rate is directly dependent on the oxygen partial pressure. Five FAs from different sources were used to compare the effects on the photodegradation of phenanthrene.

7.3.6 Carboxylic Acid

Leitner and Dore (1997) studied the oxidation of aliphatic acids by hydroxyl radicals resulting from the photolysis of hydrogen peroxide to CO_2. In order to examine the mechanism and the oxidation kinetics of aliphatic acids, experiments were conducted on mono- and dicarboxylic acids with 1 to 4 carbon atoms of various oxidation degrees. Dilute aqueous solutions under oxygenated and deoxygenated conditions and at low pH (2.9–3.5) were used. In the presence of dissolved oxygen, the weak consumption of hydrogen peroxide observed during the oxidation of formic acid and of carboxylic acids with hydroxyl groups indicates that oxygen consumption is accompanied by $HO_2^{\bullet}/O_2^{\bullet-}$ generation followed by their recombination to form H_2O_2.

The by-products identified and the oxygen consumption that occurred during elimination of unsubstituted carboxylic acids showed that the oxidation occurs via a C–C bond cleavage. Total organic carbon (TOC) measurements indicated that complete mineralization of carboxylic acids to CO_2 is slow in oxygen-free solutions. The hydroxyl group of hydroxylated acids activates the adjacent C–H bond and leads to O_2 addition and $HO_2^{\bullet}$ removal to the carbonyl group after hydrogen atom abstraction. Except for oxalic acid, the mechanism of oxidation of the other carboxylic acids to CO_2 involved oxygen consumption and $O_2^{\bullet-}/HO_2^{\bullet}$ release.

7.3.7   Ether

Wagler and Malley (1994) explored the alternative oxidation processes that consist of UV light, H_2O_2, and UV/H_2O_2 to remove methyltertiary butyl ether (MTBE) from simulated groundwater by batch oxidation systems. Kinetic studies were performed to determine the effectiveness of the selected processes and estimate the reaction order and rate constants for the effective removal processes. The results indicated that the UV/H_2O_2 process was the most promising oxidation method. A number of experiments were conducted to determine the most efficient pH range and the effects of varying the molar ratio of MTBE to H_2O_2 with the UV/H_2O_2 process.

All reaction rates appeared to be pseudo first order with respect to the removal of MTBE. The reaction rate peaked in a slightly acidic environment of pH 6.5. The theoretical dose of H_2O_2 calculated using stoichiometry and redox chemistry proved to be approximately two times the level necessary to produce similar results (14.9:1.0 vs. 7.9:1.0) because H_2O_2 is produced in the UV/H_2O_2 process when oxygen is present. An analysis of an MTBE sample that was not completely oxidized suggests that methanol is a major by-product of the UV/H_2O_2 process along with trace amounts of formaldehyde and 1,1-dimethylethyl formate. The molecular structure of these compounds and proposed pathways based on hydroxyl radical oxidation of organics were discussed, and the results for natural water and simulated groundwater using the UV/H_2O_2 process were also compared.

7.3.8   Halide

Fluorinated aromatics and one-side-chain fluorinated aromatics such as α,α,α-trifluorotoluene were degraded in either the presence or absence of dissolved oxygen by Leitner and Dore (1996). During oxidation, the dissolved oxygen was gradually depleted. In solutions containing H_2O_2, the absorbance of the trifluorobenzene derivatives gradually increased upon irradiation. During the photooxidation of α,α,α-trifluorotoluene, the production of F^- ions and the increase of absorbance was considerable more noticeable when O_2 was present. In contrast, for oxygen-free solutions the effects were not so pronounced. The analysis of fluoride ions showed the formation of about 2 mol of F^- per mole of trifluorobenzene reacted and of 1 mol of F^- per mole of trifluorotoluene reacted when irradiated in the presence of H_2O_2. Similar values were obtained in the presence and in the absence of oxygen, thus demonstrating the relative insignificance of oxygen.

Direct photolysis was assumed to occur via a polarized excited state. Kinetic studies led to determination of the quantum yield for the photolysis of studied compounds. For the oxidation of fluorinated compounds using UV/H_2O_2 process, high concentrations of hydrogen peroxide were used. Under these conditions, direct photolysis of the organic molecule was negligible. The disappearance of the fluorinated compound was due mainly to the reaction of hydroxyl radicals. Based upon gas chromatography/mass spectrometry (GC/MS) analyses, the by-products are mostly hydroxylated and dehalogenated compounds. In addition, some dimers can also be observed during photolysis. However, under oxygen-free conditions, the polymerization of by-products is favored.

Kochany and Bolton (1992) studied the primary rate constants of the reactions of hydroxyl radicals, benzene, and some of its halo derivatives based on spin trapping using detection by electron paramagnetic resonance (EPR) spectroscopy. The competitive kinetic scheme and the relative initial slopes or signal amplitudes were used to deduce the kinetic model. Based on a previously published rate constant (4.3×10^9 M^{-1} s^{-1}) in the pH range of 6.5 to 10.0 for the reaction of hydroxyl radicals with the spin trap compound 5,5'-dimethylpyrroline N-oxide (DMPO), rate constants for the reaction of hydroxyl radicals with benzene and its halo derivatives were determined.

The effect of pH on the hydroxyl radical reaction rate constants was studied in buffered solution at pH = 3.5, 7.0, and 11.0. As the pH increases above 9, the presence of alternate scavenger HO_2^- will react with hydroxyl radicals, and the rate constant is nearly 300 times that of H_2O_2. The data for halobenzenes showed that the reactivity of hydroxyl radical with these derivatives does not vary much among the compounds studied, and an average rate constant of ~5.0×10^9 M^{-1} s^{-1} could be used for these compounds.

7.3.9 Ketone

UV/H_2O_2 process has low reaction rates when acetone is treated. Stefan et al. (1996) studied the kinetics and mechanism of acetone degradation in dilute aqueous solution to understand the slow kinetics. Photodegradation was carried out in a batch tank with a UV reactor (6.5 L) in the wavelength range of 200 to 300 nm. In dilute aqueous solutions of acetone, H_2O_2 is the principal absorbent of UV light so that the direct photolysis of acetone can be neglected. The acetone degradation was followed concomitantly with an increase and then decay of reaction intermediates such as acetic, formic, and oxalic acids, until complete mineralization was achieved. The relative concentrations, pH, pK_a of each by-product, and rate constants of various components govern the kinetics of complete destruction of each component with hydroxyl radicals. Treatment of acetone is slow because the intermediates formed in low concentrations can efficiently compete with acetone for hydroxyl radicals even at relatively short irradiation times. The results showed that the initial concentrations of acetone and hydrogen peroxide strongly affected the initial rate of acetone degradation; however, no acid-catalyzed reaction occurred in the photochemical oxidation of acetone in the presence of hydrogen peroxide when the pH range was 2 to 7.

7.3.10 Chlorophenol

Apak and Hugul (1996) studied a number of chlorophenols, including 2-, 3-, and 4-chlorophenol; 2,4-dichlorophenol; and 2,4,6-trichlorophenol. All of these compounds were decomposed in aqueous solution by H_2O_2 process in a Pyrex glass column photoreactor fitted with a thermostat and a 400-W high-pressure mercury lamp. During the decomposition of chlorophenol, the organic-bound chlorine substituents (C–Cl) were stepwise replaced by alcohol (C–OH) and (C=O) groups, yielding the corresponding hydroquinone and quinone compounds as intermediates.

When H_2O_2:chlorophenol mole ratios are between 1:1 and 16:1, especially at relatively higher ratios, the photooxidation reaction kinetics can be fitted to a pseudo first-order rate model. A pseudo first-order approximation of the reaction scheme does not totally reflect the physical reality of the oxidation system but does have practical significance for environmental process design and comparison. Apak and Hugul (1996) found that the rate constants increased with the increasing ratio of the oxidant. The degradation rates decreased in the order of trichlorophenol $\geq$ dichlorophenol > monochlorophenol, especially at a high H_2O_2 ratio. In 2,4-dichlorophenol and 2,4,6-trichlorophenol compounds, the strong mesomorphic effect of the hydroxyl group makes the *ortho* and *para* positions vulnerable to hydroxyl radical attack and causes faster conversion. Because of the complex reaction sequence processing via a radical mechanism involving the photodegenerated hydroxyl radicals, no simple reaction order was found.

Ku and Ho (1990) conducted a feasibility study and established empirical reaction rate expressions for the decomposition of chlorophenols by UV irradiation combined with various oxidation processes, including O_2, H_2O_2, and sodium hypochlorite (NaOCl). Increasing dissolved oxygen (DO) level decreased the destruction of chlorophenols by UV irradiation, because the oxygen present competes for photons to form weak oxidants, such as peroxide radicals, rather than hydroxyl radicals. The presence of NaOCl and H_2O_2 significantly increased the removal efficiency until a certain threshold dosage was reached. Comparing the results for NaOCl oxidation with and without UV radiation, the removal efficiencies were almost the same. However, the reaction intermediates and mechanisms involved may be different for these two processes due to the color change of solution. The expression of the overall reaction rate was subdivided into three parts: the reaction rates caused by H_2O_2 oxidation only, by UV irradiation only, and by both UV and H_2O_2.

Shen et al. (1995) investigated the effect of light absorbance on the decomposition of aqueous chlorophenols (CPs) by UV/H_2O_2. The photoreaction system was batch annular photoreactors with 254-nm, low-pressure UV lamps at 25°C. The light absorbance and photolytic properties of chlorophenols and H_2O_2 were found to be highly dependent on the solution pH and can be adequately described with the linear summation of the light absorbance of undissociated and dissociated species of chlorophenols:

$$Abs_{total} = Abs_{OH} \frac{1}{1 + K_a / (H^+)} + Abs_{O^-} \frac{1}{1 + (H^+) / K_a} \qquad (7.37)$$

where Abs_{OH} and Abs_{O^-} are absorbances for undissociated and dissociated species, respectively. Due to the complexity of the decomposition of chlorophenols, a simplified two-step kinetic model was used:

$$(\text{Chlorophenols})_c \xrightarrow{k_{1c}} (\text{intermediates})_c \xrightarrow{k_{2c}} (CO_2)_0 \qquad (7.38)$$

where k_{1c} and k_{2c} are pseudo first-order rate constants. Similarly, the reaction rate was found to be highly dependent on the species distribution, which is a function of pH. For UV/H_2O_2 processes, the decomposition of chlorophenols was attributed to direct photolysis by UV light and indirect oxidation by HO$^\bullet$. The oxidation of chlorophenols by H_2O_2 alone was found to be negligible during the reaction. The k_{1c} values of various chlorophenols obtained for UV/H_2O_2 treatment were much larger than those for UV treatment alone under acidic solutions. For neutral and alkaline solutions, the differences of k_{1c} for the two systems were much closer for the decomposition of chlorophenols, implying that direct photolysis would gradually become dominant at higher pH conditions. The decomposition rate constants of organic intermediates, k_{2c}, were fairly constant under acidic and neutral conditions but decreased when pH values were greater than 9. The H_2O_2 concentration decreased with increasing solution pH values because the dissociated HO_2^- has higher photolytic decomposition rates than H_2O_2.

The efficiencies of some AOPs on the rate of degradation of six selected CPs were investigated in laboratory-scale studies (Trapido et al., 1997). Because the molar extinction coefficient of hydrogen peroxide is low, a relatively high molar ratio of hydrogen peroxide to CP was used in the experiments to ensure that the hydroxyl radical-mediated route is the major pathway for the degradation of compounds in UV/H_2O_2 treatment. In the control experiments, the complete mixing of the CPs with hydrogen peroxide in the absence of radiation did not lead to the degradation of the compounds.

UV/H$_2$O$_2$ treatment at pH 2.5 degraded CPs in the following order: 2,4,6-trichlorophenol (TCP) > 2,4-dichlorophenol (DCP) > 2-CP. However, differences between the degradation efficiencies of CPs studied in basic media (pH = 9.5) were less. The decrease of the initial pH decreased the overall degradation rate of CP.

Moreover, the degree of mineralization in acidic media was considerably lower than that in basic media. The differences between the degree of mineralization of CPs at different pH values can be explained partly by the more stable chlorinated intermediates at low pH. Intermediates with aromatic structures were formed soon after the beginning of the oxidation. The concentration of intermediates, however, slowly decreased during extended treatment time. By increasing the dose of oxidants or UV intensity, improved process efficiency could have been obtained, but the total mineralization of the compounds is likely to cost more in remediating waters contaminated with CPs.

To define the effectiveness of the UV/H$_2$O$_2$ process on a wide range of priority pollutants in water, Sundstrom et al. (1989) conducted experiments in a recirculating flow reactor system with low-pressure UV lamps at 254 nm. The temperature of the solution was maintained at 25°C, and pH was maintained at 6.8 by a phosphate buffer. Molar ratio of peroxide to pollutant was varied during the experiments. As the molar ratio of peroxide to pollutant increased, the reaction rates increased. Three monosubstituted benzenes were selected to examine the effect of a single substituent group on the rate of reaction of benzene. The rates of reaction were of similar magnitude for benzene and monosubstituted benzenes (toluene, chlorobenzene, and phenol) at the ratio of 7 for peroxide to pollutant.

The effect of increasing chlorine atoms to a ring was explored through a series of chlorophenols. With both UV light and hydrogen peroxide, phenol, 2,4-dichlorophenol, and 2,4,6-trichlorophenol had similar reaction rates. For the phthalate esters studied, the reaction rates of dimethyl phthalate and diethyl phthalate were virtually identical under any given set of operating conditions. The rate constants show that the phthalates were the slowest reacting compounds, while 2,4,6-trichlorophenol had the fastest reaction rate.

7.3.11 Xenobiotics

The effect of nitrate on the photochemical degradation kinetics of hydrophilic was studied by Sörensen and Frimmel (1997). A number of common pollutants were examined, including amino-polycarboxylates and aromatic sulfonates. Kinetic experiments were performed with or without H$_2$O$_2$. Quantum yield was used as a tool for examining whether a substance acts as an inner filter. Under the assumption that water compounds do not take part in the chemical reaction but exert influence only in a physical way by the absorption of light, the true integral quantum yield (Φ_{wi}) is independent of the concentration of the inner filter. Φ_{wi} is expressed as:

$$\Phi_{wi} = \Phi_{si} \times \frac{P_{Abs,\ Tot}}{P_{Abs,\ Tot} - P_{Abs,\ NO_3^-}}$$
(7.39)

where Φ_{si} is the apparent integral quantum yield, $P_{Abs,Tot}$ and $P_{Abs,\ NO_3^-}$ are the absorbed photon flux of solution and nitrate, respectively. The values of Φ_{wi} for the photochemical degradation of micropollutants in the absence of H_2O_2 increase drastically with increasing concentration of nitrate. In the UV/H_2O_2 process, the values of Φ_{wi} decrease with increasing nitrate concentration.

7.3.12 Mixture of Chemical Compounds

Beltran et al. (1996c) used trichloroethylene (TCE) and 1,1,1-trichloroethanes (1,1,1-TCA) as model compounds treated by UV/H_2O_2. UV irradiation alone was proven appropriate only when the compound to be irradiated shows noticeable level of absorptivity as is the case of TCE; however, TCAS are refractory to this kind of treatment and their elimination requires more stringent conditions. The destruction rates depend on the initial hydrogen peroxide concentration. The maximum degradation rates of TCA and TCE occurred when the H_2O_2 concentrations were 10^{-2} *M*, which represents about 5 to 250 times higher rates than those obtained from volatilization alone. When concentrations of hydrogen peroxide were below 10^{-2} *M*, the oxidation rates of TCA were close to those from volatilization alone, presumably because the hydrogen peroxide immediately consumes most of the generated radicals. On the contrary, free-radical oxidation continues to be the principal removal step in the case of TCE, as the rate constant of its reaction with hydroxyl radicals is approximately 65 times higher than that of the reaction of TCA with hydroxyl radicals.

When the concentration of bicarbonates is lower than 10^{-3} *M*, which is typical for many natural waters, CO_3^{2-} and HCO_3^- do not significantly affect the oxidation rate of these pollutants, especially in the case of TCE. On the other hand, the oxidation kinetics of both TCA and TCE in natural waters are considerably slower compared to data from laboratory experiments. A kinetic model was proposed for the compounds studied with the UV/H_2O_2 process. The elimination rate of volatile pollutants depends on the rates of volatilization (r_V), direct photolysis (r_{UV}), and hydroxyl free-radical oxidation (r_{OH}):

$$-\frac{dC_M}{dt} = r_V + r_{UV} + r_{OH}$$
(7.40)

Kinetic models for UV/H_2O_2 were developed based on known chemical and photochemical principles by Glaze et al. (1992), who examined the oxidation of nitrobenzene, naphthalene, and pentachlorophenol to illustrate some features of the UV/H_2O_2 process. The model took into account the effects of radical scavengers including bicarbonate, dose ratios of the oxidants or UV intensity, pH, and the presence of generic radical scavengers.

The direct photolysis of nitrobenzene was found to be slow and it did not occur at a significant rate. The naphthalene/pentachlorophenol system showed that the degradation rate of each compound was affected by the presence of the other compounds. In addition, both compounds when subjected to direct photolysis produced significant amounts of hydrogen peroxide. Thus, shortly after the initiation of the photolysis, the system became a UV/H_2O_2 process.

Kang and Lee (1997) evaluated the effectiveness of a UV/H_2O_2 process at relatively high concentrations of organics. TCE, PCE, benzene, and toluene were used as model compounds. The UV source used was a low-pressure, mercury-vapor lamp and primarily 254 nm of radiation. In regard to the direct photolytic rates of those model compounds, PCE was degraded fastest, while the degradation rate of TCE was the slowest. Benzene and toluene were decomposed at a similar rate, in between the rate of PCE and TCE. PCE decomposition occurring at a faster rate than the others presumably was due to the high value of the molar extinction coefficient at 254 nm and quantum yield, but decomposition of benzene and toluene by direct photolysis was ineffective due to the low value of quantum yield. The by-products produced after benzene and toluene oxidation have shown an inhibiting effect by acting as a competing UV absorbent, thus substantially decreasing the photolysis rate of the target compound. Kang and Lee (1997) developed a kinetic model for the rate of direct photolysis of organics by measuring the quantum yields of photolysis in the presence of by-products. The overall rate of the UV/H_2O_2 system was found to depend on the relative importance of the hydroxyl radical reaction pathway, the direct photolysis reaction rate, and the absorptivity of by-products at 254 nm.

Prengle et al. (1996) studied the photooxidation by UV/H_2O_2 of waterborne hazardous C_1–C_6 compounds in drinking water. Their work was conducted in a photochemical batch stirred-tank reactor, with medium pressure mercury arc immersion lamps of 100 and 450 W, covering the visible UV range, (578.0 to 222.4 nm). Tetrachloromethane, tetrachloroethane, dichloroethane, dichloroethene, trichloroethane, trichloroethene, and benzene were the compounds studied. Dark oxidation rates and photooxidation rates were determined. The latter rate constants were 10^4 to 10^5 greater than those under dark conditions.

A mathematical model assumed that photochemical reactions do not occur by conventional thermal kinetics but are driven by photon flux; therefore, the instantaneous rate equations for the disappearance of parent compound A and oxidizer B are directly proportional to the photon flux and the amounts of A and B:

$$-\dot{A} = f(A, B, \Sigma h\nu) = k_a \Phi_a A_a A_b \qquad (7.41)$$

$$-\dot{B} = g(t)\dot{A} \tag{7.42}$$

where k_a (µmol/min) is the rate constant, Φ_a (W/µmol) is the photon mass flux, and A_a (N_a/N_a^0) and A_b (N_b/N_b^0) are the dimensionless amounts of A and B. N_a^0 (µmol/L) is the standard state moles of a compound. A molecular composition structure (MCS) model was used to successfully correlate the rate constants. Because the constants are in the MCS model, it can be used to estimate $\ln k_a$ values for other C_1–C_6 chlorinated hydrocarbon compounds.

7.3.13 Chlorinated Aliphatic Compounds

Sundstrom et al. (1986) investigated the destruction of typical halogenated aliphatics, including alkanes and alkenes, by UV/H_2O_2 reaction using a batch reactor equipped with a low-pressure UV lamp at 254 nm. The degradation rate of trichloroethylene increased with increasing hydrogen peroxide concentration and temperature and was highly dependent on chemical structure. The reaction time will be approximately reduced by half when temperature increases 10°C. Chlorinated compounds with unsaturated bonds, such as trichloroethylene and tetrachloroethylene, degraded at a much faster rate than did the other compounds. Apparently, this double bond is readily attacked by UV light and reactive species. The synergistic effect of H_2O_2 and UV light was the most pronounced for trichloroethylene and dichloromethane; however, the presence of hydrogen peroxide had less effect on the disappearance of the other compounds studied. All of the organic chlorine was converted to chloride ion, indicating that the chlorinated structures were destroyed by UV catalyzed oxidation.

7.3.14 Textile Wastewater

Yang et al. (1998) evaluated whether a new UV/H_2O_2 photochemical oxidation treatment system can efficiently decompose and decolor textile dyes in a cost-effective manner, particularly when treating highly concentrated dyebaths using high-intensity ultraviolet lamps. The results showed that dyebath wastewater containing direct, basic, acid, and reactive dyes could be totally decolorized by the photochemical oxidation process. Commercial dyes, including diazo, azo, polyazo, oxazine, triarylmethane, diazovinyl sulfone, triazine, anthraquinone, and indigoid dyes, were studied. Both disperse and vat dyes have very low solubility in water and were not a part of the experiment.

7.4 QSAR Models

Quantitative structure–activity relationship (QSAR) models for kinetic rate constants and molecular descriptors, such as dipole moment, E_{HOMO}, E_{LUMO}, log *P*, and Hammett constants were presented for the UV/H$_2$O$_2$ process for selected organic pollutants (Jung, 1999). A number of properties, including dipole moments, E$_{HOMO}$, and E$_{LUMO}$, have been calculated using the computer program SPARTAN. The octanol/water partition coefficient and substituent constant are taken from Hansch et al. (1995), and the kinetic rate constants are collected from the literature.

Based on their chemical structure, the organic chemicals were divided into a number of categories: alkanes, alkenes, amines, aromatic hydrocarbons, benzenes, carboxylic acids, halides, phenols, and sulfonic acid. Linear regression analysis has been applied using the method of least-squares fit. Each correlation required at least three datapoints, and the parameters chosen were important to ensure comparable experimental conditions. Most vital parameters in normalizing oxidation rate constants for QSAR analysis are the overall liquid volume used in the treatment system, the source of UV light, reactor type, specific data on substrate concentration, temperature, and pH of the solution during the experiment.

7.4.1 Dipole Moment

Dipole moment (μ) measures the internal charge separation of a molecule and is important in evaluating how the solvent molecules will cluster around a solute particle having a dipole moment. The solvent dipoles will tend to orient themselves around the solute in the manner indicated in Figure 7.2. The orientation will be most pronounced in the innermost shell of the solvent molecules and will become increasingly random as distance from the solute particle increases. Therefore, the magnitudes of the dipole moments of studied compounds will be enclosed by more solvent particles to prevent attack from hydroxyl radicals. Furthermore, the presence of halogen atoms that have high electronegativity can increase the dipole moment. By contrast, compounds with identical opposing atoms or functional groups and compounds with larger structures may cancel dipole moments. As dipole moment increases, the correlation between rate constants and dipole moments decreases.

Alkenes ($R^2 < 0.69$), amines ($R^2 < 0.64$), carboxylic acids ($R^2 < 0.001$), and sulfonic acids ($R^2 < 0.20$) show a poor correlation with dipole moment. With dipole moment as descriptor, benzene and halide classes have a predictable trend for linear correlations. The correlation coefficients of multicarbon alkanes or combinations of one-carbon and multicarbon halogenated alkanes are smaller than 0.67. They all have the same trend of correlations in that the rate constants decrease as the dipole moment increases.

In the case of aromatic hydrocarbon derivatives, the dataset used was fluorene, acenaphthene, and phenanthrene:

$$\text{Log } k = -0.1679\mu - 0.3287; \ n = 3, \ R^2 = 0.9818 \qquad (7.43)$$

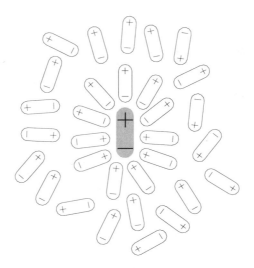

FIGURE 7.2
Interaction of dipole moments.

The oxidation of benzene class included benzene, chlorobenzene, bromobenzene, and iodobenzene; a good linear correlation is found at pH = 11.

$$\text{Log } k = -0.0788\mu + 9.7699; \; n = 4, \; R^2 = 0.985 \qquad (7.44)$$

The dipole moment correlation of trifluorobenzenes (dataset consisting of 1,2,3-trifluorobenzene, 1,2,4-trifluorobenzene, and 1,3,5-trifluorobenzene) is better than that for dichlorobenzenes of the halide class:

$$\text{Log } k = -0.0112\mu + 9.6132; \; n = 3, \; R^2 = 0.9988 \qquad (7.45)$$

Table 7.4 summarizes the QSAR models using dipole moment as the molecular descriptor.

7.4.2 E_{HOMO}

A molecular orbital calculation yields a set of given values or energy levels in which all the available electrons are accommodated. Based on the frontier molecular orbital theory, one can conclude that the strongest interactions will be highest occupied molecular orbital (HOMO)/lowest unoccupied molecular orbital (LUMO) orbital interactions that are close to each other. The interactions contribute to the energy of the interaction and hence to the energy of the transition state; therefore, the energies of the HOMO and LUMO for hydroxyl radicals and the reactive organic compounds must be compared with each other.

TABLE 7.4

QSAR Models with Dipole Moment

Chemical Class	Dataset	QSAR Model	R^2
Aromatic hydrocarbon	Fluorene, acenaphthene, phenanthrene	$\text{Log } k = -0.1679\mu - 0.3287$	0.9818
Benzene	Benzene, chlorobenzene, bromobenzene, iodobenzene	$\text{Log } k = -0.0788\mu + 9.7699$	0.9850
Halide	1,2,3-Trifluorobenzene, 1,2,4-trifluorobenzene, 1,3,5-trifluorobenzene	$\text{Log } k = -0.0112\mu + 9.6132$	0.9988
Phenol	Phenol, 2-chlorophenol, 2,4-dichlorophenol	$\text{Log } k = 0.6287\mu - 2.8023$	0.9481
	2-Chlorophenol, 2,4-dichlorophenol, 2,4,6-trichlorophenol	$\text{Log } k = -0.3262\mu - 0.9097$	0.9642

For multicarbon halogenated alkanes, no correlation exists between predicted rate constants and the values of the computed E_{HOMO} ($R^2 < 0.71$). E_{HOMO} also exhibits perfect correlation with the alkene class. The compounds contained in the dataset are trichloroethene, 1,1-dichloroethene, and tetrachloroethene:

$$\text{Log } k = 74.795\ E_{HOMO} + 28.364;\ n = 3,\ R^2 = 1 \qquad (7.46)$$

The result of QSAR models can be rationalized in terms of oxidation necessary for the activation energy of these chemicals, because HOMO is a measure of the ability of a molecule to release electrons, and hydroxyl radicals serve as oxidants that accept electrons. As the energy of the HOMO increases, the ability of organic compounds to behave as nucleophiles increases; therefore, the increased oxidation activity of compounds with hydroxyl radicals increases and leads to higher kinetic rate constants.

For halogenated phenols, a negative relationship between k and E_{HOMO} was found. The dataset includes 2-chlorophenol, 2,4-dichlorophenol, 2,4,6-trichlorophenol, and pentachlorophenol. The rate constants decrease as E_{HOMO} increases at pH = 2.5:

$$\text{Log } k = -34.65\ E_{HOMO} - 14.542;\ n = 4,\ R^2 = 0.9347 \qquad (7.47)$$

No statistically significant correlation was found for amine ($R^2 < 0.27$), aromatic hydrocarbon ($R^2 < 0.84$), benzene ($R^2 < 0.54$), carboxylic acid ($R^2 < 0.002$), halide ($R^2 < 0.87$), and sulfonic acid ($R^2 < 0.30$). The QSAR models using E_{HOMO} as the molecular descriptor are summarized in Table 7.5.

TABLE 7.5

QSAR Models with E_{HOMO}

Chemical Class	Dataset	QSAR Model	R^2
Alkene	Trichloroethene, 1,1-dichloroethene, tetrachloroethene	Log k = 74.795 E_{HOMO} + 28.364	1.0000
Phenol	2-Chlorophenol, 2,4-dichlorophenol, 2,4,6-trichlorophenol, pentachlorophenol	Log k = –34.65 E_{HOMO} – 14.542	0.9932

7.4.3 E_{LUMO}

The energy of the LUMO of a molecule can be approximated as its electron affinity. It measures the affinity of a molecule to accept electrons, and it acts as an electrophile or undergoes reduction.

For multicarbon halogenated alkanes, the dataset containing 1,2-dichloroethane, 1,1,1-trichloroethane, and tetrachloromethane demonstrates a positive relationship:

$$\text{Log } k = 13.755 \, E_{LUMO} - 2.0854; \; n = 3, \; R^2 = 0.9373 \qquad (7.48)$$

Halogenated benzenes and halides also exhibit good predictive correlations. The rate constants increase as E_{LUMO} does. The dataset for the benzene class is composed of benzene, chlorobenzene, and toluene:

$$\text{Log } k = 7.9487 \, E_{LUMO} - 2.2434; \; n = 3, \; R^2 = 0.9094 \qquad (7.49)$$

The dataset for the halide class is comprised of *o*-dichlorobenzene, *m*-dichlorobenzene, and *p*-dichlorobenzene at pH = 11:

$$\text{Log } k = 2.1831 \, E_{LUMO} + 9.394; \; n = 3, \; R^2 = 0.9876, \qquad (7.50)$$

Because LUMO is an approximate measure of the ability of a molecule to accept electrons, this result can be understood in terms of the electrophilicity necessary for the direct attack of hydroxyl radicals. As the energy of LUMO decreases, the ability of organic compounds to behave as an electrophile increases; therefore, the increased reduction activity of compounds causes decreased oxidation activity that leads to lower kinetic rate constants.

The remaining chemical classes such as alkene ($R^2 < 0.51$), amine ($R^2 < 0.90$), aromatic hydrocarbon ($R^2 < 0.29$), carboxylic acid ($R^2 < 0.06$), and sulfonic acid ($R^2 < 0.77$) do not correlate well. Table 7.6 shows all the QSAR models using E_{LUMO} as the molecular descriptor.

TABLE 7.6

QSAR Models with E_{LUMO}

Chemical Class	Dataset	QSAR Model	R^2
Alkane	1,2-Dichloroethane, 1,1,1-trichloroethane, tetrachloromethane	Log k = 13.755 E_{LUMO} – 2.0854	0.9373
Benzene	Benzene, chlorobenzene, and toluene	Log k = 7.9487 E_{LUMO} – 2.2434	0.9094
Halide	o-Dichlorobenzene, m-dichlorobenzene, and p-dichlorobenzene	Log k = 2.1831 E_{LUMO} + 9.394	0.9876
Phenol	2-Chlorophenol, 2,4-dichlorophenol, 2,4,6-trichlorophenol, pentachlorophenol	Log k = –17.463 E_{LUMO} – 0.9796	0.9700

7.4.4 Octanol/Water Partition Coefficient

The octanol/water coefficient (log P) is the standard molecular descriptor used to provide the chemical property of the hydrophobicity of a molecule. Compounds with high partition coefficients usually have very low aqueous solubility. This will decrease the chance of attack by hydroxyl radicals and lead to a lower rate constant. Nevertheless, the linear relationships with log P do not reproduce similar trends of several chemical classes such as alkane and phenol. They could be either positive or negative linear relationships. Alkene ($R^2 < 0.67$), benzene ($R^2 < 0.78$), carboxylic acid ($R^2 < 0.74$), and halide ($R^2 < 0.55$) classes do not provide significant correlations.

A good relationship is displayed for halogenated alkanes. The dataset contains 1,2-dichloroethane, 1,1,1-trichloroethane, and tetrachloromethane with negative correlation:

$$\text{Log } k = -0.9418 \log P + 1.7985; \, n = 3, R^2 = 0.9951 \qquad (7.51)$$

The amine group includes bentazone, carbofuran, and atrazine, which correlate well with negative correlation:

$$\text{Log } k = -0.4571 \log P + 10.922; \, n = 3, R^2 = 0.9019 \qquad (7.52)$$

A nearly equally good linear correlation is observed for the aromatic hydrocarbon class consisting of fluorene, acenaphthene, and phenanthrene. The rate constants exhibit a positive correlation to log P:

$$\text{Log } k = 0.2618 \log P - 1.5056; \, n = 3, R^2 = 0.9399 \qquad (7.53)$$

The best correlation of halogenated phenols is demonstrated with positive correlation at pH = 2.5. The dataset is 2-chlorophenol, 2,4-dichlorophenol, 2,4,6-trichlorophenol, and pentachlorophenol (see Figure 7.3):

$$\text{Log } k = 0.3253 \log P - 3.845; \; n = 4, \; R^2 = 0.9612 \qquad (7.54)$$

All the QSAR models using log P as the molecular descriptor can be found in Table 7.7.

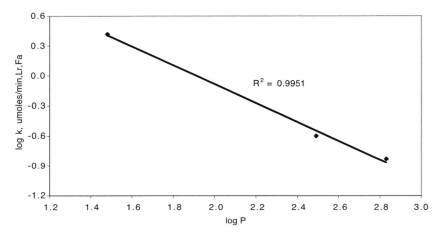

FIGURE 7.3
Correlation of oxidation rate constants of halogenated phenols with log P.

TABLE 7.7

QSAR Models with Log P

Chemical Class	Dataset	QSAR Model	R^2
Alkane	1,2-Dichloroethane, 1,1,1-trichloroethane, tetrachloromethane	Log $k = -0.9418\log P + 1.7985$	0.9951
Amine	Bentazone, carbofuran, atrazine	Log $k = -0.4571\log P + 10.922$	0.9019
Aromatic hydrocarbon	Fluorene, acenaphthene, phenanthrene	Log $k = 0.2618\log P - 1.5056$	0.9399
Phenol	2-Chlorophenol, 3-chlorophenol, 4-chlorophenol, 2,4-dichlorophenol, 2,4,6-trichlorophenol	Log $k = 0.2557\log P - 4.334$	0.9102
	Phenol, 2-chlorophenol, 2,4-dichlorophenol	Log $k = -0.5528\log P - 0.6764$	0.9664
	2-Chlorophenol, 2,4-dichlorophenol, 2,4,6-trichlorophenol, pentachlorophenol	Log $k = 0.3253\log P - 3.845$	0.9567

7.4.5 Hammett's Constants

Usually the most convenient way to use Hammett's equation is to plot log k/k_0 or just log k of the reaction of interest on the vertical axis and σ values for the substituents on the horizontal axis. The slope of the plot of log k vs. σ may change as a result of changes in the mechanism and in response to the variable electron demand of each substituent. It is anticipated that the various substituents will exert the same kinds of effects on the rate constants as they do on the benzoic acid rate constants, but the greater separation between substitution site and reaction site in the same chemical family makes the reaction less sensitive to the substituent effects. Many different oxidation reactions have been treated using Hammett's equation. The consistent trend of linear relationship with all of the Hammett's constants holds for all chemical classes except halogenated phenols.

The aliphatic systems are well correlated by Taft's constant (σ^*). Halogenated alkanes show a negative linear relationship with four datapoints (trichloromethane, 1,1,2-trichloroethane, 1,2-dichloropropane, and 1,2-dibromo-3-chloropropane), as shown in Figure 7.4.

A negative slope is observed, as anticipated, for an electron-seeking reagent. A negative slope indicates that electron-withdrawing groups have reduced the reaction constant, and a small value of slope often means that the mechanism of the reaction involves radical intermediates or a cyclic transition state with little charge separation:

$$\text{Log } k = -0.1761\sigma^* + 9.2796; \quad n = 4, \quad R^2 = 0.9934 \tag{7.55}$$

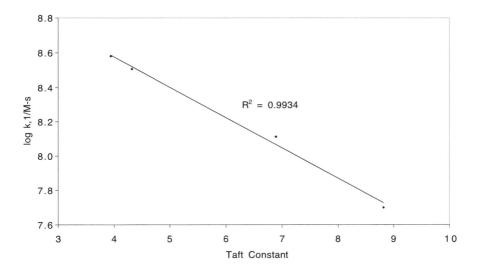

FIGURE 7.4
Correlation between the rate constants of substituted alkanes and Taft's constants.

The resonance parameter, σ_R^0, is the resonance contribution of a substituent when it is not directly conjugated with the reaction center. The dataset is composed of benzene, chlorobenzene, bromobenzene, and iodobenzene:

$$\text{Log } k = 0.8922\ \sigma_R^0 + 9.8772;\ n = 4,\ R^2 = 0.9629 \tag{7.56}$$

Although other chemical classes such as alkene ($R^2 < 0.53$), aromatic hydrocarbon ($R^2 < 0.32$), carboxylic acid ($R^2 < 0.04$), and halide ($R^2 < 0.89$) do not display significant correlations, the same trends for each class make Hammett's σ constants potential molecular descriptors, as shown in Table 7.8. The best correlation results within all the studied chemical classes with different molecular descriptors are compiled in Table 7.9.

7.5 Engineering Applications

7.5.1 Process Description

A typical UV/H_2O_2 plant consists of three major components: (1) stainless steel, titanium, or PVDF UV reactor; (2) electrical supply and UV lamp controller; and (3) dosage equipment to add H_2O_2. Usually, the contaminated water is run continuously through a tubular UV reactor that contains a UV

TABLE 7.8

QSAR Models with Hammett's Substituent Constant

Chemical Class	Dataset	QSAR Model	R^2
Alkane	Trichloromethane, 1,1,2-trichloroethane,	$\text{Log } k = -0.1761\sigma^* + 9.2796$	0.9934 / 0.9577
	1,2-Dichloropropane, and 1,2-dibromo-3-chloropropane	$\text{Log } k = -0.1645\sigma^* + 1.0086$	0.9989
	1,2-Dichloroethane, 1,1,1-trichloroethane, tetrachloromethane Dichlorobromomethane, chlorodibromomethane, bromoform	$\text{Log } k = -1.4996\sigma^* + 10.452$	
Benzene	Benzene, chlorobenzene, bromobenzene, iodobenzene	$\text{Log } k = 0.8922\ \sigma_R^0 + 9.8772$	0.9629
Phenol	3-Chlorophenol, 4-chlorophenol, 3,4-dichlorophenol, 3,5-dichlorophenol 2-Chlorophenol, 2,4-dichlorophenol, 2,4,6-trichlorophenol, pentachlorophenol	$\text{Log } k = -0.8668\sigma - 5.2609$ $\text{Log } k = 0.5142\sigma - 3.4594$	0.9411 0.9249

TABLE 7.9

Correlation Coefficients and the Rate Constants vs. Molecular Descriptors for Each Chemical Class

Chemical Class	Dipole Moment	Data Points	E_{HOMO}	Data Points	E_{LUMO}	Data Points	Log P	Data Points	Hammett Constant	Data Points
Alkane	—	—	—	—	—	—	0.9951	3	0.9989	3
Alkene	0.6895	3	1.0000	3	0.5089	3	0.6663	3	0.5295	3
Amine	0.6347	3	0.2639	3	0.8947	3	0.9019	3	—	—
Aromatic hydrocarbon	0.9818	3	0.8300	3	0.2885	3	0.9399	3	0.3119	3
Benzene	0.9850	4	0.5364	3	0.9094	3	0.7743	4	0.9629	4
Carboxylic acid	0.0004	3	0.0013	3	0.0546	3	0.7321	3	0.0347	3
Halide	0.9988	3	0.8685	3	0.9876	3	0.5474	3	0.8871	3
Phenol	0.9642	3	0.9932	3	—	—	0.9664	3	—	—
Sulfonic acid	0.1936	3	0.2921	3	0.7696	3	—	—	—	—

lamp with an optimized emission spectrum. Photons participate in three major reactions as follows:

1. The UV photons can directly excite the molecule of the organic contaminants, thus leading to their direct photochemical destruction.
2. Light activates the oxidant, such as H_2O_2, that is added to the system to produce strong oxidants such as OH.
3. Light can be absorbed by the photocatalyst, and the photocatalyst can either directly participate in the oxidation of organics or initiate the formation of OH radicals from water and the added hydrogen peroxide.

The Perox-Pure™ chemical oxidation treatment technology is one of a number of commercialized UV/H_2O_2 technologies available. The technology produces no air emissions, generates no sludge, and requires no further processing, handling, or disposal. This system has been applied for the remediation of contaminated groundwater and some specialized chemical waste treatment. It can treat approximately 64 different organic compounds and be used for the removal of biological oxygen demand (BOD), chemical oxygen demand (COD), color, and TOC. Yost (1989) presented two case studies: a pilot system and full-scale operation of the Perox-Pure system. The projected annual costs of operating a full-scale unit were reported to be $2.59/1000 gal of groundwater, operating at a flow rate of 600 gpm for a stream contaminated with aromatic compounds. The capital investment for three 200-gpm Perox-Pure systems was estimated to be $515,000 (Yost, 1989). Over 40 full-scale Perox-Pure systems are in operation today.

Another photochemical oxidation system has been developed by Sun River Innovations, Ltd. According to the company, their SR2000 system combines air stripping, oxidation, ultraviolet light, hydroxyl destruction, and self-cleaning activated carbon in one small treatment unit. The flow rate and the electrical power required are shown in Table 7.10.

The commercial success of UV/H_2O_2 processes has increased the demand for other applications. For example, UV/H_2O_2 systems can also be used for gas-phase treatment of air pollutants generated by industrial emissions. Compared to aqueous degradation the gas-phase treatment demonstrates higher degradation rates due to the lower UV absorption by air and higher mobility of the reactants and intermediates (Wekhof, 1991). However, UV/H_2O_2 systems are not suitable for the treatment of contaminated soils primarily because UV light is not capable of sufficiently penetrating soil particles. Another limiting factor is the presence of other energy sinks in the soil matrices. Additionally, in order to maximize the oxidizing efficiency, it is desirable to mobilize the contaminants into solution prior to UV/H_2O_2 oxidation.

7.5.2 Radiation Intensity

When a molecule absorbs UV photons, its internal energy increases. This increase of internal energy may be in the form of electronic, vibrational, or rotational energy. With sufficient energy, a return to the ground state, or original energy, may occur through the release of excess energy in the form of heat or fluorescent radiation. The energy loss occurs when the molecule undergoes homolytic dissociation or ionization. In other words, the correct input of energy will cause molecular bonds to break in favor of a more stable state. UV dose is simply the product of UV intensity and exposure time. UV intensity is typically measured by using ferrioxalate actinometry. Because the overall quantum yield is known to be unity, hydrogen peroxide actinometry can also be used to measure the UV intensity at 254 nm. By increasing the UV dose, the extent of oxidation may also

TABLE 7.10

Technical Data for a Typical UV/H_2O_2 Reactor

Reactor Type	Electrical Power (kW)	Water Throughput	
		(m³/hr)	(U.S.gal/min)
UX-1	1	0.5–2	2–9
UX-2	2	2–4	9–18
UX-4	4	5–8	22–35
UX-10	10	10–20	44–88
UX-20	20	20–40	88–175
UX-40	40	40–80	175–350

Source: www.uv-system.de/presse01_engl.htm, 2003.

be increased. Therefore, two methods of increasing the UV dose are to increase exposure time and UV intensity.

A radiation flow rate model can be used to quantitatively calculate UV light intensity within a reactor. If a photochemical reaction is promoted by polychromatic sources, the rate of absorbed radiation energy for the solution is given by:

$$W_{Abs} = \Sigma W_{Abs_i} = \Sigma \int_V \mu_i I_i dV \tag{7.57}$$

Equation (7.57) takes into account the individual contributions of flow rates absorbed for each wavelength; μ_i and I_i are the absorbance of the solution and the radiation intensity, respectively, for each individual wavelength. In a kinetic study, this rate of absorbed energy by the reaction medium (W_{Abs}) must be determined for each reaction time. The evaluation is carried out by solving the radiation energy balance with the help of a radiation source model that can describe the radiant energy distribution within the reactor. To account for the energy utilization, Jacob et al. (1977) developed the line source spherical emission model, which considered the geometric characteristics of the reactor. According to this model, the radiation energy flow rate absorbed by the solution for each wavelength is:

$$W_{Abs_i} = 2\pi\mu_i \int_0^H \int_{R_1}^{R_0} I_i(r,z) r dr dz \tag{7.58}$$

where $I_i(r, z)$ is the radiation intensity for each point of the reactor and for every wavelength, and r and z are the cylindrical coordinates (radial and axial respectively) of the general points considered.

If a global term that includes absorbance μ_i and all geometric parameters N_i is considered, the radiation flow rate absorbed in the entire wavelength range can finally be expressed as:

$$W_{Abs} = \Sigma W_{Abs_i} = \frac{\Sigma W_i \mu_i N_i}{2L} = W_L \frac{\Sigma \mu_i P_i N_i}{2L} \tag{7.59}$$

where L is the length of the lamp, and P_i is the fraction emitted for each wavelength. In addition to the geometric characteristics, W_{Abs} also depends on the solution absorbance μ_i and the flow rate emitted by the radiation source, W_L.

Ultraviolet intensity is easily controlled through the selection of the lamp. Low-pressure mercury cathode lamps emit most of their energy at the optimal wavelength of 253.7 nm. The electrical input ranges from 15 to 30%. The advantage of medium-pressure, mercury-vapor lamps is their high electrical rating from 0.1 to 20 kW. However, medium-pressure mercury lamps have an efficiency yield rarely exceeding 5% and only part of the light is in the region of 200 to 280 nm.

A high-power lamp that uses antimony iodide offers several advantages. First, the electrical capacities of such a lamp range from 0.5 to 5 kW, and the lamp emits light between 200 to 400 nm with a maximum wavelength of 260 nm. Additionally, the radiation yields relative to electrical rating are 12% for 200 to 280 nm, 4% for 280 to 315 nm, 7% for 315 to 400 nm, and 3% for 400 to 600 nm. The high-intensity lamps are superior due to their better water penetration and broad UV spectral emission, resulting in these lamps being better suited for most applications.

The ultimate radiation intensity to use should be based on a priority ranking of variables that are most important to the plant (e.g., cost, color removal, COD removal). Before making that decision, an extensive series of trials should be run on actual plant effluent. As a result, a cost-to-performance ratio can be assessed according to daily operating parameters.

For radiation intensity, an intensive in-plant optimization study will be required to determine the optimal level of peroxide. Ultimately, the overall system configuration should be based on a priority ranking of variables that are most important to the plant. It should be noted that in these trials, when references are made about stoichiometric mole ratio quantities of hydrogen peroxide, the assumption is made that the target contaminants being treated are 100% pure. In reality, the contaminants are not 100% pure; therefore, more peroxide should be used than would theoretically be necessary.

The relationship between radiation intensity and effective treatment rate might not be universally applicable to all substrates, especially when treatment parameters change. For example, when a more concentrated substrate solution was treated at various peroxide concentrations, a higher radiation level did not increase the decomposition rate; however, when a less concentrated substrate solution was treated, there appeared to be some treatment system efficiency improvement at higher radiation intensity. Equation (7.1) shows that UV intensity is proportional to the concentration of hydroxyl radicals produced at constant hydrogen peroxide concentration.

The sources of UV radiation in the process can be generated by low- or high-pressure mercury lamps. A low-pressure mercury lamp that has a power rating below 100 W is usually used at the laboratory scale for basic study. The high-pressure mercury lamp has a power rating that is usually larger than 25,000 W and can provide enough energy to activate the target pollutant and decompose hydrogen peroxide to hydroxyl radical effectively (Li, 1996). Table 7.11 provides a comparison of UV lamps.

7.5.3 Hydrogen Peroxide Dose

Concentration of organic substrates directly influences the required hydrogen peroxide dose. When H_2O_2 is below the optimum dose, increasing H_2O_2 dose will improve oxidation. If the H_2O_2 dose is higher than the optimum

TABLE 7.11

Comparison of Ultraviolet Lamps

Feature	Low Intensity	High Intensity
Output efficiency ($\Sigma 300$ nm)	30–40%	15–35%
Spectrum	254 nm	180–400 nm
Temperature effects	Sensitive	Not sensitive
Rated power[a]	1 W/in	200–600 W/in
Typical 6-ft lamp	65 W	15 KW
Lamp cost/KW output	$200–$3300	$90–$220

[a] Total power consumed per inch lamp length as rated by the manufacturer.

Source: Fronelich, E.M., *Water Pollut. Res. J. Can.*, 27(1), 169–183, 1992. With permission.

dose, the overdosed hydrogen peroxide may act as a hydroxyl radical scavenger, resulting in lower oxidation efficiency.

The concentration of target contaminants decreased with increasing $[H_2O_2]_0$ but then stabilized for $[H_2O_2]_0$ beyond optimal dosage. This phenomenon could be explained by two opposing effects of H_2O_2 on the steady-state concentration of $HO^\bullet$; that is, $HO^\bullet$ is generated through photolysis of H_2O_2 but is also consumed by H_2O_2. At high concentrations, excess peroxide can react with hydroxyl radicals to form water and oxygen. The entire oxidation capability of the treatment system is reduced. These opposing effects suggest a double role of hydrogen peroxide as initiator and inhibitor of oxidation. This implies that an optimal dosage of the chemical oxidant H_2O_2 can provide maximum removal of the contaminant per unit of H_2O_2. If the concentration of H_2O_2 exceeds this optimal level, no performance-to-cost benefit can be expected.

7.5.4 Temperature

Temperature will affect the degradation rate of different organic pollutants. Weir et al. (1987) reported that benzene and hydrogen peroxide are insensitive to temperature because photochemically induced reactions often have low activation energies. Koubek (1975) stated that temperature has little effect on the oxidation of refractory organics; however, Sundstrom et al. (1986) observed that the decomposition rates of some halogenated aliphatics increased with temperature.

7.5.5   Carbonate/Bicarbonate Ions

Carbonate species such as HCO_3^- and CO_3^{2-} are expected to affect the photolysis of hydrogen peroxide in aqueous solution through various pathways. These ions are considered to be scavengers of hydroxyl radicals and in natural waters are one of the main inconveniences in the applications of AOPs. Furthermore, the reaction of $HO^\bullet$ with HCO_3^- and CO_3^{2-} generates carbonate radicals ($HCO_3^\bullet$ and $CO_3^{\bullet-}$) as oxidation transients. These two forms of carbonate radicals exhibit similar reactivities toward other species. It was suggested that these radicals may disappear rapidly by attacking hydrogen peroxide, resulting in the formation of hydroperoxyl radical ($HO_2^\bullet/O_2^{\bullet-}$). Ultimately, $HO_2^\bullet/O_2^{\bullet-}$ may react to regenerate H_2O_2 through disproportionation reaction, or they may react with organic substances if their reaction rate is relatively high. For pH values above 8, the radical $CO_3^{\bullet-}$ becomes dominant (the $pK_{a,2}$ of HCO_3^- is 7.9) and reacts with $O_2^{\bullet-}$ to terminate the reaction. The carbonate ion radical, $CO_3^{\bullet-}$, is not an inert species but shows different degrees of reactivity toward organic compounds in water, as reported by Chen et al. (1975). This reactivity is highly dependent on the nature of the organics, contrary to the unselective character of the hydroxyl radical. These reactions of the carbonate species are summarized below:

$$HO^\bullet + HCO_3^-/CO_3^{2-} \rightarrow HCO_3^\bullet/CO_3^{\bullet-} + OH^-, \text{ where}$$
$$k = 8.5 \times 10^6 + 3.9 \times 10^8 \times 10^{pH-10.3} \; M^{-1} \, s^{-1} \qquad (7.60)$$

$$HCO_3^\bullet \Leftrightarrow CO_3^{\bullet-} + H^+, \text{ where } pK_a = 7.9 \qquad (7.61)$$

$$HCO_3^\bullet/CO_3^{\bullet-} + H_2O_2/HO_2^- \rightarrow HO_2^\bullet/O_2^{\bullet-} + HCO_3^-/CO_3^{2-}, \text{ where}$$
$$k = 4.3 \times 10^5 + 3 \times 10^7 \times 10^{pH-11.8} \; M^{-1} \, s^{-1} \qquad (7.62)$$

$$CO_3^{\bullet-} + O_2^{\bullet-} \rightarrow CO_3^{2-} + O_2, \text{ where } k_{19} = 6.5 \times 10^8 \; M^{-1} \, s^{-1} \qquad (7.63)$$

Significant amounts of bicarbonate in natural water may compete with organic matter and peroxide for hydroxyl radicals. Equation (7.60) indicates that an increase of bicarbonate concentration will first lower the hydroxyl radical concentration and then reduce the destruction rate of target organics. The effect of bicarbonate on the oxidation rate will be significant, particularly when the concentration of target organics is low. In fact, the concentration of substrates and the rate constant of substrate with hydroxyl radical will determine the effect of carbonate/bicarbonate. However, the effect on the decomposition rate of H_2O_2 is relatively negligible because the reaction between carbonate radical anions and hydrogen peroxide produces $HO_2^\bullet$, which then will regenerate hydrogen peroxide via termination reactions (De Laat et al., 1994).

The lack of any significant effect of carbonate ions on H_2O_2 photolysis, on the other hand, indicates that the overall quantum yield (Φ_T) remains unchanged in the presence of carbonate ions. Thus, the quantum yield for decomposition of hydrogen peroxide should be unity in bicarbonate-containing water as well as in pure water. The carbonate radicals generated by reaction of $HO^\bullet$ with HCO_3^-/CO_3^{2-} (Equation 7.60) react further with H_2O_2 (Equation 7.62) and thus regenerate the chain-propagating radicals, $HO_2^\bullet/O_2^{\bullet-}$. Accordingly, the carbonate ions behave differently than organic radical scavengers, such as acetic acid or allyl alcohol, which can reduce the quantum yield to 0.5 by quenching the chain reaction of hydrogen peroxide photolysis.

7.5.6 Natural Organic Matter

The degradation of organic pollutants can take place by direct photolysis reaction of UV irradiation. Because several UV absorbers are present, the fraction of radiation absorbed by the peroxide should be taken into consideration for the kinetic model. The presence of absorbers might have significant effects not only on the peroxide photolysis rate, but also on the direct photolysis rate of substrate due to the photon shielding on the substrate. In addition, when relatively high concentrations of M are present, intermediate reaction products may act as strong UV absorbers as well. For the case of substrate M, which is both a radical scavenger and UV absorber, assume that the rate of change of $HO^\bullet$ concentration of radicals and other radicals produced in this process is constant. The equation for $[OH]_{ss}$ may be stated as follows:

$$[OH]_{ss} = \frac{2\Phi_{PER}f_{PER}I_0[1-\exp(-2.3A_t)]}{k_{M,OH}[M]+\sum k_{HM_i,OH}[HM_i]+\sum k_{BYP_i}[BYP_i]+k_{PER,OH}[H_2O_2]}$$

$$+k_{HCO_3,OH}[HCO_3^-]+k_{CO_3,OH}[CO_3^{2-}]$$

(7.64)

where A_t is the total radiation absorbed by the solution:

$$A_t = k_M b[M] + k_{PER} b[H_2O_2] + \sum k_{HM} b[HM_i] + \sum k_{BYP} b[BYP_i] \qquad (7.65)$$

The numerator of Equation (7.64) represents the production rate of hydroxyl radicals by hydrogen peroxide photolysis. When several UV absorbers are present, the fraction of radiation absorbed by peroxide to produce the hydroxyl radicals will be smaller. The denominator of Equation (7.64) represents the consumption rate of hydroxyl radicals by key scavengers. The value of f_M and f_{PER} denotes the relative importance of the destruction pathways (i.e., the direct photolysis reaction and the hydroxyl radical reaction).

Absorption of radiation by absorbers other than peroxide can be beneficial for treatment under the following conditions: (1) M is the absorber and direct photolysis is applied; and (2) the absorber causes M to decompose by some other process such as photosensitization. In general, the overall rate of decomposition of M can be given by Equation (7.66).

$$-\frac{d[M]}{dt} = k_{M,OH}[M][OH]_{ss} + \Phi_M f_M I_0[1 - \exp(-2.3A_t)] + k_{PS}[PS][M] \quad (7.66)$$

where the first term on the right-hand side of Equation (7.66) is the rate due to reaction with the hydroxyl radicals; second term is the direct photolysis rate; and k_{PS} is the rate constant for induced decomposition by photosensitizer PS.

Natural organic matter might affect the H_2O_2/UV oxidation process through several mechanisms. First, the process might be inhibited due to absorption of UV light by the organic matter. The major source for $HO^\bullet$ formation is the photolytic decomposition of H_2O_2, the rate of which is controlled by light intensity available for H_2O_2. The fraction of the incident light intensity available for H_2O_2 is a major concern when UV-absorbing species such as humic material (HM) are present in the solution. In addition to this light-filtering effect, humic material is also known to be an effective $HO^\bullet$ scavenger. Humic material may affect the chain reactions by consuming the radicals $HO_2^\bullet/O_2^{\bullet-}$. These possible reactions involving humic material are incorporated into the reaction scheme as follows:

$$HM + h\nu \rightarrow \text{excited HM} \quad (7.67)$$

The amount of photons adsorbed can be calculated as follows:

$$A_{254} \text{ (cm}^{-1}) = \varepsilon_{HM}*DOC \text{ (mg/L)} \quad (7.68)$$

$$HO^\bullet + HM \xrightarrow{k_{HM}} HM_{oxi} \quad (7.69)$$

$$HO_2^\bullet / O_2^{\bullet-} + HM \xrightarrow{k_{HM}^*} HM_{oxi}^* \quad (7.70)$$

where A_{254} is the absorbance of light by humic material, ε_{HM} is the absorptivity of humic material at 254 nm, and DOC represents the concentration of the humic material as dissolved organic carbon. The k_{HM} and k_{HM^*} (mg/L/s are the rate constants for reaction of humic material, in terms of DOC, with HO$^\bullet$ and HO$_2$$^\bullet$/O$_2$$^{\bullet-}$, respectively. These constants are not available in the literature; however, the extent of radical scavenging by humic material is believed to remain constant during the reaction period.

In the presence of humic material, the transmitted light intensity available for absorption for wavelength of 254 nm by H$_2$O$_2$ after traveling a distance of x cm can be expressed as follows:

$$I_x = I_0 e^{-2.303A_{254}} \tag{7.71}$$

Accordingly, the average photon intensity available for absorption by H$_2$O$_2$ within the reactor can be estimated as

$$\bar{I} = \frac{\int_0^b I_x dx}{b} = \frac{\int_0^b I_0 e^{-2.303A_{254}} dx}{b} = (\frac{1-10^{-bA_{254}}}{2.303bA_{254}})I_0 = \eta' I_0 \tag{7.72}$$

where η', which represents the fraction of light available for absorption by H$_2$O$_2$, is given by:

$$\eta' = \frac{1-10^{-bA_{254}}}{2.303bA_{254}}, \text{ where } 0 < \eta' < 1 \tag{7.73}$$

In the presence of humic materials, $k_{HM}[HO^\bullet]$ is the major rate contributor while $k_{HM}[HO_2^\bullet/O_2^{\bullet-}]$ is negligible. Humic material primarily functions as a competitor to the target contaminants through the dual effects of HO$^\bullet$ scavenging and light absorption; however, humic material might also promote the radical chain reactions to generate H$_2$O$_2$, as reported for sunlight-induced reactions in natural water. But, compared to the experimental results obtained in the presence of H$_2$O$_2$, these numbers are still too small, implying that the role of the humic material as a promoter of the chain reaction is only minor, and humic material primarily functions as an effective inhibitor in this process. The slight difference between the measured and predicted values of H$_2$O$_2$ might be due to this chain-promoting effect of the humic substances. This effect has not been incorporated into the kinetic model because of difficulty in defining the interaction of the UV light with the ambiguous chemical structure of the humic material.

7.5.7 Inorganic Hydroxyl Radical Scavengers

Scavenger is a term used for substrates reacting with hydroxyl radicals that do not yield species that propagate the chain reaction. Scavengers can hinder oxidation by consuming hydroxyl radicals, hydrogen peroxide, and UV light.

Scavengers include overdosed hydrogen peroxide, humic substances, carbonate ions, bicarbonate ions, and oxidation by-products. Free-radical scavengers are compounds that consume chemical species that have at least one unpaired electron.

The most common effect of inorganic matter is caused by dissolved solids, particularly iron, calcium, and magnesium. Iron will reduce the effectiveness of the UV/H_2O_2 process by absorbing UV light, thus limiting the dissociation of H_2O_2. Dissolved iron will also undergo chemical oxidation, decreasing the amount of oxidant available to remove the target contaminant. In hard water, magnesium and calcium salts have been found to crystallize on the quartz sheaths that enclose UV lamps submerged in the wastewater. The resulting scale can reduce the intensity of the light and shield the water altogether from the UV light. For this reason, pretreatment or softening of the influent may be required. Suspended solids and some inorganic and organic species exhibiting color have a negative effect on the performance of UV/H_2O_2 systems and consequently increase operational costs. Interference due to colored compounds may have been the cause of the incomplete removal (70%) of color from bleaching wastewaters reported by Prat et al. (1990).

Pretreatment involving filtration and clarification for removal of suspended solids and turbidity may improve treatment efficiency. To reduce the interference of inorganic and organic compounds, other treatment processes may have to be combined with UV/H_2O_2 systems for effective treatment. In some situations, pH control may be required to prevent precipitation of metal salts during the oxidation process and to avoid a loss in efficiency due to the precipitates. Generally, metal hydroxide precipitation can be avoided for pH less than 6. Alkaline pH can adversely affect the reaction rate, possibly due to the base-catalyzed decomposition of H_2O_2.

7.5.8   Substrate Concentration

As substrate concentration increases, the time required for the same percent reduction of substrate might also increase. The kinetic order of a UV/H_2O_2 reaction with respect to substrate M will be zero order at high concentrations where M is the most efficient $HO^{\bullet}$ scavenger. As the reaction proceeds to completion, the order of the reaction may shift from zero order with respect to substrate to first order, scavenging effects will become pronounced, and the overall efficiency of $HO^{\bullet}$ utilization will decrease. On the other hand, more concentrated wastewaters may not be subject to scavenger effects by contaminants such as inorganic carbon unless the latter are present at very high concentrations relative to the target substrate. Photolysis of hydrogen peroxide is an effective AOP if the substrate does not absorb significant amounts of UV light.

7.5.9 pH

The highest oxidation efficiencies are usually obtained at acidic pH values and are independent of pH when pH values are below 5, where carbonic acid constitutes the major fraction of C_T ($[CO_3^{2-}] + [HCO_3^-] + [H_2CO_3^*]$). However, the oxidation efficiency is reduced drastically with increasing pH above pH 5. At this pH, the equilibrium shifts toward the bicarbonate ion, which is expected to successfully scavenge $HO^•$. Increasing pH beyond 7 transforms the bicarbonate to the carbonate ion, which has an even higher reactivity toward $HO^•$ (the rate constants for carbonate and bicarbonate ions are, respectively, 3.9×10^8 and 8.5×10^6 M^{-1} s^{-1}). As a result, the oxidation rate of target contaminants will be reduced as pH increases.

An improvement of hydroxyl radical formation occurs at 254 nm because the extinction coefficient (240 M^{-1} cm^{-1}) of the ionic form of hydroperoxide at pH 12, which is the major peroxide species, is higher than the extinction coefficient of 19 M^{-1} cm^{-1} for the nonionic form. On the other hand, the consumption rate of hydroxyl radicals by hydrogen peroxide also increases because the hydroperoxide ion reacts faster than the nonionic form. Hence, at higher pH, less $HO^•$ is available for oxidation of target contaminants. Furthermore, in the presence of bicarbonate alkalinity, a pH increase will also increase the kinetic rate constant of hydroxyl radicals toward inorganic carbon and consequently decrease the efficiency of the UV/H_2O_2 process for the degradation of organic chemicals. These results indicate that acidification of water prior to treatment by UV/H_2O_2 may prove to be quite beneficial in practical applications.

The H_2O_2 concentration was basically independent of pH, except that $[H_2O_2]$ decreased slightly at the high end of the pH range, probably because of more pronounced attack of carbonate radicals on H_2O_2/HO_2^-, as shown by Equation (7.62). It should be noted that solutions containing humic acid might change the substrate structure as a function of pH. The reaction products of target contaminants and humic acid might differ at different pH levels.

In some situations, pH control may be required to prevent precipitation of metal salts during the oxidation process and to avoid a loss in efficiency due to the precipitates. Generally, metal hydroxide precipitation is avoided for pH less than 7. Alkaline pH can adversely affect the reaction rate, possibly due to the base catalyzed decomposition of H_2O_2.

7.5.10 Nitrate

Nitrate presented in natural waters is known to interact with UV light with the formation of hydroxyl radicals. Sörensen and Frimmel (1997) used the inner filter model to describe the nitrate influence on the degradation rate of the micropollutants. In the absence of H_2O_2, the degradation rate of the photolysis is enhanced in the presence of nitrate. It can be assumed that this is caused by the formation of hydroxyl radicals during the nitrate photolysis, as shown in Equation (7.74) to Equation (7.78).

$$NO_3^- \xrightarrow{h\nu} NO_2^- + O \tag{7.74}$$

$$NO_3^- \xrightarrow{h\nu} O^{\bullet-} + NO_2^{\bullet} \tag{7.75}$$

$$2NO_2^{\bullet} + H_2O \rightarrow NO_2^- + NO_3^{\bullet} + 2H^+ \tag{7.76}$$

$$O + H_2O \rightarrow 2HO^{\bullet} \tag{7.77}$$

$$O^{\bullet-} + H_2O \rightarrow HO^{\bullet} + HO^- \tag{7.78}$$

In the presence of H_2O_2, the formation of hydroxyl radicals generated through the photolysis of H_2O_2 is much more effective than in photolysis of nitrate. So, the production of hydroxyl radicals by nitrate photolysis does not have a decisive effect on the degradation rate in the UV/H_2O_2 process; however, nitrate acts as an inner filter and reduces the UV light intensity in the photoreactor. As a consequence, the degradation rate of the micropollutants in the UV/H_2O_2 process decreases with increasing nitrate concentration. The problem of the inner filter effect caused by nitrate could be solved by using irradiation close to $\lambda = 254$ nm, because in this case photodegradation is still very fast and nitrate absorbs only weakly at this wavelength.

For some compounds such as EDTA, the photolytic degradation shows a slower rate with increasing nitrate concentration in the absence of H_2O_2. The reason is the instability of compounds themselves in UV light; they degrade very rapidly when irradiated by the light. As a result, nitrate seems to work mainly as a UV light filter rather than as a hydroxyl radical generator.

The influence of nitrate on the degradation rates in both UV photolysis and UV/H_2O_2 processes was significantly more distinct for the irradiation at the shorter wavelength. Also, the irradiation of solutions with higher nitrate concentration leads to higher nitrite concentrations, caused primarily by some orders of magnitude higher UV absorption of nitrate at the shorter wavelength. At the same time, problems with this technology can arise because nitrite and peroxynitrite can be formed during UV irradiation of nitrate. Both substances are known to be toxic, and the level of nitrite is particularly critical. In the presence of H_2O_2, the concentrations of nitrite formed are slightly but not decisively lower.

7.6   Summary

Ultraviolet/hydrogen peroxide is a promising technology for UV-catalyzed destruction of toxic or hazardous wastes. Its cost is much less sensitive to the scale of operation than ozone and the process components can be stored for use on an intermittent basis according to process demand. The hydrogen peroxide solution can be readily mixed with wastewaters, whereas ozone gas must be transferred into the water by mass transfer from ozone bubbles. Recent investigations regarding the use of UV/H_2O_2 processes for oxidative degradation of organic pollutants dissolved or dispersed in aqueous systems reveal a number of advantages when compared to other methods for chemical photochemical water treatment:

- Commercial availability of the oxidant
- Thermal stability and storage on-site; infinite solubility in water
- No mass-transfer problems associated with gases
- Formation of two hydroxyl radicals for each molecule of H_2O_2 photolyzed
- Generation of peroxyl radicals after $HO^{\bullet}$ attack on most organic substrates
- Subsequent thermal oxidation reactions occur
- Minimal capital investment
- Very cost-effective source of hydroxyl radicals
- Mobility and installation in a short time frame at different sites
- Simple operation procedure

UV/H_2O_2 systems have primarily been used to treat low levels of pollutants in the parts-per-million range typical of groundwater concentrations and may not be applicable for the treatment of high-strength wastes. Initial treatment by UV/H_2O_2 systems, followed by the use of alternative technologies, is a possible approach for highly concentrated wastes. Gray et al. (1992) pointed out that the treatment of high-strength wastes would require more efficient UV lamps with better hydroxyl radical generation efficiency.

To assess the economical viability of the technology, calculations projecting the scaled-up operating costs were also reported. As with all new and innovative treatment systems, UV/H_2O_2 oxidation will initially be compared to the industry's reference treatment process such as activated sludge. At high feed concentrations, this cost is more than the industry average for an activated sludge system; however, at low concentrations, the costs are more comparable.

Obstacles to using UV/H_2O_2 processes do exist. In fact, the rate of chemical oxidation of the contaminant is limited by the rate of formation of hydroxyl radicals, and the rather narrow absorption spectrum of H_2O_2 at 254 nm is a real disadvantage, particularly in cases where organic substrates will act as inner filters. Higher rates of hydroxyl radical formation may nevertheless be realized by the use of xenon-doped mercury arcs exhibiting a strong emission in the spectral region of 210 to 240 nm, where H_2O_2 has a higher molar absorption coefficient.

Special care in process and reactor design must be taken in order to ensure optimal oxygen concentration in and near the irradiated reactor volume. The main disadvantages of all oxidative degradation processes based on the reactivity of hydroxyl radicals, however, can be attributed to the efficient trapping of hydroxyl radicals by HCO_3^- and CO_3^{2-} ions. Although the generated carbonated radical anion has been shown to be an oxidant itself, its oxidation potential is less positive than that of the hydroxyl radical, thus introducing selectivity as far as the compounds to be degraded are concerned.

The other disadvantages of this process include the fact that radiation lamps can foul and require cleaning; lamps have finite lifetimes and require replacement; operating costs are high; special safety precautions are necessary because of the use of very reactive chemicals (hydrogen peroxide) and high-energy sources (UV lamps); and possible intermediates from incomplete oxidation may be present (Prengle et al., 1996). This process is not applicable to treatment of contaminated soil because UV light does not penetrate soil particles and also because of the presence of other energy sinks in the soil which reduce efficiency. Surfactants are required to transfer the contaminants from within soil metrics to the surface and to mobilize the contaminants into the solution prior to UV/H_2O_2 oxidation for this process to be effective. Compounds such as calcium and iron salts may precipitate during UV treatment and coat the lamp tubes, thereby reducing UV light penetration. For this reason, pretreatment or softening of the influent may be required. Suspended solids and some inorganic and organic species exhibiting color have negative effects on the performance of the UV/H_2O_2 system and consequently increase operation costs. Interference due to colored compounds may have been the cause of the incomplete removal (70%) of color from bleaching wastewater. Pretreatment involving filtration and clarification for removal of suspended solids and turbidity may improve efficiency.

References

Apak, R. and Hugul, M., Photooxidation of some mono-, di-, and tri-chlorophenols in aqueous solution by hydrogen peroxide/UV combination, *J. Chem. Tech. Biotechnol.*, 67, 221–226, 1996.

Baxendale, J.H. and Wilson, J.A., Photolysis of hydrogen peroxide at high light intensities, *Trans. Faraday Soc.*, 53, 344–356, 1957.

Beck, Detection of charged intermediate of pulse radiolysis by electrical conductivity measurements, *Int. J. Radiat. Phys. Chem.*, 1(3), 361–371, 1969.

Beltran, F.J., Ovejero, G., and Acedo, B., Oxidation of atrazine in water by ultraviolet radiation combined with hydrogen peroxide, *Water Res.*, 27(6), 1013–1021, 1993.

Beltran, F.J., Ovejero, G., and Rivas, J., Oxidation of polynuclear aromatic hydrocarbons in water, *Indust. Eng. Chem. Res.*, 35, 883–890, 1996a.

Beltran, F.J., Gonzalez, M., and Rivas, F.J., Advanced oxidation of polynuclear aromatic hydrocarbons in natural waters, *J. Environ. Sci. Health*, A31(9), 2193–2210, 1996b.

Beltran, F.J., Gonzalez, M., Acedo, B., and Jaramillo, J., Contribution of free radical oxidation to eliminate volatile organochlorine compound in water by ultraviolet radiation and hydrogen peroxide, *Chemosphere*, 32(10), 1949–1961, 1996c.

Beltran, F.J., Ovejero, G., and Rivas, J., Industrial wastewater advanced oxidation. 1. UV radiation in the presence and absence of hydrogen peroxide, *Water Res.*, 27(6), 2405–2414, 1997.

Beltran-Heredia, J., Benitez, F.J., Gonzalez, T., Acero, J.L., and Rodriguez, B., Photolytic decomposition of bentazone, *J. Chem. Tech. Biotechnol.*, 66, 206–212, 1996.

Beltran-Heredia, J., Torregrosa, J., Dominguez, J.R., and Peres, J.A., Comparison of the degradation of p-hydroxybenzoic acid in aqueous solution by several oxidation processes, *Chemosphere*, 42, 351–359, 2001.

Benitez, F.J., Heredia, J.B., Gonzalez, T., and Real, F., Photooxidation of carbofuran by a polychromatic UV irradiation without and with hydrogen peroxide, *Indust. Eng. Chem. Res.*, 34, 4099–4105, 1995.

Benitez, F.J., Beltran-Heredia, J., Acero, J.L., and Gonzalez, T., Degradation of protocatechuic acid by two advanced oxidation processes: ozone/UV radiation and H_2O_2/UV radiation, *Water Res.*, 30(7), 1597–1604, 1996.

Bevan, D.R., *QSAR and Drug Design*, 2003. http://www.NETSCI.ORG/Science/Compchem/feature12.html.

Bielski, H.J., Benon, H.J., Cabelli, D.E., Ravindra, L.A., and Alberta, A.B., Reactivity of perhydroxyl/superoxide radicals in aqueous solution, *J. Phys. Chem. Ref. Data*, 14(4), 1041–1100, 1985.

Braun, A.M. and Oliveros, E., How to evaluate photochemical methods for water treatment, *Water Sci. Technol.*, 35(4), 17–23, 1997.

Braun, W., Herron, J.T., and Kahanar, D.K., ACUCHEM: a computer program for modeling complex reaction systems, *Int. J. Chem. Kinetics*, 20, 51–62, 1988.

Buxton, G.V., Greenstock, C.L., Helman, W.P., and Ross, A.B., Critical review of rate constants for reactions of hydrated electrons, hydrogen atoms and hydroxyl radicals in aqueous solution, *J. Phys. Chem. Ref. Data*, 17(2), 513–886, 1988.

Cater, S.R., Bircher, K.G., Stevens, R.D.S., and Safarzadeh-Amiri, A., Rayox® — a second generation enhanced oxidation process for process and groundwater remediation, *Water Pollut. Res. J. Can.*, 27(1), 151–168, 1992.

Chen, S.N. and Hoffman, M.Z., Effect of pH on the reactivity of the carbonate radical in aqueous solutions, *Radiat. Res.*, 62, 18–27, 1975.

Christensen, H.S., Sehested, K., and Corftizan, H., Reactions of hydroxyl radicals with hydrogen peroxide at ambient temperatures, *J. Phys. Chem.*, 86, 15–88, 1982.

Crittenden, J.C., Hu, S., Hand, D.W., and Green, S.A, A kinetic model for H_2O_2/UV process in a completely mixed batch reactor, *Water Res.*, 33(10), 2315–2328, 1999.

Davenport, J. and Mill, T., An SAR for estimating hydroxyl radical rate constants for organic chemicals, *Natl. Mtg. Am. Chem. Soc., Div. Environ. Chem.*, 26(2), 156–161, 1986.

De, A.K., Bhattacharjee, S., and Dutta, B.K., Kinetics of phenol photooxidation by hydrogen peroxide and ultraviolet radiation, *Indust. Eng. Chem. Res.*, 36, 3607–3612, 1997.

De, A.K., Chaudhuri, B., Bhattacharjee, S., Dutta, B.K., Estimation of OH radical reaction rate constants for phenol and chlorinated phenols using UV/H_2O_2 photo-oxidation, *J. Haz. Mat.*, 64, 91–104, 1999.

De Laat, J., Tace, E., and Dore, M., Degradation of chloroethanes in dilute aqueous solution by H_2O_2/UV, *Water Res.*, 28(12), 2507–2519, 1994.

Dobson, R., Make access and the web work together, *Byte*, 21(10), 71–72, 1996.

Draganic, Z.D., Negro-Mendoza, A., Sehested, K., Vujosevic, S.I., Navarro-Gonzales, R., Albarran-Sanchez, M.G., and Draganic, I.G., Radiolysis of aqueous solutions of ammonium bicarbonate over a large dose range, *Radiat. Phys. Chem.*, 38(3), 317–321, 1991.

Eriksen, T.E., Lind, J., and Merenyi, G., On the base-acid equilibrium of the carbonate radical, *Radiat. Phys. Chem.*, 26(2), 197–199, 1985.

Ertl, P., Simple quantum chemical parameters as an alternative to the Hammett sigma constants in QSAR studies, *QSAR*, 16(5), 377–382, 1997.

Ertl, P., World Wide Web-based system for the calculation of substituent parameters and substituent similarity searches, *J. Molec. Graph. Model.*, 16(1), 11–13, 1998.

Famini, G.R., *Using Theoretical Descriptors in Quantitative Structure Activity Relationships and Linear Free Energy Relationships*, 2003. http://www.NETSCI.ORG/Science/Compchem/feature08.html.

Finar, I.L., *Organic Chemistry*, Longman, London, 1973.

Fleming, I., *Frontier Orbitals and Organic Chemical Reactions*, Wiley, New York, 1976.

Froelich, E.M., Advanced chemical oxidation of organics using the Perox-Pure™ oxidation system, *Water Pollut. Res. J. Can.*, 27(1), 169–183, 1992.

Glaze, H.W. and Kang, J.W., Advanced oxidation process. Description of a kinetics model for the oxidation of hazardous materials in aqueous media with ozone and hydrogen peroxide in a semi-batch reactor, *Ind. Eng. Chem. Res,.* 28, 1573–1579, 1989a.

Glaze, H.W. and Kang, J.W., Advanced oxidation process. Test of a kinetics model for the oxidation of hazardous materials in aqueous media with ozone and hydrogen peroxide in a semi-batch reactor, *Ind. Eng. Chem. Res.*, 28, 1580, 1989b.

Glaze, W.H., Beltran, F., Tuhkanen, T., and Kang, J.W., Chemical models of advanced oxidation processes, *Water Pollut. Res. J. Can.*, 27(1), 23–42, 1992.

Glaze, W.H., Lay, Y., and Kang, J.W., Advanced oxidation processes: a kinetic model for the oxidation of 1,2–dibromo-3–chloropropane in water by the combination of hydrogen and UV radiation, *Indust. Eng. Chem. Res.*, 34, 2314–2323, 1995.

Gray, L.W., Adamson, M.G.A., Hickman, R.G., Farmer, J.C., Chiba, Z., Gregg, D.W., and Wang, F.T., Aqueous phase oxidation techniques as an alternative to incineration, Lawrence Livermore National Laboratory, UCRL-JC-108867, Albuquerque, NM, 1992.

Haag, W.R. and Yao, C.C.D., Rate Constants for reaction of hydroxyl radicals with several drinking water contaminants, *Environ. Sci. Technol.*, 26(5), 1005–1013, 1992.

Hammett, L.P., *Physical Organic Chemistry: Reaction Rates, Equilibria, and Mechanisms*, McGraw-Hill, New York, 1970.

Hansch, C. and Leo, A., *Exploring QSAR: Fundamentals and Applications in Chemistry and Biology*, American Chemical Society, Washington, D.C., 1995.

Hansch, C., Leo, A., and Hoekman, D., *Exploring QSAR: Hydrophobic, Electronic, and Steric Constants*, American Chemical Society, Washington, D.C., 1995.

Hansch, C., Leo, A., and Taft, R.W., A survey of Hammett substituent constants and resonance and field parameters, *Chem. Rev.*, 91(2), 165–195, 1991.

He, J., Databases on the internet for engineers, *Exp. Tech.*, 22(4), 38–39, 1998.

Hendrix, T.H., Adsorption Kinetics, Web Database, and QSAR Models for Organic Degradation by TiO_2/UV, Master's thesis, Florida International University, Miami, 1998.

Heredia, J.B., Benitez, F.J., Gonzalez, T., and Acero, J.L., Photolytic decomposition of bentazone, *J. Chem. Tech. Biotechnol.*, 66, 206–212, 1996.

Hine, J., *Structural Effects on Equilibria in Organic Chemistry*, Wiley, New York, 1975.

Hirsch, P.M., Exercise the power of the World Wide Web, *IEEE Computer Appl. Power*, 8(3), 25–29, 1995.

Hirvonen, A., Tuhkanen, T., and Kalliokoski, P., Formation of chlorinated acetic during UV/H_2O_2 oxidation of ground water contaminated with chlorinated ethylenes, *Chemosphere*, 32(6), 1091–1102, 1995.

Hoigne, J, Inter-calibration of radical sources and water quality parameters, *Water Sci. Technol.*, 35(4), 1–8, 1997.

Holcman, J., Bjergbakke, E., and Sehested, K., The importance of radical–radical reactions in pulse radiolysis of aqueous carbonate/bicarbonate, *Proc. Tihany Symp. Radiat. Chem.*, 6(1), 149–153, 1987.

Huie, R.E. and Clifton, C.L., Temperature dependence of the rate constants for reactions of the sulfate radical, SO_4^-, with anions, *J. Phys. Chem.*, 94(23), 8561–8567, 1990.

Jacob, N., Balakrishnan, I., and Redd, M.P., Characterization of the hydroxyl radical in some photochemical reactions, *J. Phys. Chem.*, 81, 17–22, 1977.

Kang, J.W. and Lee, K.H., A kinetic model of the hydrogen peroxide/UV process for the treatment of hazardous waste chemicals, *Environ. Eng. Sci.*, 14(3), 183–192, 1997.

Karelson, M., Lobanov, V.S., and Katritzky, A.R., Quantum-chemical descriptors in QSAR/QSPR studies, *Chem. Rev.*, 96, 1027–1043, 1996.

Kawaguchi, H., Oxidation efficiency of hydroxyl radical in the photooxidation of 2-chlorophenol using ultraviolet radiation and hydrogen peroxide, *Environ. Technol.*, 14, 289–293, 1993.

Kim, S.M., Geissen, S.U., and Vogelpohl, A., Landfill leachate treatment by a photo-assisted Fenton reaction, *Water Sci. Technol.*, 35(4), 239–248, 1997.

Kochany, J. and Bolton, J.R., Mechanism of photodegradation of aqueous organic pollutants. 2. Measurement of the primary rate constants for reaction of HO• radicals with benzene and some halobenzenes using an EPR spin-trapping method following the photolysis of H_2O_2, *Environ. Sci. Technol.*, 26(2), 262–265, 1992.

Koppenol, W.H., Butler, J., and Van Leeuwen, J.W.L., The Haber–Weiss cycle, *Photochem. Photobiol.*, 28, 655–660, 1978.

Koubek, E., *Ind. Eng. Chem. Process Design and Development*, 14, 348, 1975.

Kruk, A., Westerland, T., and Heller, P., Database management systems, *Chem. Eng.*, 103(5), 82–85, 1996.

Ku, Y. and Ho, S.C., The effect of oxidants on UV destruction of chlorophenols, *Environ. Progr.*, 9(4), 218–221, 1990.

Legrini, O., Oliveros, E., and Braun, A.M., Photochemical processes for water treatment, *Chem. Rev.*, 93, 671–698, 1993.

Leifer, A., *The Kinetics of Environmental Aquatic Photochemistry: Theory and Practice*, American Chemical Society Professional Reference Book, Washington DC, 1988.

Leitner, N.K.V. and Dore, M., Mechanism of the reaction between hydroxyl radicals and glycolic, glyoxylic, acetic and oxalic acids in aqueous solution: consequence on hydrogen peroxide consumption in the H_2O_2/UV and O^{-3}/H_2O_2 systems, *Water Res.*, 31(6): 1383–1397, 1997.

Leitner, N.K.V. and Dore, M., Hydroxyl radical induced decomposition of aliphatic acids in oxygenated and deoxygenated aqueous solutions, *J. Photochem. Photobiol., A*, 99(2–3): 137–143, 1996.

Leitner, N.K.V., Gombert, B., Ben Abdessalem, R., and Dore, M., Kinetics and mechanisms of the photolytic and OH· radical induced oxidation of fluorinated aromatic compounds in aqueous solution, *Chemosphere*, 32(5), 893–906, 1996.

Li, Y., Photolytic hydrogen peroxide oxidation of 2-chlorophenol wastewater, *Arch. Environ. Contam. Toxicol.*, 31, 557–562, 1996,

Liao, C.H. and Gurol, M.D., Chemical oxidation by photolytic decomposition of hydrogen peroxide, *Environ. Sci. Technol.*, 29(12), 3007–3014, 1995.

Lowry, T.H. and Richardson, K.S., *Mechanism and Theory in Organic Chemistry*, Harper & Row, New York, 1987.

Lyman, W.J., Reehl, W.F., and Rosenblatt, D.H., *Handbook of Chemical Property Estimation Methods: Environmental Behavior of Organic Compounds*, McGraw-Hill, New York, 1982.

Majcen-Le Marechal, A., Slokar, Y.M., and Taufer, T., Decoloration of chlorotriazine reactive azo dyes with H_2O_2/UV, *Dyes Pigment*, 33(4), 281–298, 1997.

March, J., *Advanced Organic Chemistry: Reactions, Mechanisms, and Structure*, John Wiley & Sons, New York, 1985.

Martianov, I.N., Savinov, E.N., and Parmon, V.N., A comparative study of efficiency of photooxidation of organic contaminants in water solutions in various photochemical and photocatalytic systems. 1. Phenol photooxidation promoted by hydrogen peroxide in a flow reactor, *J. Photochem. Photobiol. A: Chem.*, 107, 227–231, 1997,

Maruthamuthu, P. and Neta, P., Phosphate radicals: spectra, acid–base equilibria, and reactions with inorganic compounds, *J. Phys. Chem.*, 82, 710–713, 1978.

McWeeny, R., *Methods of Molecular Quantum Mechanics*, Academic Press, London, 1989.

Miller, R.M., Singer, G.M., Rosen, J.D., and Bartha, R., Sequential degradation of chlorophenols by photolytic and microbial treatment, *Environ. Sci. Technol.*, 22(10), 1215–1219, 1988.

Nakashima, M. and Hayon, E., Rates of reaction of inorganic phosphate radicals in solution, *J. Phys. Chem.*, 74(17), 3290–3291, 1970.

Namboodri, C.G. and Walsh, W.K., Ultraviolet light/hydrogen peroxide for decolorizing spent reactive dyebath waste water, *Am. Dyestuff Reporter*, 85(3), 15–25, 1996.

Nirmalakhandan, N. and Speece, R.E., Structure–activity relationships, *Environ. Sci. Technol.*, 22(6), 606–615, 1988.

Pace, G, Berton, A., Calligaro, L., Mantovani, A., and Uguagliati, P., Elucidation of the degradation mechanism of 2–chloroethanol by hydrogen peroxide under ultraviolet irradiation, *J. Chromatogr. A*, 706, 345–351, 1995.

Perry, R.H., Green, D.W., and Maloney, J.D., *Chemical Engineer's Handbook,* 5th ed., McGraw-Hill, New York, 1981.

Pignatello, J.J. and Chapa, G., Degradation of PCBs by ferric ion, hydrogen peroxide and UV light, *Environmental Toxicol. Chem.,* 13(3), 423–427, 1994.

Prat, C., Vincente, M., and Esplugas, S., Treatment of bleaching waters in the paper industry by hydrogen peroxide and ultraviolet radiation, *Water Res.,* 663–668, 1988.

Prengle, H.W., Experimental rate constant and reactor considerations for the destruction of micropollutants and trihalomethane precursors by ozone with UV-radiations, *Environ. Sci. Technol.,* 17, 743-747, 1983.

Prengle, H.W., Jr., Symons, J.M., and Belhateche, D., H_2O_2/VisUV process for photooxidation of waterborne hazardous substances $-C_1-C_6$ chlorinated hydrocarbons, *Waste Manage.,* 16(4), 327–333, 1996.

Rance, P.J.W. and Skelton, R.L., The combination of iron salts with ultraviolet light and hydrogen peroxide for the decomposition of organic materials, *1994 Icheme Res. Event,* 1221–1223, 1994.

Roe, B.A. and Lemley, A.T., Treatment of two insecticides in an electrochemical Fenton system, *Environ. Sci. Health B,* 32(2), 261–281, 1997.

Safarzaden-Amiri, A., Bolton, J.R., and Cater, S.R., Ferrioxalate-mediated photooxidation of organic pollutants in contaminated water, *Water Res.,* 31(4), 787–798, 1997.

Sapach, R. and Viraraghavan, T., An introduction to the use of hydrogen peroxide and ultraviolet radiation: an advanced oxidation process, *J. Environ. Sci. Health,* A32(8), 2355–2366, 1997.

Sehested, K., Rasmussen, O.L., and Fricke, H., Rate constants of OH with HO_2, and O_2^- and H_2O_2 from hydrogen peroxide formation in pulse-irradiated oxygenated water, *J. Phys. Chem.,* 72, 626–631, 1968.

Shen, Y.S., Ku, Y., and Lee, K.C., The effect of light absorbance on the decomposition of chlorophenols by ultraviolet radiation and UV/H_2O_2 processes, *Water Res.,* 29(3), 907–914, 1995.

Sörensen, M. and Frimmel, F.H., Photochemical degradation of hydrophilic xenobiotics in the UV/H_2O_2 process: influence of nitrate on the degradation rate of EDTA, 2-amino-1-naphthalenesulfonate, diphenyl-4-sulfonate and 4,4'-diaminostilbene-2,2'-disulfonate, *Water Res.,* 31(11), 2885–2891, 1997.

Stefan, M.I., Hoy, A.R., and Bolton, J.R., Kinetics and mechanism of the degradation and mineralization of acetone in dilute aqueous solution sensitized by the UV photolysis of hydrogen peroxide, *Environ. Sci. Technol.,* 30, 2382–2390, 1996.

Stumm, W. and Morgan, J.J., *Aquatic Chemistry,* Prentice Hall, Englewood Cliffs, NJ, 1972.

Sundstrom, D.W., Klei, H.E., Nalette, T.A., Reidy, D.J., and Weir, B.A., Destruction of halogenated aliphatics by ultraviolet catalyzed oxidation with hydrogen peroxide, *J. Hazardous Waste Hazardous Mater.,* 3(1), 101–110, 1986.

Sundstrom, D.W., Weir, B.A., and Klei, H.E., Destruction of aromatic pollutants by UV light catalyzed oxidation with hydrogen peroxide, *Environ. Progr.,* 8(1), 6–11, 1989.

Symons, J.M. and Worley, K.L., An advanced oxidation process for DBP control, *J. Am. Water Works Assoc.,* 87(11), 66–75, 1995.

Symons, J.M. and Zheng, M.C.H., Technical note: does hydroxyl radical oxidize bromide to bromate? *J. Am. Water Works Assoc.,* 89(6), 106–109, 1997.

Tang, W.Z. and Huang, C.P., The effect of chlorine position of chlorinated phenols on their dechlorination kinetics by Fenton's reagent, *Waste Manage.*, 15(8), 615–622, 1995.

Tang, W.Z. and Tassos, S., Oxidation kinetics and mechanism of trihalomethanes by Fenton's reagent, *Water Res.*, 31(5), 1117–1125, 1997.

Tosato, M.L., Chiorboli, C., Eriksson, L., and Jonsson, J., Multivariate modelling of the rate constant of the gas-phase reaction of haloalkanes with the hydroxyl radical, *Sci. Total Environ.*, 109/110, 307–325, 1991.

Trapido, M., Hirvonen, A., Veressinina, Y., and Hentunen, J., Ozonation, ozone/UV and UV/H_2O_2 degradation of chlorophenols, *Ozone: Sci. Eng.*, 19(1), 75–96, 1997.

Tratnyek, P.G. and Hoigne, J., Kinetics of reactions of chlorine dioxide (OCIO) in water. II. Quantitative structure–activity relationships for phenolic compounds, *Water Res.*, 28(1), 57–66, 1994.

Tratnyek, P.G., Hoigne, J., Zeyer, J., and Schwarzenbach, R.P., QSAR analyses of oxidation and reduction rates of environmental organic pollutants in model systems, *Sci. Total Environ.*, 109/110, 327–341, 1991.

Udell, J., Microsoft's Windows database, *Byte*, 17(14), 51–52. 1992.

USEPA, Ultrox International Ultaviolet Radiation/Oxidation Technology: Applications Analysis Report, EPA/540/A5-89/012, Risk Reduction Engineering Laboratory, USEPA, Cincinnati, OH, 1990.

Venkatadri, R. and Peters, R.W., Chemical oxidation technologies: ultraviolet light/hydrogen peroxide, Fenton's reagent, and titanium dioxide-assisted photocatalysis, *J. Hazardous Waste Hazardous Mater.*, 10(2), 107–149, 1993.

Wagler, J.L. and Malley, J.P., Jr., The removal of methyl tertiary-butyl ether from a model ground water using UV/peroxide oxidation, *J. NEWWA*, Sept., 236–260, 1994.

Wang, C.X., Yediler, A., Peng, A., and Kettrup, A., Photodegradation of phenanthrene in the presence of humic substances and hydrogen peroxide, *Chemosphere*, 30(3), 501–510, 1995.

Wang, J. and Reid, E.O.F., Developing WWW information systems on the Internet, *Microcomputers Inform. Manage.: Global Internetworking Libraries*, 13(3/4), 237–252, 1996.

Weir, B.A. and Sundstrom, D.W., Destruction of trichloroethylene by UV Light-catalyzed oxidation with hydrogen peroxide, *Chemosphere*, 27(7), 1279–1290, 1993.

Weir, B.A., Sundstrom, D.W., and Klei, H.E., Destruction of benzene by ultraviolet catalyzed oxidation with hydrogen peroxide, *Haz. Waste Haz. Mat.*, 4, 165–176, 1987.

Weistein, J., Benon, H.J., and Bielski, H.J., Kinetics of the interaction of HO2 and O2– radicals with hydrogen peroxide: the Haber–Weiss reaction, *J. Am. Chem. Soc.*, 101, 58–62, 1979.

Wekhof, A., *Environ. Prog.*, 10, 241, 1991.

Worthington, S.S, Internet: a critical resource for today's professional, *Contr. Eng.*, 43(7), 81–82, 1996.

Yang, Y., Wyatt II, D.T., and Bahorsky, M., Dechlorization of dyes using UV/H_2O_2 photochemical oxidation, *Textile Chem. Colorist*, 30(4), 27–35, 1998.

Yao, C.C.D., Haag, W.R., and Mill, T., Kinetic features of advanced oxidation processes for treating aqueous chemical mixtures, in *Chemical Oxidation: Technology for the Nineties, Vol. 2* Eckenfelder, W.W., Bowers, A.R., and Roth, J.A. Eds., Technomic Publishing Company, Inc., Lancaster, PA, 1994, pp 112–139.

Yost, K.W., *Proc. of the 43rd Purdue Indus. Conf.*, 43, 441, 1989.

Yue, P.L., Degradation of organic pollutants by advanced oxidation, *Trans. IchemE*, 70 (pt. B), 145–148, 1992.

Yue, P.L., Oxidation reactions for water and wastewater treatment, *Water Sci. Technol.*, 35(4), 189–196, 1997.

Yue, P.L., and Legrini, O., Photochemical degradation of organics in water, *Water Pollut. Res. J. Can.* 27(1): 123–137, 1992.

8

Ultraviolet/Ozone

8.1 Introduction

Three dominant reactions during ultraviolet (UV)/ozone (O_3) treatment processes that effectively decompose organic pollutants are photolysis, ozonation, and reactions of hydroxyl radicals. The generation of hydroxyl radicals is essential in this oxidation process as it is the reaction between these radicals and organic compounds that can ultimately destroy organic pollutants. Physical parameters, such as temperature, pH, initial compound and ozone concentrations, UV intensity, and ozone partial pressure will also have considerable effects on the kinetic rate constants and removal efficiency of any compound.

The UV/ozone process is commonly used to degrade toxic organic compounds often found in surface and groundwater. Many of these compounds originate from the chemical, petrochemical, pesticide, and herbicide industries. The molecular structure of organic pollutants to be oxidized has a significant impact on kinetic rate constants and the removal efficiency of the compound. Parent compounds can be partially oxidized to form by-products during oxidation treatment. These intermediates can further react with hydroxyl radicals, creating a "scavenging" effect that often reduces the degradation rates of parent compounds.

8.2 Decomposition Kinetics of UV/Ozone in Aqueous Solution

Ozonation processes are rather complex, due to the high instability of ozone in aqueous solutions. Ozone absorbs UV photons with the maximal absorption at 253.7 nm. The decomposition of ozone under UV radiation typically occurs through three reactions: direct photolysis, direct ozonation, and reactions between hydroxyl radicals and hydrogen peroxide as shown in the following reactions:

Photolysis (slow):

$$O_3 + H_2O \xrightarrow{h\nu} H_2O_2 + O_2 \tag{8.1}$$

$$H_2O_2 \xrightarrow{h\nu} 2OH^\bullet \tag{8.2}$$

Ozonation:

$$HO_2^- + O_3 \rightarrow HO_2^\bullet + O_3^- \tag{8.3}$$

Hydroxyl radical reactions:

$$H_2O_2 + OH^\bullet \rightarrow HO_2^\bullet + H_2O \tag{8.4}$$

$$HO_2^- + OH^\bullet \rightarrow HO_2^\bullet + OH^- \tag{8.5}$$

In a gaseous phase enriched with water vapor, the mechanism of photolysis involves the release of a molecule of oxygen and an atom of oxygen (^1D). The latter may react with water to produce hydroxyl radicals:

$$O_3 + h\nu \rightarrow O_2 + O\ (^1D) \tag{8.6}$$

$$O\ (^1D) + H_2O \rightarrow HO^\bullet + HO^\bullet \tag{8.7}$$

where the kinetic rate constants for Equation (8.6) and Equation (8.7) are 2.7×10^7 and $7.5 \times 10^9\ M^{-1}\ s^{-1}$, respectively (Beltran et al., 1994); however, if two hydroxyl radicals are prevented from escaping the solvent cage, they may recombine in solution to form hydrogen peroxide. The overall photolysis of ozone in solution is therefore likely to be represented by the following reaction:

$$O_3 + H_2O \xrightarrow{h\nu} [2HO^\bullet]_{cage} + O_2 \rightarrow H_2O_2 + O_2 \tag{8.8}$$

Prousek (1996) showed that hydrogen peroxide is in fact the primary product of ozone photolysis and also summarized the chemistry involved in the generation of $HO^\bullet$ radicals by the UV/ozone process as follows:

$$H_2O_2 \leftrightarrow HO_2^- \xrightarrow{O_3} HOO^\bullet + O_3^- \tag{8.9}$$

$$O_3^- + H^+ \rightarrow HO_3 \rightarrow HO + O_2 \tag{8.10}$$

$$O_3^- + H_2O \rightarrow HO^\bullet + HO^- + O_2 \tag{8.11}$$

$$HO^\bullet + RH \rightarrow R^- \rightarrow ROO^\bullet \tag{8.12}$$

Reactions of ozone can be initiated by $HO^\bullet$ or $HOO^\bullet$ or by photolysis of hydrogen peroxide. Ozone can also be decomposed through the following reaction pathways:

$$O_3^- \rightarrow O_2 + O^- \tag{8.13}$$

$$O^- + H_2O \rightarrow HO^\bullet + HO^- \tag{8.14}$$

Decomposition rates of ozone can be influenced by pH, UV irradiation, and the presence of free-radical scavengers generated from anion species (Ku et al., 1996a). For example, anions such as chloride, carbonate, and nitrate in aqueous solutions tend to scavenge hydroxyl radicals produced during UV/ozone oxidation processes, subsequently reducing the decomposition rates of ozone. An increase in alkalinity will rapidly increase the decomposition rate of ozone, due to hydroxyl radicals being consistently formed by the reaction between ozone and hydroxide ions. UV irradiation also assists in increasing the decomposition rate of ozone. The rates of reaction expressed by second-order kinetics achieved by hydroxyl radicals are typically 10^6 to 10^9 times faster than the corresponding rates by molecular ozone, as shown in Table 8.1. The rate constants are useful in estimating $HO^\bullet$-induced oxidation rates of organic compounds in a variety of aqueous systems, including atmospheric water droplets, sunlit surface waters, and room-temperature radical oxidation processes.

TABLE 8.1

Reaction Rates of Ozone and Hydroxyl Radicals with Classes of Organic Compounds

Compound	k (M^{-1} s^{-1})	
	O_3	$HO^\bullet$
Olefins	$1–450 \times 10^3$	$10^9–10^{11}$
S-containing organics	$10–1.6 \times 10^3$	$10^9–10^{10}$
Phenols	10^3	10^9
N-containing organics	$10–10^2$	$10^8–10^{10}$
Aromatics	$1–10^2$	$10^8–10^{10}$
Acetylenes	50	$10^8–10^9$
Aldehydes	10	10^9
Ketones	1	$10^9–10^{10}$
Alcohols	$10^{-2}–1$	$10^8–10^9$
Alkanes	10^{-2}	$10^6–10^9$
Carboxylic acids	$10^{-3}–10^{-2}$	$10^7–10^9$

According to Hayashi et al. (1993), refractory compounds, such as saturated alcohols and carboxylic acids are built up in the system after a certain amount of ozonation. Kusakabe (1990, 1991) applied UV irradiation to the preozonation of humic acid and found that the UV/ozone process accelerated the decomposition of volatile organic compound (VOC) and nonvolatile organic carbon (NVOC) precursors. Vollmuth and Niessner (1997) argued that, if organic compounds in dilute aqueous solutions are oxidized only by a direct reaction with ozone, the accelerated decomposition of ozone should retard the oxidation rate. This is not the case at a pH greater than 7, when decomposition of ozone actually accelerates the degradation of organic pollutants. Benitez et al. (1997), Glaze et al. (1982), and Peyton and Glaze (1988) have argued that a synergistic effect must exist between ozone and UV photons that cannot be accounted for as shown in Figure 8.1.

Yue (1993) suggested a mechanism of destruction of an organic pollutant that begins with photolysis of ozone in the solution, which produces hydrogen peroxide. The deprotonated peroxide reacts with ozone to produce ozonide and hydroxyl radicals that attack the organic substrate to form an organic carbon-centered radical, which reacts quickly with oxygen to form a peroxyl radical, which decomposes to produce superoxide or hydrogen peroxide. The cyclic reaction pathway is completed with the superoxide reacting with ozone to produce ozonide.

Chemical reactions in UV/ozone process are a series of slow and fast reactions. The reaction time is determined by the time taken to complete the sequence of reactions. In the presence of $OH^{\bullet}$ radical scavengers, the oxidative efficiency of $OH^{\bullet}$ radicals will be reduced. For example, bicarbonate and carbonate ions usually play a dominant role as $OH^{\bullet}$ radical scavengers, because they present at concentrations of several millimoles per liter and react with $OH^{\bullet}$ radicals with rate constants as high as 1.1×10^7 L/mol/cm and 3.9×10^8 L/mol/cm for bicarbonate and carbonate, respectively.

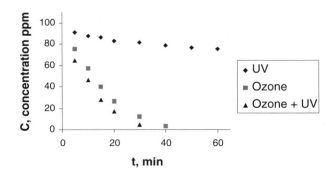

FIGURE 8.1
Comparison of the oxidation power of the different oxidants used on syringic acid degradation. $T = 20°C$, pH = 7, pressure = 0.43 kPa. (From Benitez F.J. et al., *Indust. Eng. Chem. Res.*, 33, 1264–1270, 1997. With permission.)

Mokrini et al. (1997) proposed that the photolytic ozonation kinetics of substrates is a linear combination of purging, ozonation, photolysis, and photolytic ozonation:

$$d[S]/dt = k_{purge}\,[S] + k_{photo}\,I^a[S]^b + k_{ozon}[O_3]_l^c\,[S]^d + k_{po}\,I^eD^f[S]^g \quad (8.15)$$

where I is the flux of radiation input into a reactor; D is the dose rate of ozone; $[O_3]_l$ is the concentration of ozone in the liquid phase; and $[S]$ is the substrate concentration (Figure 8.2).

Four major factors influence the oxidation rate of organic pollutants: (1) pH, (2) relative concentration of oxidants (O_3/H_2O_2), (3) photon flux in the UV/O_3 system, and (4) radical scavenger concentration.

8.2.1 pH Effect

Beltran et al. (1998) reported that the oxidation kinetics of nitroaromatic hydrocarbons at different pH levels (between 2 and 12) was similar to those found in O_3/H_2O_2 oxidation — for example, the positive effect of pH on removal rate between pH 2 and 7 and partial inhibition at pH 12. The ozone efficiency increased with pH, from 30 to 40% (pH = 4) to 95% (pH 9 or 11). At pH 4, about a 10% difference was observed between the ozone efficiencies obtained during UV/ozone radiation oxidation and ozonation alone, while no difference was observed at pH 9 or 11. Figure 8.3 shows this effect for the degradation of vanillic acid, as reported by Benitez et al. (1997).

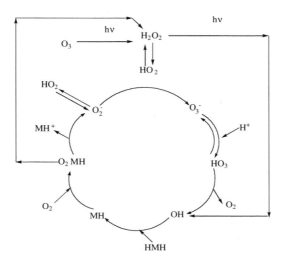

FIGURE 8.2
Ozone decomposition process by photolysis at 253.7 nm. (From Peyton, R. and Glaze, W., *Environ. Sci. Technol.*, 22, 761–767, 1988. With permission.)

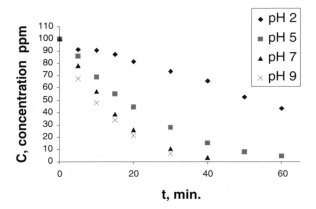

FIGURE 8.3
UV/ozone: influence of pH on the degradation of vanillic acid. $T = 20°C$, pressure = 0.43 kP$_a$. (From Benitez, F.J. et al., *Indust. Eng. Chem. Res.*, 33, 1264–1270, 1997. With permission.)

8.2.2   Concentration of Oxidants

In UV/ozone processes, the increase of ozone feed rate leads to an increase of oxidation rate at a given time. This increase of ozone feed rate leads to an increase of the ozone partial pressure and hence to an increase of ozone driving force and thus the ozone absorption rate. When ozone is combined with UV radiation, dinitrotoluene (DNT) is removed at a faster rate than by ozonation alone; therefore, a synergism exists between ozone and UV radiation (Beltran et al., 1998). Shen and Young (1997) reported that the addition of ozone slightly improved (4 to 6%) on the removal rate of trichloroethylene (TCE) until the dosage reached 480 ppmv, at which point further addition of ozone decreased the removal of TCE. Bhowmick and Semmens (1994) concluded that the absorption of UV photons by ozone increased significantly with ozone concentration and thus inhibited photolysis. The chlorinated intermediates generated from the decomposition of TCE by the UV/ozone process were found to be much fewer than for direct photolysis, indicating that hydroxyl radicals could significantly promote the decomposition of chlorinated intermediates. The optimum ozone dosage was found to be 480 ppmv; however, excessive ozone would reduce the treatment efficiency of TCE by the UV/ozone process.

8.2.3   Effect of Photon Flux in the UV/Ozone System

The effect of UV light intensity on the decomposition of organics by the UV/ozone process was studied by Ku and Shen (1999). The removal efficiency of three chloroethenes increased with increasing light intensity and could be as high as 95%. The treatment efficiency of chloroethenes by the UV/ozone process was found to be much higher than for direct photolysis under various UV light intensities.

8.2.4 Radical Scavengers

Inhibitors of free-radical reactions are compounds capable of consuming OH radicals without regenerating the superoxide anion O_2^-. Some of the common inhibitors include bicarbonate and carbonate ions, alkyl groups, tertiary alcohols, and humic substances. Natural waters contain varying concentrations of numerous organic and inorganic natural compounds. These compounds are present in dissolved and suspended forms. In the presence of algae and degraded animal and vegetable residues, physical, chemical, and biological reactions in water and soil will produce natural organic matter. The main organic constituents in natural water are a collection of polymerized organic acids called *humic acids*.

The inhibitive reactions of bicarbonate and carbonate ions with hydroxyl radicals are similar to those described in Chapter 7. During the oxidation of atrazine, UV light accelerates the decomposition of dissolved ozone in aqueous solutions (Ku et al., 1996a). Furthermore, in solutions of pH greater than 3, the decomposition rates of ozone are usually constant. UV light causes the consumption rates of ozone (r_c) to increase, while an increase in solution pH value reduces the degree to which UV irradiation affects ozone consumption rates. UV irradiation will not significantly affect the consumption rate of hydroxide ions (r_{OH^-}). Decomposition rates generally increase linearly as UV light intensity increases, regardless of pH. The main contribution of ozone decomposition in acidic solutions would be UV photolysis, because the decomposition of ozone was found to be at a minimum at a pH of 2 without the influence of UV irradiation (Ku et al., 1996a). The overall decomposition rate equation of ozone in the presence of UV light in a range of solution pH of 2 to 10 is given as follows:

$$-d[O_3]/dt = 23.47\,[O_3]^{1.5}\,[OH^-]^{0.395} + 0.1414[O_3]^{1.5}[OH^-]^{0.064}[I]^{0.9} \quad (8.16)$$

where I is the UV light intensity (W/m^2). The UV light required to decompose ozone decreases with increasing solution pH values, indicating that the decomposition of ozone by OH^- could be the major reaction in alkaline solutions. Photolysis is the main reaction pathway in acidic solutions in the presence of UV light (Ku et al., 1996a). The presence of anion species did not significantly change the decomposition rates of ozone (r_c) or hydroxide ions (r_{OH^-}). However, hydroxide ion consumption increased considerably under UV irradiation; ozone decomposition increased only slightly under UV light. Cl^- and NO_3^- ions were found to be very weak scavengers of hydroxyl radicals and had only a minimal effect on the decomposition of ozone. With the presence of CO_3^{2-}, the consumption of hydroxide ions was not detected during the decomposition of ozone without UV irradiation, possibly due to the carbonate-buffer effect; however, carbonate is the most effective scavenger among the three anions studied.

Changes in pH value and UV irradiation intensity usually have the greatest effect on ozone decomposition rates, while the selected anion species such as chloride, carbonate, and nitrate have virtually no influence on the decomposition rates. UV irradiation decomposes chemicals through photolysis, where photons in the far UV region are capable of breaking down hydrogen peroxide to form hydroxyl radicals, which oxidize organic pollutants (Esplugas et al., 1994). Ozone itself can be decomposed to oxygen radicals by its reaction with hydroperoxide ions (HO_2^-). However, the effect of the combination of the two treatments is often synergistic, with a subsequent increase in the rate of hydroxyl radical formation. The radicals produced from these reactions are responsible for the complete degradation of organic pollutants.

The rate at which ozone decomposes during photolysis can be given by the following equation:

$$-r = d[O_3]/dt = k\,I\,([O_3]^m \times [COMP]^n) \tag{8.17}$$

where $-r$ is the decomposition rate of ozone; k is the kinetic rate constant; I is UV irradiation intensity; $[O_3]$ and $[COMP]$ are initial ozone and compound concentrations, respectively; and m and n are the rate orders, which can be found according to mathematical models defined by Benitez et al. (1996). In the model, UV intensity is directly proportional to the ozone decomposition rate. Both ozone and compound initial concentrations are also proportional to degradation rate r. The actual amount of any particular compound destroyed during UV/ozone oxidation is known as the removal percentage or efficiency. This efficiency is dependent upon the kinetic rate constant, k. The amount of compound that must be removed during UV/ozone oxidation depends upon actual standards, advisory concentrations, or toxicity levels for a given organic pollutant.

The degradation kinetic rate constants are often dependent upon four factors: physical parameters, pH, scavengers, and molecular structure of the organic pollutant to be oxidized. Operating conditions, such as temperature, pH, and ozone partial pressure, are directly proportional to kinetic rates and removal efficiency. For example, low pH tends to lower removal efficiency and reaction rate constants because molecular ozone is the dominant species. At pH greater than 7, hydroxyl radicals are the dominant species; therefore, a high pH is required to achieve high removal efficiency. However, scavenger formation tends to develop more readily in alkaline solutions, which will reduce the removal rates, so optimum removal efficiency and highest kinetic rate constants usually occur when pH is near or slightly higher than neutral, which is consistent with what would be expected theoretically.

The formation of scavenger substances can also retard removal efficiency and kinetic reaction rates. Scavengers are ions such as bicarbonate, carbonate, chloride, and humic acid, etc. These scavengers subsequently react with hydroxyl radicals produced during the degradation process. Therefore, the removal efficiency will be reduced significantly. In the presence of scavenging substances, 100% removal of organic pollutants from an influent stream is often impossible.

It is important to note that the molecular structure of organic compounds has a determined effect on the oxidation rate constants. For example, if a compound is "saturated" with four chlorine atoms per carbon atom such as tetrachloroalkane, its reactivity rate with hydroxyl radicals is expected to be significantly lower than that of an "unsaturated" compound with only two or three chlorine atoms such as di- and trichloroalkanes.

8.3   Degradation Kinetics of Organic Pollutants

The kinetic rate of an organic pollutant, k, is a function of the decomposition rate of ozone, UV irradiation intensity, pH, and initial ozone concentration. The removal efficiency usually increases with kinetic rate constants, while physical parameters, such as temperature, pH, UV intensity, and ozone partial pressure have significant influences on these rates. For most organic pollutants, the optimum solution pH is near or slightly higher than neutral pH. UV intensity, temperature, and ozone partial pressure are all usually directly proportional to kinetic rate constants and removal efficiencies; however, removal efficiency has generally been found to be inversely proportional to scavengers generated during degradation processes. Kinetic rate constants can be used to determine whether the UV/ozone process is suitable for treating a particular organic pollutant. For example, if the rate constant for a certain compound were high, then, theoretically, more of the compound would be removed from the influent in less time. This would be a major advantage, because less operating time would reduce operating costs at the treatment plant. Another advantage is that most organic contaminants will respond to UV/ozone treatment, and usually to an appreciable extent. However, a significant disadvantage is that many compounds have special requirements in order for their degradation to occur. High UV intensities and initial ozone concentrations, for example, may be necessary to achieve the desired results, which will increase operating costs. This is because the generation and storage of ozone can be very expensive because ozone is unstable and highly reactive.

8.3.1   Atrazine

Ozone combined with ultraviolet radiation ($\lambda = 254$ nm) has been shown to oxidize atrazine in water. The process can be used to oxidize different organic compounds such as volatile organochlorine substances (e.g., pesticides). Mass transfer and kinetic data have been applied to the mass balance equations of atrazine to obtain corresponding concentrations under varying physical conditions (Beltran et al., 1994). Increasing the pH leads to an increase of the oxidation rate of atrazine. This is due to radical reactions involving ozone and hydrogen peroxide rather than to direct photolysis or ozonation (Beltran et al., 1994). If pH decreases, a higher concentration of dissolved ozone will be required. The increase in ozone concentration reaches a maximum value followed by a gradual decrease in ozone concentration required. This pattern is due to the competition between atrazine and ozone for hydroxyl radicals (Beltran et al., 1994).

Hydrogen peroxide is a product typically formed during direct photolysis of ozone (Beltran et al., 1994). During the first 5 min of oxidation, Beltran et al. (1994) found that the concentrations of hydrogen peroxide at pH 2 and 7 remained constant. After 5 min, the H_2O_2 at pH 7 tended to level off, while the H_2O_2 concentration at pH 2 continued to increase. This difference was due to the fact that hydrogen peroxide (pK = 11.7) remains in its non-dissociating form at pH 2, which cannot directly react with ozone (Beltran et al., 1994).

Beltran et al. (1994) reported that the higher the temperature at any given time, the greater the elimination rate of atrazine will be. The mass balance equation for atrazine is as follows:

$$-dC_A/dt = k\,C_A\,C_{O3} - I_A + k_{OH,A}\,C_A\,C_{OH} \tag{8.18}$$

where I_A is the degradation rate of atrazine due to direct photolysis, which is given as (Beltran et al., 1994):

$$I_A = -\varphi_A\,I_0 f_A\,[1 - \exp\,(-2.3L\,\Sigma\varepsilon_I\,C_i)] \tag{8.19}$$

where φ_A is the quantum yield of atrazine; I_0 is the effective intensity of incident radiation in the water; f_A is the fraction of radiation absorbed by atrazine; L is the effective path of radiation; and ε_i and C_i are the extinction coefficient and concentration of species present in water, respectively.

At pH less than 3, the reaction will occur slowly enough for kinetic models to be true. Thus, a pH 2 or 7 can be utilized for the slow kinetics of atrazine oxidation by UV/ozone processes, while rapid reactions will take place at pH 12. All of the three reaction mechanisms will be affected by other variables such as temperature, pH, and bicarbonate ion concentrations.

8.3.2 Humic Acids

Chlorinated organic compounds present in water, due to their carcinogenic nature, have become a great concern with respect to human health. Such substances are formed when humic acids react with chlorine in disinfection processes. Ozonation alone is generally not suited for the complete oxidation of chlorinated compounds because scavenger compounds such as acetic acid, formic acid, and oxalic acid can form and accumulate as by-products in the system. However, when UV irradiation is combined with ozone in the treatment process, reactions of hydroxyl radicals generated are quite fast and nonselective (Kusakabe et al., 1990). Therefore, with UV/ozone treatment, the destruction rate of the organic compounds increases because the by-products formed can also be completely mineralized in the UV/ozone process.

Kusakabe et al. (1990) reported that total organic carbon (TOC) concentrations decrease rapidly during the first 100 min of treatment with ozone, leveling off somewhat after 100 min. Furthermore, the decomposition rate of TOC was accentuated by UV irradiation; however, no direct correlation between UV intensity and TOC concentration was found. Low concentrations of TOC were still detected even after a 5-hour sampling period, indicating that the destruction of humic substances produces refractory compounds that are oxidized quite slowly (Kusakabe et al., 1990).

To estimate the destruction rate of TOC, it is assumed that the decomposition rate of humic substances is proportional to C_1, which is the TOC of refractory components produced (Kusakabe et al., 1990). Consequently, the decomposition rate of humic compounds measured in the recirculating system is expressed as the sum of the rate in the bubble column and in the UV reactor ($V_R + V_B$) as follows:

$$-(dC/dt)(V_R + V_B) = k_B[O_3]_B(C - C_1)\,V_B + k_P\,(I_{avg})[O_3]_P(C - C_1)V_R \quad (8.20)$$

where k_B and k_P are the destruction rate coefficients in the bubble column and UV reactor, respectively; I_{avg} is the average UV intensity; and $[O_3]_B$ and $[O_3]_P$ are the dissolved ozone concentrations in the bubble column and UV reactor, respectively (Kusakabe et al., 1990). C_1, k_B, and k_P are assumed to remain constant within the 30- to 250-min sampling range.

Kusakabe et al. (1990) reported that the destruction rate coefficients increase as temperature increases. UV light intensity of 8.7 W/m^2 yielded a slightly more than tenfold increase in the decomposition rate. The decomposition rate of ozone increases with UV intensity. These results imply that, under UV irradiation, radical chain reactions are predominant over molecular ozone reactions. When light intensity is greater than 3 W/m^2, the degradation rate of TOC by UV/O$_3$ can be expressed as follows:

$$-dC/dt = 1.1 \times 10^4 \times \exp[(-39/RT)I_{avg}(C - 0.056C_0)[O_3]] \quad (8.21)$$

where the concentrations are expressed in mol/m^3; t is in seconds; T is temperature (K); R is the ideal gas constant; and I_{avg} is UV intensity (W/m^2) (Kusakabe et al., 1990). Oxalic acid concentration is greatly reduced when exposed to UV irradiation, acidic acid is only slightly reduced (after about $T > 150$ min), and formic acid does not seem to be affected. This is partially because the absorption coefficient of oxalic acid at $\lambda = 253.7$ nm is higher than that of the other organic compounds (Kusakabe et al., 1990).

Chloroform compounds can be adequately degraded after 100 min. Although in the study by Kusakabe et al. (1990) UV irradiation only slightly reduced the chloroform concentration as compared to the use of ozone only, total organic halide (TOX) concentrations after about 4 hours decreased to about one third of those when ozone alone was used. Although decomposition rates are increased when UV and ozone are used together, ozone has a maximal utilization efficiency at certain UV intensities. The destruction rate of humic acids by UV/O_3 was significantly higher than by ozone alone. Additionally, as ozone concentration in the presence of UV light increased, TOX concentrations decreased accordingly.

8.3.3 Volatile Organic Compounds

Volatile organic compounds (VOCs), especially trihalomethanes, are frequently found in drinking water due to the chlorination of humic acids. When UV irradiation is applied to the pre-ozonation of humic acids, the decomposition of VOC precursors increases (Hayashi et al., 1993). The ozonation rates of compounds such as trichloroethylene, tetrachloroethylene, 1,1,1-trichloroethane, 1,2-dichloroethane, and 1,2-dichloropropane were found to be dependent on UV intensity and ozone concentration in the aqueous phase by Kusakabe et al. (1991), who reported a linear relationship between the logarithmic value of $[C]/[C_0]$ and $[O_3]_t$ for 1,1,1-trichloroethane, trichloroethylene, and tetrachloroethylene. The other two organochlorines followed the same first-order kinetics with and without UV irradiation (Kusakabe et al., 1991). Thus, the decomposition rate can be expressed as:

$$-d[C]/dt = k\,[C]\,[O_3] \qquad (8.22)$$

where k is the reaction rate coefficient. The decreasing order of rate constant k can be arranged as trichloroethylene > tetrachloroethylene > 1,1,1-trichloroethane (Kusakabe et al., 1991). Under UV irradiation of 8 to 9 W/m^2, the rate coefficients for 1,1,1-trichloroethane and tetrachloroethylene were about 30 times larger than those without UV irradiation (Kusakabe et al., 1991). The destruction rate of organochlorine compounds under UV irradiation can be expressed by the following equation:

$$-d[C]/dt = k_1[C][O_3] + k_2[C][O_3] \qquad (8.23)$$

where k_1 and k_2 are the rate coefficients in the presence and absence of UV irradiation, respectively (Kusakabe et al., 1991). Under UV intensities of 1 W/m^2, $k_2[C][O_3]$ can be neglected. Kusakabe et al. reported that when the UV intensity is higher than about 5 W/m^2, the rate coefficient increases linearly with UV intensity, suggesting that the ozonation is controlled by photolytic reactions. The overall decomposition rate of ozone in water under UV irradiation around neutral pH is given by Kusakabe et al. (1991):

$$-d[O_3]/dt = k_1[OH^-]^{0.28}[O_3]^{1.5} + k_2[OH^-][O_3] + 2.8\exp$$
$$[(-16\,kJ/mol/RT)[OH^-]^{0.07}[O_3]I] \tag{8.24}$$

where $[O_3]$ and $[OH^-]$ are expressed in terms of mol/m^3; R is the ideal gas constant; T is temperature (K); and I is UV intensity (W/m^2). The destruction rates of 1,1,1-trichloroethane, trichloroethylene, and tetrachloroethylene by UV/ozone were proportional to the dissolved ozone concentration and the UV intensity in the reaction chamber.

In the gaseous phase, neither ozone nor any of the organochlorines can be decomposed substantially without UV irradiation. This result implies that active species such as hydroxyl or oxygen radicals formed by the decomposition of ozone under UV irradiation are indispensable for the destruction of the organic compounds in gaseous phase (Hayashi et al., 1993). Trichloroethylene was decomposed to the largest extent; tetrachloroethylene was decomposed to a lesser extent; and 1,1,1-trichloroethane was nearly nonreactive. A general linear relationship exists between $\log(C_s/C_s^0)$ and $I \cdot C_{O3}t$; therefore, with respect to the concentrations of the organic substances in the presence of gaseous ozone, the decomposition rate can be described as first-order kinetics as follows:

$$-dC_S / dt = k_{SG}IC_{O_3}C_S \tag{8.25}$$

where the values of k_{SG} (destruction rate coefficient under UV irradiation) at 293 K were 0.055 and 0.003 J/mol/m^5 for trichloroethylene and tetrachloroethylene, respectively (Hayashi et al., 1993). At temperature of 293 K and UV intensity of 10 W/m^2, the destruction rate coefficient of 1,1,1-trichloroethane was found to be about 30 times greater than without UV, while the destruction rate coefficient of 1,2-dichloropropane was about 10 times greater than without UV (Hayashi et al., 1993). The rate coefficient increases linearly with increased UV intensity. This leads to the corresponding rate equation in aqueous phase as follows (Hayashi et al., 1993):

$$-d[S]/dt = (k_{SL0} + k_{SL1}I)[S][O_3] \tag{8.26}$$

where $[S]$ is the organochlorine concentration (mol/m^3); k_{SL0} is the destruction rate coefficient without UV irradiation (mol/m^3/s); k_{SL1} is the destruction rate coefficient under UV irradiation (J/mol/m^5); I is the UV intensity (W/m^2); and $[O_3]$ is the ozone concentration (mol/m^3). k_{SL1} represents the slope of the lines for each organochlorine analyzed. The different rate coefficients under UV irradiation indicate that k_{SL1} is typically much greater than k_{SG}; for tetrachloroethylene, k_{SL1}/k_{SG} is about 50 at 293 K (Hayashi et al., 1993).

The general equation for the decomposition rate of ozone under UV light, in the gas phase, is given as follows:

$$-d[O_3]/dt = k_{G(O3)}I[O_3] \qquad (8.27)$$

where I = 240 W/m^2; λ = 253.7 nm, $[O_3]$ = 0.1–2.0 mol/m^3; and $k_{G(O3)}$ is 1.26 $\times$ 10^{-3} $J/mol/m^5$ when $[O_3]$ is 1.0 mol/m^3 (Hayashi et al., 1993). Similarly, the overall decomposition rate for ozone under UV irradiation, in the aqueous phase, is given as:

$$-d[O_3]/dt = k_a[O_3]^{1.5}[OH^-]^{0.28} + k_b[O_3][OH^-] + k_{L(O3)}[O_3]I \qquad (8.28)$$

where $k_{L(O3)}$ = 2.8 $\exp[(-16 \text{ kJ/mol/RT})[OH^-]^{0.07}]$; all concentrations are in units of mol/m^3; and the following parameters must fall in these ranges: pH = 2 to 9, T = 279 to 303 K, I = 2 to 40 W/m^2, and $[O_3]$ = 0.03 to 0.4 mol/m^3 (Hayashi et al., 1993). At pH = 7 and T = 293 K, the value of $k_{L(O3)}$ is 2.06 $\times$ 10^{-3} $J/mol/m^5$. The value of $k_{L(O3)}$ is larger than $k_{G(O3)}$, but $k_{L(O3)}/k_{G(O3)}$ is 1.63 (much less than k_{SL1}/k_{SG}), verifying that radicals produced in the gas phase by UV light contribute to the destruction of the organochlorines, but with less efficiency than those in the aqueous phase (Hayashi et al., 1993).

Ozone utilization efficiency (R_L) will vary with changes in average UV intensity. The value of R_L represents a ratio between the amount of decomposed ozone in the aqueous phase to the amount of ozone in the gas phase initially introduced into the treatment chamber. Additionally, R_L corresponds to the fraction of ozone actually utilized for the destruction of solutes (Hayashi et al., 1993). R_L values are maximized at UV intensities from about just over 10 W/m^2 to about 100 W/m^2.

None of the five compounds analyzed can be destroyed in the gas phase without UV irradiation; however, both trichloroethylene and tetrachloroethylene were degraded under UV irradiation. In the aqueous phase, UV irradiation destroyed the five compounds tested. The degradation rates increased linearly with UV intensity. Finally, the utilization efficiency of ozone, as well as the corresponding destruction rates of organic compounds, is influenced by UV intensity. The maximum efficiency roughly occurred in light intensity ranging from 10 to 100 W/m^2.

8.3.4   Chlorophenol

Chlorinated phenols such as 4-chlorophenol (4-CP) constitute a large portion of halogenated volatile organic pollutants. It is reported that UV/ozone produced significant intermediates during the destruction of 4-CP. These intermediates were resistant to oxidation by UV irradiation and by ozone alone; however, they were susceptible to oxidation by hydroxyl radicals formed in the UV/ozone process (Esplugas et al., 1994). The degradation rate of TOC destruction ($-r_{TOC}$) is first order with respect to the rate of light energy absorption ($\mu_\lambda[q_\lambda]$) and TOC concentration. The rate equation can be expressed as follows:

$$-r_{TOC} = k_p \Sigma_\lambda \mu_\lambda [q_\lambda][TOC] + k_d[TOC] \tag{8.29}$$

where the summation is extended to the entire range of wavelengths of radiation absorbed; [TOC] is the concentration of total organic carbon; k_p is the rate constant of the photolytic oxidation reactions; and k_d is the rate constant for "dark" reactions (Esplugas et al., 1994).

Assuming complete mixing, the mass balance for TOC in the photoreactor under the initial conditions of $t = 0$ and $[TOC] = [TOC]_0$ when operated in the batch mode is given as:

$$d[TOC]/dt = -[TOC][k_p W_{Abs} + k_d] \tag{8.30}$$

where W_{Abs} is the radiation flow rate (Einstein/s) absorbed by ozone in the liquid phase (Esplugas et al., 1994). Integration of the previous equation yields:

$$[TOC] = [TOC]_0 \exp[(-k_p W_{Abs} + k_d)t] \tag{8.31}$$

where the value of k_d, experimentally determined to be 2.65×10^{-4} s^{-1}, can be used for the determination of the rate constant of photolytic oxidation (k_p) (Esplugas et al., 1994). In this equation, the destruction of TOC follows first-order kinetics with concentration decreasing exponentially with time.

8.3.5 Protocatechuic Acids

When UV/ozone was applied to the destruction of protocatechuic acid, the remaining concentration of protocatechuic acid decreases with increasing O_3 and protocatechuic acid concentrations (Benitez et al., 1996). The total ozone absorbed in the liquid phase was considered to react with the solute present in the solution by two parallel chemical reactions — namely, direct reaction between ozone alone and the solute (B):

$$O_3 + B \rightarrow \text{products} \tag{8.32}$$

and the combined reaction of ozone plus UV irradiation and the solute:

$$O_3 + B \xrightarrow{h\nu} \text{products} \tag{8.33}$$

where the rate constant for the ozone plus UV irradiation combined reaction is represented by:

$$-r_C = k_C I C_A^p C_B^q \tag{8.34}$$

where r_C is the reaction rate (mol/L/s), k_C is the rate constant (Einstein/ mol^2·m^8), I is the radiation intensity (E/m^2/s); C_A is the ozone concentration (mol/L); C_B is the protocatechuic acid concentration (mol/L); and q and p are reaction orders with respect to C_A and C_B, respectively (Benitez et al., 1996). Values of k_C can be calculated according to the following equation:

$$k_2 = k_C(W_{Abs}/\varepsilon V)C_B[\exp(q - 1)] \qquad (8.35)$$

where W_{Abs} is the radiation flow rate (Einstein/s), ε is the extinction coefficient for protocatechuic acid at a given λ (mol/L/m); and V is liquid volume (L) (Benitez et al., 1996). Similarly, values of k_2 can be calculated according to the following equation:

$$(M_1)^2 = (M_D)^2 + [(2/p + 1)] [k_2(C_A)^{p-1}D_A/(k_L)^2] \qquad (8.36)$$

where M_1 is the total Hatta number; M_D is the Hatta number for direct ozonation; D_A is the ozone diffusivity in liquid phase (m^2/s); and k_L is the liquid-phase mass transfer coefficient (m/s) (Benitez et al., 1996).

Both temperature and pH have significant effects on the rate constant, k_C. An increase in either parameter will lead to a higher rate constant. The highest pH values of 7 and 9 corresponded to the two highest k_C values of 12.19 and 20.42 Einstein/mol^2·m^8, respectively. Furthermore, when the temperature increased to 40°C, the rate constant k_C increased to 14.41 Einstein/ mol^2·m^8. Conversely, a low pH value of 5 combined with a low temperature of 10°C yielded the lowest k_C value of 1.34 Einstein/mol^2·m^8.

When the effects of UV/ozone treatment on the degradation of protocatechuic acid were compared with the rates of six other advanced oxidation processes (AOPs), the UV/ozone process ranked second behind only the UV/O$_3$/H$_2$O$_2$ process in terms of oxidation kinetics. The combination of UV irradiation and ozone was more effective than either UV or ozone alone in terms of degradation rates of protocatechuic acid. The contribution of different reaction pathways in a combined system will be discussed in detail in Chapter 14.

8.3.6 Propoxur

The UV/ozone process can treat pesticides without little generation of refractory products. One of the most common pesticides found in water supplies is Propoxur. The oxidation kinetics was developed in terms of the reaction orders and apparent kinetic constants (Benitez et al., 1994). The amount of chemical removed (Xp) is inversely proportional to the amount of initial Propoxur concentration (C_p), ozone partial pressure (kP$_a$), temperature, and pH. When Propoxur is degraded by UV/ozone, the reactions can be represented by the following general reactions:

$$O_3 + \text{products} \rightarrow B' \tag{8.37}$$

$$h\nu + \text{products} \rightarrow B'' \tag{8.38}$$

where the first equation describes direct ozonation and the second presents the photochemical oxidation of the pesticide (Benitez et al., 1994). Using Equation (8.17), developed by Benitez et al. (1996), m was found to be 0, and n was found to be 1. Therefore, with values for m and n now known, the equation needed to obtain k_C is given as follows:

$$H_{aC} = (1/k_L)(2D_A k_C\, C_P/C_A)^{0.5} \tag{8.39}$$

where H_{aC} is the Hatta number for UV/ozone; k_L is the liquid-phase mass-transfer coefficient (m/s); D_A is the ozone diffusivity in liquid phase (m^2/s); and C_P and C_A are the Propoxur and ozone concentrations (mol/L), respectively (Benitez et al., 1996). The apparent constant, k_C, can be expressed as a function of both pH and temperature in the following equation:

$$k_C = k_O \exp[(-E_a/RT)[\text{OH}^-]^p] \tag{8.40}$$

where T is the temperature (K), and R is the ideal gas constant. Multiple-regression analysis of k_C values yielded $k_O = 7.1 \times 10^{15}$, $E_a = 72.76$ kJ/mol, and $p = 0.26$ (Benitez et al., 1996). k_C is proportional to pH and temperature.

8.3.7 Chlorinated Benzenes

Masten et al. (1996) investigated the oxidation of chlorinated benzenes such as 1,2-dichlorobenzene (1,2-DCB), 1,3,5-trichlorobenzene (1,3,5-TCB), and pentanoic acid (PA). TCB is often generated as a by-product of pesticide manufacturing, while DCB is commonly manufactured as an insecticide or a fumigant for industrial odor control. Due to their resistance to biological treatments, PA is usually nonreactive with ozone but can react with hydroxyl radicals (Masten et al., 1996).

The degradation kinetics of the target chemical compounds (TCB, DCB, or PA) was assumed to be first order in terms of the concentration of target compound (C). The conditional rate constant (k) can be expressed as:

$$k = 1/\theta[([C_0] - [C])/[C]] \tag{8.41}$$

where $[C]$ is the target chemical concentration in the reactor (continuous-flow stirred tank); $[C_0]$ is the target chemical concentration before oxidation; and θ is the hydraulic retention time (Masten et al., 1996). The UV/ozone process removed 86.4% of the influent PA at pH of 6.8. Upon further increasing the pH to 11, the removal efficiency decreased slightly to 83.4% (Masten et al., 1996). The UV/ozone process destroyed nearly all the influent TCB when the pH was increased up to 9; however, the efficiency decreased considerably at higher pH values. Virtually all the DCB was destroyed when the pH was less than about 9. The efficiency decreased only slightly at higher pH values. For all three compounds, low pH did not significantly change degradation rates; however, efficiency was notably reduced at high pH for all three substances. This decrease was probably partially due to the scavenging of OH$^\bullet$ radicals by carbonate or hydrogen phosphate (Masten et al., 1996). The humic acid scavenges OH$^\bullet$ or other radicals responsible for the degradation of both TCB and DCB. Bicarbonate ions were found to decrease the decomposition rates of both TCB and DCB as concentration increased due to the scavenging effect of OH$^\bullet$ radicals.

Photolysis, molecular ozone reactions, or hydroxyl radical reactions with the target compound can be quantified by:

$$k = k_{O3}[O] + k_{photo}\, \phi I + k_{OH\bullet}\, [OH^\bullet] \qquad (8.42)$$

where k is the first-order rate constant; I is UV intensity; and k_{O3} and $k_{photo}\phi$ are the reaction by ozonation rate constant and reaction by direct photolysis rate constant, respectively. The rate constants for DCB, PA, and TCB were found to be 4×10^9, 2.9×10^9, and $4 \times 10^9\ M^{-1}\ s^{-1}$, respectively (Masten et al., 1996). At low pH values, DCB, TCB, and PA exhibited negligible changes in degradation rates. At high pH values, degradation efficiencies decreased significantly due to scavengers such as humic acids and bicarbonate ions.

8.3.8   Polycyclic Aromatic Hydrocarbons

The U.S. Environmental Protection Agency identified 16 polycyclic aromatic compounds (PAHs) as primary pollutants, eight of which are known to be carcinogenic (Trapido et al., 1995). These substances are often formed as the by-products of incomplete combustion of fossil fuels and have been identified in many emission sources such as vehicle exhausts, power plants, and the chemical and oil shale industries (Trapido et al., 1995). PAHs such as fluorene, phenanthrene (PHEN), and acenaphthene (ACEN) can be degraded by UV/ozone. Beltran et al. (1995) reported that all three compounds were almost completely removed from their respective solutions after about 4 min of treatment, with PHEN being decomposed at the fastest rate. Degradation rates for all three chemicals continually increased during about the first 3 min, suggesting that little or no scavengers were present to inhibit these rates during this time period. The degradation of these compounds was due to the direct photolysis of ozone because no dissolved ozone was detected (Beltran et al., 1995).

The highest bicarbonate concentration (0.01 M) produced the greatest decrease in the degradation rate due to the competition for hydroxyl radicals by bicarbonate ions; however, a 0.001-M concentration of bicarbonate ion barely affected the oxidation rate of fluorene (Beltran et al., 1995). Similar tests were conducted by Beltran et al. (1995) on PHEN and ACEN, and the presence of the bicarbonate species did not significantly decrease their oxidation rates, either, indicating that the radical pathway is negligible compared with direct photolysis and ozonation reactions.

When pH increased from 2 to 7, the removal rate of fluorene increased; however, a subsequent increase in pH from 7 to 12 reduced the removal efficiency back to about the rate at the pH of 2. The increase of pH leads to an increase in the hydroxyl-ion-catalyzed decomposition of ozone into hydroxyl radicals; however, the amount of ozone available to undergo direct photolysis and produce hydrogen peroxide will decrease with increasing pH. Eventually more hydroxyl radicals will be produced, which is particularly important at pH 12 (Beltran et al., 1995). The rate of oxidation of fluorene is given by:

$$r_F = -dC_F/dt = \Phi F_F I_a + k_F C_{O3} C_t + k_{RF} C_t \qquad (8.43)$$

where Φ is the quantum yield (mol/photon); F_F is the fraction of absorbed light of fluorene; I_a is the total flow of absorbed radiation (Einstein/L/s); k_F is the rate constant of fluorine (M^{-1} s^{-1}); C_t is the fluorene concentration in water (M); and k_{RF} is the rate constant for radical reactions between hydroxyl radical and fluorene (M/s) (Beltran et al., 1995). The contributions of direct ozonation and photolysis to the degradation of fluorene can be estimated by the following equations:

$$\gamma_{O3} = [(k_F C_{O3} C_F)/r_F] \times 100 \qquad (8.44)$$

$$\gamma_{UV} = [(\Phi_F F_F I_a)/r_F] \times 100 \qquad (8.45)$$

where γ is the percentage contribution of direct reactions in the oxidation of the PAH (Beltran et al., 1995). The oxidation of fluorene with UV/ozone is due almost exclusively to direct photolysis and radical attack (Beltran et al., 1995). Through the use of similar equations, ozonation can be shown to be the main pathway for the oxidation of ACEN, while both photolysis and ozonation contribute nearly evenly to the elimination of PHEN (Beltran et al., 1995).

The UV/ozone process destroyed about 75% of the influent anthracene within about 2 min. The degradation rate was generally steady for about the first minute of treatment but decreased somewhat thereafter, perhaps indicating the buildup of scavengers reacting with hydroxyl radicals. The UV/ozone process decelerates the chemical reaction of the ozonolysis of anthracene with molecular ozone as compared with ozone treatment alone (Trapido et al., 1995).

The UV/ozone process removed nearly 100% of the influent phenanthrene after 5 min at pH around 7; furthermore, the UV/ozone process removal rate was virtually identical to that of ozone alone, with only slight variations occurring after about 3 min. Evidence of the buildup of scavengers can be seen due to the continual decrease in the degradation rate, especially after about 3 min. When the pH is about 7, almost 90% of a 40-μg/L sample tested after 1.5 min can be removed. The degradation rate for pyrene was found to be substantially greater than for phenanthrene. The rate remained fairly constant throughout the treatment, indicating few scavengers competing for hydroxyl radicals; however, ozone alone again resulted in slightly better decomposition rates of pyrene than for the UV/ozone process. The effects of scavenger buildup were minimal, except in the treatment of phenanthrene, where buildup appeared to be significant toward the end of the treatment period.

Beltran et al. (1995) concluded that: (1) the UV/ozone oxidation process can achieve high removal rates of fluorene, phenanthrene, and acenaphthene, with total efficiencies being near 100% in some cases; (2) neutral pH of 7 yields the highest removal rate of fluorene in solution; and (3) the greater the bicarbonate concentration added to fluorene, the lower the removal efficiency becomes. When an excess of ozone is present in the reaction mixture, the degradation rate of anthracene, phenanthrene, and pyrene can be given as:

$$dC_{PAH}/dt = -k_1(C_{PAH}{}^n) \tag{8.46}$$

where $k_1 = k_2(C_{O3}{}^m)$; m is the reaction order with respect to ozone; n is the reaction order with respect to the given PAH; C_{PAH} is the PAH concentration (M); and k_2 is the second-order rate constant for a given PAH/ozone reaction (Trapido et al., 1995).

8.3.9 Halogenated VOCs

Halogenated VOCs such as CCl_4, Cl_3C, and CCl_3 in natural water are refractory to O_3 and UV/ozone. Their removal is affected only by stripping. However, Sebastian et al. (1996) reported that organochlorides can be destroyed by the UV/ozone combination, and the reactivity depends on the type of halogen in the molecule ($CHBr_3 > CHBr_2Cl > CHCl_3$). In addition, Francis (1987) was able to remove CCl_4 in experiments conducted with deionized water. If no radical scavengers exist, phenol, p-cresol, 2,3-xylenol, and 3,4-xylenol can be completely mineralized to CO_2 and H_2O by ozone alone.

Bhowmick and Semmens (1994) studied the photooxidation kinetics of five halogenated VOCs, including chloroform (CHL), carbon tetrachloride (CTC), trichloroethylene (TCE), tetrachloroethylene (PCE), and 1,1,2-trichloroethane (TCA) by UV/ozone. In the low to intermediate concentration range of 0.07 to 1.11 mg/L, the reactivities of the organics rank in the following order: TCE > PCE > CTC > CHL > TCA. This order indicates that the reaction between ozone and saturated hydrocarbons is much slower than that of ozone with unsaturated compounds such as alkenes. For CHL, TCA, and CTC, an increase in ozone concentration had little effect on the rate constant. By comparison, the rates of oxidation for TCE and PCE increased with added ozone concentration. The rate increases for TCE and PCE were about 15 to 30% and 2%, respectively (Bhowmick and Semmens, 1994).

When ozone concentration was 2.86-mg/L, changes in the oxidation rate for all three compounds were barely detectable for about the first 45 min of treatment; however, after about 50 min of treatment, the rate of oxidation increased considerably for both TCA and CHL. The rate for CTC degradation, however, showed no such increase. Oxidation rates increased with increasing UV intensity for all five substances; for CHL, TCA, and CTC, the increase was in direct proportion to intensity, but the oxidation rates of TCE and PCE were less sensitive and the increase was marginal (Bhowmick and Semmens, 1994).

Phosgene was identified as the principal intermediate during the photo-oxidation of all five compounds. Chloroaldehyde and chloroacetylchloride were also detected in low concentrations (Bhowmick and Semmens, 1994). The concentrations of these intermediates reached a maximum value but subsequently fell with time as the oxidation process continued. The rates of oxidation of the VOCs tested are proportional to UV irradiation intensity. The addition of ozone to UV irradiation improved the kinetics of TCE and PCE oxidation; however, no significant change in the oxidation rates of CHL, TCA, and CTC was observed.

8.4   QSAR Models

Molecular descriptors can be used to correlate with the oxidation kinetic rate constants; therefore, the treatability of different classes of organic pollutants by UV/ozone can be evaluated using quantitative structure–activity relationship (QSAR) models. Chapter 4 demonstrated that the rate constants of selected compounds reacting with hydroxyl radicals were greatly dependent upon molecular structure. In general, saturated compounds such as chloro-ethanes produce much lower reaction rates compared with unsaturated compounds such as dichlorobenzene and chloroethylenes. The same distinction was noted between fluorene or pyrene and phenanthrene. Of all the compounds investigated, atrazine, chlorendic acid, and 4-chlorophenol were removed at a rate of near or at 100%, while both dichloroethane and dichloropropane were virtually nonreactive. Table 8.2 summarizes the treatability of different classes of organic pollutants to show that qualitative structure and reactivity relationships exist for different classes of organic compounds.

8.4.1   Amine Herbicides

Deethylatrazine, deisopropylatrazine, and acetamido-*s*-triazines are the primary oxidation by-products of atrazine by UV/O_3. UV photolysis of atrazine

TABLE 8.2

Treatability of Different Classes of Organic Pollutants

Classes of Compounds	Degradation Order
Halogenated aliphatic	Trichloroethylene > tetrachloroethylene
Halogenated aliphatic	Trichloroethylene > tetrachloroethylene > tetrachloride > chloroform > trichloroethane
Halogenated aromatic	Dichlorobenzene > trichlorobenzene
Polynuclear aromatic	Fluorene > acenaphthene > phenanthrene
Polynuclear aromatic	Pyrene > anthracene > phenanthrene

at the monochromatic radiation of 253.7 nm yields hydroxyatrazine as the major product. The s-triazine ring is found to be resistant to chemical and photochemical oxidation. Kinetic rate constants for these compounds were correlated with the energy of the highest occupied molecular orbital (HOMO) and the energy of the lowest unoccupied molecular orbital (LUMO). The QSAR model developed for these compounds and correlations is plotted in Figure 8.4. The correlation shows a fit of 0.9122 between kinetic rate constants and E_{HOMO}. The kinetic rate constant and E_{LUMO} have a correlation coefficient of only 0.3624. This suggests that nucleophilic reactions are the dominant mechanism during oxidation of these amines. This also indicates that E_{HOMO} can be used to predict kinetic rate constants for amine herbicides. The QSAR model for amine herbicides such as desipropylatrazine, simazine, atratone, hydroxyatrazine, terbutylazine, and desethylatrazine is as follows:

$$k = (1 \times 10^{11})E_{HOMO} + (6 \times 10^{10}); \ n = 6, \ r = 0.9122 \tag{8.47}$$

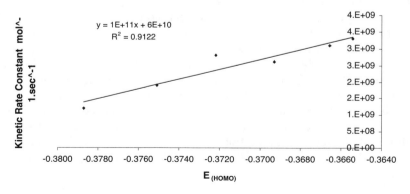

Kinetic Rate Constant vs. E_{HOMO} (Herbicides-Amines)

FIGURE 8.4

Correlation between E_{HOMO} and kinetic rate constant of amine herbicides. Experimental conditions: $T = 20°C$, pH = 7.5. (Data from De Laat, J. et al., *J. Water Sci.*, 8, 23–42, 1995.)

8.4.2 Chloroethanes

The mechanism that accounts for the oleofinic oxidation by hydroxyl radicals is the hydrogen abstraction. Moreover $Cl^•$ radicals may also be an important mechanism for chlorinated organics. When the C–Cl bond is broken by photolysis, a $Cl^•$ radical is released and can initiate additional oxidation reactions through a chain mechanism as follows:

$$Cl^• + \text{Oleofins} \rightarrow R^• \qquad (8.48)$$

Ozone molecules can also react with chlorine radicals to induce the chain reaction and generate oxygen molecules and chlorine oxide radicals. The reaction mechanism is very much the same as that for ozone layer depletion by CFC compounds:

$$Cl^• + O_3 \rightarrow O_2 + ClO^• \qquad (8.49)$$

Table 8.3 shows QSAR models between the degradation rate constants of chloroethanes and E_{HOMO} and E_{LUMO} under different ozone dosages and UV intensities. Table 8.3 also shows that E_{HOMO} exhibits better correlation with the kinetic rate constants than does E_{LUMO}. The correlation increases as the UV intensity increases. Because E_{HOMO} is a measure of the energy necessary

TABLE 8.3

QSAR Models for Chloroethanes: 1,2-Dichloroethane, 1,1,1-Trichloroethane, and 1,1,2,2-Tetrachloroethane

QSAR Model	Slope	Experimental Conditions	Correlation Coefficient
$k = -4.2024\ E_{HOMO} - 1.8389$	-4.2024	UV = 68 W/m^2	0.7451
$k = -5.9265\ E_{HOMO} - 2.5951$	-5.9265	UV = 87 W/m^2	0.8211
$k = -7.3872\ E_{HOMO} - 3.2354$	-7.3872	UV = 105 W/m^2	0.8409
$k = -8.794\ E_{HOMO} - 3.8514$	-8.794	UV = 117 W/m^2	0.8461
$k = -11.32\ E_{HOMO} - 4.9611$	-11.32	UV = 128 W/m^2	0.8717
$k = -13.21\ E_{HOMO} - 5.7911$	-13.21	UV = 141 W/m^2	0.8742
$k = 0.163\ E_{LUMO} - 0.0208$	0.163	[O$_3$] = 37 ppmv	0.8753
$k = 0.23\ E_{LUMO} - 0.0286$	0.23	[O$_3$] = 159 ppmv	0.888
$k = 0.234\ E_{LUMO} - 0.0273$	0.234	[O$_3$] = 360 ppmv	0.828
$k = 0.13\ E_{LUMO} - 0.0105$	0.13	[O$_3$] = 779 ppmv	0.49
$k = 0.162\ E_{LUMO} - 0.0164$	0.162	[O$_3$] = 1174 ppmv	0.746
$k = 0.149\ E_{LUMO} - 0.0156$	0.149	[O$_3$] = 1589 ppmv	0.822

to lose an electron from the highest occupied molecular orbital (Peijnenburg et al., 1991, 1993), the increase in kinetic rate constants that occurs with increasing UV light intensity suggests that UV photons provide additional energy for electrons to escape the orbital.

8.4.3   Chloromethanes

The oxidation rate constants of chloromethanes seem to have good fit with both E_{HOMO} and E_{LUMO}. Figure 8.5 demonstrates that the slopes of the correlations decrease with increasing UV intensity. For different ozone dosages, the correlations do not follow those for UV intensity. The degradation rate of dichloromethane is greater than chloroform, while the degradation rate of trichloromethane is greater than that for carbon tetrachloride (Shen and Ku, 1999). This indicates that the number of chlorines contained in a class of organic pollutants plays a critical role in terms of treatability by the UV/ ozoene process. Table 8.4 summarizes the findings for the chloromethanes studied.

8.4.4   Chlorophenols

Ozonation of aromatic compound has been considered to be initiated by electrophilic reaction; hence, in the ozonation of chlorophenol, the electronegativity of aromatic ring governs the degradation rate (Abe and Tanaka, 1997). Thus, the graphs that seem to give the best fit are the ones correlating kinetic rate constant with E_{LUMO} (Abe and Tanaka, 1997; Trapidoet al., 1995). The kinetic rate constants were found to correlate well with Hammett's constants as well as log P, which is a measure of hydrophobicity of compounds (see Figure 8.6 and Figure 8.7). Table 8.5 summarizes the QSAR models for the chlorophenols studied.

8.4.5   Substituted Phenols

At a pH less than 6, molecular ozone directly attacks the phenolic ring. Then, the ozone further oxidizes the dihydric phenol to either *o*- or *p*-quinone. Because ozone is a relatively less powerful oxidant, the selectivity can be clearly demonstrated, as in Figure 8.8. The Hammett plot was first reported by Hoigne (1982) and confirmed by Gurol and Nekoulnalni (1984) (Figure 8.9). At a pH greater than 6, however, ozone is decomposed as hydroxyl radicals, and substituted phenols are ionized to form phenolate anions, which are much stronger electrophilic species than the protonated forms at low pH. As a result, the measured rate constants for some substituted phenolates approach the diffusion-controlled limits.

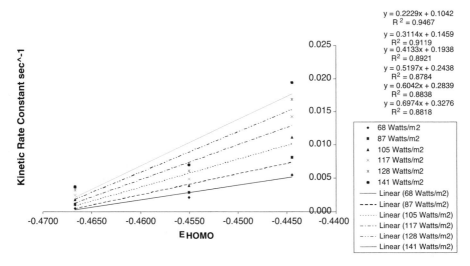

FIGURE 8.5

Correlation between kinetic rate constants and E_{HOMO} for chloromethanes with different UV intensities. Experimental conditions: $T = 40°C$, pH = 7, $[O_3]$ = 360 ppmv. (Data from Shen, Y.S. and Ku, Y., *Chemosphere*, 38(8), 1855–1866, 1999.)

TABLE 8.4

QSAR Models for Chloromethanes: 1,2-Dichloroethane, 1,1,1-Trichloroethane, 1,1,2,2-Tetrachloroethane

QSAR Model	Slope	Ultraviolet Intensity (W/m²)	Correlation Coefficient
$k = 0.222\ E_{HOMO} + 0.104$	0.222	68	0.9467
$k = 0.311\ E_{HOMO} + 0.145$	0.311	87	0.9119
$k = 0.413\ E_{HOMO} + 0.1938$	0.413	105	0.8921
$k = 0.519\ E_{HOMO} + 0.2438$	0.519	117	0.8784
$k = 0.604\ E_{HOMO} + 0.284$	0.604	128	0.884
$k = 0.697\ E_{HOMO} + 0.327$	0.697	141	0.881
$k = 0.083\ E_{LUMO} - 0.0077$	0.083	68	0.9394
$k = 0.116\ E_{LUMO} - 0.0105$	0.116	87	0.9028
$k = 0.154\ E_{LUMO} - 0.0137$	0.154	105	0.882
$k = 0.194\ E_{LUMO} - 0.0171$	0.194	117	0.868
$k = 0.226\ E_{LUMO} - 0.0195$	0.226	128	0.873
$k = 0.26\ E_{LUMO} - 0.0226$	0.26	141	0.8715

8.4.6 Chlorinated Alkanes and Alkenes

Kinetic rate constants, E_{HOMO}, and Hammett's constants can be correlated with reaction rate constants. Table 8.6 summarizes the QSAR models for these volatile compounds.

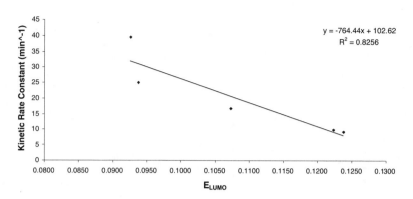

FIGURE 8.6
Correlation between kinetic rate constants and E_{LUMO} of different chlorophenols. Experimental conditions: $T = 20°C$, pH = 2.5. (Data from Trapido, M. et al., *Environ. Technol.*, 16, 729–740, 1995.)

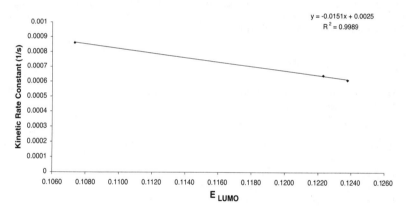

FIGURE 8.7
Correlation between kinetic rate constant and E_{LUMO} of different chlorophenols. Experimental conditions: $T = 25°C$, pH = 7, $[O_3] = 16.8$ mg/L. (Data from Abe, K. and Tanaka, K., *Chemosphere*, 35, 2837–2847, 1997.)

TABLE 8.5

QSAR Models for Chlorophenols

QSAR Model	Slope	Correlation Coefficient
$k = -1408\ E_{HOMO} - 440.39$	−1408	0.8037
$k = -764.44\ E_{LUMO} + 102.62$	−764.44	0.8256
$k = 22.153\sigma_{res} - 0.7456$	22.153	0.8689
$k = 0.0003 \log P + 1 \times 10^{-5}$	0.0003	0.869

Dataset: 2-Chlorophenol, 4-chlorophenol, 2,4-dichlorophenol, 2,4,5-trichlorophenol, 2,4,6-trichlorophenol.

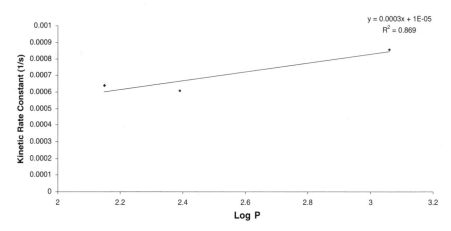

FIGURE 8.8
Correlation between kinetic rate constant and log P of different chlorophenols. Experimental conditions: $T = 20°C$, pH = 2.5. (Data from Trapido, M. et al., *Environ. Technol.*, 16, 729–740, 1995.)

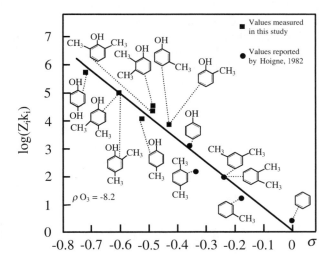

FIGURE 8.9
Correlation between rate constant for reaction with ozone and Hammett's σ constant for substituted phenols. (From Gurol, M.D. and Nekoulnalni, S., *Indust. Eng. Chem. Fundam.*, 23, 1–54, 1984. With permission.)

8.4.7 Halogenated Aliphatics

Francis (1987) and Hayashi et al. (1993) studied the oxidation kinetics of the mixtures of alkanes and alkenes. In regard to oxidation of the mixtures, the kinetic rate constants have the best correlation with E_{HOMO}. Hayashi et al.

TABLE 8.6

QSAR Models for Volatile Alkanes and Alkenes

QSAR Model	Slope	Ultraviolet Intensity (W)	Correlation Coefficient
$k = 2.5644\ E_{HOMO} + 1.1988$	2.5644	48	0.916
$k = 2.5921\ E_{HOMO} + 1.2243$	2.5921	60	0.8904
$k = 2.6153\ E_{HOMO} + 1.256$	2.6153	100	0.9119
$k = -0.0213\sigma^* + 0.2489$	-0.0213	48	0.8984
$k = -0.0215\sigma^* + 0.2638$	-0.0215	60	0.8705
$k = -0.0217\sigma^* + 0.2872$	-0.0217	100	0.8949

Dataset: Trichloroethylene, trichloroethane, tetrachloroethylene, chloroform, carbon tetrachloride.

(1993) reported that the kinetic rate constants correlate strongly with Hammett's constant, suggesting the nucleophilic character of these compounds during UV/ozone oxidation (see Figure 8.10 and Figure 8.11). Table 8.7 summarizes the findings for these halogenated aliphatics. At low and neutral pH conditions, the kinetic rate constants increased, but not so at alkaline solutions. The correlations between reaction rate constants and E_{HOMO} are better at a lower (not acidic) pH of 5 and neutral pH of 7; for the other pH values, no correlation could be found, as shown in Table 8.8, which summarizes the QSAR models for these halogenated aliphatics at different pH levels.

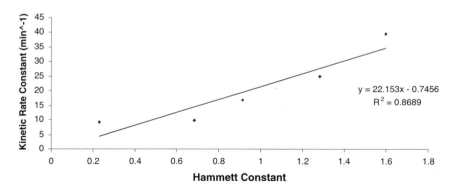

FIGURE 8.10

Correlation between kinetic rate constant and Hammett's constant of different chlorophenols. Experimental conditions: $T = 25°C$, pH = 7, ozone concentration = 16.8 mg/L. (Data from Abe, K. and Tanaka, K., *Chemosphere*, 35, 2837–2847, 1997.)

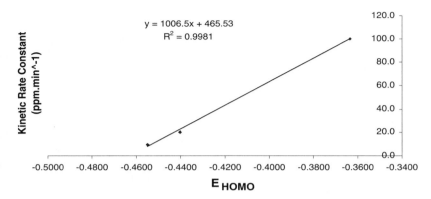

$$y = 1006.5x + 465.53$$
$$R^2 = 0.9981$$

FIGURE 8.11

Correlation between kinetic rate constant and E_{HOMO} for a mixture of alkanes and alkenes. Experimental conditions: $T = 25°C$, ozone rate 10 g/hr, UV intensity = 0.24 Einstein/hr. (Data from Francis, P.D., *Ozone Sci. Eng.*, 9, 369–390, 1987.)

TABLE 8.7

QSAR Models for Halogenated Aliphatics

QSAR Model	Slope	Ultraviolet Intensity (W/m²)	Correlation Coefficient	Ref.
$k = 0.008\sigma^* + 0.01$	0.008	68	0.9983	Shen and Ku (1998)
$k = 0.0087\sigma^* + 0.0126$	0.0087	87	0.9192	Shen and Ku (1998)
$k = 0.0169\sigma^* + 0.0204$	0.0169	105	0.9916	Shen and Ku (1998)
$k = 0.0205\sigma^* + 0.027$	0.0205	117	0.9209	Shen and Ku (1998)
$k = 0.0496\sigma^* + 0.0536$	0.0496	128	0.9705	Shen and Ku (1998)
$k = 0.0312\sigma^* + 0.0483$	0.0312	141	0.9997	Shen and Ku (1998)
$k = 1006.5\, E_{HOMO} + 465.5$	1006.5	—	0.9981	Francis (1987)
$k = 0.4716\, E_{HOMO} + 0.222$	0.4716	—	0.9157	Hayashi et al. (1993)
$k = -0.0034\sigma^* + 0.0484$	–0.0034	—	0.9435	Hayashi et al. (1993)

Dataset: Perchloroethylene, 1,1-dichloroethene, trichloroethylene.

TABLE 8.8

QSAR Models for Halogenated Aliphatics

QSAR Model	Slope	pH	Correlation Coefficient
$k = 1.402\, E_{HOMO} + 0.6164$	1.402	3	0.3357
$k = 2.8323\, E_{HOMO} + 1.1491$	2.832	5	0.8432
$k = 2.052\, E_{HOMO} + 1.0326$	2.052	7	0.9044
$k = 1.019\, E_{HOMO} + 0.7385$	1.0192	9	0.4526
$k = -0.1466\, E_{HOMO} + 0.3292$	–0.146	11	0.0257

Dataset: Perchloroethylene, 1,1-dichloroethene, trichloroethylene.

8.4.8   Benzene-Ring-Based Compounds (BTX)

For substituted benzenes such as benzene, toluene, and xylene (BTX), the best correlations can be found between the kinetic rate constants and E_{HOMO}. The correlations hold for both different UV light intensities and different ozone dosages. As reported by Shen and Ku (1999), the decomposition rates of these compounds increased with increasing number of methyl groups substituted on the benzene ring and with UV intensity. However, increasing ozone dosage did not seem to increase the decomposition rate. Shen and Ku (1999) reported that radical addition could be the dominant mechanism, as compared to hydrogen elimination, because the compound with more substituted methyl groups (i.e., fewer OH• addition sites on the benzene ring) can cause faster reactions due to the electron-donating nature of methyl groups. For substituted benzenes, the more methyl groups an organic molecule contains, the easier it is decomposed by UV/ozone (see Figure 8.12 and Figure 8.13). Table 8.9 summarizes the findings for these BTX compounds. Other substituted benzenes have also been investigated in terms of the effect of substituents on reactivity to molecular ozone. Hammett's constants, σ^+, were used in the correlation analysis shown in Figure 8.14.

8.4.9   Triazin Herbicides

Excellent correlations were obtained between the degradation rate constants of triazin herbicides and E_{HOMO}, as shown in Figure 8.15 and Figure 8.16. This suggests that amines subject to nucleophilic oxidation; therefore, E_{HOMO}

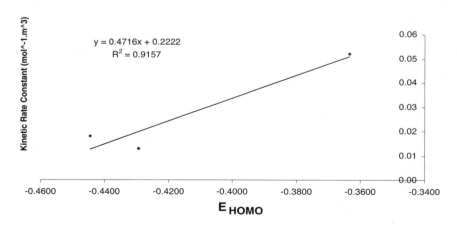

FIGURE 8.12
Correlation between kinetic rate constant and E_{HOMO} for a mixture of alkanes and alkenes. Experimental conditions: $T = 20°C$, pH = 6.9, UV output = 253.7 nm. (Data from Hayashi, J.-I. et al., *Water Res.*, 27, 1091–1097, 1993.)

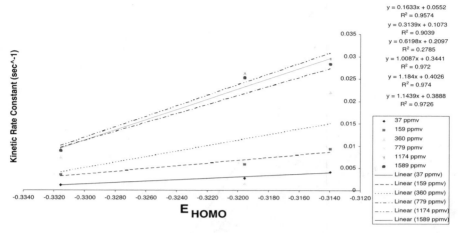

FIGURE 8.13

Correlation between kinetic rate constant and E_{HOMO} for different O_3 doses for BTX compounds. Experimental conditions: $T = 40°C$, pH = 7, UV intensity = 117 W/m². (Data from Shen, Y.S. and Ku, Y., *Chemosphere*, 38(8), 1855–1866, 1999.)

TABLE 8.9

QSAR Models for BTX Compounds

QSAR Model	Slope	Experimental Conditions	Correlation Coefficient
$k = 0.8019\ E_{HOMO} + 0.267$	0.8019	UV = 68 W/m²	0.9474
$k = 0.876\ E_{HOMO} + 0.292$	0.876	UV = 87 W/m²	0.9131
$k = 0.8938\ E_{HOMO} + 0.3004$	0.8938	UV = 105 W/m²	0.9469
$k = 0.827\ E_{HOMO} + 0.281$	0.827	UV = 117 W/m²	0.9426
$k = 1.076\ E_{HOMO} + 0.3644$	1.076	UV = 128 W/m²	0.9435
$k = 1.181\ E_{HOMO} + 0.4$	1.181	UV = 141 W/m²	0.916
$k = 0.163\ E_{HOMO} + 0.0552$	0.163	$[O_3]$ = 37 ppmv	0.9574
$k = 0.314\ E_{HOMO} + 0.1073$	0.314	$[O_3]$ = 37 ppmv	0.9039
$k = 0.6198\ E_{HOMO} + 0.2097$	0.6198	$[O_3]$ = 37 ppmv	0.2785
$k = 1.008\ E_{HOMO} + 0.344$	1.008	$[O_3]$ = 37 ppmv	0.972
$k = 1.184\ E_{HOMO} + 0.4026$	1.184	$[O_3]$ = 37 ppmv	0.974
$k = 1.144\ E_{HOMO} + 0.388$	1.144	$[O_3]$ = 37 ppmv	0.9726

Dataset: Benzene, toluene, *o*-xylene.

can be used to predict kinetic rate constants for amine herbicides. Figure 8.15 suggests that the sensitivity of reaction rate constants is proportional to the UV intensities, while the effect of ozone concentrations on this sensitivity decreases as its concentration increases, as shown in Figure 8.16. Table 8.10 summarizes the QSAR models for triazin herbicides. The dataset includes atrazine, desethylatrazine, and desethyl-desisopropylatrazine.

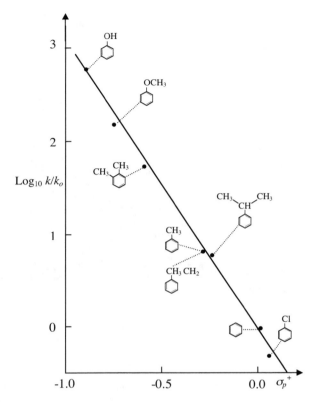

FIGURE 8.14
Relationship between rate constant for reaction with ozone and Hammett's σ^+ constant for series of substituted benzenes. (From Hoigne, J. and Bader, H., *Proc. of the International Ozone Association Symposium*, International Ozone Association, Toronto, 1977, p. 16. With permission.

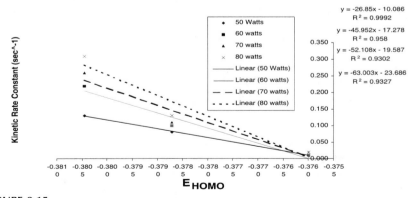

FIGURE 8.15
Correlation between kinetic rate constant vs. E_{HOMO} for different UV intensities for triazin herbicides. Experimental conditions: $T = 25°C$, pH = 7.8, ozone concentration = 38 mg/L. (Data from Zweiner, C. et al., *Vorm Wasser*, 84, 47–60, 1995.)

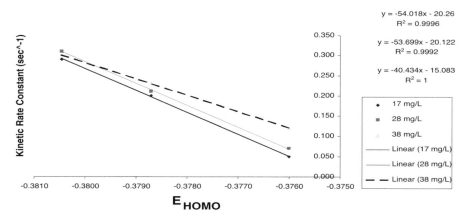

FIGURE 8.16

Correlation between kinetic rate constant vs. E_{HOMO} for different O_3 doses for triazin herbicides. Experimental conditions: $T = 25°C$, pH = 7.8, UV intensity = 80 W. (Data from Zweiner, C. et al., *Vorm Wasser*, 84, 47–60, 1995.)

TABLE 8.10

QSAR Models for Triazin Herbicides

QSAR Model	Slope	Experimental Conditions	Correlation Coefficient
$k = -26.85\ E_{HOMO} - 10.086$	−26.85	UV = 50 W	0.9992
$k = -45.952\ E_{HOMO} - 17.278$	−45.95	UV = 60 W	0.958
$k = -52.108\ E_{HOMO} - 19587$	−52.108	UV = 70 W	0.9302
$k = -63\ E_{HOMO} - 23.686$	−63	UV = 80 W	0.9327
$k = -54.018\ E_{HOMO} - 20.26$	−54.018	$[O_3]$ = 17 mg/L	0.9996
$k = -53.7\ E_{HOMO} - 20.122$	−53.7	$[O_3]$ = 28 mg/L	0.9992
$k = -40.434\ E_{HOMO} - 15.083$	−40.434	$[O_3]$ = 38 mg/L	1

Dataset: Atrazine, desethylatrazine, desethyl-desisopropylatrazine.

8.4.10 Chlorinated Dioxins and Furans

Figure 8.17 reveals that the decomposition of dioxins and furans undergoes reduction because the best fit is for the correlation between kinetic rate constants and E_{LUMO}; therefore, the compounds act as electrophilic agents. Figure 8.17 presents the correlation between the kinetic rate constants and log P for chlorinated dioxins and furans. It indicates that the higher the hydrophobicity of a given chlorinated dioxin or furan, the less reactive it will be. Table 8.11 summarizes the QSAR models for the dioxins and furans studied.

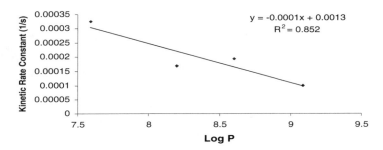

FIGURE 8.17
Correlation between kinetic rate constant and log *P* for chlorinated dioxins and furans. Experimental conditions: $T = 20°C$, pH = 6.5, ozone dose = 960 mg/hr, UV intensity = 254 nm. (Data from Vollmuth, S. and Niessner, R., *Toxicol. Environ. Chem.*, 61, 27–41, 1997.)

TABLE 8.11

QSAR Models for Dioxin and Furans

QSAR Model	Slope	Correlation Coefficient
$k = -0.0466 \, E_{HOMO} - 0.0156$	-0.0466	0.8318
$k = -0.0001 \log P + 0.0013$	-0.0001	0.852

Dataset: 1,2,3,4,6,7,8-Heptachlorodibenzodioxin, 1,2,3,4,6,7,9-heptachlorodibenzo-*p*-dioxin, octachlorodibenzofuran, 1,2,3,4,6,7,8,9-octachlorodibenzo-*p*-dioxin.

8.5 Summary

During the degradation of organic pollutants, pH produces adverse effect on the correlation until it reaches the vicinity of neutral. The increasing UV intensity accelerates the degradation rate of organic pollutants. Depending on the nature and the number of substituents, electron-donating substituents such as the methyl group will increase the degradation rate of the substituted compound. The electron-withdrawing groups such as halogens will decrease the treatability of the substituted chemical. Aromatic compounds are electron-rich species; therefore, E_{HOMO} of an organic pollutant will determine its degradation rate in the UV/ozone process (Abe and Tanaka, 1997). For example, the degradation kinetic rate constants of aromatic compounds correlate well with E_{HOMO}. Compounds with more chlorine atoms are more difficult to degrade than those with fewer chlorines; therefore, the number of halogen atoms that an organic pollutant contains determines the treatability of the halogenated compound.

References

Abe, K. and Tanaka, K., Fe^{3+} and UV-enhanced ozonation of chlorophenolic compounds in aqueous medium, *Chemosphere*, 35, 2837–2847, 1997.

Adams, C.D., Fusco, W., and Kanzelmeyer, T., Ozone, hydrogen peroxide/ozone and UV/ozone treatment of chromium- and copper-complex dyes: decolorization and metal release, *Ozone Sci. Eng.*, 17, 149–162, 1995.

Beltran, F.J., Garcia-Araya, J.F., and Acedo, B., Advance oxidation of atrazine in water. II. Ozonation combined with ultraviolet radiation, *Water Res.*, 28, 2165–2174, 1994.

Beltran, F.J., Ovejero, G., Garcia-Araya, J.F., and Rivas, J., Oxidation of polynuclear aromatic hydrocarbons in water. 2. UV radiation and ozonation in the presence of UV radiation, *Indust. Eng. Chem. Res.*, 34, 1607–1615, 1995.

Beltran, FJ., Gonzales, M., and Rivas, F.J., Advanced oxidation of polynuclear aromatic hydrocarbons in natural waters, *J. Environ. Sci. Health A: Environ. Sci. Eng. Toxic. Hazardous Substance Contr.*, A31, 2193–2210, 1996.

Beltran, F.J., Encinar, J.M., and Gonzales, J.F., Industrial wastewater advanced oxidation. Part 2. Ozone combined with hydrogen peroxide or UV radiation, *Water Res.*, 31, 2415–2428, 1997.

Beltran, F.J., Encinar, J.M., and Alonso, M.A., Nitroaromatic hydrocarbon ozonation in water. 2. Combined ozonation with hydrogen peroxide or UV radiation, *Indust. Eng. Chem. Res.*, 37, 32–40, 1998.

Benitez, F.J., Beltran-Heredia, J., and Gonzales, T., Kinetic study of propoxur oxidation by UV radiation and combined O_3/UV radiation, *Indust. Eng. Chem. Res.*, 33, 1264–1270, 1994.

Benitez, F.J., Beltran-Heredia, J., Acero J.L., and Gonzales, T., Degradation of protocatechuic acid by two advanced oxidation process ozone/UV radiation and H_2O_2/UV radiation, *Water Res.*, 30, 1597–1604, 1996.

Benitez, F.J., Beltran-Heredia, J., Acero, J.L., and Pinilla, M.L., Simultaneous photodegradation and ozonation plus UV radiation of phenolic acids: major pollutants in agro-industrial wastewaters, *J. Chem. Tech. Biotech.*, 70, 253–260, 1997.

Beschkov, V., Bardarska, G., Gulyas, H., and Sekoulov, I., Degradation of triethylene glycol dimethyl ether by ozonation with UV irradiation or hydrogen peroxide addition, *Water Sci. Technol.*, 36, 131–138, 1997.

Bhowmick, M. and Semmens, M.J., Ultraviolet photooxidation for the destruction of VOCS in air, *Water Res.*, 28, 2407–2415, 1994.

Boncz, M.A., Bruning, H., Rulkens, W.H., Sudholter, E.J.R., Harmsen, G.H., and Bijsterbosch, J.W., Kinetic and mechanistic aspects of the oxidation of chlorophenols by ozone, *Water Sci. Technol.*, 35, 65–72, 1997.

Bose, P., Glaze, W.H., and Maddox, D.S., Degradation of RDX by various advanced oxidation processes. I. Reaction rates, *Water Res.*, 32, 997–1004, 1998.

Craig D.A. et al., Effect of ozonation on the biodegradability of substituted phenols, *Water Res.*, 31, 2655–2663, 1997.

De Laat, J., Dore, M., and Suty, H., Oxidation of s-triazines by advanced oxidation processes: by-products and kinetic rate constants, *J. Water Sci.*, 8, 23–42, 1995.

Erikson, L., Vervoom, H., and Peijnenburg, W.J., Multivariate QSAR modelling of the rate of reductive dehalogenation of haloalkanes, *J. Chemometrics*, 10, 483–492, 1996.

Espuglas, S., Yue, P.L., and Pervez, M.I., Degradation of 4–chlorophenol by photolytic oxidation, *Water Res.*, 28, 1323–1328, 1994.

Francis, P.D., Oxidation by UV and ozone of organic contaminants dissolved in deionized and raw mains water, *Ozone Sci. Eng.*, 9, 369–390, 1987.

Glaze, W.H., Peyton, G.R., Lin, S., Huang, R.Y., and Burleson, J.L., Destruction of pollutants in water with ozone in combination with ultraviolet radiation. 2. Natural trihalomethane precursors, *Environ. Sci. Technol.*, 16, 454–458, 1982.

Gorchev, V.F., Sova, A.N., and Goncharuk, V.V., Influence of UV irradiation on oxidation of aqueous phenol solutions with ozone, *Soviet J. Water Chem. Tech.*, 11, 35–37, 1989.

Gurol, M.D. and Nekoulnalni, S., Kinetic behavior of ozone in aqueous solution of substituted phenols, *Indust. Eng. chem. Fundam.*, 23, 1–54, 1984.

Hansch, C., Leo, A., and Hoekman, D., *Exploring QSAR Hydrophobic, Electronic, and Steric Constants*, American Chemical Society, Washington D.C., 1995.

Hashem, T.M., Zirlewagen, M., and Braun, A.M., Simultaneous photochemical generation of ozone in the gas phase and photolysis of aqueous reaction systems using one UV light source, *Water Sci. Technol.*, 35, 41–48, 1997.

Hayashi, J.-I., Ikeda, J., Kusakabe, K., and Shigeharu, M., Decomposition rate of volatile organochlorines by ozone and utilization efficiency of ozone with ultraviolet radiation in a bubble-column contactor, *Water Res.*, 27, 1091–1097, 1993.

Hoefl, C., Sigl, G., Specht, O., Wurdack, I., and Wabner, D., Oxidative degradation of AOX and COD by different advanced oxidation processes: a comparative study with two samples of pharmaceutical wastewater, *Water Sci. Technol.*, 35, 257–264, 1997.

Hoigne, J., Mechanisms, rates, and selectivities of oxidations of organic compounds initiated by ozonation in water, in *Handbook of Ozone Technology and Applications*, Rice, R.G. and Netzer, A., Eds., Ann Arbor Science, Ann Arbor, MI, 1982.

Hoigne, J. and Bader, H., *Proceedings of the International Ozone Association Symposium*, International Ozone Association, Toronto, 1977, p. 16.

Huang, C.-R. and Shu, H.-Y., The reaction kinetics, decomposition pathways and intermediate formations of phenol in ozonation, UV/O_3 and H_2O_2 processes, *J. Hazardous Mater.*, 41, 47–64, 1995.

Kozai, S. and Matsumoto, H., Decomposition of ketones in water by ozone treatment and UV irradiation, *Jpn. J. Toxicol. Environ. Health*, 43, 25–34, 1997.

Ku, Y. and Shen, Y.-S., Treatment of gas-phase volatile organic compounds (VOCs) by the UV/O_3 process, *Chemosphere*, 8, 1855–1866, 1999.

Ku, Y., Su W.J., and Shen, Y.S., Decomposition kinetics of ozone in aqueous solution, *Indust. Eng. Chem. Res.*, 35, 3369–3373, 1996a.

Ku, Y., Su, W.-J., and Shen, Y.-S., Destruction rate of volatile organochlorine compounds in water by ozonation with ultraviolet radiation, *Indust. Eng. Chem. Res.*, 35, 3369–3373, 1996b.

Ku, Y., Su, W.-J., and Shen, Y.-S., Decomposition of phenols in aqueous solution by UV/O_3 process, *Ozone Sci. Eng.* 18, 443–460, 1996c.

Kusakabe, K., Aso, S., Hayashi, J.-I., Kazuaki, I., and Morooka, S., Decomposition of humic acid and reduction of trihalomethane formation potential in water by ozone with UV irradiation, *Water Res.*, 24, 781–785, 1990.

Kusakabe, K., Aso, S., Wada, T., Hayashi, J.-I., Morooka, S., and Isomura, K., Destruction rate of volatile organochlorine compounds in water by ozonation with ultraviolet radiation, *Water Res.*, 25, 1199–1203, 1991.

Langlais, B., Reckhow, D., and Brink, D., *Ozone in Water Treatment, Application and Engineering,* Lewis Publishers, Boca Raton, FL, 1991.

Li, L., Chen, D.H., and Li, K.Y., UV/ozone-enhanced forced oxidation and simulations NO_x and SO_2 removal, *Waste Manage.,* 13, 518–519, 1993.

Mansour, M., Feicht, E.A., Behechti, A., and Scheunert, I., Experimental approaches to studying the photostability of selected pesticides in water and soil, *Chemosphere,* 35, 39–50, 1997.

Masten, S.J. and Hoigne, J., Comparison of ozone and hydroxyl radical-induced oxidation of chlorinated hydrocarbons in water, *Ozone Sci. Eng.,* 14, 197–213, 1992.

Masten, S.J., Shu, M., Galbraith, M.J., and Davies, S.H.R., Oxidation of chlorinated benzenes using advanced oxidation processes, *Hazardous Waste Hazardous Mater.,* 13, 265–281, 1996.

Masten, S.J., Galbraith, M.J., and Davies, S.H.R., Oxidation of 1,3,5–trichlorobenzene using advanced oxidation processes, *Ozone Sci. Eng.,* 18, 535–547, 1997.

Mokrini, A., Ousse, D., and Espuglas, S., Oxidation of aromatic compounds with UV radiation/ozone/hydrogen peroxide, *Water Sci. Technol.,* 35, 95–102, 1997.

Nagamany N.N. and Speece, R.E., QSAR model for predicting Henry's constant, *Environ. Sci. Technol.,* 22, 1349–1357, 1988.

Olson, T.M. and Barbier, P.F., Oxidation kinetics of natural organic matter by sonolysis and ozone, *Water Res.,* 28, 1383–1391, 1994.

Peijnenburg, W.J., The use of quantitative structure–activity relationships for predicting rates of hydrolysis processes, *Pure Appl. Chem.,* 63, 1667–1676, 1991.

Peijnenburg, W.J., Hart, M.J., den Hollander, H., de Meent, D., Verboom, H., and Wolfe, N., QSARs for predicting biotic and abiotic reductive transformation rate constants of halogenated hydrocarbons in anoxic sediment systems, *Sci. Total Environ.,* 109/110, 283–300, 1991.

Peijnenburg, W.J., De Beer, K., den Hollander, H., Stegeman, M.H.L., and Verboom, H., Kinetics, products, mechanisms and QSARs for the hydrolytic transformation of aromatic nitriles in anaerobic sediment slurries, *Environ. Toxicol. Chem.,* 12, 1149–1161, 1993.

Peyton, R. and Glaze, W., Destruction of pollutants in water with ozone in combination with ultraviolet radiation. 3. Photolysis of aqueous ozone, *Environ. Sci. Technol.,* 22, 761–767, 1988.

Prado, J., Arantegui, J., Chamarro, E., and Espuglas, S., Degradation of 2,4–D by ozone and light, *Ozone Sci. Eng.,* 16, 235–245, 1994.

Preis, S., Kamenev, S., Kallas, J., and Munter, R., Advance oxidation process against phenolic compounds in wastewater treatment, *Ozone Sci. Eng.,* 17, 399–418, 1995.

Prousek, J., Advanced oxidation processes for water treatment. photochemical processes, *Chemicke listy,* 90, 307–315, 1996.

Rorije, E. and Peijnenburg, W.J., QSARs for oxidation of phenols in the aqueous environment, suitable for risk assessment, *J. Chemometrics,* 10, 79–93, 1996.

Sebastian, J.H., Weber, A.S., and Jensen, J.N., Sequential chemical/biological oxidation of chlorendic acid, *Water Res.,* 30, 1833–1843, 1996.

Shen, Y.S. and Ku, Y., Treatment of gas-phase chloroethenes by UV/O_3 process, *Water Res.,* 32(9), 2669–2679, 1998.

Shen, Y.S. and Ku, Y., Treatment of gas-phase volatile organic compounds (VOCs) by UV/O_3 process, *Chemosphere,* 38(8), 1855–1856, 1999.

Shen, Y.-S. and Young, K., Treatment of gas-phase trichloroethene in air by the UV/ O_3 process, *J. Hazardous Mater.*, 54, 189–200, 1997.

Shu, H.-Y. and Huang, C.-R., Degradation of commercial azo dyes in water using ozonation and UV enhanced ozonation process, *Chemosphere*, 31, 3813–3825, 1995.

Tang, W. Z., Advanced Oxidation Processes for Treatment of Organic Pollutants, Dept. of Civil and Environmental Engineering, Florida International University, Miami, 1997.

Tang, W.Z. and Hendrix, T., An Internet Database of Kinetic Rate Constants and QSAR Models for Hydroxyl Radical Reactions and TiO_2/UV, Dept. of Civil and Environmental Engineering, Florida International University, Miami, 1998.

Trapido, M., Veressinina, J., and Munter, R., Ozonation and AOP treatment of phenanthrene in aqueous solutions, *Ozone Sci. Eng.*, 16, 475–485, 1994.

Trapido, M., Veressinina, Y., and Munter, R., Ozonation and advanced oxidation processes of polycyclic aromatic hydrocarbons in aqueous solutions: a kinetic study, *Environ. Technol.*, 16, 729–740, 1995.

Trapido, M., Hirvonen, A., Veressinina, Y., Hentunen, J., and Munter, R., Ozonation, ozone/UV and UV/H_2O_2 degradation of chlorophenols, *Ozone Sci. Eng.*, 19, 75–96, 1997.

Tratnyek, P.G., Hoigne, J., Zeyer, J., and Schwarzenbach, R.P., QSAR analyses of oxidation and reduction rates of environmental organic pollutants in model systems. *Sci. Total Environ.*, 109/110, 327–341, 1991.

Vollmuth, S. and Niessner, R., Degradation of PCDD, PCDF, PAH, PCB, and chlorinated phenols during the destruction-treatment of landfill seepage water in laboratory model reactor (UV, ozone, and UV/ozone), *Chemosphere*, 30, 2317–2331, 1995.

Vollmuth, S. and Niessner R., Degradation of polychlorinated dibenzo-*p*-dioxins and polychlorinated dibenzofurans during the UV/ozone treatment of pentachlorophenol-containing water, *Toxicol. Environ. Chem.*, 61, 27–41, 1997.

Vollmuth, S., Wenzel, A., and Niessner, R., Purification of organic contaminants in seepage water of a landfill by UV/ozone technique, *Proc. SPIE*, 2504, 520–530, 1995.

Weichgrebe, D. and Vogelpohl, A., Comparative study of wastewater treatment by chemical wet oxidation, *Chem. Eng. Process.*, 33, 199–203, 1994.

Yue, P.L., Modelling of kinetics and reactor for water purification by photo-oxidation, *Chem. Eng. Sci.*, 48, 1–11, 1993.

Yue, P.L. and Legrini, O., Photochemical degradation of organics in water, *Water Pollut. Res. J. Can.*, 27, 123–137, 1992.

Zeff, J.D. and Barich, J.T., UV/oxidation of organic contaminants in ground, waste and leachate waters, *Water Pollut. Res. J. Can.*, 27, 139–150, 1992.

Zwiener, C., Weil, L., and Niessner, R., Atrazine and parathion-methyl removal by UV/O_3 in drinking water treatment, *Int. J. Environ. Analyt. Chem.*, 58, 247–264, 1993.

Zweiner, C., Weil, L., and Niessner, R., UV- and UV/ozone degradation of triazine herbicides in a pilot plant: estimation of UV-photolysis rate constants and quantum yields, *Vorm Wasser*, 84, 47–60, 1995.

9

UV/Titanium Dioxide

9.1 Introduction

The photochemical reactions of titanium dioxide (TiO_2) have drawn much attention in the field of advanced oxidation processes for over 30 years. Research on photocatalysis using semiconductors began in the late 1940s with the work of Laidler et al., who investigated the chemical reactions of zinc oxide under illumination. At the same time, Mashio and Kato were studying the photocatalytic oxidant action of alcohols by TiO_2. In the late 1960s, Honda and Fujishima made significant discoveries that led to the development of TiO_2 as a semiconductor photocatalyst (Fujishima and Honda, 1971). The irradiation of a TiO_2 electrode coupled with a platinum electrode was found to be able to split water into hydrogen and oxygen (Fujishima et al., 1969); the water was simultaneously oxidized and reduced. This discovery laid the foundation for further research on TiO_2 to generate hydrogen as a combustible fuel. It soon became apparent that redox reactions of organic and other inorganic compounds could be induced by band gap irradiation of a variety of semiconductor particles. The photocatalytic oxidation of simple organic compounds such as alcohols and carboxylic acids was further studied by Kraeutler and Bard (1998). Through the 1970s and by the end of the 1980s, photocatalytic reactions of TiO_2 were studied for almost all classes of organic compounds. In the 1990s, a UV/TiO_2 process for treating wastewaters containing organic pollutants was commercialized.

Titanium dioxide is widely used in the production of plastics, enamels, artificial fibers, electronic materials, and rubber (Hadjiivanov and Klissurski, 1996). Its ability to photocatalyze the oxidation of organic materials has been known for years in the paint industry. For this reason, TiO_2 is used as a white paint pigment (Stafford et al., 1996). TiO_2 is also known as an excellent catalyst for semiconductor photocatalysis due to its nonselectivity for environmental engineering applications; it is nontoxic, insoluble, reusable, photostable, readily available, and inexpensive (Tang and Chen, 1996). UV/TiO$_2$ is an appealing advanced oxidation process (AOP) because it has been demonstrated that the process can mineralize a wide variety of organic pollutants.

Titanium dioxide is the most widely used catalyst in photocatalytic degradation of organic pollutants due to its suitable band-gap energy of 3.05 eV over a wide range of pH (Tang and Huang, 1995). The chemical viability of UV/TiO$_2$ has been established, but the future of photocatalysis depends on new designs for photochemical reactors and fixation of TiO$_2$ onto supports to eliminate its separation from the treated effluent. For this purpose, catalyst supports, glass beads, glass plates, fiberglass mesh, and porous films on glass substrates have been studied (Peil and Hoffmann, 1996).

9.2 Fundamental Theory

9.2.1 Photoexcitation

A semiconductor has a band structure that is characterized as the valance band (VB), which has a series of closely spaced energy levels associated with covalent bonding between atoms. The valence band is composed of a crystallite structure. The *conduction band* (CB) is a series of spatially diffuse, energetically similar levels at a higher energy. The magnitude of energy between the electron-rich valence band and the electron-deficient conduction band is responsible for the extent of thermal distribution of the conduction band and, ultimately, the electrical conductivity of the particle. This band gap also defines the sensitivity of the semiconductor to irradiation by photons at different wavelengths.

Photocatalytic oxidation by UV/TiO$_2$ involves the excitation of TiO$_2$ particles by UV light from the valance band of the solid to the conduction band.

$$TiO_2 + h\nu \rightarrow e^-_{CB} + h^+_{VB} \qquad (9.1)$$

The equation can also be illustrated in Figure 9.1. When a semiconductor such as TiO$_2$ absorbs photons, the valence band electrons are excited to the conduction band. For this to occur, the energy of a photon must match or exceed the band-gap energy of the semiconductor. This excitation results in the formation of an electronic vacancy or positive hole at the valence band edge. A positive hole is a highly localized electron vacancy in the lattice of the irradiated TiO$_2$ particle. This hole can initiate further interfacial electron transfer with the surface bound anions.

The primary steps in photoelectrochemical mechanism are as follows: (1) formation of charge carriers by a photon; (2) charge-carrier recombination to liberate heat; (3) initiation of an oxidative pathway by a valence-band hole; (4) initiation of a reductive pathway by a conduction-band electron; (5) further

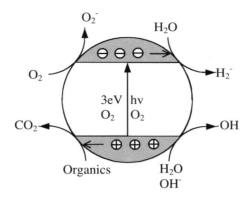

FIGURE 9.1
Excitation of electron from valance band to conduction band; semiconductor potential of TiO_2.

thermal and photocatalytic reactions to yield mineralization products; (6) trapping of a conduction band electron in a dangling surficial bond to yield Ti(III); and (7) trapping of a valence-band hole at a surficial titanol group.

The photogenerated electrons and holes of a semiconductor lose their energy by interactions with oxidants and reductants, respectively, in an aqueous solution. They are excited from the bottom of the valence band to the upper edge of the conduction band. Photoexcitation of TiO_2 and the generation of an electron/hole pair creates the potential not only for oxidation but also for reduction. The energy level of the bottom of the conduction band can be considered to be a measure of the reduction strength of the photoexcited electrons. The energy level of the upper level of the valance band is a measure of the oxidation strength of the holes. The positions of the valence and conduction bands of TiO_2 at pH = 1, relative to standard carbon electrode (SCE), are –0.1 and +3.1 V, respectively (O'Shea and Cardona, 1994). Conduction band electrons generated within TiO_2 molecules have chemical reactivity patterns that can be monitored. The potential of semiconductors for oxidation and reduction can be classified into four groups according to the water-splitting reaction shown in Table 9.1.

This classification based on water splitting is important to understanding the redox potential of a given semiconductor. Although this classification is simple, it is convenient in selecting a semiconductor that is appropriate for a desired reaction. For a more detailed reactor design, factors such as the lifetimes of carriers; energy levels of surface states; adsorption and desorption of molecules on the surface; kinetic nature of the surface; and electron kinetics must be considered (Serpone and Pelizzetti, 1989).

9.2.2 Hydroxyl Radical Formation

The production and reactivity of the radical intermediates are the main factors limiting the entire oxidation reaction rate. The formation of the

TABLE 9.1

Classification of Semiconductors Based on Water Splitting Reaction

Classification	Reaction
OR type	The oxidation and reduction power is strong enough to promote hydrogen and oxygen production. Examples include TiO_2, $SrTiO_3$, and CdS. OR indicates a strong ability for both oxidation and reduction.
R type	Only the reduction power is strong enough to reduce water; the oxidation power is too weak. Examples include CdTe, CdSe, and Si.
O type	The valence band is located deeper than the O_2/H_2O level so the oxidation power is strong enough to oxidize water, but the reduction power is not strong enough to reduce water. Examples include Fe_2O_3, MoS_2, and Bi_2O_3.
X type	The conduction and valence bands are located between the H^+/H_2 and O_2H_2O levels; therefore, both the oxidation and reduction powers are so weak that neither oxygen nor hydrogen can be produced.

hydroxyl radicals on irradiated TiO_2 in an aqueous system has been probed by pulse radiolysis. The detection of singly oxidized transients suggests that surface-bound hydroxyl radicals initiate the oxidation of the surface-bound substrate rather than diffuse into the bulk solution. Several other studies on this topic have also reached the same conclusion (Fox and Dulay, 1993).

Superoxide and perhydroxyl radicals are other radicals formed from the reactions of electrons with adsorbed oxygen. The generated radicals can then oxidize organic pollutants at the solid–liquid interface. Direct oxidation of organic pollutants may also be possible by photogenerated holes and may proceed in competition with hydroxyl radical oxidation, as proposed for benzene oxidation, although Okamoto et al. (1985) did not report the existence of this pathway for phenol. Oxidation by holes was suggested for some acids, such as trichloroacetic acid, which are formed as intermediates from the oxidation of chlorinated ethanes by hydroxyl radicals. Photogeneration of radical species can be represented by the following reactions:

$$h^+_{VB} + OH^- \text{ (surface)} \rightarrow OH^\bullet \tag{9.2}$$

$$h^+_{VB} + H_2O \text{ (adsorbed)} \rightarrow OH^\bullet + H^+ \tag{9.3}$$

$$e^-_{CB} + O_2 \text{ (adsorbed)} \rightarrow O_2^\bullet \tag{9.4}$$

$$e^-_{CB} + h^+_{VB} \rightarrow \text{heat} \tag{9.5}$$

where $h\nu$ represents UV radiation, h^+_{VB} represents valence-band holes, e^-_{CB} represents conduction-band electrons, and $O_2^\bullet$ represents a superoxide ion. H_2O_2, which can also be generated from the superoxide ion by various mechanisms, can also be a source for hydroxyl radicals by further reacting with electrons or superoxide ions. It may be decomposed to hydroxyl radicals by photolytic disassociation. Kormann et al. (1991) determined that the reaction of surface-bound hydroxyl radicals with the adsorbed organic compound was the rate-determining step. The recombination of electrons and holes was reportedly the main factor in limiting oxidation rates of organic substrates.

The holes and the electrons at the surface of the TiO_2 molecule can then form hydroxyl radicals from the oxidation of oxygen, water, or hydroxide ions (Venkatadri and Peters, 1993); however, the oxidative degradation rates of organic pollutants are based upon the energy needed to cleave a given chemical bond and the concentration of dissolved molecular oxygen.

The formation of hydroxyl radicals has been observed when TiO_2 is irradiated with ultraviolet light. Hydroxyl radical is the most powerful oxidizing species after fluorine. In the UV/TiO_2 process, if organic compounds are rich in π electrons, hydroxylation will proceed as previously described in Equation (4.1). Hydrogen abstraction usually occurs in reaction with unsaturated organic compounds. Peroxyl radicals are produced by the reaction between the organic radicals and the molecular oxygen:

$$R^\bullet + O_2 \rightarrow RO_2^\bullet \tag{9.6}$$

9.2.3 The Role of Adsorption in the UV/TiO$_2$ Process

Photocatalytic oxidation is a surface-catalyzed reaction; therefore, a chemical must first be adsorbed onto the TiO_2 surface before it can undergo photocatalytic oxidation. One of the requirements of the Langmuir–Hinshelwood model is the adsorption of the compound has to occur before oxidation takes place. For UV/TiO_2 systems, Fox et al. (1990) proposed that strong adsorption of substrates on the TiO_2 surface is required due to the very fast recombination of electron/hole pairs. Photocatalytic degradation has been demonstrated to follow the Langmuir–Hinshelwood kinetic model (Matthews, 1991; Davis and Huang, 1990). It provides strong evidence of substrate preadsorption onto the TiO_2 surface.

Langmuir–Hinshelwood (LH) kinetics are widely used to quantitatively delineate substrate preadsorption in both solid–gas and solid–liquid reactions. The model assumptions are stated in Table 9.2. Under these

TABLE 9.2

Langmuir–Hinshelwood Model Assumptions

The number of surface adsorption sites is fixed at equilibrium.

Only one substrate may bind at each surface site.

The heat of adsorption by the substrate is identical for each site and is independent of surface coverage.

No interaction occurs between adjacent adsorbed molecules.

The rate of surface adsorption of the substrate is greater than the rate of any subsequent chemical reaction.

No irreversible blocking of active sites by binding to product occurs.

---

assumptions, the relationship between surface coverage, θ; initial substrate concentration, C; and adsorption equilibrium constant, K, can be expressed by the following equation:

$$\theta = KC/(1 + KC) \tag{9.7}$$

The initial degradation rate of a substrate can be described by its initial concentration and adsorption characteristics:

$$r_{LH} = -dC/dt = kKC/(1 + KC) \tag{9.8}$$

where k is the rate constant, and t is the reaction time. For two or more species competing for one adsorption site, the following expression was suggested by Davis and Huang (1990):

$$r_{LH} = kKC/(1 + KC + \Sigma_i \, K_iC_i) \tag{9.9}$$

where i represents a competitively adsorbed species. Good linearity of a plot $1/r_{LH}$ vs. $1/C$ reflects the validity of the LH model, which demonstrates the preadsorption of a target compound during photocatalytic degradation. Oxidation rate constants and adsorption equilibrium constants can be obtained from the slope $(1/kK)$ and intercept $(1/k)$ of the straight line.

9.2.4   Characteristics of TiO$_2$ Surface

The characteristics of the TiO$_2$ surface determine its adsorptive capacity and photocatalytic activity. TiO$_2$ exists as anatase, rutile, and brookite; however, the catalytic activities of each are significantly different. The rutile form has the most practical applications, as it is the most stable of the three. The effectiveness of TiO$_2$ as a catalyst is determined by its surface properties. The TiO$_2$ surface has many active sites that contribute to its high catalytic activity. The anatase surface is highly heterogeneous and has three kinds of Lewis acid sites, due to the differently coordinated Ti^{4+} ions. The surface also has at least two kinds of hydroxy groups present on the surface (Hadijiivanov and Klissurski, 1996). Rutile and anatase have similar crystal structures, both tetragonal (Stafford et al., 1996).

The difference in the catalytic activities of the anatase and rutile form is due to differences in lattice structure. It has been reported that the reducing properties of conduction-band electrons are dependent on lattice structures (Stafford et al., 1996). Anatase has the highest energy for the lowest unoccupied molecular orbital (LUMO), making it the least reactive of the three forms of TiO$_2$ (Gratzel and Rotzinger, 1985).

Titanium dioxide is synthesized by various methods according to the structure that is desired; therefore, the surface properties depend on preparation methods. Two methods that are used to produce TiO_2 are known as the sulfate method and vapor-phase oxidation of $TiCl_4$. Vapor-phase oxidation is the most widely used preparation technique, and the product formed is anatase. The most popular commercial TiO_2 is Degussa P25, which is a fumed TiO_2 that consists of both anatase and rutile in proportions of 80:20. The product is mostly anatase with a coating of rutile on the surface. Degussa TiO_2 usually has a surface area of 50 m^2/g and an average particle size of 30 nm.

The preparation of TiO_2 has been studied to enhance its photocatalytic properties (Scalfani et al., 1990). For example, it has been found that the fastest photocatalytic degradation rates exist for anatase samples that have been prepared by precipitation of titanium isoperoxide at 350°C. Anatase samples that have been heated to 800°C will form rutile TiO_2.

Titanium dioxide particles that are colloidal in size have been proposed for improving the efficiency of photocatalysis. Decreasing the particle size is known to cause a widening of the bandgap (Brus, 1990). This quantum size effect may also contribute to the increased reactivity of surface holes and electrons (Stafford et al., 1996). The particle size of TiO_2 has also been correlated to its charge distribution. In photocatalytic experiments with UV light, illumination of TiO_2 has shown the zeta potential to become more positively charged, thus yielding better adsorption and degradation rates of organic anions (Kim and Anderson, 1996).

Titanium dioxide can be used in the stationary phase attached to a support medium such as silica gel or fiberglass, to glass beads, or in ceramic membranes. Use of TiO_2 in the stationary phase avoids the need for separation. TiO_2 can also be suspended in a heterogeneous system, which requires separation by microfiltration or centrifugation. TiO_2 has also been used as a coating on glass tubes as a catalyst support. In reactions using TiO_2 suspensions, degradation rates are two to six times faster than when the TiO_2 is fixed on a support.

Titanium dioxide photodegradation rates can be significantly enhanced with H_2O_2. With the addition of H_2O_2, degradation times of trichloroethylene (TCE) dropped from 75 to 20 min in a study by Tanaka et al. (1989). This enhancement was most likely due to an increase of hydroxyl radicals. The half-lives of pesticides DDVP and DEP were demonstrated to be shortened with the addition of H_2O_2 (Harada et al., 1990). Similar enhancements were shown for the photodegradation of chloral hydrate, phenol, and chlorophenols (Venkatadri and Peters, 1993).

The catalytic activity of TiO_2 can be increased with the loading of metals such as silver or platinum. The loading of silver onto the surface of TiO_2 has been shown to increase the removal of chloroform from 35 to 45% and the removal of urea from 16 to 83% (Kondo and Jardim, 1991). The drawback of this treatment is the dissolution of silver into solution at a level of 0.5 ppm, which is 10 times the regulatory limit (Venkatadri and Peters, 1993).

9.2.5   Adsorption of Organic Compounds on TiO$_2$

In the past, the degradation of most classes of organic compounds has been studied using an UV/TiO$_2$ system. Detailed mechanisms and kinetic data are presented in several recent literature reviews by Legrini et al. (1993), Mills et al. (1993), and Hoffmann et al. (1995). Adsorption has great effects on photocatalytic oxidation kinetics. For example, a strong correlation between the adsorptive capacity of thiocarbamates and the extent of photo-catalytic oxidation has been demonstrated (Sturini et al., 1996). The reaction rate of thiocarbamate has been found to be governed by its adsorption kinetics. Thiocarbamates are not soluble in water and were observed to be adsorbed onto the TiO$_2$ surface. It was found that only substrates adsorbed onto the surface of the TiO$_2$ molecule are photodegraded (Sturini et al., 1996).

Tunesi and Anderson (1991) studied the influence of chemisorption on the photodecomposition of salicylic acid and several other compounds. The role of adsorption may be even more important for compounds that exhibit pH-dependent adsorption behavior in aqueous solutions. The presence of water in TiO$_2$ suspensions has been shown to strongly affect the bonding behavior of compounds to be oxidized. It has been observed that water and hydroxide groups readily adsorb onto the TiO$_2$ surface. In chemisorption, these ligands are exchanged with absorbing solutes, while in physisorption no interactions between the TiO$_2$ surface and the compound of interest are observed (Tunesi and Anderson, 1991).

The degradation kinetics of fenuron is affected by the pH solution (Richard and Benagana, 1996). The products of the hydroxyl radical attack on fenuron are easily identified. It can also be concluded that the pH effect on the degradation of fenuron can be explained by differences in adsorption with changing pH. Because the point of zero charge of Degussa P-25 TiO$_2$ is equal to 6.3, the surface charge is neutral in neutral medium; however, at pH lower than 6.3, the surface is positively charged and molecules are attracted to the surface by their electronegative component (Richard and Benagana, 1996).

As shown in Figure 9.2, the mechanism for the adsorption of fenuron onto TiO$_2$ can be explained in two different ways. In the first scheme, the positively charged TiO$_2$ surface repels the positively charged nitrogen atoms. This prevents the hydroxyl radicals attacking the methyl groups. In the second scheme, the carbon–nitrogen bonds are planar to the surface. This position seems more favorable because the lone electron pairs of the nitrogen and oxygen are close to the positive TiO$_2$ surface and the methyl groups can be oxidized.

The degradation of fenuron in neutral medium is accomplished by oxida-tion of the methyl groups and the aromatic ring. In acidic TiO$_2$ suspensions, the degradation of fenuron is accomplished primarily by oxidation of the methyl groups and not oxidation of the aromatic ring. The oxidation of methyl groups can be explained by the position in which fenuron was adsorbed at the surface of TiO$_2$ (Richard and Benagana, 1996).

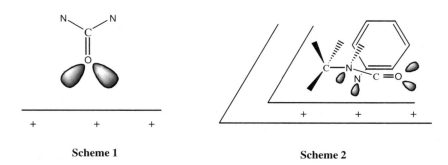

FIGURE 9.2
Adsorption of fenuron onto TiO_2. (From Richard, C. and Benagana, S., *Chemosphere*, 33, 635, 1996. With permission.)

Tunesi and Anderson (1991) reported that the electronic environment of the Ti^{4+} cation is affected by the number of bonds with oxygen and by the coordination number of these oxygen ligands. In aqueous media, Ti^{4+} cations are capable of bonding to water molecules. For salicylic acid, benzoic acid, phenol, and 4-chlorophenol, the stereochemical configurations of the compounds are critical to adsorptive capacity. Salicylic acid has the stereochemical configuration that makes ring formation possible, while it is impossible for benzoic acid, phenol, and 4-chlorophenol to have this structure. When this TiO_2–salicylate chelate is formed, the salicylate bonds to the O surface orbitals of the TiO_2 molecule. Upon irradiation, the O orbitals become the source of holes and attract electrons from the ligand, leading to oxidation of the salicylate (Tunesi and Anderson, 1991). The photocatalytic oxidation mechanisms of phenol and 4-chlorophenol were not observed to be affected by surface conditions. Phenol and 4-chlorophenol do not significantly adsorb on TiO_2 surfaces, so the degradation rates are independent of adsorption.

The adsorption of organic ligands onto metal oxides and the parameters that have the greatest effect on adsorption were also studied (Stone et al., 1993). The extent of adsorption was measured by determining the loss of the compound of interest from solution. The physical and chemical forces that control adsorption into two general categories were classified as either specific or nonspecific adsorptions. Specific adsorption involves the physical and chemical interaction of the adsorbent and adsorbate. Under specific adsorption, the chemical nature of the sites influences the adsorptive capacity. Nonspecific adsorption does not depend on the chemical nature of the sites but on characteristics such as surface charge density (Stone et al., 1993). The interactions of specific adsorption can be explained in two ways. The first approach uses activity coefficients to relate the electrochemical activity at the oxide/water interface to its electrochemical activity in bulk solution (Stone et al., 1993). This approach is useful in situations where the making or breaking of covalent bonds is not occurring. Long-range electrostatic forces may be described using the Poisson–Boltzmann term:

$$\{I\}_x = \{I\}_{bulk}\, exp^{(-zF\Psi/RT)} \tag{9.10}$$

where z is the ion charge, F is the Faraday constant, Ψ is the electrical potential, R is the universal gas constant, and T is the temperature of the reaction.

The second approach postulates a new chemical species and an equilibrium constant that relates the activity of the new species to the established species. The mechanism of nonspecific adsorption is believed to be due to long-range electrostatic forces on counterions near the charged TiO_2 surface. The extent of nonspecific adsorption can be calculated once the electrical potential on the TiO_2 surface is known (Stone et al., 1993).

The extent of adsorption of organic ligands was further studied by Vasudevan and Stone (1996). Organic ligands have Lewis-base functional groups and are capable of forming bonds with protons, metal ions, and metal oxides. These ligands are polar and ionizable. Interaction between a ligand and the TiO_2 surface adds electrostatic forces to the interaction of a compound and the metal oxide surface. The extent of adsorption was affected by the substituents on the aromatic ring (Vasudevan and Stone, 1996). TiO_2 has a high ionic contribution to bonding, and ligands with the greatest amount of ionic bonding capability are most readily adsorbed onto the TiO_2 surface. The interactions of ligands and metal oxides have been classified in two ways: those arising from long-range electrostatic forces and those that are difficult to quantify, which are referred to as near-range physical and chemical forces.

According to the position of substituents on an aromatic ring, Lewis-base groups at the *ortho* position adsorb to a significant extent compared to compounds with groups at the *meta* and *para* positions (Vasudevan and Stone, 1996). This is due to the fact that Lewis-base groups at the *ortho* position can simultaneously coordinate a single metal ion, while the *meta* and *para* base groups cannot.

The effect of mineral surfaces on the chemical transformation of organic chemicals was studied by Torrents and Stone (1991), who used the pesticide-like compound phenyl picolinate (PHP). Metal oxides, including TiO_2, SiO_2, Al_2O_3, Fe_2O_3, and FeOOH, were used as adsorbents. Effects of adsorbents on photocatalytic, redox, polymerization, and hydrolysis reactions have been widely studied, as it is necessary to understand the role that adsorption plays in these reactions. For these surface-catalyzed reactions to occur, the compound of interest must be adsorbed onto the surface of a metal oxide. If the adsorption is non-specific, the compound does not change in structure. If the adsorption is specific, the compound will change in structure and in chemical properties, making the compound more susceptible to reaction. Torrents and Stone (1991) concluded that mineral surfaces can catalyze reactions of carboxylic acids and esters in three ways:

- Metals bound to oxides can bond with the carbonyl group and adjacent ligand donors, allowing catalysis to originate from the polarization on the oxygen–carbon bond, and later allowing nucleophilic attack.
- Hydroxy groups bound to the oxide surface can act as nucleophiles, providing hydroxyl radicals for the addition of ester molecules.
- Electrostatic and other forces in the surface region can attract reactants, which facilitate the catalytic reaction.

The amount of picolinate adsorbed was near 100%. No significant levels of adsorption for phenol were observed. The adsorption of PHP was observed to be 5%. These levels of adsorption are directly related to the structures of the compounds. Adsorption of PHP was significantly high due to the two-ligand donor groups that are capable of forming a chelate with surface-bound metals (Torrents and Stone, 1991). The position of these ligand donor groups is critical to the bond formation between the surface and the compound. Phenyl isoicotinate, the isomer of PHP, was unable to form a surface chelate and therefore did not undergo catalysis. This suggests that the formation of a surface chelate is crucial for a compound to undergo catalysis.

Dichlorvos is a commercially manufactured insecticide that is widely used to protect stored products and crops. The adsorption characteristics of dichlorvos onto TiO_2 were studied by Lu et al. (1996), along with the photocatalytic oxidation of dichlorvos by UV-illuminated TiO_2. Also, the role of adsorption in the photocatalytic oxidation of dichlorvos was evaluated using the LH model. To gain a better understanding of adsorption of dichlorvos onto TiO_2, cosolvent, temperature, and ionic strength were investigated. Adsorption of dichlorvos onto TiO_2 in the presence of organic solvents was shown to decrease adsorption. Of the two solvents (methanol and acetone), methanol was shown to inhibit dichlorvos adsorption to a greater extent than did acetone. Temperature was also shown to have an important role in controlling adsorption rates. Lu et al. (1996) found that electrostatic characteristics of the solid–liquid interface are a function of temperature, and the amount of dichlorvos adsorbed increases with increasing temperature. The effect of electrolytes on the adsorption of dichlorvos onto TiO_2 demonstrated that the adsorption rates significantly decreased in the presence of the electrolytes. $NaClO_4$ and CH_3COONa/CH_3COOH were the electrolytes chosen. A relation between dichlorvos adsorption and ionic strength shows that adsorption decreased by 26% in the presence of $NaClO_4$ and as much as 70% in the presence of CH_3COONa/CH_3COOH (Lu et al., 1996). Adsorption reached a maximum value at pH 5.5 and decreased on both sides of this value. The pH of maximum adsorption density is lower than at the pH_{zpc} of TiO_2 (6.4). Lu et al. (1996) concluded that the ion–dipole interactions between the charged surface and the nonionic adsorbate are negligible. Furthermore, coordination by exchanged and structural metal cations is not relevant because the organic ligand is not a good electron donor; therefore, adsorption may be caused by hydrophobic interaction and hydrogen bonding. The decrease in adsorption in the presence of electrolytes may result from the difference in proton affinity and solvation of the solvents. Adsorption may be favorable in less polar solutions because of the hydrophobic driving force. Dichlorvos is a nonionizable compound, and the effect of pH on the adsorption is a result of the modification of the TiO_2 surface.

Martin et al. (1996) studied the surface structures formed when 4-chlorocatechol adsorbs onto TiO_2. These surface interactions were studied to gain a better understanding of how these surface structures affect photoreactivity. Adsorption isotherms of 4-chlorocatechol demonstrate that the compound adsorbs to a greater extent at pH values 7 to 9. The interactions of protons and 4-chlorocatechol with the TiO_2 surface are explained by the double layer theory (Martin et al., 1996).

To analyze the surface structures, infrared spectra were studied. It was concluded from the similarities between the spectra and the deprotonated CT^{2-} that CT forms a bidentate structure on TiO_2 (Martin et al., 1996). The infrared (IR) peak positions also suggest that the CT adsorbate carries a negative charge. For a purely covalent bond, the charge would be zero. This suggests that the bond formed may have a 60% ionic character and a 40% covalent character. The concentration of 4-chlorocatechol apparently determines the type of surface structure formed. At concentrations below 50 μM, 4-chlorocatechol adsorbs as a bidentate structure. At concentrations above 50 μM, 4-chlorocatechol adsorbs nonspecifically in a multilayer environment.

Adsorption of 4-aminobenzoic acid (ABZA) and 3-chloro-4-hydroxybenzoic acid (CHBZA) was studied by Cunningham and Al-Sayyed (1990). Adsorption isotherm data were collected and analyzed to determine what role adsorption plays in influencing photocatalytic efficiencies. The Langmuir–Hinshelwood kinetic model was applied and limitations of this model are discussed.

Adsorption constants (K) were determined from slope/intercept relationships. It was observed that significantly lower K values resulted when the total number of adsorption sites was 4×10^{-4} mol/g TiO_2; therefore, the adsorption rate constants are observed to be dependent on the minimum cross-sectional area assigned per adsorbed solute molecule (Cunningham and Al-Sayyed, 1990).

McBride and Wesselink (1988) studied IR spectra of catechol adsorbed onto the oxide surface and found evidence that the compound was chemically altered, indicating that chemisorption was the dominant mechanism. In addition to catechols, phenols are known to adsorb onto metal oxide surfaces. This adsorption is dependent on the number and position of hydroxy substitutions on the benzene ring. Diphenolic compounds adsorb to a greater extent than monophenolic compounds, suggesting the formation of a bidentate bond with the metal oxide. This bidentate bond is formed when the two phenolic ligands coordinate with one or two surface metal ions (McBride and Wesselink, 1988).

Two photocatalysts, Hombikat UV100 and Degussa P25, were used in batch experiments to compare their ability to degrade toxic compounds in landfill leachate (Bekbolet et al., 1996). Both of the TiO_2 powders showed strong adsorption of the compounds onto the powders. The highest adsorption rates and the highest photodegradation rates were observed at pH 5. The degradation rates of compounds were shown to increase significantly when the Hombikat UV100 photocatalyst was used. To study the adsorption behavior, reactions were carried out under different pH values using a thin-film, fixed-bed reactor. When TiO_2 was added to the water, a significant reduction in total organic carbon (TOC) and absorbance was observed. At pH 3, a 43% reduction of TOC was observed. In treatment of leachate, a 23% reduction in leachate was observed, but a 68% reduction was observed after adjusting the pH to 3. Adsorption onto Hombikat UV100 was found to be greater than on Degussa P25. After 5 hours of reaction, the removal of leachate from the batch reactor was nearly the same for both catalysts.

The adsorption studies show that adsorption onto TiO_2 is greatest at pH 5 compared with adsorption at pH 3 and pH 8. Freundlich isotherms were able to give adsorption rate constants for the leachate. Adsorption rates were much faster under lower pH conditions than higher pH conditions. The photocatalytic studies also show that TOC removal is greatest at pH 5, which has the maximum adsorption. Higher photocatalytic rates could be obtained by diluting the leachate. It was also concluded that higher removal rates could be determined using the Hombikat UV 100 photocatalyst.

9.3 Degradation of Organic Pollutants

Numerous inorganic and organic pollutants have been effectively degraded or completely mineralized using TiO_2-assisted photocatalysis; however, most studies have investigated pollutants present as a single compound at low concentrations and not actual wastewaters. The pollutants investigated include chlorinated aliphatic compounds, phenols, chlorinated and fluorinated aromatic compounds (such as chlorophenol, chlorobenzenes, polychlorinated biphenyls, and dioxins), surfactants, nitroaromatics and aminorganophosphorus insecticides, colored organic, and polycyclic aromatic compounds (PAHs).

Most studies of UV/TiO_2 photocatalytic oxidation have proposed that hydroxyl radical attack on a substrate is the primary step in the oxidation of organics. Organic compounds not only undergo photocatalytic oxidation but also photocatalytic reduction. Photooxidation is the dominant reaction for the degradation of organic substrates. It has been found that almost every organic functional group, bearing a non-bonded lone pair of any π conjugated electrons, can be activated toward TiO_2-photocatalyzed oxidative reactivity by dehydrogenation, oxygenation, or oxidative cleavage (Fox and Dulay, 1993).

These reactions can be illustrated in gaseous anaerobic photodehydrogenation of ethanol to acetaldhyde by irradiated TiO_2 powder:

$$CH_3CH_2OH \xrightarrow[\quad TiO_2 \,/\, O_2 \quad]{} CH_3CHO \qquad (9.11)$$

Metal oxides, such as TiO_2, can sometimes act as high-temperature thermal oxidation catalysts, but oxidative selectivity can be observed at room-temperature photocatalytic oxidations. For example, the oxidation of cyclohexane by O_2 and TiO_2 is thermodynamically possible, but its rate at room temperature is impossibly slow without irradiation. At higher temperatures, little oxidative selectivity is obtained. With the use of TiO_2 photocatalysis, high oxidative selectivity is obtained.

Photocatalytic reductions are not observed as frequently as photocatalytic oxidations. This is due to the fact that the reducing power of a conduction band electron is significantly lower than the oxidizing power of a valence-band hole and because most reducible substrates do not compete kinetically with oxygen in trapping photogenerated conduction band electrons. Most photocatalytic reductions require a co-catalyst such as platinum.

The degradation kinetics and mechanics of several classes of organic compounds are discussed below.

9.3.1 Alcohol

Lichtin and Avudaithai (1996) compared the photocatalytic oxidation of methanol in both aqueous and gas phases. Photocatalytic efficiencies for the two phases were analyzed. Differences in the chemistry of photocatalytic oxidation for the two phases were also studied. Batch reactors were used for both aqueous- and gas-phase reactions. The reactor was a 450-mL, magnetically stirred, cylindrical vessel with an axially aligned 6-W fluorescent lamp emitting light at 360 nm. O_2 was bubbled into the reactor at 35 mL/min. Samples were irradiated for 140 min, and product concentrations were determined using gas chromatography.

Apparent initial kinetic orders of removal of organic reactants were evaluated from log (initial rate of removal of organics) vs. log (initial concentration of organic). The initial photoefficiencies of removal of organic reactant (E) are defined by Equation (9.12):

$$E = \frac{\text{No. molecules of organic reactant removed}}{\text{No. of photons incident on catalyst}} \qquad (9.12)$$

Removal of methanol was measured using COD. The degree to which the formation of CO_2 from CH_3OH lagged behind its removal decreased with increasing concentration of $O_2^{\bullet}$. The molar ratios of CO_2 produced to CH_3OH removed after 15 min of irradiation in 2.3, 6.2, 20, and 100% dry O_2 were

0.25, 0.40, 0.55, and 0.85, respectively (Lichtin et al., 1996). Two intermediates were detected by GC analysis and identified to be formaldehyde and methyl formate. The initial rate removal of CH_3OH increased with partial pressure of oxygen. When the kinetic orders in the organic reactants of initial rates of photocatalytic oxidation in aqueous phase were compared with gas phase reactions, the gas phase removal rates were significantly higher than the aqueous phase removal rates (Lichtin and Avudaithai, 1996).

9.3.2   Alkyls

Alkyl compound degradation by the UV/TiO_2 process has been widely studied. The photoreduction of carbon dioxide and bicarbonate in the system has also been studied (Willner and Willner, 1988). The reduction of aqueous CO_2 is thermodynamically favored over H_2 evolution according to the redox potentials of –0.52 V for CO_2 reduction and –0.41 V for hydrogen evolution. However, the kinetics for CO_2 reduction is unfavorable, and H_2 evolution generally predominates. Willner and Willner (1988) studied the selective reduction of CO_2/HCO_3^- by a palladium (Pd)-loaded UV/TiO_2 photocatalytic system. Palladium was loaded onto TiO_2 in order to suppress the evolution of hydrogen and to promote the reduction of carbon dioxide. The reaction is then able to proceed without an electron mediator via the direct coupling of conduction-band electrons of TiO_2 to the Pd surface catalyst.

Palladium was loaded onto TiO_2 at 9 to 11 mg of Pd per gram of TiO_2. Solutions containing 3 mL of the compounds were prepared at pH 6.8. Solutions were prepared at 0.05-M initial concentration. Oxalate was added to the solutions to act as a sacrificial electron donor. The solutions were irradiated by a 450-W xenon short-arc lamp that emitted light up to 360 nm. The samples were irradiated for 300 min, filtered, and analyzed with an HP5890 gas chromatograph.

Solutions containing 0.05-M bicarbonate and oxylate were reduced to formate and hydrogen. Upon irradiation of oxalate alone, only hydrogen evolution was observed, suggesting that formate originates from the reduction of CO_2/HCO_3. The activity of Pd-loaded TiO_2 was found to be dependent on the method of preparation of the catalyst. Addition of Pd to the TiO_2 resulted in an increase in the activity of the catalyst of 6.2 times. Temperatures used in the preparation of the catalyst also were shown to have an effect on the activity of the catalyst. Pd-loaded TiO_2 prepared at 90°C was found to have much greater activity than samples prepared at 60°C. The reduction of CO_2/HCO_3 by semiconductor particles is usually inefficient and nonselective. The reaction products are a mixture of various undesirable organic products such as formic acid, glycoxylic acid, oxalic acid, formaldehyde, and methanol. Willner and Willner (1988) were able to show that Pd–TiO_2 particles are able to promote the selective two-electron reduction of CO_2/HCO_3^- to formate. This reduction is thermodynamically favorable because the reduction potential of the TiO_2 conduction-band electrons (–0.53 V) is more negative than that required for the conversion of CO_2/HCO_3 to formate (–0.42 V). Initial reaction rates and quantum yields for the evolution of H_2 and formate are presented in the database. The reactions were studied using different aqueous suspensions of Pd–TiO_2 and TiO_2.

9.3.3 Alkyl Halides

Alkyl halides such as 2-chloroethyl methyl sulfide and 2-chloroethyl ethyl sulfide were studied using UV/TiO_2 (Fox et al., 1990). Solutions containing 0.001 to 0.025 M of the alkyl halide and 5 to 10 mg/L TiO_2 were prepared. The solutions were then sonicated and transferred to a reaction vessel. The solutions were irradiated under low-pressure mercury lamps emitting light at 350 nm. After irradiation, the solutions were filtered and the resulting organic concentrate was analyzed by gas chromatography. Reactions were carried out in acetonitrile and aqueous acetonitrile to determine the most efficient photocatalytic conditions.

The photocatalytic reactions in acetonitrile were observed to have fewer primary products and a higher chemical yield of the desired sulfoxide (Fox et al., 1990). The photocatalytic oxidation of 2-chloroethyl methyl sulfide was found to be faster than the degradation of 2-chloroethyl ethyl. UV/TiO_2, in the presence of oxygen, yields the greatest amount of removal for the alkyl halides. The rate of alkyl halide removal was found to increase with increased concentrations of TiO_2.

The photocatalyzed oxidation of trichloroethylene and methylene chloride in aqueous solution and in vapor phase was studied (Lichtin and Avudaithai, 1996). Much difference exists between the chemistries of photocatalytic oxidation in the aqueous phase and in the vapor phase. For example, higher reactivities have been observed for photocatalytic oxidation in the vapor phase when compared to those of photocatalytic oxidation in the aqueous phase for selected compounds. Lichtin and Avudaithai, (1996) performed a comparison of photocatalytic oxidation of several selected compounds in the aqueous phase and in the vapor phase. For the aqueous photocatalytic reactions, TiO_2 was prepared as a thin film on the inner surface of the photoreactor. The batch reactors were prepared using a magnetically stirred 450-mL Pyrex vessel. The light source was provided by an axially aligned 6-W lamp emitting light at 360 nm. For the aqueous solutions, 35 mL/min of O_2 was bubbled into the solution. The products of the reaction were analyzed using gas chromatography (GC).

Degradation of aqueous trichloroethylene was followed by the determination of ionic chloride. Chloroform was identified as a minor product through GC analysis. The fact that no O_2 was consumed during the degradation of TCE suggests that TCE, not O_2, is the principal electron trap. The low reactivity of TCE in aqueous solution may be due to a strong interaction between adsorbed TCE and liquid water.

Degradation of methylene chloride resulted in the formation of ionic chloride. Products from the degradation of methylene chloride were observed to be $COCl_2$, HCl, CO, and CO_2. Although the O_2 concentration in water is lower than in air, reactive adsorption sites were closer to saturation in the presence of water. The degradation kinetics of TCE and methylene chloride from aqueous and vapor phases were observed to be different (Lichtin and Avudaithai, 1996).

When reactions in aqueous phase and in vapor phase were compared, the molar concentrations and mole fractions were prepared to be equimolar. The mole fraction of a compound in 1 atm gas phase is 1350 times its mole fraction in aqueous solution; therefore, water competition with the reactants for adsorption sites is much greater in the aqueous phase.

Application of the Langmuir model suggests the following mechanism for the photodegradation of alkyl radicals. Photogeneration of the electron/hole pair occurs and the surface-confined hole is capable of generating an adsorbed thioether cation radical. Co-adsorbed oxygen traps electrons and inhibits electron/hole recombination, and further reduction of the adsorbed cation radical. The latter species is attacked by oxygen or photogenerated superoxide to generate the sulfoxide product (Fox et al., 1990). To increase efficiency of the process, the reactions were carried out in dry acetonitrile. TiO_2 was also found to be the superior catalyst when compared to SnO_2, ZrO_2, or ZnO. The co-deposition of platinum yields faster rates of photooxidation, but the small increase in rate is not significant enough to compensate the cost of this process.

Infrared spectroscopy was used to gain a better understanding of the mechanism of photocatalytic oxidation of TCE (Fan and Yates, 1996). IR spectroscopy also was used to determine intermediates formed from the reaction. Chemisorption of oxygen onto the TiO_2 surface plays an important role in the oxidation of TCE. The reaction was also temperature dependent. TCE is more easily degraded in the gas phase than in the aqueous phase. For this reason, a process that strips TCE from the groundwater and then treats the vapor containing TCE can be used.

Previous studies have shown that a significant amount of by-products are formed during the photocatalytic oxidation of TCE. The photocatalytic oxidation was studied at 300, 473, and 150 K to determine temperature effect on the process (Fan and Yates, 1996). Reactions were carried out in an infrared cell. The TCE was removed from the aqueous phase, and the gas stream containing the TCE was treated. Determination of the products and intermediates was accomplished using IR spectroscopy, GC, and mass spectrometry. High-pressure mercury lamps provided the UV light. The TiO_2 used was Degussa P25 and it was sprayed onto a 5.2-cm^2 area. Upon irradiation of TCE, IR spectra showed that all gas-phase spectral features disappeared and spectral features of the intermediates appeared. The intermediates include dichloroacetyl chloride, CO, phosgene, and HCl. The kinetics of the reaction indicated that the TCE had been completely degraded in 100 min.

9.3.4 Anisoles (Methoxybenzenes)

The photocatalytic oxidation of *para-* and *meta*-substituted anisoles with UV/ TiO$_2$ was studied by Amalric et al. (1996). A 100-mL batch reactor containing aqueous solutions of the anisoles was irradiated with UV light at 340 nm. The light was provided by mercury lamps. Initial concentrations of the anisoles were 0.1 mM/L. The pH of the solution was adjusted to 5.1. TiO$_2$ was added at 2.5 g/L. The samples were irradiated for 80 min. After irradiation, the samples were filtered and concentrations were determined by liquid chromatography.

The degradation of the anisoles was found to follow pseudo first-order degradation. The apparent first-order rate constant (k_{app}) was compared to Hammett's constant to determine if a correlation existed between the degradation of *meta*-substituted and *para*-substituted anisoles. Removal rates were greater that 80%, and for fluoroanisole and aminoanisole they were 100%.

Because molar refractivity (MR) is proportional to the electronic polarizability, which represents the ability of electrons pertaining to the compound to be polarized (Amalric et al., 1996), MR was useful in predicting the degradation kinetics of organic compounds. With the use of MR, it was determined that for a given TiO$_2$ sample under identical conditions the kinetics for the photodegradation of anisoles is faster for anisoles that are better electron donors and that are better dispersed at the TiO$_2$ surface (Amalric et al., 1996). These findings may be applied to all aromatic compounds. The established correlation equation is as follows:

$$\text{Log } k_{app} = 0.20 \log K_{OW} - 0.20\sigma^+ - 0.033 \text{ MR} - 0.43 \ (r = 0.903) \quad (9.13)$$

where k_{app} is the apparent rate constant (1/min), K_{OW} is the 1-octanol/water partition coefficient, σ^+ is Hammett's constant, and MR is the molar refractivity. When comparing chlorophenols, a good correlation was found ($r = 0.987$) when the photocatalytic degradability was related to log K_{OW} and σ. The study showed that for *meta-* and *para*-substituted anisoles, it is necessary to replace σ by σ^+ and to include MR.

9.3.5 Chlorinated Hydrocarbons

Chlorinated aromatic compounds are hazardous compounds that result from various industrial and agricultural activities. Water disinfection, waste incineration, and uncontrolled use of biocides are the major sources of chlorinated aromatics in the environment. Chlorinated compounds are also formed as subproducts of the biochemical reactions of herbicides containing chlorophenoxy compounds. Treatment of chlorinated compounds has been studied using biological treatment, adsorption, air stripping, and incineration. Biodegradation of chlorinated compounds is a slow process that is ineffective for extremely low concentrations. Air stripping and adsorption simply trans-

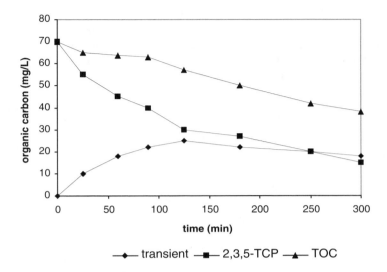

FIGURE 9.3
Photodegradation of 2,3,5-trichlorophenol by UV/TiO$_2$. (From Jardim, W. et al., *Water Res.*, 31, 1728, 1997. With permission.)

fer the contaminant from one phase to another and posttreatment is required. For these reasons, photocatalytic degradation of chlorinated compounds has been widely studied.

Figure 9.3 shows the degradation of pentachlorophenol (PCP), 2,4-dichlorophenol (2,4-DCP), 3,5-dichlorophenol (3,5-DCP), and 2,3,5-trichlorophenol (2,3,5-TCP) as reported by Jardim et al. (1997). Degradation rates as well as the toxicity of intermediates formed were investigated. Toxicity of the intermediates formed is of concern due to the fact that some of the intermediates can be more toxic than the primary compound. Intermediates formed during the oxidation of PCP and 2,4-dichlorophenol (2,4-DCP) were found to be more toxic to activated sludge than the primary compounds (Jardim et al., 1997).

Solutions containing concentrations of the chlorinated compounds and TiO$_2$ were prepared. TiO$_2$ was added to the solution at 0.1 g/L. Various initial concentrations of the compounds were prepared. The compounds were illuminated using a high-pressure, 125-W Philips HPL-N mercury lamp. The samples were irradiated for 360 min, filtered, and then analyzed using high-performance liquid chromatography (HPLC), and TOC. At various irradiation times, samples were withdrawn and analyzed using the acute toxicity test with *Escherichia coli*.

The photodegradation of all four compounds fits the pseudo first-order kinetics. Destructive efficiencies for PCP were found to be 97% when analyzed according to TOC measured at the beginning and end of the reaction. Toxicity tests on samples withdrawn at various reaction times show that, after 90 min of reaction, the intermediates formed are more toxic to *E. coli* than PCP. Samples withdrawn at 120 min were less toxic than PCP. Throughout the reaction, the toxicity of the intermediates and by-products increased; however, at the end of the reaction, residual toxicity approached zero. Degradation of 2,3,5-TCP was found to yield a higher concentration of toxic intermediates among the four compounds studied. The major by-product was found to be 2,3,5-trichloro-1,4-hydroquinone. After 60 min of reaction, the toxicity of the intermediates was less than that of 2,3,5-TCP. The toxicity of 3,5-DCP and 2,4-DCP was shown to decrease with decreasing concentration of the parent compound. After 90 min of irradiation, 2,4-DCP disappeared from solution without an increase in toxic effects from the products.

Chlorinated compounds have been shown to be successfully degraded by UV/TiO$_2$ process. Although most attention regarding the process has been focused on the mineralization of hazardous compounds, little has been paid to the formation of potentially harmful intermediates; however, the by-products of the photodegradation of DCP have been found to be harmful to human health and are present in drinking water. When detoxifying chlorinated compounds, the toxicity of intermediates and by-products should be monitored. Minero et al. (1995) reported that some of the undesired by-products that result from photocatalytic processes are trihalomethanes, chlorinated aromatics, aldehydes, ketones, and dibenzofurans. The intermediates and by-products formed may be potentially more harmful that the parent compounds. Six dichlorophenol isomers (2,3-, 2,4-, 2,5-, 2,6-, 3,4-, and 3,5-DCP) were studied by Minero et al. (1995) to identify reaction by-products. A 1500-W xenon lamp emitting light at >340 nm was used to irradiate 5-mL samples containing 20 mg/L of DCP and 250 mg/L TiO$_2$ for various time periods (1, 3, 6, 10, 20, 30, and 60 min). GC analysis was used to identify the condensation by-products from DCP photodegradation, which were present for all DCP isomers. The compound structure was shown to have an effect on the class of condensation by-products formed. The linkage between the rings, either C–C or C–O–C, was shown to yield polyhydroxy PCBs or polyhydroxypolychlorobiphenyl ethers, respectively. Reaction pathways for the formation of the condensation by-products proceed by the production of OH$^•$ radical-like species and valance-band electrons, which can concurrently oxidize or reduce the aromatic ring, forming semiquinone radicals that dimerize or condense (Minero et al., 1995).

Photocatalytic oxidation of 2,4-dichlorophenoxyacetic acid (2,4-D) was investigated (Sun and Pignatello, 1995). In addition to the dominant hydroxyl radical mechanism, Sun and Pignatello found evidence that direct hole oxidation may be the mechanism for the photocatalytic degradation of some organic compounds. The assumed mechanism for this oxidation is H$^+$ acting as an electron-transfer oxidant, while O$^•$ behaves like a free OH$^•$ and abstracts H or adds to C=C multiple bonds. Hole oxidation has been used to explain the oxidation of oxalate and trichloroacetate ions, which lack abstractable hydrogens or unsaturated C–C bonds. Whether the reaction takes place on the surface of the TiO_2 molecule or in solution is also contro-versial. Sun and Pignatello (1995) reported evidence for a surface dual hole mechanism in the photocatalytic oxidation of 2,4-dichlorophenoxyacetic acid.

Solutions containing 0.25 mM of 2,4-dichlorophenoxyacetic acid and 0.1 g/L TiO_2 were prepared. Reactions were carried out in borosilicate vessels irradiated with fluorescent blacklamps emitting light at 300 to 400 nm. pH values were adjusted with $HCLO_4$ or NaOH from pH 1 to 12. Samples were irradiated for 180 min, filtered, and analyzed by GC to determine product concentrations. The degradation of 2,4-dichlorophenoxyacetic acid obeyed first-order kinetics. Rate constant k was observed to vary with pH in the range of 1 to 12 with a maximum value at pH 3. The pK_a of 2,4-dichlorophe-noxyacetic acid is 2.9. Half-lives of 2,4-dichlorophenoxyacetic acid were stud-ied in relation to pH and were found to decrease on either side of pH 2 to 3 and increase between pH 2 and 3, as shown in Figure 9.4. These results suggest a progressive shift from a decarboxylation reaction at pH 2 to 3 to some other reaction at values on either side of pH 2 to 3.

The optimum pH for photocatalytic oxidation of 2,4-D is pH 3. The observed products from the photocatalytic oxidation of 2,4-D are carboxyl CO_2, formaldehyde, 2,4-DCP, and dichlorophenol formate. The yields of the products suggest that the first step in the oxidation of 2,4-D is one-electron oxidation of the carboxyl group (Sun and Pignatello, 1995); however, the reaction pathway is found to change with pH. On each side of pH 3, decar-boxylation increasingly lags the disappearance of 2,4-D due to decreasing half-lives of 2,4-D and declining yields of decarboxylation products.

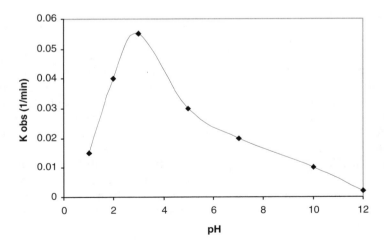

FIGURE 9.4
Photocatalytic oxidation of 2,4-dichlorophenoxyacetic acid by UV/TiO_2. (From Sun, Y. and Pignatello, J., *Environ. Sci. Technol.*, 29, 2065, 1995. With permission.)

A dual hole-radical mechanism where H⁺ oxidation of 2,4-D (Equation 9.14) exists in competition with H⁺ oxidation of surface hydroxyl groups (Equation 9.15) can be described as below.

$$H^+ + OH_s^- \rightarrow O^\bullet \qquad (9.14)$$

$$H^+ + OH_s \rightarrow R(oxidized) \qquad (9.15)$$

Holes carry out electron transfer oxidation, preferring carboxyl to other functional groups. These holes cannot abstract hydrogen atoms or add to aromatic rings. It is also assumed that trapped holes ($O^\bullet$) behave like free $HO^\bullet$ in that they abstract hydrogen atoms and/or add to the aromatic ring. Relative rates of reactions for the above equations are assumed to be pH dependent (Sun and Pignatello, 1995). The surface-bound $O^\bullet$ is the oxidant at a pH of ~3 during the reaction with the adsorbed or colliding substrate. Above and below pH 3, it is dissociated as free $HO^\bullet$ (Sun and Pignatello, 1995).

Application of the UV/TiO_2 photocatalytic process to several chlorinated hydrocarbons was studied by Martin et al. (1996). The addition of oxyanion oxidants such as ClO_2^-, IO_4^-, $S_2O_8^{2-}$, and BrO_3^- was studied to increase the rate of photodegradation. TiO_2 catalyzes the oxidation of chlorinated hydrocarbons in the presence of UV radiation at photon energies greater than the band gap energy of 3.2 eV according to the following stoichiometry:

$$C_xH_yCl_z + (x + y - z/4)O_2 \rightarrow xCO_2 + zH^+ + zCl^- + (y - z/2)H_2O \quad (9.16)$$

The degradation rates can be increased upon the addition of inorganic oxidants (Martin et al., 1996). The oxidants increase quantum efficiency by inhibiting electron–hole pair recombination through scavenging conduction-band electrons at the surface of the TiO_2 molecule. Martin et al. (1996) studied the photooxidation of 4-chlorophenol (4-CP) to describe the mechanism by which oxyanion oxidants serve as efficient electron acceptors in the UV/TiO_2 process. Aqueous solutions containing 1.0 g/L TiO_2, 100 μM 4-chlorophenol, and a particular oxidant were bubbled with O_2 prior to and during the reaction. The samples were irradiated by a 1000-W xenon arc lamp emitting light at >320 nm. The solutions were filtered and analyzed by liquid chromatography to determine product concentrations. Quantum efficiencies, Φ, were calculated from the photooxidation experimental data of 4-CP concentration vs. time. Quantum efficiency is defined by the following equation:

$$\Phi = [d(4 - CP)/dt]/I_{incident} \qquad (9.17)$$

where $d(4 - CP)/dt$ is the initial rate of disappearance of 4-CP, and $I_{incident}$ is the incident light intensity.

Irradiation of 4-chlorophenol and the effect of oxyanion oxidants were also studied in the absence of TiO_2. For irradiation of 4-CP in the absence of TiO_2, Φ was found to decrease according to oxidant added to solution in the following order: $ClO_2^- > IO_4^- > BrO_3^- > ClO_3^-$. In the absence of TiO_2, ClO_3^- was the only oxidant that did not show direct photoreactivity with respect to oxidation of 4-chlorophenol. In the presence of TiO_2, Φ was found to increase following the same order: $ClO_2^- > IO_4^- > BrO_3^- > ClO_3^-$. These quantum efficiencies were calculated over a pH range of 3 to 6.

The reaction mechanism proposed by Martin et al. (1996) is presented in Figure 9.5. 4-CP reacts with OH$^\bullet$ to form the 4-chlorodihydroxycyclodienyl radical (4-CD). After this initial hydroxylation, three parallel reaction pathways exist. In the first pathway, 4-CD is reduced by a conduction band electron to yield 4-CD$^+$, hydroquinone (HQ), and Cl$^-$. In the second pathway, 4-CD reacts with oxygen to form the molecule 4-CDO. In the third pathway, ClO_3^- facilitates the abstraction of an electron from 4-CD to yield 4-CD$^+$. 4-CD$^+$ is stabilized by a resonance interaction and by the strong electron-releasing capability of the $^-$OH substituent at the 1-position.

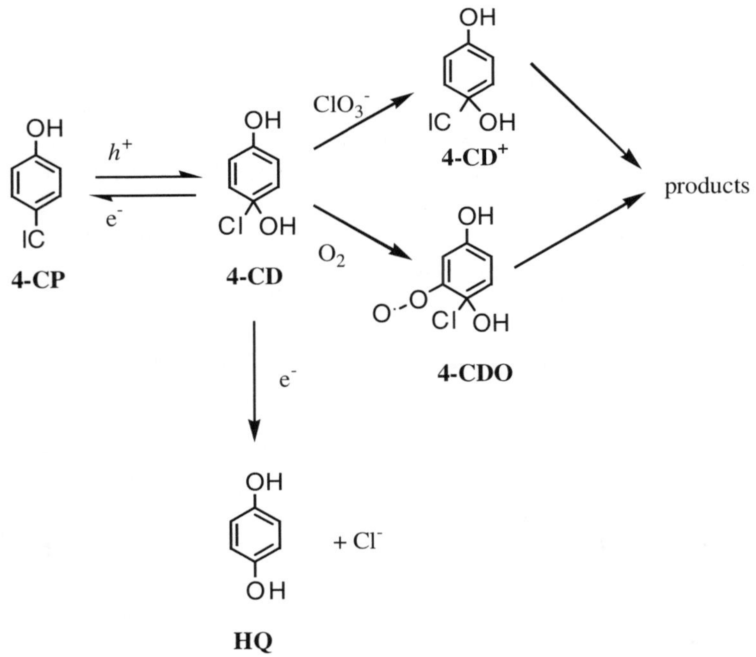

FIGURE 9.5
Reaction mechanism for the degradation of 4-chlorophenol by UV/TiO_2. (From Martin et al., *Environ. Sci. Technol.*, 30, 2535, 1996. With permission.)

9.3.6   Herbicides

Herbicides and pesticides are widely applied at agricultural sites and are often found in groundwater. The contamination of groundwater with herbicides is of concern due to the toxicity, stability, and persistence of herbicides in the environment. The photodegradation of various herbicides by UV/ TiO$_2$ has been studied by Kinkennon et al. (1995) and Richard and Benagana (1996). Herbicides that have been studied include bentazon (Basagran™), diquat (Reglone™), diuron, and fenuron (1,1-dimethyl-3-phenylurea).

Basagran has a herbicidal activity but a high bioconcentration factor (Kinkennon, 1995). Diquat is able to undergo natural photochemical decomposition after application to plant surfaces. Studies show that 50 mg of herbicide is depleted to below 6 mg/kg within 7 days after treatment. In soil, diquat is biodegraded at a rate of 10% per year. In aqueous environments, diquat was degraded to levels that were undetectable in less than 30 days.

Diuron is a highly persistent herbicide with a measured half-life of over 300 days. Diuron is easily degraded in aqueous solutions. In organic-rich environments, 90% of the compound is degraded in less than 8 months (Kinkennon, 1995). Basagran, diquat, and diuron were prepared to make a concentration of 10 ppm. TiO$_2$ was added to the 500-mL solution to create a 0.1% suspension. UV light was supplied by a 1000-W xenon lamp with a filter to allow light between 332 and 420 nm to be studied. Solar radiation was concentrated by creating a vacuum that created a parabolic shape on the surface of the reaction vessel. This concentrated the light beam on a 15-cm diameter area of the reaction vessel. Basagran, diquat, and diuron were also studied under irradiation of simulated solar irradiation and concentrated irradiation. Solutions were irradiated for 150 min and concentrations of the herbicides were determined by visible–UV spectrophotometry. Fenuron was suspended in aqueous solution to prepare a concentration of 2 g/ L. The pH of the solutions was adjusted to 7, 4.6, and 2. The solutions were then irradiated by a fluorescent lamp that emitted light in the range of 300 to 450 nm. The samples were then filtered and analyzed with HPLC to determine the concentrations of herbicide in the solutions.

Figure 9.6 shows that the degradation of Basagran, diquat, diuron, and fenuron appears to follow first-order degradation. Under the above experimental conditions, Basagran and diuron are decomposed to less than 0.5 ppm in less than 60 min. The degradation of diquat took approximately twice as long as that of Basagran and diuron. The rate of photocatalytic decomposition of Basagran, diquat, and diuron was dependent on light intensity. When a solar concentrator was applied to the reaction, the degradation rates were shown to approximately double. Therefore, increasing light intensity by adding more light-focusing elements should increase the reaction rates of herbicides (Kinkennon, 1995).

The chemical structures of herbicides are shown in Figure 9.7. Herbicides containing secondary amino moieties such as Basagran and diuron are more

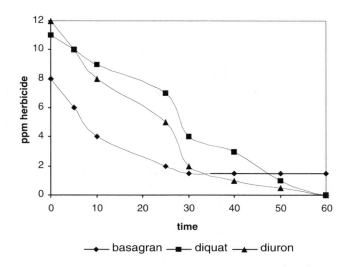

FIGURE 9.6
Degradation of Basgran, diquat, and diuron by UV/TiO$_2$. (From Kinkennon, A., *Chemosphere*, 31, 3663, 1995. With permission.)

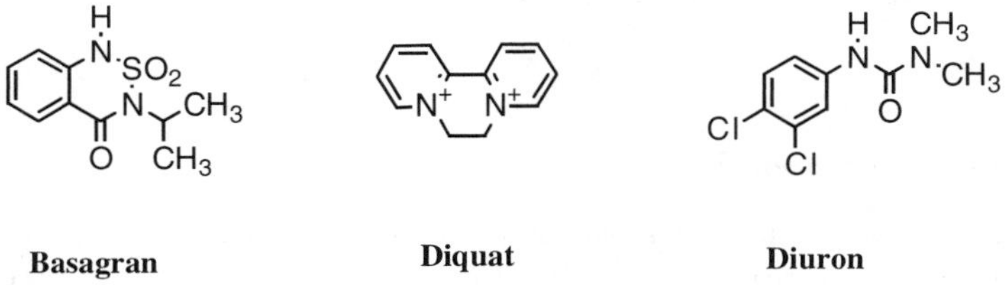

Basagran   **Diquat**   **Diuron**

FIGURE 9.7
Structures of Basagran, diquat, and diuron.

susceptible to photocatalytic oxidation. The compounds with quaternary amine structures such as diquat are more stable due to the highly stabilized pie structure. Therefore, the chemical structure of diquat has greater stability than Basagran and diuron.

Carbetamide, a pesticide that is found in groundwater, was degraded by UV/TiO$_2$ photocatalytic process by Percherancier et al. (1995). This molecule has been observed to be unaffected by the direct action of light, and its structure contains both an aromatic ring and an aliphatic chain. The degradation of carbetamide was studied using various parameters such as power of the light source, mass of the catalyst, and concentration of the pollutant. A comparison of catalysts using ZnO and TiO$_2$ was made. Solutions containing 100 mg/L carbetamide and 400 mg/L TiO$_2$ were prepared. The concentration of TiO$_2$ in solution was varied to determine which parameters would yield optimum efficiency. 60-mL samples were placed in a batch reactor and illuminated with a 125-W mercury lamp. After reaction, the samples were filtered and concentrations of the pollutant were determined using HPLC. When catalysts were compared, degradation of carbetamide was complete at 90 min with the use of TiO_2 and 60 min with the use of ZnO. Direct photolysis was found to have no significant effect on carbetamide degradation. Initial reaction rates were found to increase with the amount of catalyst present. The reaction rates reached a maximum at levels that corresponded to complete adsorption of incident light by TiO_2. The reaction rate of carbetamide was found to plateau at 100 mg/L TiO_2. The initial reaction rate was also found to increase up to 4.2×10^{-3} M. Kinetics of the reaction were modeled using the Langmuir–Hinshelwood relationship. The adsorption rate constant, K, and the rate constant, k, are reported in the database.

Percherancier et al. (1995) proposed that degradation of carbetamide occurs via formation of hydroxyl radicals. Hydroxyl radicals are formed in this reaction by the interaction of the photo-produced holes in the TiO_2 surface and adsorbed OH^- radicals or adsorbed water. The formation of superoxide ions from dioxygen present in the solution may also account for the formation of H_2O_2, which is unstable under UV light, and ultimately OH^- radicals. The cleavage of side chains appears to be carried out by hydroxylation of the ring, and direct cleavage from carbetamide is a minor process. Radiolysis with gamma rays detected $OH^\bullet$ radicals (Percherancier et al., 1995). Carbetamide is completely mineralized by both reaction mechanisms.

9.3.7 Nitro Compounds

Photocatalytic oxidation of acetonitrile was studied by Lichtin and Avudaithai. (1996). A comparison of photocatalytic efficiencies for aqueous- and gas-phase photooxidation was made. Differences in the chemistry of photocatalytic oxidation for the two phases were also studied. A batch reactor was used for both the aqueous- and gas-phase reactions. The reactor was a 450-mL, magnetically stirred, cylindrical vessel with an axially aligned 6-W fluorescent lamp emitting light at 360 nm. Oxygen was bubbled into the reactor at 35 mL/min. Samples were irradiated for 140 min, and product concentrations were determined using gas chromatography. Equation (9.18) represents a likely stoichiometry for the complete oxidation of acetonitrile.

$$CH_3CN + 4O_2 \rightarrow 2CO_2 + H_2O + HNO_3 \tag{9.18}$$

Nitric acid was observed to be an intermediate in the photocatalytic oxidation of acetonitrile. Removal rates for acetonitrile in gas phase were found to be significantly greater than rates for acetonitrile in aqueous phase (Lichtin and Avudaithai, 1996).

Nitroaromatics are often found in former military sites at which explosives were manufactured or handled. Contamination of the original explosives or their by-products is often widespread. Biotic and abiotic degradation of nitroaromatics results in the formation of aminonitrotoluenes, nitrotoluenes, nitrobenzenes, and nitrophenols (Dillert et al., 1995). These contaminants pose serious threats to ground and surface water. The most common contaminant found at these sites is 2,4,6-trinitrotoluene (TNT). The concern over contamination of ground and surface water with TNT is due to its toxicity to a wide variety of organisms, including humans. TNT is a known mutagen and is classified as a priority pollutant by the U.S. Environmental Protection Agency (Schmelling and Gray, 1995).

Photocatalytic transformation and mineralization of TNT was examined by Schmelling and Gray (1995) and Dillert et al. (1995). 2,4,6-Trinitrotoluene was prepared at a 50-mg/L initial concentration in ultrapure water. TiO_2 was added to create a concentration of 250 mg/L. The solutions were irradiated under mercury lamps with a wavelength greater than 340 nm for 120 min. The TNT concentration and by-products were analyzed with liquid chromatography. It was observed that TNT was capable of being mineralized in the presence or absence of TiO_2. The pseudo first-order degradation rate constants were shown to have significant differences depending on the presence or absence of TiO_2 in the solution. In the presence of TiO_2 the rate constant was 4.2 1/hr; in the absence of TiO_2, the rate constant was 1.2 1/hr. In the presence of TiO_2, the conversion of TNT was complete at 60 min. In the absence of TiO_2, the conversion of TNT was complete after 90 min. Reaction intermediates formed during the photocatalytic oxidation of TNT were identified as 2,4,6-trinitrobenzoic acid, 1,3,5-trinitrobenzene, 3,5 dinitroaniline, and 2,4,6-trinitrophenol (Schmelling and Gray, 1995). The sequence of reactions was proposed to be (1) oxidation of the methyl group to COOH, (2) nucleophilic substitution of the COOH by NO^{3-}, and (3) hydrolysis of the nitrate ester.

The photocatalytic degradation of TNT and 10 other nitroaromatic compounds was studied as a function of pH (Dillert et al., 1995). Solutions of TNT were prepared at 100 μm/L and TiO_2 was added to the solution at 1 g/L. The pH of the samples was adjusted with KOH. A xenon lamp was used to irradiate the 5-mL samples for 20 min. Concentrations of the compounds were determined using liquid chromatography. The results show that nitroaromatic compounds can be degraded. The reactivity of the compounds was found to decrease with increasing numbers of nitro groups on the ring. The following order of reactivity was observed: nitrotoluenes > nitrobenzenes > dinitrotoluenes > dinitrobenzenes > 2,4,6-trinitrotoluene > 1,3,5-trinitrobenzene, as shown in Figure 9.8. The degradation of nitroaromatics is believed to proceed via two pathways: oxidation of the methyl groups and hydroxyl radical attack at the aromatic ring.

In the expression $\log r_o = n \times \log I + \log(k' \times C_0)$, I is the light intensity, k' is the rate constant, and C_0 is the initial concentration of the compound. The factor n has been determined for the degradation of TNT to be 0.73 ± 0.06 and 0.75 ± 0.09 at pH 5 and pH 9, respectively. The degradation of the nitroaromatic compounds studied was not affected by the pH of the solution.

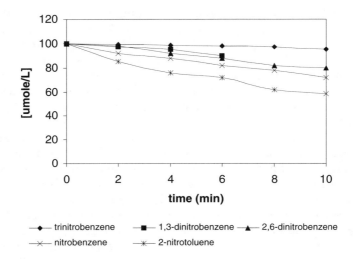

FIGURE 9.8

Degradation of nitroaromatic compounds by UV/TiO$_2$. (From Dillert, R., Brant, M., Siebers, U., and Bahnemann, D., *Chemosphere*, 30(12), 2333, 1995. With permission.)

Aniline is a compound used in the synthesis of insecticides, chemical brighteners, and dyes and is a by-product of the petroleum, paper, and coal industries. The photocatalytic oxidation of aniline was studied by Sanchez et al. (1997). The reaction was found to follow Langmuir–Hinshelwood kinetics. The adsorption rate constant and the reaction rate constants were also reported. Higher yields were reported for acidic conditions and values near the pH at the point of zero charge (pH$_{pzc}$) of TiO$_2$. The rate of photocatalytic oxidation was also found to increase with the addition of small amounts of Fe. Hydroquinone is the main intermediate formed from the reaction. Photocatalytic reactions were carried out in a 130-cm^3 cylindrical Pyrex cell. Medium-pressure mercury lamps provided UV light. Initial concentrations of 1.0×10^{-4} and 2 g/L were prepared. The pH of the solutions was adjusted and reactions were carried out for 15 min. Concentrations of aniline and by-products formed were determined by HPLC.

Adsorption of aniline onto TiO$_2$ was observed. Approximately 10% of the initial concentration of aniline was adsorbed onto the TiO$_2$ surface in the dark reactions. The presence of UV light significantly increased the rates of removal of aniline from solution. The photodegradation of aniline was found to be a function of light intensity. The reaction was found to follow Langmuir-type kinetics. The data reported showed an adsorption rate constant of 1.1×10^{-3} k/mol and a reaction rate constant of 9.86×10^{-4} mol/L/min. The solution pH was also shown to have an effect on the photodegradation rates of aniline. A maximum adsorption is reached near the pH at the point of zero charge of TiO$_2$. Photodegradation rates are also greatest near pH$_{pzc}$. Higher degradation rates were also observed at pH values less than 3 and greater than 12. The formation of polyanilines at the higher and lower pH values is suggested as a mechanism for photo-degradation at the extreme pH values.

The addition of Fe ions to the solution also increased the rate of photo-catalytic degradation. The TOC reduction was 80% in the presence of Fe ions and 10% in the absence of Fe ions. pH is the key parameter in determining the photocatalytic degradation of anilines as well as other compounds previously studied. The high photocatalytic degradation near the pH_{pzc} has also been observed for other compounds studied with Ti and Zn. The major mechanism is the attack of photogenerated OH radicals on the aniline molecule under alkaline conditions. The addition of Fe ions to the solution can enhance the formation of OH radicals and lead to higher photocatalytic rates.

Mahdavi et al. (1993) studied the effect of UV/TiO_2 photoinduced reduc-tion of 12 aliphatic and aromatic compounds. Although much effort has been put forth investigating the oxidation pathways of various compounds, little effort has been made to investigate the reduction of these compounds. The compound that donates an electron to the TiO_2 semiconductor will be oxi-dized, but when a reactant receives an electron from the semiconductor, it is reduced. Twelve aliphatic compounds were prepared to make a solution containing 0.01 M nitro compound and 5.0 g/L TiO_2. The compounds were irradiated in a Rayonet photochemical reactor, filtered, and then analyzed by gas chromatography to determine the concentration of the compound being studied. The mechanism for reduction of a nitro compound is believed to follow hydrogen production during oxidation of the solvent (Mahdavi et al., 1993). This was proposed due to the fact that when *p*-nitroacetophenone was saturated with hydrogen the compound was not reduced by the UV/TiO_2 process. In the absence of TiO_2 no significant amount of reduction was observed. Reduction rates were calculated according to the percent produc-tion of the reduced form of the nitro compound. Visible–UV absorption spectra were used to identify the reduction products and their corresponding concentration.

The efficiency of the reduction generally increases when the concentration of the nitro compound decreases. The photoredox reactions are halted when photons are absorbed by the nitro compounds. The concentration limit of this reaction is dependent on the extinction coefficient of a nitro compound in the UV region. These limits are 5×10^{-3} M for 6-nitrocoumarin and 5×10^{-2} M for *p*-nitrobenzaldehyde (Mahdavi et al., 1993). Complete reduction of a nitro compound would require six protons and would leave six electron holes behind. The oxidation of an alcohol and the reduction of a nitro com-pound are presented in Figure 9.9.

The pH of the solution containing the nitro compound does not change. This suggests that the oxidation of an alcohol and the reduction of the nitro compound are coupled (Mahdavi et al., 1993). These findings are consistent with the generation of hydrogen gas. Reduction of nitro compounds is highly dependent on the reduction potential of the compound being studied. Nitro

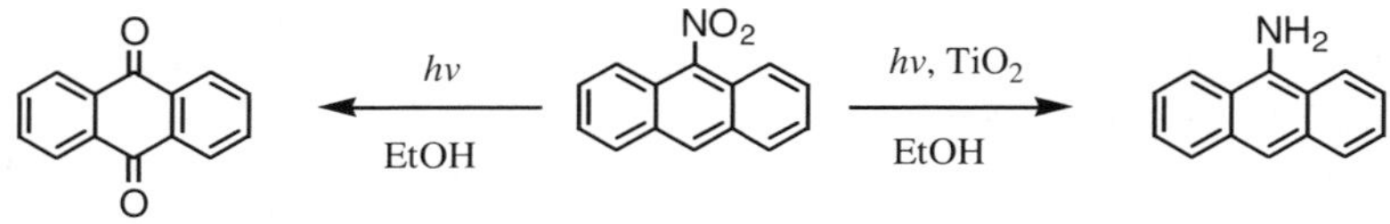

FIGURE 9.9

Reaction mechanisms of nitro compounds by UV/TiO$_2$. (From Mahdavi, F., Bruton, T., and Li, Y., *J. Org. Chem.*, 58, 744, 1993. With permission.)

compounds with reduction potentials well below the conduction band energy of TiO$_2$ are more readily reduced than compounds with higher reduction potentials.

The role of reduction in the photocatalytic degradation of 2,4,6-trinitrotoluene was studied (Schmelling et al., 1996). Earlier research by Schmelling and Gray (1995) demonstrated that TNT was degraded in the presence of TiO$_2$ and UV light. Over 90% removal of the compound was observed. The formation of ammonia during the photocatalytic degradation of TNT suggests that the degradation pathway involves reductive transformations. Mahdavi et al. (1993) concluded that nitroaromatics were reduced to amines during photocatalytic degradation by interfacial electron transfer.

The initial reductive step observed in aerated systems is the transfer of the photoexcited electron to absorbed molecular oxygen. This absorption of the electron onto oxygen creates a superoxide radical anion that is capable of forming HO$_2$• and H$_2$O$_2$. The superoxide radical is then able to react with the compound of interest. TNT is susceptible to reductive attack due to steric effects and the three deactivating nitro groups. To improve the efficiency of the photodegradation process for TNT, an understanding of the role of reduction is necessary. 2,4,6-Trinitrotoluene was prepared to form a 200-μM solution at pH 5.1. KCL (100 mM) was added as an electrolyte, and UV light with a wavelength of 340 nm was applied to the solution. TNT concentration and reduction products were determined by HPLC. The reduction and oxidation potentials of TNT were determined by cyclic voltometry. Reduction potentials ($E_{1/2}$) for removing three nitro groups sequentially were reported to be –0.22, –0.39, and –0.57 V. The conduction band energy for TiO$_2$ in aqueous solution is defined as:

$$E_{CB} = -0.1 - 0.059\text{pH} \tag{9.19}$$

This suggests that the reduction of at least two or three of the nitro groups on TNT by TiO$_2$ photocatalysis is thermodynamically feasible near pH 5. The reduction potential of TiO$_2$ decreases with increasing pH, and the reduction potential of TNT also becomes more negative with increasing pH.

A comparison of sensitized and direct photocatalysis was performed for the degradation of 4-nitrophenol (Dieckmann and Gray, 1996). The application of sensitization to the degradation of nitrophenolic compounds and azo dyes has been studied, and it has been demonstrated that colored pollutants can be used to sensitize the photocatalytic process. Sensitized photocatalysis involves the adsorption of colored compounds onto semiconducting surfaces (e.g., TiO_2). The adsorbed compound is then promoted to an excited state by the input of visible radiation. The excited molecule can then inject an electron into the conduction band of the semiconductor and then become oxidized to a cation radical (Dieckmann and Gray, 1996); therefore, sensitized photocatalysis does not involve charge separation in the semiconductor or production of OH• radicals. Instead, it involves charge injection from an excited state of the colored compound. Sensitized photocatalysis is advantageous because it can make use of excitation energies in the visible range as well as energies in the UV range.

Dieckmann and Gray (1996) investigated the sensitized photocatalytic oxidation of 4-nitrophenol because of its environmental importance and its use in the production of pesticides, herbicides, insecticides, explosives, and synthetic dyes. Because it is widely used, 4-nitrophenol and its derivatives are found in many natural water and wastewater systems. Reaction pathways for sensitized and direct photocatalytic oxidation of 4-nitrophenol were studied, as well as the effect of oxygen on the degradation products of 4-nitrophenol. Solutions containing 0.5-mM 4-nitrophenol and 2.5-g/L TiO_2 were prepared in an 800-mL reaction vessel. Before placing the solutions in the photoreactor, the solutions were saturated with either O_2 or N_2. The solution was placed in a batch reactor and irradiated with a 450-W mercury lamp. Concentrations of 4-nitrophenol and the aromatic intermediates were determined using liquid chromatography. In the direct photocatalysis of 4-nitrophenol in the presence of O_2, the degradation of 4-nitrophenol was observed to follow pseudo first-order kinetics. Complete elimination of 4-nitrophenol was achieved in less than 3 hours. This degradation rate is two to three times larger than those reported for the photocatalytic degradation of 4-nitrophenol in unbuffered aqueous solutions (Dieckmann and Gray, 1996). Aromatic intermediates 4-nitrocatechol and hydroquinone were identified in the degradation of 4-nitrophenol; 4-nitrocatechol appears immediately after irradiation, but disappears after 3 hours of irradiation.

The direct photocatalytic oxidation of 4-nitrophenol in the absence of O_2 was observed to follow pseudo first-order degradation kinetics. In the absence of O_2, the degradation rate constant decreased by 82%. Three aromatic compounds were identified as intermediates for the reaction. The intermediates were 4-nitrocatechol, 1,2,4-benzenetriol, and hydroquinone. Inorganic nitrogen species were observed to be produced by the degradation of 4-nitrophenol. NO_2^- was found to be the dominant nitrogen species produced in the reaction. Sensitized photocatalysis in the presence of O_2 and in the absence of O_2 was found to follow pseudo first-order degradation kinetics. The intermediates formed were the same as those for direct oxidation. Total degradation of 4-nitrophenol was 50% lower for the deoxygenated system than for the oxygenated system. The reaction mechanism for direct photocatalysis of 4-nitrophenol (4-NP) is summarized in Figure 9.10.

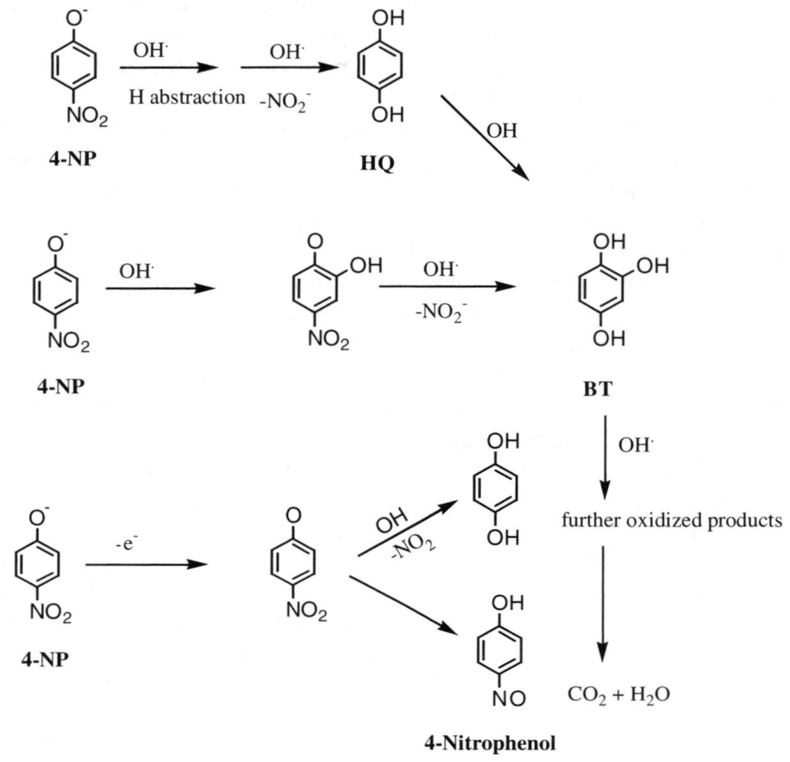

FIGURE 9.10
Degradation pathways of 4-nitrophenol via direct photocatalysis. (From Dieckmann, M. and Gray, K., *Water Res.*, 30(5) 1169, 1996. With permission.)

In the oxygenated system, $OH^{\bullet}$ abstracts hydrogen for solution phase nitrophenolate ion (4-NP$^-$), leading to denitration and $OH^{\bullet}$ substitution to form hydroquinone. Both hydroquinone and 4-nitrocatechol (4-NC) will further react with $OH^{\bullet}$ to form 1,2,4-benzenetriol (BT), followed by ring cleavage and eventual mineralization. In the deoxygenated system, the nitrophenolate ion behaves much differently. The major intermediate formed is 4-NC, suggesting that a surface reaction of nitrophenol is preferred and 4-NC production is achieved by oxidation (Dieckmann and Gray, 1996). The tendency of the nitrophenolate ion to react at the surface is due to the increased adsorption of 4-nitrophenol onto the surface of TiO_2. The major reduction reaction is proposed to be the production of 4-nitrophenol, as shown in the pathway of Figure 9.10. Hydroquinone has been identified as a minor reduction product of nitrophenolate ion that perhaps occurs via conduction-band electron reduction of nitrophenolate.

The primary intermediates are 4-NC and hydroquinone, and the secondary intermediate is benzenetriol. In both oxygenated and deoxygenated systems, the nitrophenolate ion injects an electron into the conduction band of TiO_2, creating a radical cation in the pathway as shown in Figure 9.10 The cation then reacts with water to yield 4-NC, which is then able to undergo synthesized photocatalysis to yield 1,2,4-benzenetriol. In the deoxygenated system, nitrophenolate must behave as the conduction-band electron acceptor and donator. The reductive degradation of nitrophenolate proceeds as illustrated in pathway 2 of Figure 9.11.

Direct and sensitized photocatalysis creates similar products and involves primarily oxidative steps when oxygen is present. In deoxygenated systems, both oxidative and reductive steps are involved in the degradation process. The presence of oxygen in direct and sensitized photocatalytic systems affects the extent of mineralization and the denitration of 4-NP (Dieckmann and Gray, 1996).

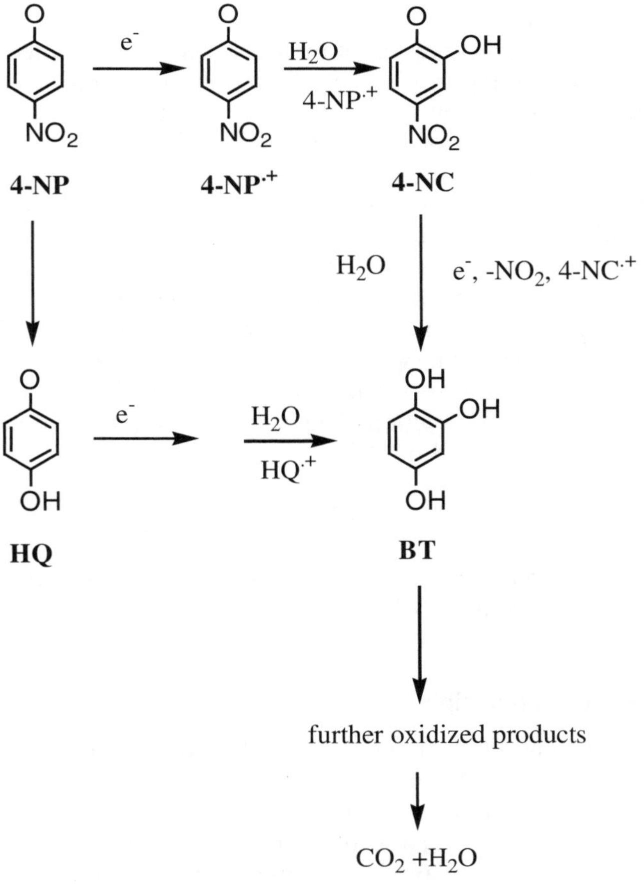

FIGURE 9.11
Degradation pathways of 4-nitrophenol via sensitized photocatalysis. (From Dieckmann, M. and Gray, K., *Water Res.*, 30(5) 1169, 1996. With permission.)

9.3.8 Phenols

Titanium dioxide photocatalytic oxidation of *para*-substituted phenols was studied by O'Shea and Cardona (1994). The phenols were irradiated with UV light from mercury lamps emitting light at a wavelength of 350 nm. The phenols were studied in aqueous solution with an initial concentration of 0.1 M. TiO_2 was added to the solution at 0.1 g/L. The phenols were placed in the photoreactor for 30 min and then analyzed by gas chromatography to determine phenol concentrations.

Removal rates greater than 90% were observed for all of the phenols studied. The decay kinetics for phenol; *p*-methoxyphenol; *p*-cresol; *p*-fluorophenol; *p*-chlorophenol; *p*-bromophenol; 4-hydroxyacetophenone; α,α,α-trifluoro-*p*-cresol; *p*-cyanophenol; and *p*-iodophenol were found to be consistent with the Langmuir–Hinshelwood kinetic model. After employing all the substituents, little variation on the Langmuir–Hinshelwood kinetic parameters was observed.

The mechanism for degradation of the monohalogenated phenols (fluoro-, chloro-bromo-, and iodophenol) may involve the formation of a positively charged reaction center. Hammett plots suggest that the mechanism for dehalogenation is the addition of a hydroxyl radical followed by elimination of a halide (O'Shea and Cardona, 1994). The major mechanism does not involve simple carbon–halide bond cleavage, but instead may involve the formation of an intermediate in which the relative bond strengths of the carbon–halide bonds are strongest to weakest in the following order: C–I > C–Br > C–Cl > C–F. This intermediate may be the result of a complex and the TiO_2 molecule. Different adsorption mechanisms were observed for halophenols and non-halophenols. It has been suggested that halophenols adsorb onto the TiO_2 molecule by charge transfer, while non-halophenols adsorb by hydrogen bonding or π electron interaction. The rate constants were determined by the Langmuir–Hinshelwood model:

$$1/\text{rate} = 1/k' + 1/k'K \qquad (9.20)$$

where k' is the apparent rate constant (M/min), and K is the apparent adsorption equilibrium constant ($1/M$).

9.3.9 Polychlorinated Biphenyls

Polychlorobiphenyls (PCBs) have been studied under the UV/TiO_2 photocatalytic process to assess the use of photocatalytic oxidation in the treatment of PCBs. These organic microcontaminants are biologically refractory; therefore, AOPs were studied for the remediation of PCBs. Haloaliphatic and haloaromatic compounds have been found to be mineralized to CO_2 and HCl by UV/TiO_2 process (Mathews, 1988).

The photocatalytic degradation of 2-chlorobiphenyl by UV/TiO_2 was studied by Huang et al. (1996) (Figure 9.12). Aroclor™ is a commercial grade of

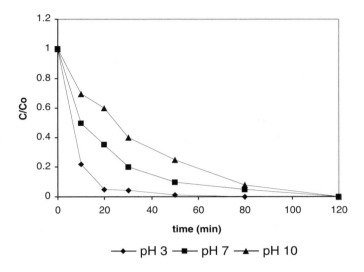

FIGURE 9.12
Degradation of 2-chlorobiphenyl by UV/TiO$_2$. (From Huang, I., Hong, C., and Bush, B., *Chemosphere*, 32, 1869, 1996. With permission.)

PCBs available for various uses. For this reason, 2-chlorobiphenyl and a mixture of Aroclor were studied. Samples (100 mL) of the mixtures were prepared with 265 µg/mL of PCB and 25 µg/mL of TiO$_2$. The pH of the samples was adjusted to 3, 7, and 10. The samples were irradiated under a UVA-340 light source for 120 min. The concentrations of the PCBs were determined using GC. The degradation rate of 2-chlorobiphenyl increased with decreasing pH values. After 1 hour of irradiation, more that 90% of the PCBs had been degraded at pH 3 and pH 7. The degradation rate constant, k, can be determined from the first-order reaction model:

$$\ln (C / C_0) = -kt \qquad (9.21)$$

where C_0 and C are the concentrations of 2-chlorobiphenyl at time 0 and t, respectively. The half-lives can also be determined:

$$t_{1/2} = \ln (2)/k \qquad (9.22)$$

Light alone and TiO$_2$ alone in the dark produce no degradation of PCB. The combination of TiO$_2$ and light shows high rates of degradation. The pathway of PCB degradation by UV/TiO$_2$ is likely to involve an attack of OH radicals on the phenyl ring. At low contaminant levels (ppb), PCB degradation is accomplished in short periods of time; however, a mixture of the catalyst and UV light shows increased rates of degradation. Less chlorinated PCBs show faster degradation rates than compounds substituted with more chlorine.

The first step of the degradation pathway is assumed to be addition of the OH radical to the PCB. Further reaction pathways are too detailed to identify, due to the complexity of degradation of PCBs on the TiO_2 surface. Data suggest that the pathway also involves an attack of OH radicals on the phenyl ring. After addition of the OH radicals to the aromatic ring, CO_2 evolution has been observed via oxidation of carbonyl compounds after the ring is ruptured (Huang et al., 1996).

In PCB degradation, photocatalytic oxidation has been observed to occur under ambient conditions. Degradation rates were found to be dependent on the concentration of TiO_2, pH of solution, and light intensity. Acidic conditions yield faster degradation rates than alkali conditions. Concentrations of PCB at parts-per-billion levels are readily degraded by UV/TiO_2 process. Economically, photocatalysis is comparable to activated carbon removal systems and more economic than UV-ozone oxidation systems (Huang et al., 1996).

9.3.10   Toluenes

The photocatalytic oxidation of alkylbenzenes and alkenylbenzenes has been widely reported. The data concerning alkylbenzenes have shown that the reactivity of toluenes is low when compared to other monosubstituted benzenes (Somarani et al., 1995). The effect of adding a zeolite, which is an acid solid catalyst, by TiO_2/UV on various 4-substituted toluenes was studied by Somarani et al. (1995). The compounds of interest were prepared at 0.03 M in solutions containing TiO_2. The effect of a zeolite was also studied by adding HY-type zeolites with various Si/Al ratios. The solutions were irradiated with a 125-W mercury lamp emitting light at 330 nm. Samples were taken at 48 hours and analyzed by GC/mass spectroscopy (MS) to determine percent conversions of the toluenes to the desired products.

In all photooxidation of the toluenes, the initial step was observed to be oxidation of the side chains on the aromatic nucleus. Two oxidation sites were observed for the alkyltoluenes, either the $-CH_3$ or the *para*-alkyl, and oxidation of these sites led to formation of alkylbenzaldehydes and α-ketones, respectively. Table 9.3 presents the compounds studied and percent conversion to the desired products.

Conversion increases from toluene to *p*-xylene to *p*-methoxytoluene due to the electron-donating effects of the substituents. The increased number of electron-donating substituents creates a positive charge on the reaction center and activates the ring for increased oxidation rates. This effect is in agreement with the radical cation oxidation mechanism proposed by Somrani et al. (1995). Electron density on the reaction center was shown to play an important role in determining oxidation rates of toluenes.

Addition of the acidic zeolite (HY_{15} or HY_{20}) led to an increase in conversion for all substrates. This effect is due to the influence of acidity on the reaction mechanism. This confirms the assumption that oxidation rates of toluenes are dependent on the acidity of the reaction.

TABLE 9.3

Selectivity of Photogenerated Products

Substrate	Conversion (%)	Selectivity of Main Photogenerated Products		
(toluene)	30	CHO 12	COOH 33.5	
(p-xylene)	50	CHO 40	COOH 20	
(p-ethyltoluene)	63	CHO 12	COOH 19	(acetyl) 33
(p-isopropyltoluene)	47	CHO 21.5	COOH 21.5	(ketone) 10.5
(p-tert-butyltoluene)	30	CHO 60	COOH 7	
(p-methylanisole)	75	CHO 60	COOH 143.5	

Source: Somarani, C. et al., *Catalysis Lett.*, 33, 395, 1995. With permission.

The primary advantage of semiconductor-assisted photodegradation as a treatment process involves its ability to degrade a wide range of pollutants with rapid rates of reaction under moderate reaction conditions. The ability to mineralize pollutants completely to CO_2 makes it an attractive treatment option, especially in cases where alternative treatment technologies are ineffective (e.g., when the pollutants are refractory toward biological degradation). TiO_2-assisted photocatalysis has an advantage over the UV/H_2O_2 oxidation process in the degradation of colored organics, as these compounds do not interfere with TiO_2-assisted photocatalytic reactions. UV/H_2O_2 oxidation is accomplished best using wavelengths in the 200- to 280-nm range. TiO_2-assisted photooxidation can be accomplished with higher wavelengths because TiO_2 has an absorption maximum at approximately 340 nm. Thus, solar radiation or simulated sunlight at wavelengths greater than 340 nm can also catalytically degrade organic compounds, albeit with a lower overall efficiency. TiO_2-assisted photocatalysis may also be applicable in the treatment of water contaminated with both organics and heavy metals.

Clearly, molecular structure influences the reaction kinetics of organic compounds during their photocatalytic oxidation. This relationship between degradability and molecular structure may be described using quantitative structure–activity relationship (QSAR) models. QSAR models can be developed to predict kinetic rate constants for organic compounds with similar chemical structures. The following section discusses QSAR models developed by Tang and Hendrix (1998) as well as those developed by other researchers.

9.4   QSAR Models

9.4.1   Substituted Phenols

Hammett correlations were developed from experimental data for substituted phenols studied under the UV/TiO_2 process (D'Oliviera et al., 1993). The mechanism for this reaction is understood to proceed via the hydroxyl radical. Experimental data from the study of dichlorophenols and trichlorophenols under UV/TiO_2 were used for QSAR analysis (D'Oliviera et al., 1993). Figure 9.13 demonstrates the QSAR model for substituted phenols formulated from experimental data. The QSAR model developed for substituted phenols shows a goodness of fit of 0.9766. A good correlation was also established for substituted phenols using Hammett's constant, σ; the correlation coefficient is 0.987 (D'Oliviera et al., 1993). Similar correlation coefficients for the constants σ and σ_{res} demonstrate that the descriptor σ_{res} can be used to accurately predict kinetic rate constants for substituted phenols.

Experimental data and calculated data for the photodegradation of chlorophenols were compared using multivariate analysis and SPSS™ statistical software. A linear expression was developed using the independent variables K_{ow} and σ_{res} and the dependent variable $\log k/k_0$. Calculated values for $\log k/k_0$ were obtained from the following linear equation:

$$\text{Log } k/k_0 = A\,\sigma_{res} + B \log K_{ow} + C \qquad (9.23)$$

The values of A and B were determined to be –0.049 and –1.022, respectively. Constant C was determined to be 0.107. Table 9.4 shows the experimental values of $\log k/k_0$ and the calculated values from the linear relationship.

Figure 9.14 is a plot of experimental vs. predicted values for $\log k/k_0$. Hammett's equation for the photocatalytic degradation of substituted phenols was able to predict kinetic rate constants with 92.9% accuracy. The above

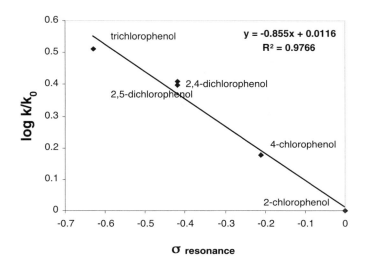

FIGURE 9.13
Hammett's equation for rate constants of substituted phenol degradation by UV/TiO_2. Experimental conditions: 2.5 g/L TiO_2; pH = 4.4–5.

TABLE 9.4

Experimental and Calculated Values of Log k/k_0 of Chlorophenols

Chlorophenol	Log K_{ow}	σ_{res}	Log k/k_0 (calc.)	Log k/k_0 (exp.)
2-Chlorophenol	2.17	0	−0.09563	0.132339751
4-Chlorophenol	2.39	−0.21	0.10821	0.174277918
2,4-Dichlorophenol	3.16	−0.42	0.2851	0.2955671
2,5-Dichlorophenol	3.42	−0.42	0.27236	0.329906123
3,4,5-Trichlorophenol	4.29	−0.63	0.44435	0.42833731

Note: k_0 = 1.6/hr; calc. = calculated; exp. = experimental results.

correlation could be used to predict degradation rate constants for the compounds of similar chemical structure given a specific value of σ_{res}.

The degradation of substituted phenols could also be modeled according to Cl^- and CO_2 formation. Hammett's equations were derived using sigma resonance as a predictor for Cl^- and CO_2 formation rates. Analysis of the Hammett correlations for Cl^- and CO_2 formation can provide support for the proposed mechanism. The rates of Cl^- formation and CO_2 formation are also dependent on the number and type of substituents attached to the benzene ring. For Cl^- formation, the rate of formation is fastest for the least substituted compound (2-chlorophenol) and slowest for the most substituted compound (3,4,5-trichlorophenol). In regard to the formation of CO_2 (Figure 9.14), when plotted vs. the molecular descriptor, σ_{res} was random. This suggests the formation of intermediates in the reaction pathway, consistent with the findings of D'Oliviera et al. (1993).

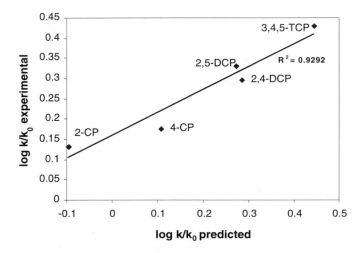

FIGURE 9.14
Experimental vs. predicted rate constants for degradation of chlorinated phenols by UV/TiO$_2$.
Experimental conditions: 20-mL reactor volume; initial concentration, 20 mg/L; 2.5 g/L TiO$_2$;
initial pH = 4.4–5.

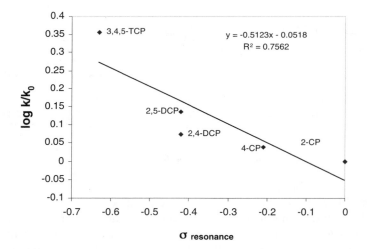

FIGURE 9.15
Hammett's equation for Cl$^-$ formation during degradation of chlorinated phenols. Experimental
conditions: 20-mL reactor volume; initial concentration, 20 mg/L; 2.5 g/L TiO$_2$; initial pH =
4.4–5.

Figure 9.15 plots the kinetics of Cl$^-$ formation by hydroxyl radical attack
on chlorinated phenols. It shows that the kinetic rate constants of Cl$^-$ formation are linear with respect to σ_{res}. This is consistent with the findings of
D'Oliviera et al. (1993) on the study of photodegradation of dichlorophenols

and trichlorophenols. The linear formation of Cl⁻ suggests that hydroxyl radical attack at the Cl site is the major pathway.

The reaction pathway of 2,4-dichlorophenol is illustrated in Figure 9.16. The Cl⁻ reaction site leads to the product 2-chloro-4-hydroxyphenol (D'Oliviera et al., 1993). Bond cleavage of the aromatic ring is observed to yield the minor products 1-hydroxy-4-chlorophenol and 2-chloro-2,5-cyclodihexene-1,4-one, as shown in Figure 9.17.

The plot shown in Figure 9.18 for the kinetics of CO_2 formation from hydroxyl radical attack on chlorinated phenols appears to be random. This can be justified by the various intermediates formed along the reaction pathway. Each of these intermediates will have different degradation rates, so CO_2 formation will be different according the reaction.

9.4.2 Substituted Benzenes

A Hammett equation was also established for substituted benzenes. A separate model was established for halogenated benzenes. It has been found that halogen substituents behave differently than substituents such as –CH_3 and NO_2 (Hansch and Leo, 1995). For this reason, an accurate comparison of halogenated substituents and other substituents could not be made.

Figure 9.19 shows the Hammett's equation for degradation rates of substituted nitrobenzenes from experimental data. Photocatalytic degradation of the nitro aromatic compounds by UV/TiO_2 from the experimental data reported by Dillert et al. (1995) are used to construct the model. The degradation of nitrobenzenes can be described according to the number of nitro substituents: nitrobenzene > dinitrobenzene > 1,3,5-trinitrobenzene (Dillert et al., 1995).

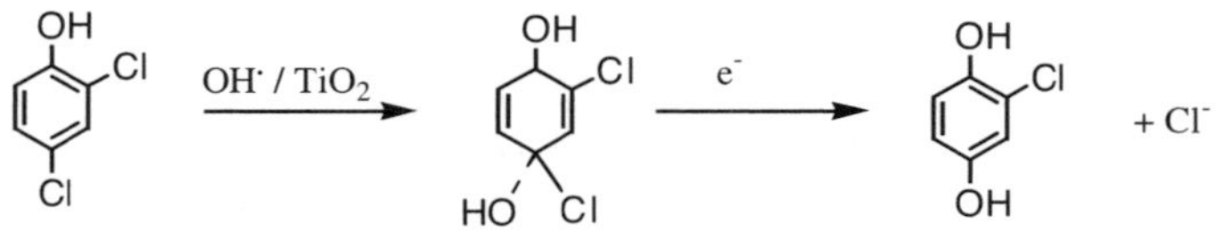

FIGURE 9.16
Degradation mechanism of dichlorophenol by UV/TiO_2.

minor products

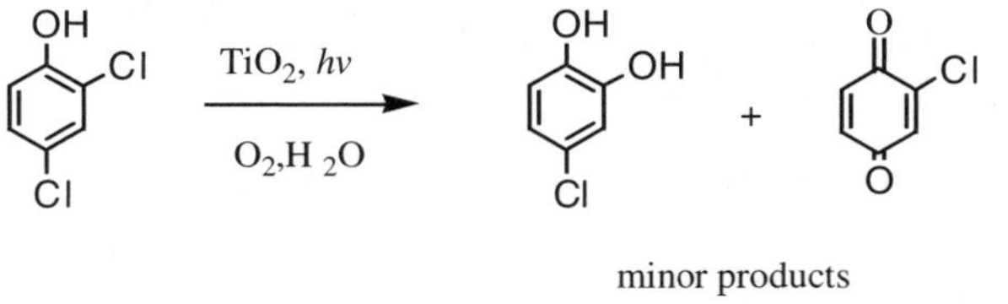

FIGURE 9.17
Minor products from degradation of dichlorophenol by UV/TiO_2.

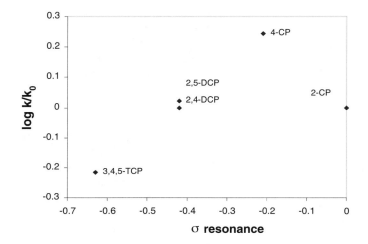

FIGURE 9.18
Hammett's equation for CO_2 formation during degradation of chlorophenols. Experimental conditions: 20-mL reactor volume; initial concentration, 20 mg/L; 2.5 g/L TiO_2; initial pH = 4.4–5.

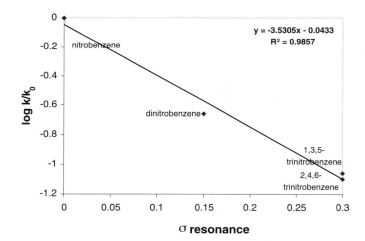

FIGURE 9.19
Hammett's equation for rate constants of substituted benzene degradation by TiO_2/UV. Experimental conditions: 5-mL reactor volume; 100-μmol/L initial concentration; 1 g/L TiO_2; 30°C; pH = 9.

In Figure 9.19, the negative slope of the line suggests a positive transition state complex. The decrease in the values of log k/k_0 for the more substituted benzenes can be attributed to the greater electron-withdrawing effect of the NO_2 groups. The presence of the NO_2 groups decreases the π electron density, making the transfer of electrons to the positive hole on the valence band of TiO_2 more difficult. This explains why the monosubstituted ben-

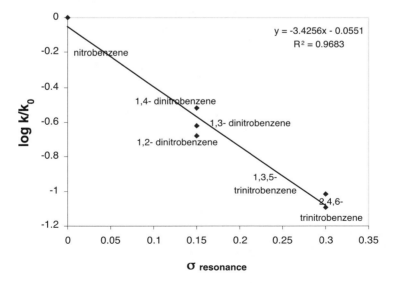

FIGURE 9.20
Reaction pathway for nitrobenzene degradation by UV/TiO$_2$.

zenes are easier to degrade than the di- and trisubstituted benzenes. Figure 9.20 demonstrates the reaction pathway for hole oxidation of nitrobenzenes.

Hole oxidation appears to be the mechanism for the degradation of nitrobenzenes. The presence of the electron-withdrawing NO$_2$ groups inhibits hydroxylation. Hydroxylation will not form a positive transition state complex, which does not correspond to the Hammett correlation shown in Figure 9.19. The reaction proceeds via the formation of a positive hole and creation of a positive transition-state complex. The positive transition-state complex will gain a free electron to form NO$_2$•$^-$. The NO$_2^-$ group will disassociate from the ring to form benzene. Degradation of nitrobenzene was also studied at pH 3. Figure 9.21 demonstrates the Hammett correlation for nitrobenzenes at pH 3.

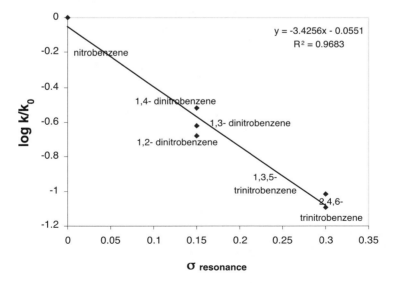

FIGURE 9.21
Hammett's equation for substituted benzenes. Experimental conditions: 5-mL reactor volume; 100-μmol/L initial concentration; 1 g/L TiO$_2$; 30°C; pH = 3.

The Hammett correlation for nitrobenzenes is almost identical to the correlation at pH 9; therefore, the degradation of substituted benzenes at pH 3 can also be described by the same hole oxidation mechanism. Figure 9.21 demonstrates the oxidation of nitrobenzene by positive hole. At pH 3, substituted benzenes are oxidized by the formation of a positive hole. An electron transfer from nitrobenzene to TiO_2 creates this positive hole.

9.4.3 Substituted Alcohols

The degradation of alcohols could be described using a derived Hammett's equation. Degradation rate constants were modeled using σ^* as a molecular descriptor. The least substituted alcohol, methanol, was used as a reference compound. Figure 9.22 demonstrates Hammett's equation for substituted alcohols. The correlation coefficient for substituted alcohols is 0.9031. The degradation of alcohols behaves similarly to the elementary hydroxyl radical under UV/TiO_2 photocatalytic oxidation. The kinetic rate constant appears to be dependent on the number of carbon atoms present. Alcohols with longer carbon chains are degraded more slowly than alcohols with shorter carbon chains. For example, Figure 9.22 demonstrates that methanol is degraded nearly three times faster than heptanol.

9.4.4 Chlorinated Alkanes

A Hammett correlation was also established for chlorinated alkanes studied by UV/TiO_2. The least substituted compound, ethane, was used as a reference compound. Figure 9.23 demonstrates the Hammett correlation for chlo-

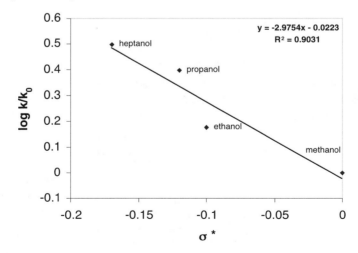

FIGURE 9.22
Hammett's equation for substituted alcohols. Experimental conditions: batch reactor with 0.1 wt% TiO_2; 10–100 ppm initial concentration; pH = 9.

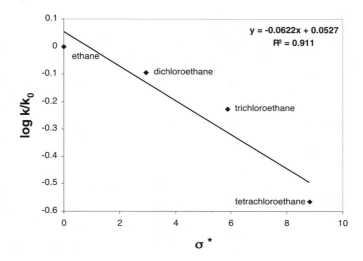

FIGURE 9.23
Hammett's correlation for chlorinated alkanes studied by UV/TiO$_2$. Experimental conditions: 0.1 wt% TiO$_2$; 10 ppm initial concentration; 20-mL reactor volume; pH ≈ 7.

rinated alkanes. The degradation rate appears to be a function of the number of chlorine substituents present. The least substituted compound, ethane, was used as a reference compound and is degraded at the fastest rate of the alkanes studied. The most substituted compound, tetrachloroethane, is degraded at the slowest rate of the alkanes studied. This is similar to the Hammett correlation established for chlorinated phenols, with the least substituted phenol being degraded at the fastest rate and the most substituted phenol being degraded at the slowest rate.

9.4.5   Chlorophenols

The apparent first-order degradation rate constants of six dichlorophenols and three trichlorophenols were studied by D'Oliviera et al. (1993). It was determined that the first-order rate constant, k, correlated with the Hammett's constant σ and the octanol/water partition coefficient. The Hammett correlation for the cholorophenols was developed as follows:

$$K_{app} = -10\sigma + 5.2 \log K_{ow} - 7.5 \qquad (9.24)$$

This linear relationship was used to predict first-order rate constants based on log K_{ow} and Hammett's constants for chlorophenols. The experimental and calculated values for the chlorophenols studied are listed in Table 9.5.

The correlation coefficient for the Hammett correlation is 0.987 if the value for 2,4,6-trichlorophenol is omitted. This experimental value is too small and should be disregarded. This correlation successfully demonstrates that some

TABLE 9.5

Predicted and Experimental First-Order Rate Constants for Substituted Chlorophenols

Chlorophenol	Log K_{ow}	s	$k_{predicted}$ (1/hr)	$k_{experimental}$ (1/hr)
2-Chlorophenol	2.17	0.227	1.5	1.6
3-Chlorophenol	2.48	0.373	1.7	1.6
4-Chlorophenol	2.39	0.227	2.6	2.4
2,4-Dichlorophenol	3.16	0.454	4.4	4.0
2,5-Dichlorophenol	3.42	0.600	4.3	4.1
2,6-Dichlorophenol	2.8	0.454	2.5	2.7
3,4-Dichlorophenol	3.47	0.600	4.5	4.2
3,5-Dichlorophenol	3.72	0.746	4.4	4.1
2,4,5-Trichlorophenol	3.72	0.827	3.6	3.7
3,4,5-Trichlorophenol	4.29	0.973	5.1	5.2
2,4,6-Trichlorophenol	3.37	0.681	3.2	2.0

characteristics of aromatic compounds, such as log K_{ow} and σ, can be used as descriptors to predict the photocatalytic degradation of organic pollutants.

9.4.6 Substituted Anisoles

A correlation for less structurally related compounds was also developed by Amalric et al. (1996). To develop this correlation, *meta-* and *para-*substituted anisoles were studied. These aromatic compounds were substituted with F, Cl, NO_2, OH, and NH_2 groups. The first-order degradation rate constant, k_{app}, was predicted with the octanol/water partition coefficient (log K_{ow}), Brown's constant (σ^+), and molar refractivity (MR) used as descriptors. The following correlation was developed:

$$K_{app} = 0.2 \log K_{ow} - 0.2\sigma^+ - 0.033 \text{ MR} - 0.43 \qquad (9.25)$$

The correlation was developed with a correlation coefficient of 0.903. The correlation was able to accurately predict the first-order degradation rate constants as a function of Brown's constant, the octanol/water partition coefficient, and molar refractivity. This correlation held for a broader class of aromatic compounds substituted at the *meta* and *para* positions, as compared to a simple substitution at one position on the aromatic ring. Table 9.6 lists the experimental and predicted values for the first-order degradation rate constants of substituted anisoles.

This correlation of anisoles using the above descriptors demonstrates that broader ranges of compounds can be compared with QSAR models. Selecting descriptors that allow for the comparison of these compounds across wider classes is the key to developing useful QSAR models. By experimenting, it was also found that a better correlation could be achieved by replacing

TABLE 9.6

Experimental and Predicted Values of *Meta*- and *Para*-Substituted Anisoles

Anisole	Log K_{ow}	σ^+	MR	$k_{experimental}$ (10^2/min)	$k_{predicted}$ (10^2/min)
An	2.11	0	32.89	7.8	8.1
m-An-Cl	2.94	0.399	37.84	6.4	6.7
p-An-Cl	2.84	0.114	37.19	7.6	7.7
m-An-F	2.42	0.352	32.92	8.6	7.9
p-An-F	2.32	−0.073	32.60	10.0	9.4
m-An-NO$_2$	2.16	0.674	39.93	3.6	3.5
p-An-NO$_2$	2.03	0.790	39.44	4.9	3.3
m-An-OH	1.58	0.121	35.09	5.4	5.1
p-An-OH	1.34	−0.92	34.93	10.4	7.4
m-An-NH$_2$	0.93	−0.16	37.37	2.7	3.6
p-An-NH$_2$	0.95	−1.3	37.50	6.2	6.1

Hammett constant's (σ) with Brown's constant (σ^+). It was also found that molar refractivity was proportional to the electronic polarizability and is a good descriptor of dispersion forces (Amalric et al., 1996).

9.4.7 Comparison of Hammett's Correlations for Elementary Hydroxyl Radical Reactions and UV/TiO$_2$

The Hammett correlation for substituted alcohols studied by elementary hydroxyl radical reaction can be compared to the correlation for alcohols studied by UV/TiO$_2$. The slopes of both correlations are negative, suggesting a similarity in the reaction mechanism. In both cases, the longer carbon chain alcohol was degraded at a slower rate. Figure 9.24 presents a comparison between the kinetic rate constants for elementary hydroxyl radical reaction and UV/TiO$_2$. The correlation coefficient of 0.9637 suggests that the reaction mechanisms for alcohols studied under elementary hydroxyl radical reaction and UV/TiO$_2$ are similar. The mechanism appears to be hydroxylation by OH$^\bullet$ radical attack.

A comparison of the kinetic rate constants for elementary hydroxyl radical and UV/TiO$_2$ is provided in Figure 9.24. The slopes of the Hammett correlations for the hydroxyl radical data and UV/TiO$_2$ are both negative. This suggests a similarity in the reaction mechanism that is supported by a comparison of these correlations.

Figure 9.25 shows the correlation coefficient of 0.9636 between the hydroxyl radical rate constants and UV/TiO$_2$ rate constants for chlorinated alkanes. Therefore, the reaction mechanism for the degradation of chlorinated alkanes by UV/TiO$_2$ is similar to the reaction with hydroxyl radical. This correlation suggests that the reaction proceeds via hydroxyl radical attack on the chlorinated alkane.

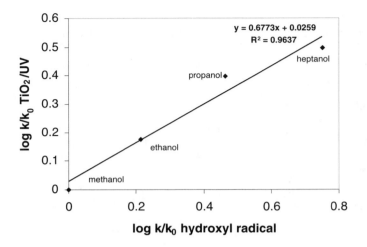

FIGURE 9.24
Correlation between hydroxyl radical rate constants and UV/TiO$_2$ rate constants for substituted alcohols.

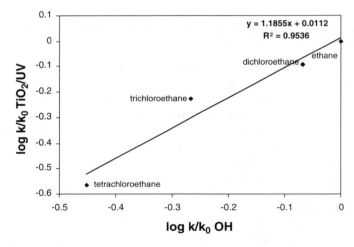

FIGURE 9.25
Correlation between hydroxyl radical rate constants and UV/TiO$_2$ rate constants for chlorinated alkanes.

The Hammett correlations for substituted benzenes were similar for elementary hydroxyl radical reaction and UV/TiO$_2$ because both correlations have negative slopes. The similarity of the correlations may be due to the same oxidation species, such as hydroxyl radical. For both reactions, the benzenes substituted with electron withdrawing groups were difficult to degrade.

In the case of phenols, Hammett correlations for elementary hydroxyl radical reactions and UV/TiO₂ both had negative slopes. The halogenated phenols showed that the halogen substituents decreased the rate of degradation. Electron-withdrawing groups on phenols also had the same effect as on benzenes. The phenols with electron-withdrawing groups had the slowest degradation rates among the substituted phenols. The electron-withdrawing effect of the substituents decreases the electron density on the ring, making hydroxyl radical attack more difficult. Figure 9.26 demonstrates a correlation between kinetic data for elementary hydroxyl radical reactions and UV/TiO₂ for chlorophenols.

The correlation coefficient of 0.8897 suggests that the mechanism for chlorophenol degradation by UV/TiO₂ is the hydroxyl radical attack on the aromatic ring. This correlation is consistent with the reaction mechanism for the degradation of chlorophenols proposed by O'Shea and Cardona (1994) and D'Oliviera et al. (1993). In other words, no correlation would suggest that degradation by UV/TiO₂ occurs via a different mechanism.

Quantitative structure–activity relationship models can be used to predict kinetic rate constants for compounds similar in chemical structure under UV/TiO₂ photocatalytic oxidation. In formulating the Hammett correlations, the descriptors, such as σ_{res} and σ^*, proved to be successful in deriving equations for aromatic and aliphatic compounds respectively. For the case of aromatic compounds, the σ_{res} descriptor formulated better Hammett correlations than the σ_m descriptor.

For each class of compounds studied by UV/TiO₂, reaction mechanisms agree with the slope of the Hammett's equations developed. A comparison

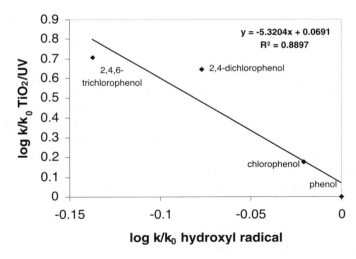

FIGURE 9.26
Correlation between hydroxyl radical rate constants and UV/TiO₂ rate constants for chlorophenols.

of the kinetic data for elementary hydroxyl radical data and UV/TiO$_2$ suggests that hydroxyl radicals are the dominant species. A good correlation coefficient between the two processes suggests that the reaction mechanism proceeds via elementary hydroxyl radical attack on the aromatic ring. For nitrobenzenes, however, positive hole oxidation appears to be the dominant mechanism due to the negative slope of the Hammett plot.

9.5 Engineering Applications

The need for highly efficient new methods for the treatment of toxic and biologically persistent compounds has led to increased interest in semiconductor photocatalysis. The term *photodegradation* is used to refer to the complete mineralization or conversion of organic compounds to CO$_2$, H$_2$O, NO$_3$, or other oxides, halides, phosphates, etc. Extensive work with UV/TiO$_2$ has shown that many organic wastes can be completely mineralized with irradiated TiO$_2$ suspended in water. Semiconductor photocatalysis using TiO$_2$ is appealing due to the fact that TiO$_2$ is capable of extended use without the loss of catalytic activity. TiO$_2$ particles recovered by filtration or centrifugation retain much of their catalytic activity after repeated catalytic reactions. Essentially all chlorinated aliphatics, chlorinated aromatics, most pesticides, herbicides, and surfactants are completely mineralized to water, carbon dioxide, and mineral acids. This makes UV/TiO$_2$ an attractive AOP, as almost all of the pollutants are not simply transferred from one media to another. UV/TiO$_2$ has also been applied in the destruction of microbes, viruses, inactivation of cancer cells, odor control, production of hydrogen gas, and remediation of oil spills.

Titanium-dioxide-assisted photocatalytic degradation using solar illumination has been developed into a process for compounds (such as TCE) with high rates of degradation. However, the process may be unsuitable for contaminants that have slower rates of degradation, such as chlorinated methanes and ethanes, because of the need for a large surface area collector in the solar-driven processes. Based on longer half-lives of degradation, a large number of aromatic compounds may also be unsuitable target compounds for TiO$_2$-assisted photodegradation by solar energy. Field tests of TiO$_2$-assisted solar detoxification of groundwaters containing TCE and other volatile organic compounds has been reported. A pilot-scale system that does not use concentrated sunlight and thereby eliminates the costs associated with solar collectors was found to yield superior TCE degradation rates. The cost of the process was estimated to be about $8/1000 gal of contaminated water. Systems that do not use concentrated sunlight and yield high rates of degradation are likely to be extended to the treatment of compounds that have low rates of degradation. The economics of TiO$_2$-assisted photodegradation using UV lamps for large-scale systems is not known, although it is possible that solar-illuminated systems are more likely to be applied commercially. In addition, TiO_2-assisted photodegradation may also be applicable for the treatment of industrial wastewaters.

Gaseous pollutants present in air emissions from various chemical plants are primarily inorganic compounds, such as nitrogen and sulfur oxides, and odor-causing compounds such as sulfides and various sulfur-containing compounds can be degraded by TiO_2-assisted photodegradation.

Groundwater contamination is most commonly caused by pollutants such as unsaturated or saturated chlorinated aliphatic compounds (e.g., TCE, PCE, and various other chlorinated alkanes and alkenes) and some aromatic compounds (e.g., gasoline and diesel hydrocarbons). These contaminants are usually present at low concentrations in the low parts-per-million range and require treatment to the parts-per-billion level. Low concentrations of dissolved salts and low concentrations of other organic compounds are typical of contaminated groundwater. The treatment of groundwater can be readily accomplished by UV/H_2O_2 systems as evidenced by numerous full-scale units in operation. The application of the TiO_2-assisted photocatalytic process in groundwater treatment appears promising because of the numerous reports of complete mineralization of a wide range of contaminants.

High concentrations of organic compounds, a range of mixtures from a few components to complex mixtures, high chemical oxygen demand (COD) values, and high dissolved-salt concentrations are all typical characteristics of industrial wastewaters. Pollutants commonly present in industrial wastewaters, such as phenols, chlorinated aromatic compounds, colored organic compounds, pesticides, surfactants, and nitroaromatics, have been readily photodegraded by UV/TiO_2. However, the requirement for higher illumination intensities, longer treatment time periods, and the possible interference of suspended solids and dissolved salts will likely make the application of TiO_2-assisted photodegradation processes economically or technically impractical for the treatment of high-strength industrial wastewaters.

UV/TiO_2 process has been used to treat azo dyes found in the wastewaters of the dyeing industry. Textile dyes and commercial colorants are one of the largest groups of industrial chemicals produced. The wastewaters are highly colored with these dyes. Removal efficiency has been reported as high as 100% for various dyes (Tang and Huang, 1995). This technology is attractive, because physical and biological treatment methods are ineffective for the removal of these commercial dyes.

Photocatalytic oxidation using UV/TiO_2 has been commercialized for water treatment. Due to the extent of research done in this area, most chemical functionalities have been investigated (Amalric et al., 1996). Photodegradation of pollutants can be estimated based on structural relationships to compounds whose photocatalytic degradability has already been investigated. Photodegradation correlations between degradation rate constants and molecular descriptors of many organic pollutants have also been reported.

UV/TiO_2 process has been shown to be successful in the treatment of trichloroethylene (TCE), which is a widely known contaminant found in gaseous emissions, landfills, hazardous waste sites, and groundwater. Photocatalytic oxidation using UV/TiO_2 has been shown to remove 99.4% of TCE from aqueous solutions (Hung and Marinas, 1997). When fluoroalkanes or fluoroaromatics were exposed to suspension of TiO_2 and UV light, CO_2 and HF were formed (Fox and Dulay, 1993). Trinitrotoluene (TNT) is another widespread environmental contaminant that is found in soil and groundwater samples of former U.S. Department of Defense sites. Investigations of degradation of TNT by UV/TiO_2 have demonstrated that more than 90% of TNT can be oxidized to CO_2.

During the oxidation of TCE and PCE, several chlorinated by-products are formed, at least one of which, dichloroacetic acid, is a known carcinogen and has greater toxicity than either TCE or PCE (Glaze et al., 1992); however, several researchers have shown quantitative conversion of chlorinated organics to innocuous end products such as inorganic chloride and CO_2. Also, PCP and 2,4-dichlorophenol have been completely detoxified (as judged by toxicity to activated sludge microorganisms) by photocatalytic oxidation; however, toxicity increased during the treatment, most likely due to the formation of intermediate compounds more toxic than the parent compound. Because toxic by-products are formed during the treatment process, the effluent from such a process needs to be carefully analyzed for residual compounds and toxicity, as incomplete treatment may cause more of a treatment problem.

In most cases, the half-lives for mineralization of pollutants are on the order of a few hours. Substantial irradiation times are required for some pollutants; therefore, TiO_2-assisted photodegradation may not be economical in some cases. For example, cyanide, 2,7-dichlorobenzo-*p*-dioxin, and benzamide require about 30, 90, and 150 hr of irradiation time, respectively. In some studies, the pollutants have only been tested as single compounds at low concentrations ranging from parts-per-billion to low parts-per-million levels, typical of those encountered during treatment of groundwater and drinking water. Few studies have evaluated the efficiency of TiO_2-assisted photodegradation in the treatment of mixtures of contaminated wastewaters. The process is unlikely to be used in the treatment of high-strength industrial wastewaters or for large-scale, direct clean-up of contaminated soils. Loss of efficiency may result in cases where substrates compete with mixtures containing dissolved anions and cations, which cause significant reductions in rates of photodegradation. Loss of efficiency may also result when light interference by high concentrations of pollutants occurs; however, soils spiked with 2-chlorophenol, dioxin, and atrazine could be treated using soil slurries up to 60 g/L. The economics of this application for soil decontamination is not known.

References

Amalric, L., Guillard, C., Brude, E., and Pichat, P., Correlation between the photocatalytic degradability over TiO_2 in water of *meta*- and *para*-substituted methoxybenzenes and their electron density, hydrophobicity and polarizability properties, *Water Res.*, 30, 5, 1137, 1996.

Bekbolet, M., Lidner, M., Weichgrebe, D., and Bahnemann, D., Photocatalytic detoxification with the thin-film fixed bed reactor: clean up of highly polluted landfill effluents using a novel TiO_2 photocatalyst, *Solar Energy*, 56(5), 455, 1996.

Borman, S., New QSAR techniques eyed for environmental assessments, *Chem. Eng. News*, 68, 20, 1990.

Brus, L., Size-dependent development of band structure in semiconductor crystallites, *New J. Chem.*, 11, 123, 1990.

Connors, K., *Chemical Kinetics: The Study of Reaction Rates in Solution*, VCH, New York, 1990.

Cunningham, J. and Al-Sayyed, G., Factor influencing efficiencies of TiO_2-sensitized photodegradation, *J. Chem. Soc. Faraday Trans.*, 86, 3935, 1990.

Davis, A. and Huang, C., Adsorption of some substituted phenols on hydrous CdS, *Langmuir*, 6, 857, 1990.

Dieckmann, M. and Gray, K., A comparison of the degradation of 4-nitrophenol via direct and sensitized photocatalysis in TiO_2 slurries, *Water Res.*, 30(5) 1169, 1996.

Dillert, R., Brant, M., Siebers, U., and Bahnemann, D., Photocatalytic degradation of trinitrotoluene and other nitroaromatic compounds, *Chemosphere*, 30(12), 2333, 1995.

Doherty, S., Guillard, C., and Pichat, P., Kinetics and products of the photocatalytic degradation of morpholine in TiO_2 suspensions, *J. Chem. Soc. Faraday Trans.*, 91(12), 1853, 1995.

D'Oliviera, J., Minero, C., Pelizzetti, E., and Pichat, P., Photodegradation of dichlorophenols and trichlorophenols in TiO_2 aqueous suspensions: kinetic effects of the positions of the Cl atoms and identification of the intermediates, *J. Photochem. Photobiol. A: Chem.*, 72, 261, 1993.

Ehmayer, G., Kappel, G., and Reich, S., Connecting databases to the Web: a taxonomy of gateways, *Database and Expert Systems Applications 8th International Conference Proceedings*, 1997.

Fan, J. and Yates, J., Mechanism of photooxidation of trichloroethylene on TiO_2: detection of intermediates by infrared spectroscopy, *J. Am. Chem. Soc.*, 118, 4686, 1996.

Fox, M. and Dulay, M., Heterogeneous photocatalysis, *Chem. Rev.*, 93, 341, 1993.

Fox, M., Kim, Y., Wahab, A., and Dulay, M., Photocatalytic decontamination of sulfur containing alkyl halides on irradiated semiconductor suspensions, *Catalysis Lett.*, 5, 369, 1990.

Fujishima, A. and Honda, K., *Bull. Chem. Soc. Jpn.*, 44, 1148, 1971.

Fujishima, A. and Honda, K., *Nature*, 238, 38, 1972.

Glaze, W.H., Beltran, F., Tuhkanen, T., and Kang, J.W., Chemical models of advanced oxidation processes, *Water Poll. Res. J. Can.*, 27(1), 23–42, 1992.

Goren, Z., Willner, I., Nelson, A., and Frank, A., Selective photoreduction of CO_2/ HCO_3^- to formate by aqueous suspensions and colloids of Pd–TiO_2, *J. Phys. Chem.*, 94, 3784, 1990.

Gratzel, M. and Rotzinger, F., *Chem. Phys. Lett.*, 118, 474, 1985.

Hadjipavlou, D. and Hansch, C., Quantitative structure–activity relationships of the benzodiazepines, *Chem. Rev.*, 94, 1483, 1994.

Hadjivanov, K.I. and Klissurski, D., Surface chemistry of titania and titania-supported catalysts, *Chem. Soc. Rev.*, 25, 61–69, 1996.

Hansch, C. and Leo, A., *Exploring QSAR: Fundamentals and Applications in Chemistry and Biology*, American Chemical Society, Washington, D.C., 1995.

Hansch, C., Leo, A., and Hoekman, D., *Exploring QSAR: Hydrophobic, Electronic, and Steric Constants*, American Chemical Society, Washington, 1995.

Hansch, C., Hoekman, D., and Gao, H., Comparative QSAR: toward a deeper understanding of chemicobiological interactions, *Chem. Rev.*, 96, 1045, 1996.

Harada, H., Ueda, T., and Sakata, T., *J. Phys. Chem.*, 93, 1542, 1990.

Hisanga, T., Harada, K., and Tanaka, K., Photocatalytic degradation of organochlorine compounds in suspended TiO_2, *J. Photochem. Photobiol. A*, 54, 113, 1990.

Hoffmann, M., Martin, S., Choi, W., and Bahnemann, D., Environmental applications of semiconductor photocatalysis, *Chem. Rev.*, 95, 69, 1995.

Huang C., Dong, C., and Tang, Z., Advanced chemical oxidation: its present and potential future in hazardous waste treatment, *Waste Manage.*, 13, 361,1993.

Huang, I., Hong, C., and Bush, B., Photocatalytic degradation of PCBs in TiO_2 aqueous suspensions, *Chemosphere*, 32, 1869, 1996.

Hung, C. and Marinas, B., Role of chlorine and oxygen in the photocatalytic degradation of trichloroethylene vapor on TiO_2 films, *Environ. Sci. Technol.*, 31, 562, 1997.

Jardim, W., Moraes, S., and Takiyama, M., Photocatalytic degradation of aromatic chlorinated compounds using TiO_2: toxicity of intermediates, *Water Res.*, 31, 1728, 1997.

Karelson, M. and Lobanov, V., Quantum chemical descriptors in QSAR/QSPR studies, *Chem. Rev.*, 96, 1027, 1996.

Kim, D.H. and Anderson, M.A., Solution factors affecting the photocatalytic and photoelectrocatalytic degradation of formic acid using supported TiO_2 thin films, *J. Photochem. Photobiol. A—Chemistry*, 94(2–3): 221–229, 1996.

Kinkennon, A., The use of simulated of concentrated natural solar radiation for the TiO_2 mediated photodecomposition of Basagran, diquat, and diuron, *Chemosphere*, 31, 3663, 1995.

Kondo, M.M. and Jardim, F., Photodegradation of chloroform and urea using Ag-loaded titanium dioxide as catalyst, *Water Res.*, 25, 823–827, 1991.

Kormann, C., Bahnemann, D.W., and Hoffmann, M.R., Photolysis of chloroform and other organic molecules in aqueous TiO_2 suspensions, *Environ. Sci. Technol.*, 25, 494, 1991.

Kraeutler, B. and Bard, A.J., Heterogeneous photocatalytic decomposition of saturated carboxylic acids on TiO_2 powder. Decarboxylative route to alkanes, *J. Amer. Chem. Soc.*, 100, 5985–5992, 1978.

Legrini, O., Oliveros, E., and Braun, A., Photochemical processes for water treatment, *Chem. Rev.*, 93, 671, 1993.

Lichtin, N. and Avudaithai, M., TiO_2 Photocatalyzed oxidative degradation of CH_3CN, CH_3OH, C_2HCl, and CH_2Cl_2 supplied as vapors and in aqueous solution under similar conditions, *Environ. Sci. Technol.*, 30, 2014, 1996.

Lu, M., Roam, G., Chen, J., and Huang, C., Adsorption characteristics of dichlorvos onto hydrous titanium dioxide surface, *Water Res.*, 30, 1670, 1996.

Mahdavi, F., Bruton, T., and Li, Y., Photoinduced reduction of nitro compounds on semiconductor particles, *J. Org. Chem.*, 58, 744, 1993.

Martin, S., Kesselman, J., Park, D., Lewis, N., and Hoffman, M., Surface structures of 4-chlorocatechol adsorbed on titanium dioxide, *Environ. Sci. Technol.*, 30, 2535, 1996.

Matthews, R.W., Kinetics of photocatalytic oxidation of organic solutes over titanium dioxide, *J. Catalysis*, 111, 264, 1988.

Matthews, R.W., Photooxidative degradation of colored organics in water using supported catalysts TiO_2 on sand, *Water Res.*, 25, 1169–1176, 1991.

McBride, M. and Wesselink, L., Chemisorption of catechol on gibbsite, boehmite, and noncrystalline alumina surfaces, *Environ. Sci. Technol.*, 22, 703, 1988.

Mills A., Davies, R., and Worsley, D., Water purification by semiconductor photocatalysis, *Chem. Soc. Rev.*, 22, 417–425, 1993.

Minero, C., Pelizzetti, E., Pichat, P., Sega, M., and Vicenti, M., Formation of condensation products in advanced oxidation technologies: the photocatalytic degradation of dichlorophenols on TiO_2, *Environ. Sci. Technol.*, 29, 2226, 1995.

Okamoto, K., Yamamoto, Y., Tanaka, K., and Itaya, A., Heterogeneous photocatalytic decomposition of phenol over TiO_2 powder, *Bull. Chem. Soc. Japan*, 58, 2015–2022, 1985.

O'Shea, K. and Cardona, C., Hammett study on the TiO_2-catalyzed photooxidation of *para*-substituted phenols: a kinetic and mechanistic analysis, *J. Org. Chem.*, 59, 5005, 1994.

Ollis, D., Hsiao, C.Y., Budiman, L., and Lee, C., Heterogeneous photoassisted catalysis: conversions of perchloroethylene, dichloroethane, chloroacetic acids, and chlorobenzenes, *J. Catalysis*, 88, 89, 1984.

Peil, N.J. and Hoffmann, M.R., Chemical and physical characterization of a TiO_2–coated fiber optic cable reactor, *Environ. Sci. Technol.*, 30(9): 2806–2812, 1996.

Percherancier, J., Chapelon, B., and Pouyet, B., Semiconductor-sensitized photocatalysis of pesticides in water: the case of carbetamide, *J. Photochem. Photobiol. A*, 87, 261, 1995.

Richard, C. and Benagana, S., pH effect in the photocatalytic transformation of a phenyl-urea herbicide, *Chemosphere*, 33, 635, 1996.

Sakata, T. and Kawai, T., Photosynthesis and photocatalysis with semiconductor powders, in *Energy Resources through Photochemistry and Catalysis*, Gratzel, M., Ed., Academic Press, New York, 1983.

Sanchez, L., Peral, J., and Domenech, X., Photocatalyzed destruction of aniline in UV-illuminated aqueous TiO_2 suspensions, *Electroquimica Acta*, 42(12), 1877, 1997.

Scalfani, A., Palmisano, L., and Davi, E., Photocatalytic degradation of phenol by TiO_2 aqueous dispersions: rutile and anatase activity, *New J. Chem.*, 14, 265, 1990.

Schmelling, D. and Gray, K., Photocatalytic transformation and mineralization of 2,4,6-trinitrotoluene in TiO_2 slurries, *Water Res.*, 29, 2651, 1995.

Schmelling, D., Gray, K., and Kamat, P., Role of reduction in the photocatalytic degradation of TNT, *Environ. Sci. Technol.*, 30, 2547, 1996.

Serpone, N. and Pelizzetti, E., *Photolysis. Fundamentals and Applications*, John Wiley & Sons, New York, 1989.

Shaul, G., Dempsey, C., and Dostal, K., *Fate of Water Soluble Azo Dyes in the Activated Sludge Process*, EPA/600/S2-88/030, U.S. Environmental Protection Agency, Research and Development Office, Washington, D.C., 1988.

Somarani, C., Finiels, A., Graffin, P., Guida, A., Klaver, M., Olive, J., and Saaedan, A., Photocatalytic oxidation of substituted toluenes with irradiated TiO_2 semiconductor: effect of a zeolite, *Catalysis Lett.*, 33, 395, 1995.

Stafford, U., Gray, K., and Prashat, K., Photocatalytic degradation of organic contaminants: halophenols and related model compounds, *Heterogeneous Chem. Rev.*, 3, 77, 1996.

Stone, A., Torrents, A., Smolen, J., Vasudevan, D., and Hadley, J., Adsorption of organic compounds possessing ligand donor groups at the oxide/water interface, *Environ. Sci. and Technol.*, 27, 895, 1993.

Sturini, M., Fasani, E., Prandi, C., Casaschi, A., and Albini, A., Titanium dioxide-photocatalysed decomposition of some thiocarbamates in water, *J. Photochem. Photobiol. A: Chem.*, 101, 251, 1996.

Sun, Y. and Pignatello, J., Evidence for a surface dual hole-radical mechanism in the TiO_2 photocatalytic oxidation of 2,4-dichlorophenoxyacetic acid, *Environ. Sci. Technol.*, 29, 2065, 1995.

Taft, R., *Steric Effect in Organic Chemistry*, Wiley, New York, 1956.

Tanaka, K., Hisanga, T., and Harada, K., Efficient photocatalytic degradation of chloral hydrate in aqueous semiconductor suspensions, *J. Photochem. Photobiol. A. Chem.*, 48, 155, 1989.

Tang, W.Z. and An, H., UV/TiO_2 photocatalytic oxidation of commercial dyes in aqueous solutions, *Chemosphere*, 31(9), 4157, 1995a.

Tang, W.Z. and An, H., Photocatalytic degradation kinetics and mechanism of acid blue 40 by TiO_2/UV in aqueous solution, *Chemosphere*, 9, 4171, 1995b.

Tang, W.Z. and Chen, R., Decolorization kinetics and mechanisms of commercial dyes by H_2O_2/iron powder system, *Chemosphere*, 32, 947, 1996.

Tang, W.Z. and Huang, C., Photocatalyzed oxidation pathways of 2,4-dichlorophenol by CdS in basic and acidic aqueous solutions, *Water Res.*, 29, 745, 1995.

Tang, W.Z. and Hendrix, T., Development of QSAR models to predict kinetic rate constants for TiO_2/UV photocatalytic oxidation, Water and Wastewater Industry and Sustainable Development, in honor of Prof. Xu Baojiu's 80th anniversary, Tsinghua Press, Beijing, China, 1998, pp. 427–439.

Theurich, J., Linder, M., and Bahnemann, D., Photocatalytic degradation of 4-chlorophenol in aerated aqueous titanium dioxide suspensions: a kinetic and mechanistic study, *Langmuir*, 12, 6368, 1996.

Torimoto, T., Ito, S., Kuwabata, S., and Yoneyama, H., Effects of adsorbents used as supports for titanium dioxide loading on photocatalytic degradation of propyzamide, *Environ. Sci. Technol.*, 30, 1275, 1996.

Torrents, A. and Stone, A., Hydrolysis of phenyl picolinate at the mineral/water interface, *Environ. Sci. Technol.*, 25, 143, 1991.

Tunesi, S. and Anderson, M., Photocatalysis of 3,4-DCB in TiO_2 aqueous suspensions: effects of temperature and light intensity, *Chemosphere*, 16, 1447, 1987.

Tunesi, S. and Anderson, M., Influence of chemisorption on the photodecomposition of salicylic acid and related compounds using suspended TiO_2, ceramic membranes, *J. Phys. Chem.*, 95(8), 3399, 1991.

USDOC, *Selected Specific Rates of Reactions of Transients from Water in Aqueous Solutions*, U.S. Department of Commerce, Washington, D.C.,1977.

Vasudevan, D. and Stone, A., Adsorption of catechols, 2-aminophenols, and 1,2-phenyldiamines at the metal hydroxide/water interface: effect of ring substituents on the adsorption of TiO_2, *Environ. Sci. Technol.*, 30, 1604, 1996.

Venkatadri, R. and Peters, R., Chemical oxidation technologies: ultraviolet light/ hydrogen peroxide, Fenton's reagent, and titanium dioxide-assisted photocatalysis, *J. Hazardous Wastes Hazardous Mater.*, 10, 107, 1993.

Willner, I. and Willner, B.S., *Int. J. Hydrogen Energy*, 13, 593, 1988.

10

Supercritical Water Oxidation

10.1 Introduction

The destruction of concentrated toxic and hazardous organic wastes is a major problem facing the nation. For example, the U.S. Navy and Army generate about 10,000 tons/year of concentrated hazardous organic materials. The U.S. pulp and paper industry generates 2.8 million tons of dry sludge each year. Currently, incineration and supercritical water oxidation (SCWO) are the two primary options for complete destruction of concentrated toxic wastewater and organic sludge; however, incineration process might cause air pollution if incomplete combustion occurs.

SCWO process is one of the best physicochemical processes for destruction of concentrated toxic and hazardous organic wastes. Organic-laden wastes such as waste oil, solvent, cleaners, paint, sludge, pulp sludge, and municipal and refinery slurries can be treated economically with high destruction efficiencies due to the high content of thermal energy. After these wastes are treated, residuals including gases, liquids, and solids are usually nontoxic.

Remediation of contaminated soils is another application of the SCWO process. Bench-scale testing has shown that hazardous and toxic organics can be extracted from soil and simultaneously oxidized in an SCWO reactor. Soil slurries containing 1 to 20% of organic wastes can be continuously fed to an SCWO reactor system. The effluents are water, CO_2, and NO_2, and the soil is sterilized and free of organics. Concentrated organic wastes and dilute aqueous waste frequently present at the same site can be blended to provide a feed with the appropriate heating value in a concentration range of 1 to 20 wt%. Table 10.1 summarizes historical aspects of SCWO research and development.

10.2 Fundamental Theory

10.2.1 Characteristics of Supercritical Water

Water may exist in three phases: solid, liquid, and gas; however, at elevated temperatures and pressures above its critical point (374°C and 221 atm),

TABLE 10.1

Historical Aspects of SCWO

Author	Topics of SCWO Investigated
Wightman (1981)	Initial studies of phenol SCWO
Thomason and Modell (1984)	Involved in extensive SCWO study
Frank (1987)	Investigated the unique features of supercritical water in terms of density, dielectric constant, viscosity, diffusivity, electric conductance, and solvating ability
Modell (1989)	Treatment of hazardous organic compounds
Gloyna et al. (1990)	Application of SCWO to the decomposition of sludges
Gloyna (1992)	Found that sludge readily decomposes at near-critical water conditions with O_2 or H_2O_2 as an oxidant in a batch or continuous flow reactor
Shanableh and Gloyna (1991)	Treatment of sludges
Barner et al. (1992)	Involved in extensive study of SCWO processes
Modell et al. (1992)	Investigated SCWO process for pulp mill sludges
Thornton and Savage (1992)	Explored kinetics of SCWO of phenol
Rice et al. (1995)	Explored supercritical water reactor
Tester et al. (1993a,b)	Investigated the unique features of supercritical water in terms of density, dielectric constant, viscosity, diffusivity, electric conductance, and solvating ability
Tufano (1993)	Explored multistep kinetic model of phenol in SCWO
Gloyna et al. (1995)	Involved in extensive SCWO study of priority pollutants
Blaney et al. (1995)	Treatment of sludges
Savage et al. (1995)	Reviewed previous SCWO research with model pollutants and demonstrated that phenolic compounds are the model pollutants studied most extensively under SCWO conditions
Modell and Mayr (1995)	Studied supercritical water oxidation of aqueous waste
Gopalan and Savage (1994, 1995)	Explored reaction pathways in SCWO of phenol
Ding et al. (1998)	Studied catalytic oxidation in supercritical water
Kranjc and Levec (1997a,b)	Explored metal oxides as catalysts in SCWO
Goto et al. (1997)	Studied decomposition of municipal sludge by SCWO
Martino and Savage (1997)	Investigated the SCWO kinetics, products, and pathways for CH_3- and CHO-substituted phenols
Rice (1998)	Determined oxidation rates of common organic compounds in SCWO

water enters another phase and is referred to as *supercritical water*. A pressure/temperature diagram for pure water is shown in Figure 10.1. The critical point is the highest value for both temperature and pressure at which the equilibrium of water vapor/liquid can exist.

Supercritical water is neither a liquid nor a gas, but it has properties between the liquid and gas phases (i.e., density approaching its liquid phase and diffusivity and viscosity approaching its gas phase). At the critical point, hydrogen bonds disappear, and water becomes similar to a moderately polar solvent. Oxygen and almost all hydrocarbons become completely miscible

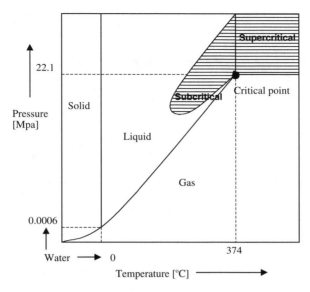

FIGURE 10.1
Pressure–temperature phase diagrams for water.

in water, mixing occurs almost instantaneously, and the solubility of inorganic salt drops to parts per million (ppm).

The properties of supercritical water are illustrated in Figure 10.2. As the temperature of water increases, it becomes less dense due to thermal expansion from 100 to 410°C. Beyond 410°C, density decreases with both temperature and pressure. The high dielectric constant of water, at ambient conditions, is due to strong hydrogen bonds that result in the high solubility of inorganic compounds. As the density decreases the dielectric constant of water decreases. At supercritical conditions, the dielectric constant of water dramatically drops from 80 to 2 due to the disappearing hydrogen bonds. At this point, the number of hydrogen bonds is about one third the number found under ambient conditions. The effect of reduced hydrogen bonding and the dielectric constant are responsible for the change in solubility of organic compounds in supercritical water.

The dielectric constant of supercritical water is in the range of 2 to 3. This range is similar to the range of nonpolar solvents such as hexane or heptane, which have dielectric constants of 1.8 and 1.9, respectively. When hazardous wastes are heated to high temperature and pressure, physical properties such as density, dielectric constant, viscosity, diffusivity, electric conductance, and solubility are optimum for destroying organic pollutants. Table 10.2 lists the characteristics of supercritical water, and Figure 10.3 illustrates the influence of temperature and pressure on the dielectric constants and density of water. As both temperature and pressure increase, the dielectric constants and density of water decrease dramatically.

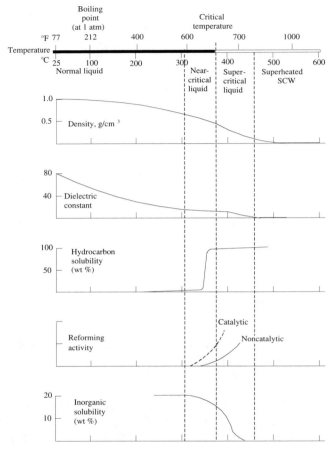

FIGURE 10.2
Properties of water at 250 atm. (From Modell, M., in *Standard Handbook of Hazardous Waste Treatment and Disposal*, Freeman, Ed., McGraw-Hill, New York, 1989, pp. 8.153–8.168. With permission.

TABLE 10.2

Characteristics of Supercritical Water

Property	Characteristics
Temperature (374–450°C)	Density of supercritical water decreases
Dielectric constant	Less than 80
Density	Higher diffusivity, rapid mass transfers
Low density and dielectric constant	Change the solvation characteristics of water
Solubility	High for organic compounds and low for inorganic salts
Miscibility (gases)	Complete
Viscosity	Decrease
H-bond	Decrease in numbers and force

Source: Data from Modell (1989), Shanableh and Gloyna (1991), and Blaney et al. (1995).

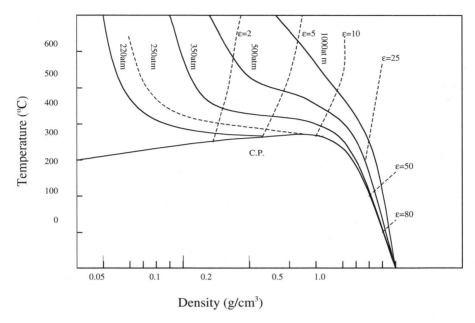

FIGURE 10.3
The temperature–pressure dependency of the dielectric constant of water.

Table 10.3 provides some examples of organic compounds and their respective dielectric constants. Many organic compounds become miscible in supercritical water because they behave almost as a nonaqueous fluid. For example, at 25°C, benzene is barely soluble in water (solubility, 0.07 wt%); however, at 260°C, the solubility is about 7 to 8 wt% and is fairly independent of pressure. At 287°C, the solubility is somewhat pressure dependent, with a maximum of solubility of 18 wt% at 20 to 25 MPa. In this pressure range and at 295°C, the solubility rises to 35 wt%. At 300°C, the critical point of

TABLE 10.3

Dielectric Constants of Organic Compounds

Organic Compound	Dielectric Constant
Propane	1.6
Hexane	1.8
Heptane	1.9
Carbon tetrachloride	2.2
Benzene	2.3
Acetone	20.7
Ethanol	24.5
Methanol	32.6

the benzene/water mixture is surpassed. When the mixture becomes super-critical, by definition, only a single phase exists. Thus, the components are miscible in all proportions.

Supercritical water has very low viscosity. For organic constituents, the diffusivities in supercritical water are higher in order of magnitude than under normal conditions. The low viscosity of supercritical water and higher diffusivity of organic compounds permit a much higher mass-transfer rate of organic contaminants in supercritical water. As a result, organic compounds are highly soluble in supercritical water.

The unique physical properties of supercritical water make it an ideal medium for the oxidation of organic compounds. When organic compounds and oxygen are dissolved in supercritical water, they are immediately brought into intimate molecular contact in a single homogeneous phase at high temperature. With no mass-transport limitations between the inter-phase, kinetic reactions in oxidizing organic pollutants proceed rapidly to form end products such as CO_2 and H_2O. Heteroatoms are converted to inorganic compounds, usually acids, salts, or oxides in high oxidation states that can be precipitated from the mixture along with other unwanted inorganic compounds that might be present in the feed. Phosphorus is converted to phosphate and sulfur to sulfate; nitrogen-containing compounds are oxidized to N_2 with some NO_2. Inorganic materials, if present in the wastewater, are precipitated due to the reduced solubility of salts under supercritical water conditions. The advantage of this process is the high efficiency (99.9999% or greater) of waste destruction; residual pollutants are well within the limits set by regulatory agencies.

10.3 SCWO Processes

Supercritical water oxidation technology has undergone extensive testing, both at the bench scale and at the pilot-plant scale, to determine its ability to treat a wide variety of waste compounds. A typical steady-state operating temperature for both the bench-scale and pilot-scale units is in the range of 600 to 640°C. SCWO process is capable of completely destroying undesirable organic constituents, converting the carbonaceous and nitrogenous compounds into non-noxious materials. In an enclosed environment, hydrothermal energy can be efficiently recovered. The application of SCWO to wastewater and sludge treatment provides a practical solution to treat high concentrated hazardous wastes. Organic material is dissolved in oxygen or hydrogen peroxide or in an air-rich environment, where conversion occurs rapidly due to the high temperature and pressure.

Salt formation, inorganic solubility, mass and heat transfer, transformation product identification, and effects of catalysts and additives are all important factors that affect operation of SCWO processes. In addition, process development has also focused on the materials of construction, reactor design, heat exchange and heat recovery, solid/liquid separation, gas/liquid separation, control systems, effluent handling, ash disposal, safety requirements, and process system integration. One method for reducing corrosion and plugging during SCWO process is to cool the walls of the reactor and to install a recirculation zone that sustains the reaction via cold-feed injection. This fundamental concept is to envelop the hot reaction zone with cool material, which protects the metal walls from the hot corrosive fluid. The buildup of sticky solids on the walls can also be reduced. A recirculation zone provides heat and mass transfer to the cold inflowing stream, thus heating it to ignition temperature and stabilizing the reaction zone.

Kriksunov et al. (1995b) studied corrosive SCWO environments under such conditions as extreme pH, high concentration of dissolved oxygen, ionic and inorganic species, and abnormally high temperature/pressure variations. Corrosive products in the effluent stream may cause two problems. First, the presence of metals such as chromium in the reactor effluent might affect the quality of the effluent and the ash. Second, excess corrosive products, usually metal oxides, are likely to plug up pressure-regulating devices (Gloyna and Li, 1993). A variety of corrosion-resistant alloys, corrosion rates, and corrosion by-products have been investigated.

The SCWO process is able to achieve destruction efficiencies for organic waste comparable with those attained by incineration technology, without the requirement of expensive dewatering equipment. The key to a successful SCWO process is a design that integrates various unit operations. Important design considerations include:

- Reactor residence times and associated temperatures
- System pressures and related temperatures
- Materials of construction for each unit operation
- Control and removal of solid, either from the supercritical fluid or the treated effluent
- Operations and maintenance of the facility, including safety, analytical support, and regulatory monitoring/disposal requirements

Because an SCWO process can quickly and completely oxidize a wide range of organic compounds in a reaction system, it can achieve a greater destruction efficiency for organic waste more economically than controlled incineration or activated carbon treatment. An SCWO process is also more efficient than wet oxidation, without the added expense of dewatering equipment. It also offers a viable solution for sludge disposal and reduced water usage. The scale of an SCWO application is also adaptable. Portable benchtop or trailer-mounted units are capable of treating from 1 gallon up to several thousands of gallons of toxic waste per day. Large stationary plants are capable of processing 10,000 to 100,000 gallons of toxic waste per day, with 8 to 10% organic contents.

10.3.1 Process Description of SCWO

A flow chart of a generic SCWO process is shown in Figure 10.4. It illustrates the feed stream of a typical aqueous waste. Oxidants such as air, oxygen, or hydrogen peroxide must be provided unless the waste itself is an oxidant. A supplemental fuel source should also be available for low-heat-content wastes. The streams entering the SCWO reactor must be heated and pressurized to supercritical conditions. Influent streams are frequently heated by thermal contact with the hot effluent. Both influent pressure and back pressure must be provided. The influent streams are then combined under supercritical conditions where oxidation occurs. Certain properties of supercritical water make it an excellent medium for oxidation. Acetic acid is generally considered one of the most refractory by-products of the SCWO process of industrial waste.

From the reactor, the effluent flows through a separator (for the separation of inorganic salts precipitated during the process), a heat exchanger (with the influent), a heat recovery unit, an effluent cooler, and a letdown system to return the effluent to the ambient pressure. Then, a liquid/gas separator sep-

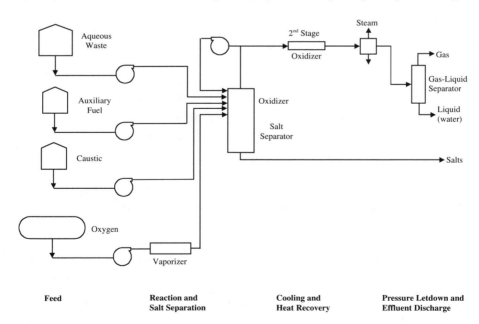

| Feed | Reaction and Salt Separation | Cooling and Heat Recovery | Pressure Letdown and Effluent Discharge |

FIGURE 10.4
SCWO flow chart (From USEPA, Supercritical water oxidation, *Eng. Bull.,* EPA/540/S-92/006, 1–7, 1992.)

arates the effluent into liquid (pure water) and gas phases (CO_2, O_2, N_2). The heat-recovery system utilizes the stream by-product to heat up the influent.

The SCWO process brings together water, organics, and oxygen at high temperatures above 400°C and high pressures of about 25 MPa. Under these conditions, a single, fluid-phase reaction system exists in which many of the inherent transport limitations of multiphase contacting are absent. The temperatures are high enough to induce spontaneous oxidation of the organics. The heat of the reaction raises the mixture temperature to levels as high as 600°C. At 550°C, organic materials are rapidly oxidized with destruction efficiency greater than 99.99% for reactor residence times of 1 min or less. Heteroatomic groups, such as chlorine, sulfur and phosphorus, are oxidized into their corresponding acids, which can be neutralized and precipitated as salts by adding a base to the feed. Table 10.4 shows the applicability of the SCWO process to treating various classes of hazardous wastes.

10.3.2   Effects of Operating Parameters of SCWO

Table 10.5 provides performance data regarding the SCWO process. Typical destruction efficiencies (DEs) for a number of compounds are also summarized in Table 10.5, which indicates that the DE could be affected by various parameters such as temperature, pressure, reaction time, oxidant type, and feed concentration. Feed concentrations can slightly increase the DE in supercritical oxidation processes. For SCWO, the oxidation rates appear to be first order and zero order with respect to the reactant and oxygen concentration, respectively. Depending upon reaction conditions and reactants involved, the rate of oxidation varies considerably. Pressure is another factor that can affect the oxidation rate in supercritical water. At a given temperature, pressure variations directly affect the properties of water, and in turn change the reactant concentrations. Furthermore, the properties of water are strong functions of temperature and pressure near its critical point.

TABLE 10.4

Applications of SCWO

Application	SCWO
Potentially treatable compounds	Liquid waste, sludge, and slurred solid wastes
	Halogenated and nonhalogenated aliphatic and aromatic hydrocarbons, aldehydes; ketones; esters; carbohydrates; organic nitrogen compounds; polychlorinated biphenyl (PCBs), phenols, and benzene; aliphatic and aromatic alcohol; pathogens and viruses; mercaptans, sulfides, and other sulfur-containing compounds; dioxins and furans; leachable metals; and propellant components
Treatability	Municipal and industrial sludges (a total organic carbon destruction efficiency of 99.3%)

Source: Adapted from USEPA, Supercritical water oxidation, *Eng. Bull.,* EPA/540/S-92/006, 1–7, 1992.

TABLE 10.5

Hazardous Waste Compounds Results

Reactant	Reaction Temperature (°C)	Time (min)	Concentration In (mg/L)	Concentration Out (mg/L)	Reaction Pressure (psi)	Destruction Efficiency (%)
2-Nitrophenol *	515	10	104	10	6500	90
2-Nitrophenol**	530	15	104	<1	6300	>99
2,4-Dimethylphenol**	580	10	135	<1	6500	>99
Phenol*	490	1.5	1100	60	6100	95
Phenol**	490	1	1650	130	5700	92
Phenol*	535	10	150	<1	6100	>99
2,4-Dinitrotoluene*	385	5	84	18	4900	79
2,4-Dinitrotoluene	410	3	84	14	6500	83
2,4-Dinitrotoluene	460	10	180	3	5300	98
2,4-Dinitrotoluene	528	3	180	<1	4200	>99

Note: Compounds followed by (*) were treated with pure oxygen, and compounds followed by (**) were treated with hydrogen peroxide and pure oxygen.
Source: Lee, D.S. and Gloyna, E.F., *Water Poll. Control Fed. Specialty Conf.*, 1988.

10.3.2.1   Reaction Time

Reaction time is the standard parameter used to measure the efficiency of the reaction at a given initial concentration. Figure 10.5 illustrates the change of total organic carbon (TOC) as a function of the reaction time. In the figure, the TOC decreases as reaction time increases. Goto et al. (1997) used a batch reactor at a temperature of 673 K and a pressure of 30 MPa, with 100 and 200% hydrogen peroxide (oxidant) in excess of stoichiometric oxygen demand.

10.3.2.2   Oxidants

Oxidation using hydrogen peroxide has different reaction mechanisms than oxidation using oxygen or air. Lin et al. (1998) showed that the use of hydrogen peroxide as an oxidant in SCWO systems produces a DE that is significantly higher than those obtained from air or oxygen. In addition, aqueous hydrogen peroxide is easier to pump, requires a less expensive feed system, and may be combined with the influent more readily than oxygen. However, other factors may influence the choice between hydrogen peroxide and oxygen as an oxidant; for example, hydrogen peroxide is significantly more expensive than oxygen. Table 10.6 illustrates that the DE is proportional to both temperature and residence time. Also, the DE slightly increases with operating pressure. Recent studies also indicate that the addition of catalysts such as potassium permanganate, manganese sulfate, copper, and iron can enhance the DE.

Figure 10.6 shows the effect of 100% hydrogen peroxide and 200% hydrogen peroxide on TOC in the liquid phase. Temperatures range from 658 to

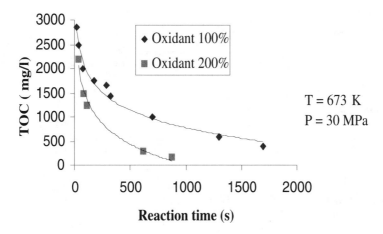

FIGURE 10.5
Effect of reaction time on TOC in liquid-phase product.

873 K, with an estimated pressure of 30 MPa. TOC in the liquid phase dramatically decreases with an increase in the amount of hydrogen peroxide. TOC also decreases with increasing temperature.

10.3.2.3 Temperature

The TOC in the liquid phase is plotted as a function of the reaction temperature in Figure 10.7. The amounts of hydrogen peroxide that are used to break TOC in the liquid are 0, 50, and 70%. TOC is almost zero when the oxidant is saturated. TOC in the liquid phase decreases with increasing temperature. At insufficient amounts of oxidant, TOC decreases with increasing temperature. When oxidant is insufficient, hydrolysis and thermolysis become more important than oxidation for the reduction of TOC.

10.3.2.4 Pressure

Figure 10.8 shows the effect of the reaction pressure on TOC in the liquid phase. The reaction pressure was estimated from the water density in the reactor. The water density was varied to give corresponding pressures from 22 to 35 MPa. In the supercritical pressure region, lower pressure favors a reduction of the TOC in the liquid phase.

10.3.2.5 Catalysts

Catalytic oxidation has been used in many wastewater treatment processes. Catalysts are now being applied to enhance supercritical water oxidation operations (Ding et al., 1998; Krajnc and Levec, 1997a). Hazardous organic pollutants can be destroyed by supercritical water oxidation at temperatures around 500°C and reactor residence times of less than 1 min, with the

TABLE 10.6

SCWO Performance Data

Pollutant	Temperature (°C)	Pressure	Destruction Efficiency (DE)	Reaction Time		Feed Concentration
1,1,1-Trichloroethane	495	—	99.99	4	Oxygen	—
1,1,2,2-trichloroethylene	495	—	99.99	4	Oxygen	—
1,2-Ethylene dichloride	495	—	99.99	4	Oxygen	—
2,4-Dichlorophenol	400	—	33.7	2	Oxygen	2000
2,4-Dichlorophenol	400	—	99.440	1	H_2O_2	2000
2,4-Dichlorophenol	450	—	63.3	2	Oxygen	2000
2,4-Dichlorophenol	450	—	99.950	1	H_2O_2	2000
2,4-Dichlorophenol	500	—	78.2	2	Oxygen	2000
2,4-Dichlorophenol	500	—	>99.995	1	H_2O_2	2000
2,4-Dimethylphenol	580	443	>99	10	H_2O_2 + O_2	135
2,4-Dinitrotoluene	410	443	83	3	Oxygen	84
2,4-Dinitrotoluene	528	287	>99	2	Oxygen	180
2-Nitrophenol	515	443	99	10	Oxygen	104
2-Nitrophenol	530	430	>99	15	Oxygen	104
Acetic acid	400	—	3.10	5	Oxygen	2000
Acetic acid	400	—	61.8	5	H_2O_2	2000
Acetic acid	450	—	34.3	5	Oxygen	2000
Acetic acid	450	—	92.0	5	H_2O_2	2000
Acetic acid	500	—	47.4	5	Oxygen	2000
Acetic acid	500	—	90.9	5	H_2O_2	2000
Activated sludge (COD)	400	272	90.1	2	—	62,000
Activated sludge (COD)	400	306	94.1	15	—	62,000
Ammonium perchlorate	500	374	99.85	15	None	12,000
Biphenyl	450	—	99.97	7	Oxygen	—
Cyclohexane	445	—	99.97	7	Oxygen	—
DDT	505	—	99.997	4	Oxygen	—
Dextrose	440	—	99.6	7	Oxygen	—
Industrial sludge (TCOD)	425	—	>99.8	20	Oxygen	—
Methyl ethyl ketone	505	—	99.993	4	Oxygen	—
Nitromethane	400	374	84	3	None	10,000
Nitromethane	500	374	>99	0.5	None	10,000
Nitromethane	580	374	>99	0.2	None	10,000
o-Chlorotoluene	495	—	99.99	4	Oxygen	—
o-Xylene	495	—	99.93	4	Oxygen	—
PCB 1234	510	—	99.99	4	Oxygen	—
PCB 1254	510	—	99.99	4	Oxygen	—
Phenol	490	389	92	1	Oxygen	1650
Phenol	535	416	99	10	Oxygen	150

Source: Adapted from USEPA, Supercritical water oxidation, *Eng. Bull.*, EPA/540/S-92/006, 1–7, 1992.

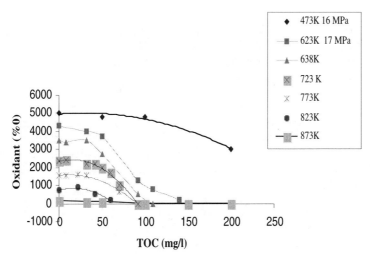

FIGURE 10.6
Effects of reaction temperature and amounts of oxidant on TOC in liquid-phase product.

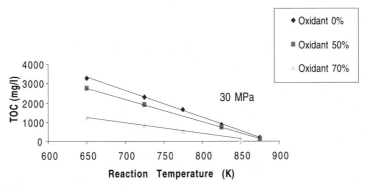

FIGURE 10.7
Relations of TOC and reaction temperature.

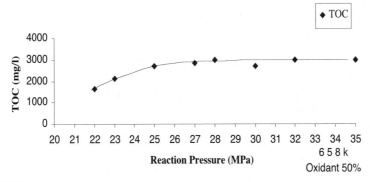

FIGURE 10.8
Effect of reaction pressure on TOC in liquid-phase product.

formation of refractory organic intermediates such as acetic acid and ammonia. In addition, SCWO of aromatic and substituted aromatic compounds forms dimers and other condensation products. At higher process temperature (higher than 600°C) and longer reaction residence times, these refractory compounds and condensation products will be further oxidized to end products, such as water, carbon dioxide, molecular nitrogen, and mineral acids.

The major incentive for using catalysts in SCWO is the reduction in energy and processing costs. The addition of catalysts such as potassium permanganate, manganous sulfate, copper, and iron can enhance destruction efficiency of organic pollutants. Different catalysts have been developed to operate at lower temperatures and pressures than thermal processes such as incineration. Fe^{2+} and Cu^{2+} are good catalysts, but the dissolved ions need to be separated at the end of process. Manganese–cerium and cobalt–bismuth are active for oxidation of many lower carboxylic acids or polyethyleneglycol. Copper–zinc is effective in the oxidation of substituted phenols and of p-coumaric acid, and RU/CeO_2 has a much higher activity than copper salts for formaldehyde or formic acid oxidation (Beziat et al., 1997).

10.4 Degradation of Hazardous Wastes in SCWO

The goal of SCWO kinetics research is to establish detailed elementary mechanisms for the oxidation of rate-limiting compounds such as carbon monoxide and methane. Simplified global reaction expressions are also needed in the design of commercial SCWO processes. To evaluate and classify the kinetic constants of organic degradation by SCWO, the destruction efficiencies and kinetic parameters of toxic wastewater, phenolic compounds, hazardous waste, some organic substances, and sludge resulting from batch and continuous flow reactors operating under a variety of environmental conditions will be discussed. The DEs of specific compounds and sludge will be presented in terms of the percentage reduction of total organic carbon (TOC) and chemical oxygen demands (COD). Supercritical water conditions can be used for the efficient destruction of sludge from biological treatment plants, 2-nitrophenol, 2,4-dimethylphenol, and 2,4-dinitrotoluene with short residence time.

Early experiments were primarily concerned with establishing destruction efficiencies (percent of parent material destroyed in a given residence time) to identify compounds that are appropriate for treatment with the SCWO technology (Modell, 1989). Since then, researchers have extensively studied using plug flow reactors for a variety of simple organic and inorganic compounds including carbon monoxide (Helling and Tester, 1987, 1994), methane (Webley and Tester, 1991), methanol (Webley and Tester, 1989; Webley et al., 1991; Tester et al. 1993b), and ammonia (Helling and Tester, 1988; Webley et al., 1991). Thornton and Savage (1992a,b) examined the kinetics of phenol at near-critical and supercritical conditions in a flow reactor. Wightman

(1981) performed flow reactor experiments and extracted global rate expressions for acetic acid and phenol. Lee and Gloyna (1988) also examined the kinetics of acetic acid and phenol in a batch reactor.

At Sandia National Laboratories, experiments in an SCWO flow reactor provided data on a number of organics, including methanol, phenol, and other industrial chemicals, as well as military munitions (Rice, 1994). Commercial SCWO processes are designed to operate at temperatures typically less than 700°C. The development of SCWO technology depends on understanding the reaction kinetics of a wide variety of compounds at SCWO conditions. Predictive chemistry models, as they become available, will play an important role in finding answers to such design problems as:

- Predicting reaction rate dependency on temperature, pressure, and species concentrations
- Calculating heat release rate and temperature during reaction
- Predicting reaction completeness and by-product profiles
- Estimating catalysis effects
- Scaling laboratory mechanisms for the oxidation of rate limiting to commercial scale reactors

Regardless how different are the reaction parameters such as pressure, temperature, organic wastes, oxygen (Lin et al., 1998), the global chemical reactions that occur during SCWO can be simplified as follows.

$$\text{Organic wastes} + O_2 \rightarrow CO_2 + H_2O \tag{10.1}$$

$$\text{N–organic wastes} + O_2 \rightarrow N_2 \tag{10.2}$$

$$\text{Cl-organic or S-organic wastes} + O_2 \rightarrow SO_4^{-2} \text{ or } Cl^- + CO_2 + H_2O \tag{10.3}$$

There are three general trends in SCWO kinetic data. First, in most cases, the oxidation rate is independent of or weakly dependent on the concentration of oxidants. As a general rule, the oxidant concentration is at least 200% in excess of stoichiometric oxygen demand. Second, the pseudo first-order kinetics with respect to the concentration of starting compounds is usually a reasonable assumption for SCWO. Third, the activation energy ranges from about 30 to 480 kJ/mol.

Generalized kinetic models, as based on simplified reaction schemes involving the formation and destruction of rate-controlling intermediates, have been developed by Li et al. (1993a,b). The rate-controlling intermediates are three common types of organic compounds: hydrocarbons and oxygenated hydrocarbons; nitrogen-containing organics; and chlorinated organics. The global oxidation rate depends on the final product formation rate as well as the formation and destruction rates of the more stable intermediates.

Activation energies of intermediate compounds such as acetic acid, methanol, and other low-molecular-weight compounds (170 to 350 kJ/mol) are greater than high-molecular-weight compounds (20 to 100 kJ/mol).

Studies on reaction mechanisms and by-product analyses have indicated that short-chain carboxylic acids, ketones, aldehydes, and alcohols are the major oxidation intermediates under wet air-oxidation conditions. Kinetic studies have been conducted on a number of refractory compounds, including acetic acid, methanol, ammonia, and carbon monoxide. For nitrogen-containing organic compounds, nitrogen gas is the predominant SCWO end product, regardless of the oxidation state of nitrogen in the initial material. Ammonia might also be formed due to incomplete oxidation

The reaction scheme for SCWO of organic compounds can be further simplified as shown in Figure 10.9, where C represents the oxidation end products, B is the rate-controlling intermediate, and A represents the initial and intermediate organic compounds other than B. Application of this generalized reaction scheme requires the definition of the groups. The concentration of Groups A and B may be expressed in terms of TOC, COD, total oxygen demand (TOD), or biological oxygen demand (BOD). The concentration units for these measurements are expressed in mass per unit volume, usually mg/L.

By solving the rate equations of A and B, the following simplified expression can be obtained:

$$\frac{[A+B]}{[A+B]_0} = \frac{k_2}{k_1+k_2-k_3}e^{-k_3 t} + \frac{(k_1-k_3)}{k_1+k_2-k_3}e^{-(k_1+k_3)t} \qquad (10.4)$$

where $[A+B] = [A] = [B]$; $[A+B]_0 = [A]_0 + [B]_0$; and $[\ ]_0$ and $[\ ]$ are the reactant concentrations at time = 0 and t, respectively. The three rate constants (k_1, k_2, and k_3) require further description. If k_2 is much smaller than k_1, the organic compounds in the waste stream may be oxidized more easily to the end products. If k_2 becomes larger, more acetic acid will be formed. Values of k_1 may be determined from the initial reaction rate data based on the lumped

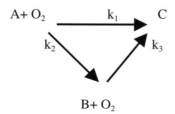

FIGURE 10.9
Simplified reaction scheme for SCWO of organic compounds.

parameters, such as COD and TOC, because group A is a collective term encompassing the initial compounds as well as unstable intermediates. Conversion of group A to acetic acid does not change the effluent COD or TOC concentration significantly; therefore, the kinetic parameters based on COD or TOC are directly related to k_1. The existing kinetic parameters for acetic acid can be substituted for k_3. Values of k_2 may be derived from further manipulation of Equation (10.4).

In Equation (10.5), α_1 is defined as the formation rate of acetic acid having the oxidation end product carbon monoxide. If first-order reactions are assumed, Equation (10.5) may be used to define α_1 as follows:

$$\alpha_1 = \frac{r_2}{r_1} = \frac{k_2[A]}{k_2[A]} = \frac{k_2}{k_2} = \frac{k_1^0}{k_1^0} e^{-\frac{(E_{a2} - E_{a1})}{RT}} \tag{10.5}$$

The global reaction rate depends on activation energy levels for the three-step reactions. Because acetic acid is a refractory intermediate, E_{a3} is greater than either E_{a1} or E_{a2}. For most organic wastes, E_{a2} should approximate E_{a1} because groups B and C represent series degradation products derived from similar reactions. For this reason, α_1 should be a weak function of temperature and should merely become a ratio of the frequency factors of the two parallel reactions. The value of α_1 for a feed stream matching the above assumptions falls in between 0 and 1. For example, assuming that a wastewater contains a high concentration of short-chain organic compounds, including those with two-carbon atoms (other than acetic acid), the value of α_1 may be large. The values of α_1 generally can be used to characterize the "strength" of the waste stream in a SCWO process.

Martino and Savege (1997) provided SCWO global kinetics of cresol and hydrobenzaldehydes. Li et al. (1993a) analyzed water oxidation reaction mechanisms and evaluated existing water oxidation kinetic models encompassing a range of temperatures and pressures from 150 to 550°C and from 20 bar to 440 bar to propose a simplified water oxidation reaction scheme and to develop a generalized kinetic model.

10.4.1 Carbon Monoxide

Helling and Tester (1987) reported the oxidation kinetics of carbon monoxide over the temperature range 420 to 570°C at a pressure of 246 bar (24.6 MPa). Holgate and Tester (1994) examined oxidation kinetics of carbon monoxide. In addition to direct oxidation with oxygen, Helling and Tester found that the reaction of carbon monoxide with water was significant, as the following equations show:

$$\text{Direct oxidation } CO + 1/2O_2 \rightarrow CO_2 \tag{10.6}$$

$$\text{Dissolved } CO + 1/2O_2 \rightarrow CO_2 \tag{10.7}$$

The effects of temperature and concentration on direct and indirect oxidation kinetics of carbon monoxide were correlated with global models. In 1986, Helling and Tester found the oxidation of carbon monoxide to be globally first order in carbon monoxide and independent of oxygen concentration over the range investigated.

10.4.2   Aliphatic Organic Compounds

Helling and Tester (1987) determined Arrhenius parameters for ethanol oxidation over the temperature range 480 to 540°C, assuming the reaction was first order in ethanol and zero order in oxygen. The major products of the reaction were carbon monoxide, carbon dioxide, and acetaldehyde. The reaction exhibited apparent activation energy of 340 kJ/mol. The oxidation kinetic of methane over the temperature range 560 to 650°C was determined by Webley and Tester (1991) at a pressure of 246 bar. The major products of the oxidation were carbon monoxide and carbon dioxide. No methanol was detected in the effluent. The oxidation was first order in methane and 0.66 order in oxygen over the range of concentrations investigated, and the activation energy was 179.1 kJ/mol. The oxidation kinetics of methanol over the temperature range 450 to 550°C was determined at a pressure of 246 bar. The oxidation was found to be highly activated, with apparent activation energy of 408.8 kJ/mol. The products of oxidation were carbon monoxide, carbon dioxide, and hydrogen. The oxidation was found to be approximately first order in methanol and zero order in oxygen over the concentration range investigated.

10.4.3   Methane and Methanol

The reaction pathways from Helling and Tester (1988) and Tester et al. (1993a,b) for methane, methanol, and hydrogen are summarized in Figure 10.10. The reaction network shown in the figure demonstrate several important points. First, the mechanisms of methane and methanol destruction are intimately linked and would be sub-mechanisms for the oxidation of larger hydrocarbons. The final steps of destruction are common to all hydrocarbons, with the last step being the oxidation of carbon monoxide. Second, hydroxyl (•OH) radical plays an extremely important role in the oxidation process, as it is the oxidant in many of the destruction steps. Molecular oxygen, on the other hand, plays only a limited role. Third, the mechanism for hydrogen oxidation demonstrates how oxygen is converted in a sequence of steps to the "macroscopic" oxidant OH. It is this sequence of steps that apparently masks the role of oxygen in the overall oxidation process and leads to the experimental observation of an oxidation reaction whose rate is independent of oxygen concentration.

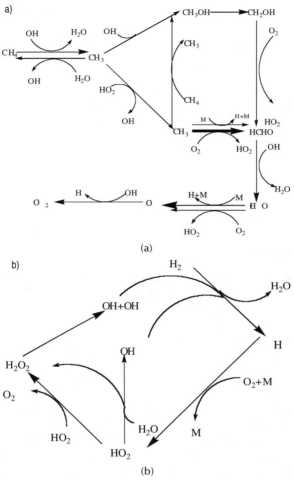

FIGURE 10.10

Elementary reaction pathway in SCWO (methane, methanol, and hydrogen). (From Tester, J.W. et al., *Indust. Eng. Chem. Res.*, 32(1), 236–139, 1993. With permission.)

10.4.4   Cresol

Figure 10.11 offers the general reaction network for the SCWO of cresol. This network shows three parallel paths for the oxidation of cresol by SCW. Three reaction intermediates are a hydroxybenzaldehyde via oxidation of the methyl substitute; ring-opening products; and phenol via demethylation. The end products are CO_2 and H_2O. The relative importance of the parallel pathways depends on the specific cresol isomer being oxidized. Figure 10.11 shows that phenol and hydroxybenzaldehydes are key organic intermediates in the reaction network, so the reaction network should also include the reaction paths for these compounds. Two parallel primary paths produce

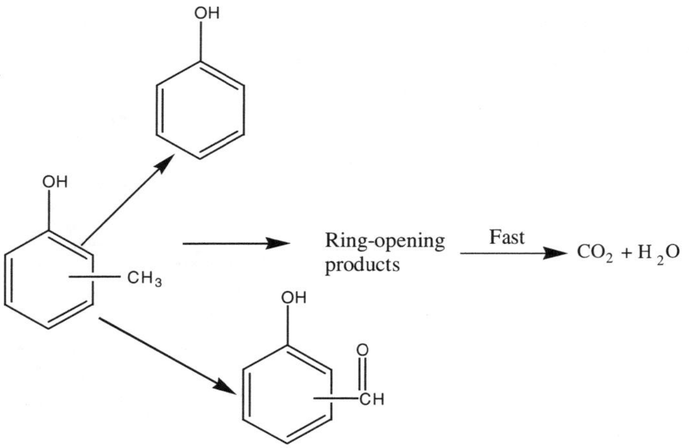

FIGURE 10.11
Cresol SCWO reaction network.

either phenol dimers or ring-opening products. Both the dimers and the ring-opening products are ultimately oxidized to CO_2 and H_2O.

10.4.5 Hydroxybenzaldehydes

CO, CO_2, and phenol were the major products resulting from the SCWO of hydroxybenzaldehydes. In addition, some trace amounts of the by-products such as xanthone and fluororenone are produced during its oxidation. The reaction network for SCWO of hydroxybenzaldehyde has two pathways: one leads to phenol and the other leads to ring-opening reactions and ultimately CO_2. The reaction network for cresols resulted from combining the reaction network for hydroxybenzaldehydes and the reaction network for phenol (Gopalan and Savage, 1995) as shown in Figure 10.12. This complete reaction network shows how cresols and important intermediate by-products are converted to CO_2.

10.4.6 Phenol

Phenolic compounds are the model pollutants studied most extensively under SCWO conditions. Equation (10.8) and Equation (10.9) describe the formation of free radicals in the absence of initiators. They are formed by the reaction of oxygen with the weakest C–H bonds of the organic compound:

$$RH + O_2 \rightarrow R^{\bullet} + HO_2 \tag{10.8}$$

$$RH + HO_2^{\bullet} \rightarrow R^{\bullet} + H_2O_2 \tag{10.9}$$

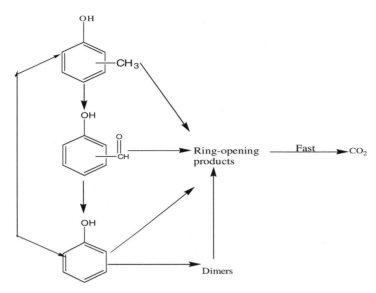

FIGURE 10.12
Combined reaction network for the SCWO of cresol.

where R is the organic functional group. In wet air oxidation of phenol, the reaction of oxygen with the O–H bond may be involved in the first reactions as shown in Equation (10.10) and Equation (10.11):

$$PhOH + O_2 \rightarrow PhO^\bullet + HO_2^\bullet \qquad (10.10)$$

$$PhOH + HO_2^\bullet \rightarrow PhO^\bullet + H_2O_2 \qquad (10.11)$$

The hydrogen peroxide formed in the two preceding equations will decompose to generate hydroxyl radicals:

$$H_2O_2 + M \rightarrow 2HO^\bullet \qquad (10.12)$$

where M can be either a homogeneous or heterogeneous catalyst.

The oxidation of organic compounds by hydroxyl radicals follows the same reaction through hydrogen abstraction or electrophilic addition, depending upon the nature of the organic compound:

$$RH + HO^\bullet \rightarrow R^\bullet + H_2O \qquad (10.13)$$

In the presence of oxygen, the organic radical (R) reacts with oxygen to form an organic peroxy radical ($ROO^\bullet$):

$$R^{\bullet} + O_2 \rightarrow ROO^{\bullet} \qquad (10.14)$$

The organic peroxy radical abstracts a hydrogen atom from the organic compound, producing an organic hydroperoxide (ROOH) and another organic radical:

$$ROO^{\bullet} + RH \rightarrow ROOH + R^{\bullet} \qquad (10.15)$$

The organic hydroperoxides formed are relatively unstable, and the decomposition of intermediates often leads to molecular breakdown and formation of subsequent intermediates with lower carbon numbers. These reactions continue rapidly until the formation of acetic or formic acid. Both acids will be converted into carbon dioxide and water as final products (Gloyna and Li, 1993).

The kinetics of phenol oxidation in near-critical and supercritical water has also been investigated by Thornton and Savage (1992a), Gopalan and Savage (1995), Rice et al. (1995), Kranjc and Levec (1996), and Tufano (1993).

Figure 10.13 shows parallel–consecutive oxidation of phenol in supercritical conditions. During the conversion of the compound, at higher residence time (>100 s), more intermediates were found in the form of highly oxidized compounds. Also, it may be assumed that phenol mainly disappears via phenoxy-phenols. According to Thornton and Savage (1992), 4-phenoxy-phenol oxidizes about three orders of magnitude faster than 2-phenoxy-phenol. The main phenol oxidation takes place via the 4-phenoxy-phenol route. Thornton and Savage (1992) also reported a relatively low rate of 2-2'-biphenol formation, which is much more difficult to oxidize than phenol.

Phenol is commonly present in industrial streams and is classified as a priority pollutant. At temperatures of 380 to 440°C and pressures of 190 to 270 atm, oxidation rates were calculated from kinetic Equation 10.16 by Minok et al. (1997). Their results showed that, under the designed system conditions, the rate of phenol destruction was dependent only on temperature, concentration of water, oxygen, and phenol but not on pressure. Water acts in the system as a reactant and was considered to be a reactive radical producer. The destruction rate of phenol can be expressed as follows:

$$d[phenol]/dt = k[phenol][H_2O]^{1.38} \{\chi/(2.89 + \chi)\}$$

$$k = 10^{4.95} \exp(-23.8/RT) M^{-1.38} \ (s^{-1}) \qquad (10.16)$$

where χ is the ratio of oxygen to phenol defined as $[O_2]/([phenol])$; k is the rate constant; R is the universal gas constant; and T is the absolute temperature.

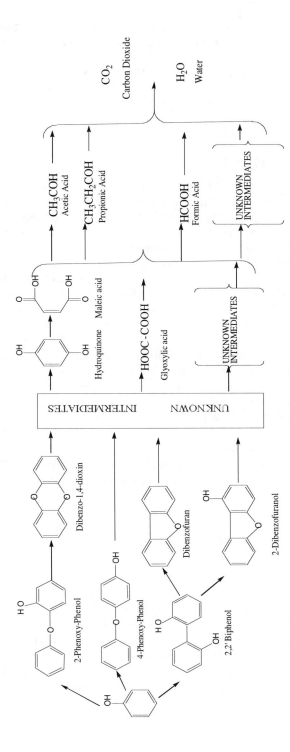

FIGURE 10.13
Proposed reaction pathway of phenol in supercritical water. (From Krajnc, M. and Levec, J., *AIChE J.*, 42(7), 1977–1984, 1996. With permission.)

When the oxygen concentration is equal to 300% in excess of stoichiometric oxygen demand, the global rate of phenol destruction can be expressed as:

$$d[phenol]/dt = k[phenol][H_2O]^{1.38}[H_2]^{0.42} \qquad (10.17)$$

$$k = 10^{2.34}\exp(-12.4/RT) \qquad (10.18)$$

10.4.7 Substituted Phenols

In general, the oxidation of CH_3 and CHO-substituted phenols in supercritical water proceeds more rapidly than the oxidation of phenol itself; also, *ortho*-substituted phenols are the most reactive and *meta*-substituted phenols are the least reactive. Moreover, for a given point of substitution, CHO-substituted phenols are more reactive than CH_3-substituted phenols.

The oxidation of cresol in supercritical water proceeds through three parallel paths: demethylation (to form phenol); oxidation of the methyl group (to form hydroxybenzaldehyde); and ring opening. The oxidation of hydroxybenzaldehydes in supercritical water proceeds through two parallel paths, with one path leading to phenol and the other involving ring opening and CO_2 formation

Table 10.7 illustrates the results of Gloyna and Li (1993) for treatment of hazardous wastes by SCWO. Pressure is given in psi, where 1 bar = 14.50 psi. The following equations describe global reaction rates for the SCWO of six different substituted phenols. The rates of phenolic compound expressions were obtained at 460°C and 250 atm with [organic] = 100 μmol/L and $[O_2]$ = 7 mmol/L. This rate was calculated as pseudo first order for SCWO of each phenolic compound.

These rates were calculated using Equation (10.16) to Equation (10.18) and divided by the reactant concentration. The global equations are:

$$r_{oHB} = 10^{-0.78\pm1.24}[_oHB]^{0.47\pm0.11}[O_2]^{0.57\pm0.53} \qquad (10.19)$$

$$r_{mHB} = 10^{-0.31\pm1.23}[_mHB]^{0.98\pm0.13}[O_2]^{-0.33\pm0.38} \qquad (10.20)$$

$$r_{pHB} = 10^{-1.81\pm0.62}[_pHB]^{0.77\pm0.11}[O_2]^{-0.02\pm0.21} \qquad (10.21)$$

Table 10.8 presents the kinetic rates that have been studied by Martino and Savage (1997). The rate constants for compounds followed by a superscript a ([a]) indicates the degradation rate via both thermal and oxidative reactions.

For compounds followed by a superscript b ([b]) in Table 10.8, the rate constants describe the degradation rates due to oxidative reaction only. These rates were also calculated from Equation (10.16) to Equation (10.18) and were divided by the reactant concentration:

$$r_{oHB} = 10^{-1.17}[_oHB]^{0.86} + 10^{-0.98}[_oHB]^{0.39}[O_2]^{0.68} \qquad (10.22)$$

$$r_{mHB} = 10^{-3.57}[_mHB]^{0.52} + 10^{-0.99}[_mHB]^{1.37}[O_2]^{0.28} \qquad (10.23)$$

$$r_{pHB} = 10^{-2.83}[_pHB]^{0..80} + 10^{-2.01}[_pHB]^{0.72}[O_2]^{0.02} \qquad (10.24)$$

Table 10.9 lists kinetic parameters for various organic compounds and wastes. In addition to the kinetic parameters listed, the experimental conditions were also listed. These conditions are from data that were generated in pilot and bench-test studies. The kinetics are defined by Equation (10.25) and Equation (10.26).

$$-d[C]/dt = k[C]^m[O]^n \qquad (10.25)$$

$$k = k'\exp{(-E_a/RT)} \qquad (10.26)$$

where:

[C] is the concentrations of organic reactants.

[O] is the concentrations of organic oxidant.

E_a is the activation energy (kJ/mol).

T is the temperature (degrees K).

R is the constant of gases (8.314 J/mol-K).

k' is the pre-exponential factor (first order) (1/s).

The concentrations of compounds labeled with "COD" were quantified by the chemical oxygen demand method. The concentrations of other compounds were quantified by chromatographic techniques. Excess oxidants were used in all tests. The SCWO system used either oxygen or air as oxidants, and has the same DE results. If liquid hydrogen peroxide had been used as an oxidant, it would have produced an apparent variation in the DE. The oxidation rates would be significantly higher than those obtained from the use of air or oxygen. The kinetic models presented in Table 10.10 are derived from a simplified reaction scheme involving the formation and destruction of rate-controlling intermediates.

10.4.8 Mixture of Organic Pollutants

Table 10.10 summarizes the kinetic data on SCWO of organic compounds, organic mixtures, ammonia, and carbon monoxide. M and n are the constants in Equation (10.25). These kinetic data can fit the pseudo first-order reaction models proposed by Wightman (1981). The activation energy from

TABLE 10.7

Global Kinetic Models for Supercritical Water Oxidation of Organic Substances

Compound	Oxidant	Reactor Type	Kinetic Parameters *				Temperature (K)	Pressure (atm)	$[C_A]_0$ (g/L)
			k'	E_a	m	n			
Acetamide	H_2O_2	Flow	2.75×10^5	88.3	1.15	0.05	673–803	240–350	1.5–4.0
Acetamide**	H_2O_2	Flow	5.01×10^4	94.7	1	0.17	673–803	240–350	1.5–4.0
Acetic acid	H_2O_2	Flow	2.63×10^{10}	167.1	1	0	673–803	240–350	1.3–3.3
Acetic acid	H_2O_2	Flow	9.23×10^7	131	1	0	673–773	240–350	1.0–5.0
Acetic acid	O_2	Flow	9.82×10^{17}	231	1	1	611–718	394–438	0.525
Acetic acid	O_2	Flow	2.55×10^{11}	172.7	1	0	611–718	394–438	0.525
Activated sludge (COD)	O_2	Batch	$\sim 1.5 \times 10^2$	~54	1	0	573–723	240–350	46.5
Ammonia	O_2	Flow	3.16×10^6	157	1	0	913–973	246	0.03–0.11
2-Butanone	O_2	Batch	1.20×10^1	36.2	1	0	673–773	240–400	~6
Carbon monoxide	O_2	Flow	3.16×10^6	112	1	0	673–814	246	0.02–0.11
Carbon monoxide**	O_2	Flow	3.16×10^8	134	0.96	0.34	693–884	246	0.01–0.098
o-Cresol	O_2	Batch	3.16×10^0	28.5	1	0	673–773	240–400	~10
Digested sludge (COD)	O_2	Batch	4.36×10^3	20.4	1.86	0	573–723	240–350	46.5
2,4-Dichlorophenol	O_2	Flow	1.92×10^4	71.9	1	0.38	683–788	276	0.4–0.8
Ethanol	O_2	Flow	6.46×10^{21}	340	1	0	755–814	241	0.03–0.036
Formic acid	O_2	Flow	—	~96	1	1	683–691	408–432	1.0
Glucose (TOC)	—	Batch	—	130	0.5	1	653–683	~400	~1.0
Methane	O_2	Flow	1.26×10^7	156.8	1	0	913–973	245	—

			k'	E_a	m	n			
Methane	O_2	Flow	2.51×10^{11}	178.9	0.99	0.66	833–903	245	—
Methane	O_2	Flow	2.04×10^{7}	41.7	1	0	833–903	245	—
Methanol	O_2	Flow	2.51×10^{24}	395.0	1	0	723–823	243	—
Methanol	O_2	Flow	3.16×10^{26}	408.4	1.1	−0.02	723–823	243	0.038–0.17
Phenol	O_2	Flow	2.61×10^{5}	63.8	1	1	557–702	292–340	0.1–0.4
Phenol	O_2	Flow	—	—	0.5	0	653	188–278	0.25–1.0
Pyridine	O_2	Flow	3.44×10^{4}	227	1	0.2	698–800	276	1–3

* Kinetic parameters are defined by $-dC/dt = k[C]^m[O]^n$, $k = k'\exp(-E_a/RT)$; E_a is in kJ/mol; T is in degrees K; R = 8.314 J/mol-K; and $k' = 1$/second

** Parameters have been obtained for oxidation only (e.g., exlcuding reactions with water)

Source: Gloyna, E.F. and Li, L., *Waste Manage.*, 36, 379–394, 1993. With permission.

TABLE 10.8

Pseudo First-Order Rate Constants for SCWO

Compound	k'' (1/s)
Phenol	0.0379
o-Chlorophenol	0.0308
o-Cresol	0.179 ± 0.023
m-Cresol	0.023 ± 0.047
o-Hydrybenzaldehyde[a] (Eq. (10.5))	1.20 ± 0.20
o-Hydrybenzaldehyde[b] (Eq. (10.8))	0.939
m-Hydrybenzaldehyde[a] (Eq. (10.6))	0.111 ± 0.029
m-Hydrybenzaldehyde[b] (Eq. (10.9))	0.0842
p-Hydrybenzaldehyde[a] (Eq. (10.7))	0.147 ± 0.020
p-Hydrybenzaldehyde[b] (Eq. (10.10))	0.147

[a] Thermal oxidative reactions

[b] Oxidative reaction only

Source: Data from Martino and Savage (1997), Gopalan and Savage (1995b), and Li et al. (1993a).

TABLE 10.9

Global Kinetic Models for Supercritical Water Oxidation of Organic Compounds

Compound	Oxidant	Reactor Type	Kinetic Parameters *				Temperature (K)	Pressure (atm)	$[Cl]_0$ (g/L)
			(k′)	E_a	M	n			
Acetic acid	O_2	Batch	4.4×10^{12}	182	1	0	543–593	20–200	~30
Alcohol from distiller waste (COD)	O_2	Batch	3×10^3	45.3	1.5	0.3–0.6	423–483	2–20	33–35
2,4-Dichlorophenol	O_2	Batch	9.0×10^1	28.5	1	0	673–773	240–350	0.3–1
Methane	O_2	Flow	2.51×10^{11}	178.9	0.99	0.66	833–903	245	—
Methane	O_2	Flow	2.04×10^7	141.7	1	0	833–903	245	—
m-Xylene (COD)	O_2	Batch	—	89.5	1	0.5	300–500	70–140	0.128
Phenol	O_2	Batch	—	112	1	0.5	300–500	70–140	0.128
Phenol	O_2	Flow	1.98×10^7	56.6	1	1	418–453	26–48	0.1
Phenol	O_2	Batch	—	93	1	0.5	423–468	4–15	0.2
Phenol (COD)	O_2	Batch	—	48	1	1	413–453	4–15	0.5
Phenol (total phenolics)	O_2	Batch	—	57.5	0.5	1.5	353–383	1.5–4.5	0.01–0.05
Pyridine	O_2	Batch	1.59×10^4	91.5	1	0	673–773	240–350	0.3–1

* Kinetic parameters are defined by $-dC/dt = k[C]^m[O]^n$, $k = k′\exp(-E_a/RT)$; E_a is in kJ/mol; T is in degrees K; R = 8.314 J/mol-K; and k′ = 1/second

** Parameters have been obtained for oxidation only (e.g., excluding reactions with water)

Source: Li, L. et al., *ACS Symp. Ser.*, 514, 305–313, 1992. With permission.

TABLE 10.10

Global Kinetic Models for Supercritical Water Oxidation of Organic Substances

Compound	Oxidant	Reactor Type	Kinetic Parameters *				Temperature (K)	Pressure (atm)	$[C_a]$ (g/L)
			(k')	E_a	M	n			
Acetamide	H_2O_2	Flow	2.75×10^5	88.3	1.15	0.05	673–803	240–350	1.5–4.0
Acetamide **	H_2O_2	Flow	5.01×10^4	94.7	1	0.17	673–803	240–350	1.5–4.0
Acetic acid	H_2O_2	Flow	3.47×10^{11}	179.5	1.01	0.16	673–803	240–350	1.3–3.3
Acetic acid	H_2O_2	Flow	2.63×10^{10}	167.1	1	0	673–803	240–350	1.3–3.3
Acetic acid	H_2O_2	Flow	8.94×10^5	314	2.36	1.04	673–773	240–350	1.0–5.0
Acetic acid	H_2O_2	Flow	9.23×10^7	131	1	0	673–773	240–350	1.0–5.0
Acetic acid	O_2	Flow	9.82×10^{17}	231	1	1	611–718	394–438	0.525
Acetic acid	O_2	Flow	2.55×10^{11}	172.7	1	0	611–718	394–438	0.525
Activated sludge(COD)	O_2	Batch	$\sim1.5 \times 10^2$	~54	1	0	573–723	240–350	46.5
Ammonia	O_2	Flow	3.16×10^6	157	1	0	913–973	246	0.03–0.11
Carbon monoxide	O_2	Flow	3.16×10^6	112	1	0	673–814	246	0.02–0.11
Digested sludge(COD)	O_2	Batch	4.36×10^3	204	1.86	0	573–723	240–350	46.5
Ethanol	O_2	Flow	6.46×10^{21}	340	1	0	755–814	241	0.03–0.036
Formic acid	O_2	Flow	—	~96	1	1	683–691	408–432	1.0
Glucose(TOC)	—	Batch	—	130	0.5	1	653–683	~400	~10
Methane	O_2	Flow	1.26×10^7	156.8	1	0	913–973	245	—
Methanol	O_2	Flow	2.51×10^{24}	395.0	1	0	723–823	243	—
Methanol	O_2	Flow	2.51×10^{29}	478.6	1	0	723–823	243	—
Methanol	O_2	Flow	3.16×10^{26}	408.4	1.1	−0.02	723–823	243	0.038–0.17
Phenol	O_2	Flow	2.61×10^5	63.8	1	1	557–702	292–340	0.1–0.4
Phenol	O_2	Flow	—	—	0.5	0	653	188–278	0.25–1.0

* Kinetic parameters are defined by $-dC/dt = k[C]^m[O]^n$; $k = k'\exp(-E_a/RT)$; E_a is in kJ/mol; T is in degrees K; R = 8.314 J/mol-K; and $k' = 1$/second

** Parameters have been obtained for oxidation only (e.g., exlcuding reactions with water)

Source: Li, L. et al., Int. J. Chem. Kinet., 23, 971, 1991.

the best-fit model for SCWO of acetic acid by oxygen was 231 kJ/mol, based on a first-order reaction for both acetic acid and oxygen. An initial comparison of kinetic results obtained from SCWO of acetic acid by hydrogen peroxide using two different flow reactor systems showed considerable differences in reaction orders and activation energies (Lee, 1990). However, when the pseudo first-order reaction model is applied to the kinetic data reported by Wightman (1981) and Lee (1990), the activation energies of acetic acid, methanol, and ethanol become 172.7 kJ/mol, 167.1kJ/mol, and 131 kJ/mol, respectively. The pre-exponential factors for methanol and ethanol were much higher than that for acetic acid, indicating a much faster rate of reaction for alcohols. Thus, acetic acid is assumed to be the key rate-limiting intermediate for the SCWO of organic compounds.

10.4.9 Sludge

Table 10.11 shows the different residual sludges and their characteristics after the SCWO process. Based on characteristics presented in Table 10.11, the chemical composition of sludge residue from SCWO can be disposed of at any sanitary landfill.

10.5 QSAR Models

A quantitative structure–activity relationship (QSAR) can be expected for compounds within the same class and under similar supercritical reaction conditions such as pressure, temperature, reactor type, and oxidant. A set of organic compounds treated by SCWO technology was divided into two major classes, aliphatic and aromatic, by Vaca (1999). It was taken into consideration that aliphatic compounds are those in which the characteristics groups are linked to the straight or branched carbon chain, while the aromatic compounds have these groups linked to a particular type of six-member carbon ring that contains three double bonds. Various sets of aliphatic and aromatic compounds were analyzed by Vaca (1999). This analysis took their respective molecular descriptors (E_{HOMO}, E_{LUMO}, Hammett's constant, and Taft's constant) and a series of kinetic parameters (destruction kinetic rates such as first order and pseudo first order and activation energy) to create a database. Then, kinetic parameters and molecular descriptors were correlated. These analyses were done with the aim of obtaining better knowledge about the behavior of the chemical reactivity of organic pollutants treated by SCWO technology.

TABLE 10.11

Destruction Efficiency (DE) at Huntsman Pilot Plant Treatment Results

Waste Stream	Contaminant	Concentration in Feed (mg/L)	Destruction Efficiency (DE) (%)	Test Duration (days)
Organic-laden wastewater	TOC	50,000	99.9	3
Municipal sludge	TOC	93,230	98.3	1
	TS	66,850	71.7	
Refinery sludge	COD	68,949	97.1	1
	TS	31,715	50.5	
Chemical industrial biosludge	COD	42,239	98.9	1
	TS	39,340	70.9	
Paper mill sludge	COD	54,732	99.9	2
	TS	33,353	57.8	
Fermentation still bottoms	COD	75,471	99.9	1
	TS	19,033	98.9	
Excess Hazardous Materials				
Paint sludge	COD	500,000	99.7	1
Paint liquids	COD	299,500	99.3	1
Adhesive sludge	COD	500,000	99.8	1
Greases and lubricants	COD	1,682,000	99.9	1
Mixed solvents	COD	412,600	99.4	1
Nonchlorinated waste oil	COD	1,589,000	99.9	1
Chlorinated waste oil	COD	2,782,000	99.9	1
AFFF solution	TOC	25,280	99.9	1
TEA/citric acid solution	TOC	29,200	99.9	1
Mixture of all excess hazardous materials	COD	500,000	99.9	10

Source: http://www.epa.gov.

10.5.1 Aliphatic Compounds

The first set of aliphatic compounds included methane, acetic acid, 2-butanone, and acetamide, which were analyzed using E_{HOMO} as the molecular descriptor. The relationship between the kinetic rates of these compounds and E_{HOMO} reflected a very good correlation ($r^2 = 0.96$). Figure 10.14 shows the correlation of the set of aliphatic compounds. From this figure, it can be seen that the kinetic rate decreases as the E_{HOMO} increases. Application of the F test showed that the level of significance was 2.5%. That is, there is a 2.5% chance of erroneously concluding that they are not related. Based on Table 10.12, the calculated $F_{(1,2)}$ was 40.5, which is larger than the $F_{(1,2)}\alpha_{0.025}$

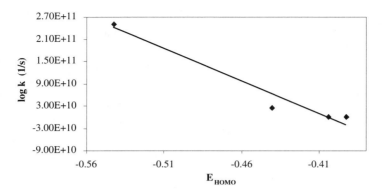

FIGURE 10.14
k (1/s) vs. E_{HOMO} for aliphatic compounds.

TABLE 10.12

F Test Results

Correlation	Calculated F	Significant Level	Degree of Freedom	F Distribution
Aliphatic				
k vs. E_{HOMO}	40.5	0.025	(1,2)	38.5
E_a vs. E_{HOMO}	2610	0.05	(1,3)	55.6
k vs. E_{LUMO}	22.2	0.025	(1,3)	17.4
E_a vs. E_{LUMO}	207	0.05	(1,3)	55.6
E_a vs. Taft's constant	10.92	0.1	(1,2)	8.53
Aromatic				
k vs. E_{HOMO}	24.92	0.025	(1,3)	17.4
k'' vs. E_{HOMO}	95.70	0.05	(1,3)	55.6
E_a vs. E_{HOMO}	179.39	0.05	(1,2)	18.6
k vs. E_{LUMO}	12.56	0.025	(1,4)	12.2
k'' vs. E_{LUMO}	52.33	0.05	(1,3)	34.1
E_a vs. E_{LUMO}	15.23	0.05	(1,3)	10.1
k vs. Hammett's constant	7.76	0.1	(1,4)	5.54
k'' vs. Hammett's constant	9.63	0.1	(1,3)	10.1
E_a vs. Hammett's constant	105.95	0.01	(1,2)	98.5

distribution value of 38.5; therefore, the null hypothesis was rejected. It was then concluded that a relationship exists between the previous parameters. Note that as the F value increases the correlation that has been achieved becomes more significant.

A set of aliphatic compounds was used for the correlation between the activation energy and E_{HOMO}. The dataset contained acetamide, methanol, and ethanol. The regression coefficient (r^2) for this relationship was 0.998. The probability of getting a correlation was 0.995 for a sample size of five. The significance of $F_{(1,3)} = 2610$ can be ascertained by consulting the F values in distribution tables. The $F_{(1,3)}\alpha_{0.005}$ distribution value is found to be 55.6. When the calculated F value and F distribution values are compared, it can be seen that the $F_{(1,3)}$ of 2610 is greater than 55.6; therefore, it can be assumed that the equation is significant at a 0.5% confidence level. In Figure 10.15 a plot of the activation energy vs. E_{HOMO} shows that activation energy decreases as E_{HOMO} increases.

Because E_{HOMO} is a measure of the ability of a compound to donate electrons (i.e., act as a nucleophilic species or undergo oxidation), the increase in kinetic rate leads to increasing destruction efficiency. In the case of activation energy, activation energy increases as E_{HOMO} decreases; therefore, E_{HOMO} is an indicator of the ability of the aliphatic compound to donate electrons. Muller and Klein (1991) studied a set of 63 aliphatic compounds, and they stated that the reaction rate was significantly correlated with the E_{HOMO}.

Figure 10.14 and Figure 10.15 show the correlation of kinetic rates and activation energy vs. E_{LUMO}. The dataset used was 2-butanone, acetic acid, methane, and acetamide:

$$k = -2E^{+12}E_{HOMO} - 7E^{+11}; \ r^2 = 0.9575, \ n = 4, \ r = 0.9785 \qquad (10.27)$$

The regression coefficient (r^2) for this relationship was 0.8685. The significance of the calculated $F_{(1,3)} = 22.2$ can be ascertained by consulting tables of F values. From such a table, it is found that the $F_{(1,3)}\alpha_{0.025}$ distribution is 17.4. Because the $F_{(1,3)}$ of 22.2 is greater than 17.4, it can be assumed that the equation is significant at the 2.5% confidence level. This relationship showed that the kinetic rate increases as E_{LUMO} increases. A set of aliphatic compounds was used for the correlation between E_{LUMO} as molecular descriptor and activation energy as a predictive molecular descriptor. The dataset of

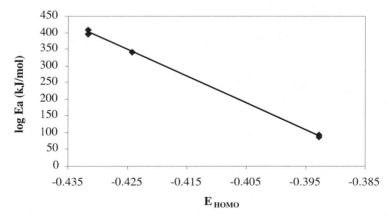

FIGURE 10.15
E_a (kJ/mol) vs. E_{HOMO} for aliphatic compounds.

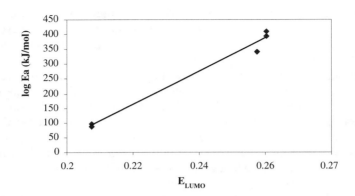

FIGURE 10.16
E_a (kJ/mol) vs. E_{LUMO} for aliphatic compounds.

aliphatic compounds that was used in Figure 10.16 included methanol, acetamide, and ethanol:

$$E_a = -7967.4E_{HOMO} - 3038.1; r^2 = 0.9988, n = 5, r = 0.9993 \quad (10.28)$$

The relationship between these two properties of the molecule showed a very good correlation (0.985). The probability of getting a correlation was 0.995 for a sample size of five. The calculated $F_{(1,3)}$ is 207, and the $F_{(1,3)}\alpha_{0.005}$ distribution is 55.6. Because $F_{(1,3)}$ of 207 is greater than 55.6, it can be assumed that the equation is significant at a confidence level of 0.5%. Thus, this relationship showed that activation energy increases as E_{LUMO} increases.

Because E_{LUMO} is a measure of the ability of a compound to accept electrons (i.e., act as an electrophilic species or undergo reduction), the above correlations show that the kinetic rates and activation energy increase as E_{LUMO} increases. Therefore, the ability of the compound to behave as an electrophilic species increases as E_{LUMO} increases. The increased reactivity of organic compounds may have a direct influence on its degradation rate constants as follows:

$$k = 2E^{+12}{}_{LUMO} - 3E^{+11}; r^2 = 0.8685, n = 5, r = 0.9319 \quad (10.29)$$

$$E_a = 5625.3E_{LUMO} - 1076.7; r^2 = 0.9857, n = 5, r = 0.992 \quad (10.30)$$

Taft's constants were used as molecular descriptors to correlate with activation energy and kinetic rates, respectively. They were used because they describe the inductive effects of the substituents in aliphatic compounds. Taft's constant showed a poor correlation ($r^2 = 0.68$) for aliphatic compounds; however, the correlation between activation energy and Taft's constant was significant ($r^2 = 0.8452$). The dataset was acetic acid, carbon monoxide, and methane. The $F_{(1,2)}$ value was calculated, and from that calculation a value of 10.92 for a set of four compounds was obtained. Because the calculated $F_{(1,2)}$ of 10.92 is greater than $F_{(1,3)}\alpha_{0.1}$ (8.53), it can be assumed that the equation is significant at the confidence level of 10%.

$$E_a = -1456.5\sigma^* + 126.85; \; r^2 = 0.8452, \; n = 4, \; r = 0.9193 \qquad (10.31)$$

10.5.2 Aromatic Compounds

Various sets of aromatic compounds and three different molecular descriptors such as E_{HOMO}, E_{LUMO}, and Hammett's constants have been used for other QSAR models. First-order kinetic rates, pseudo first-order kinetic rates, and activation energy were used to correlate with different molecular descriptors.

For a set of five aromatic compounds, a correlation of 0.6744 was obtained between the first-order kinetic rate constants and E_{HOMO}. The compounds used were phenol, *o*-cresol, 2,4-dichlorophenol, and pyridine. *F* values were used to verify the level of correlation of these data. The *F* results were obtained from calculation of the mean squares of the variation in *Y* and were divided by the corresponding degrees of freedom. Then, this value and corresponding degree of freedom were compared with the *F* distribution values from tables. The calculated $F_{(1,3)}$ was 24.92, which was then compared with the $F_{(1,3)}\alpha_{0.025}$ value of 17.4 using $(\alpha - 1) = 0.975$. Because the $F_{(1,3)}$ value of 24.92 is greater than 17.4, application of the *F* test shows that the relation is significantly different than 0 at the 2.5% confidence level. Figure 10.17 shows that the relationship between E_{HOMO} and kinetic rates is inversely proportional, thus the kinetic rates decrease as E_{HOMO} increases.

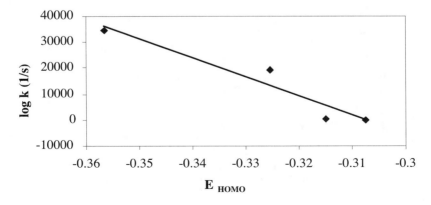

FIGURE 10.17
k (1/s) vs. E_{HOMO} for aromatic compounds.

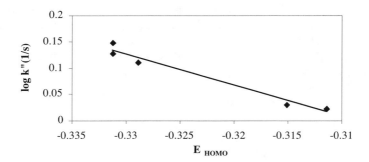

FIGURE 10.18
k'' (1/s) vs. E_{HOMO} for aromatic compounds.

A set of four aromatic compounds such as *m*-cresol, *m*-hydroxybenzalde-hyde, *p*-hydroxybenzaldehyde, and phenol was used to correlate pseudo first-order kinetic rates and E_{HOMO} descriptors. From Figure 10.18, it can be observed that the correlation between pseudo first-order kinetic rates and E_{HOMO} descriptors behaved the same as the correlations of aliphatic compounds. In other words, even though the first-order kinetic rate and pseudo first-order kinetic rate were treated as distinct kinetic parameters, these two correlations behaved the same.

The correlation between E_{HOMO} and the pseudo first-order kinetic rates showed a very good correlation (0.9696). The significance of $F_{(1,3)}\alpha$ is 95.70, and $F_{(1,3)}\alpha_{0.005}$ is 55.6. Because the $F_{(1,3)}$ of 95.70 is greater than 55.6, it can be assumed that the equation is significant at the confidence level of 0.5%. This relationship shows that the pseudo first-order kinetics rate increases as E_{HOMO} decreases.

The correlation between the activation energy of aromatic compounds and E_{HOMO} has a correlation coefficient (r^2) of 0.9676. Figure 10.19 shows the

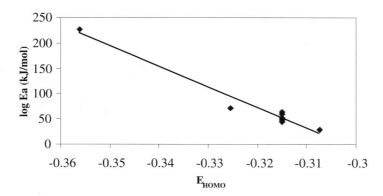

FIGURE 10.19
E_a (kJ/mol) vs. E_{HOMO} for aromatic compounds.

correlation between $\log E_a$ and E_{HOMO}. This set of aromatic compounds included pyridine, *o*-cresol, phenol, and 2,4-dichlorophenol. From the figure, it can be seen that activation energy decreases as the E_{HOMO} increases. Application of the F test showed that the level of significance was 0.005, because $F_{(1,2)0.95}$ is 179.39 and is much larger than 18.6. The value of 18.6 can be obtained from a table of F distribution values. As the F value increases, the correlation that has been achieved becomes more significant, so the null hypothesis is rejected. Therefore, there is a 0.5% chance of erroneously concluding that they are related.

The first set of aromatic compounds that were analyzed using E_{LUMO} as the molecular descriptor and the first-order kinetic rate as the kinetic parameter was composed of 2,4-dichlorophenol, pyridine, phenol, and 1,3-dichlorobenzene. The correlation between the first-order kinetic rates of the compounds and E_{LUMO} reflected a coefficient of $r^2 = 0.7834$. Figure 10.20 shows the correlation of this set of aromatic compounds. From this figure, it can be seen that the kinetic rate increases as E_{LUMO} increases. The F test showed that the level of significance was 2.5%. Because the $F_{(1,4)0.975}$ of 12.56 is larger than 12.2, it can be concluded that a relationship exists between the two variables:

$$k = -738322E_{HOMO} - 227302; \; r^2 = 0.9198, \; n = 4, \; r = 0.9590 \qquad (10.32)$$

$$k'' = -5.89E_{HOMO} - 1.8; \; r^2 = 0.9696, \; n = 5, \; r = 0.9846 \qquad (10.33)$$

$$E_a = -4094E_{HOMO} - 1238; \; r^2 = 0.9676, \; n = 8, \; r = 0.9836 \qquad (10.34)$$

For pseudo first-order kinetic rates, the aromatic compounds such as *p*-hydrobenzaldehyde, *m*-hydrobenzaldehyde, and *o*-cresol were used. Figure 10.20 shows that the correlation of aromatic compounds between the pseudo first-order kinetic rates and E_{LUMO} behaves in a directly proportional

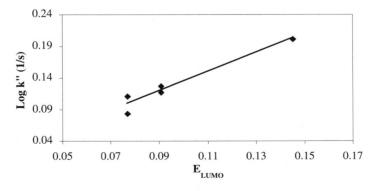

FIGURE 10.20
k'' (1/s) vs. E_{LUMO} for aromatic compounds.

manner. A very good correlation of 0.9458 was obtained. The F test showed that the level of significance was 0.5%. A comparison of the $F_{(1,3)}$ of 52.33 and $F_{(1,3)0.995}$ of 55.6 from the F test table shows that this result is not simply a chance correlation.

The correlation between activation energy and E_{LUMO} is another example of how molecular orbital can help to describe the mechanism of the molecules and characterize each of them. Figure 10.20 shows the relation between these two parameters. In this correlation, activation energy increases as E_{LUMO} increases. A significant correlation was obtained for 2,4-dichlorophenol, 1,4-dichlorobenzene, 1,3-dichlorobenzene, phenol, and pyridine. The correlation confidence between these two parameters was 0.8355. The F test showed that the level of significance between activation energy and E_{LUMO} with a degree of freedom of (1,3) was 15.23. Because $F_{(1,3)}$ is 15.23 and is larger than 10.1, there is only a 5% chance of erroneously concluding that they are related.

In general, E_{LUMO} correlated well with the kinetic parameters such as k, k″, and E_a of the aliphatic and aromatic compounds, as shown in Figure 10.20. QSAR models can be described as follows:

$$k = 2E + 6\ E_{LUMO} - 228329;\ r^2 = 0.7834,\ n = 6,\ r = 0.885 \qquad (10.35)$$

$$k'' = 1.5159\ E_{LUMO} - 0.0175;\ r^2 = 0.9458,\ n = 5,\ r = 0.9725 \qquad (10.36)$$

$$E_a = 2491\ E_{LUMO} - 230.11;\ r^2 = 0.8355,\ n = 5,\ r = 0.9140 \qquad (10.37)$$

As E_{LUMO} increases, the ability of compounds to undergo reduction decreases. Therefore, the kinetic rates and activation energy increase with E_{LUMO}. In other words, the increase in kinetic rates and activation energy increases the reactivity of the molecule and enhances the possibility of oxidizing organic pollutant compounds, thus the destruction efficiency of these compounds will be higher.

Hammett's constants can also be used to correlate with the previous kinetic parameters. The aromatic compounds that were analyzed by Vaca (1999) using Hammett's constant as the molecular descriptor and the first-order kinetic rate as the predicted variable include 2,4-dichlorophenol, 1,4-dichlorobenzene, 1,3-dichlorobenzene, pyridine, and phenol. The correlation of Hammett's constant and first-order kinetic rates of this compound reflected a coefficient of $r^2 = 0.718$. Figure 10.21 shows the correlation of this set of aromatic compounds. From the figure, it can be seen that the kinetic rate decreases as the Hammett's constant increases. The F test showed that the level of significance was 10%. Because $F_{(1,4)0.90}$ was 7.76 and is larger than 5.54, it can be concluded that there is a relationship between the two variables.

To analyze pseudo first-order kinetic rates, the aromatic compounds such as *p*-hydrobenzaldehyde, *m*-hydrobenzaldehyde, *o*-chlorophenol, *m*-cresol, and *o*-cresol were used by Vaca. Figure 10.22 shows that the correlation of

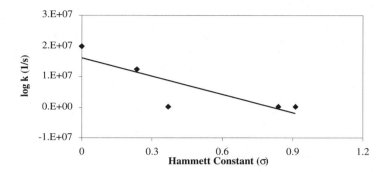

FIGURE 10.21

k (1/s) vs. Hammett's constant for aromatic compounds.

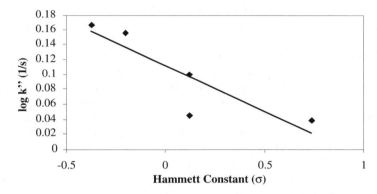

FIGURE 10.22

k'' (1/s) vs. Hammett's constant (σ) for aromatic compounds.

aromatic compounds between the pseudo first-order kinetic rates and the Hammett's constant behaves in an inversely proportional manner. In this relationship, a correlation of 0.7623 was obtained. The F test showed that the level of significance was 10%. A comparison of the calculated $F_{(1,3)0.90}$ value of 9.63 and the $F_{(1,3)}$ value of 10.1 from the F distribution table shows that this result is not merely a chance correlation.

For aromatic compounds, activation energy increases as Hammett's constants decrease, as shown in Figures 10.21 and 10.22. Figure 10.21 shows the relation between these two parameters. The dataset of 2,4-dichlorophenol, m-xylene, and phenol were used.

Hammett's constant represents 89.82% of the variance in the linear regression equation. The F test showed that the level of significance of E_{LUMO} and activation energy with a degree of freedom of 1,2 was 0.01. Because the calculated F is 105.95 and is larger than the $F_{(1,2)0.995}$ value of 98.5 from the F distribution table, the null hypothesis is rejected; therefore, the chance of erroneously concluding that they are related is only 1%:

$$k = -2E + 7\sigma + 2E + 7;\ r^2 = 0.7176,\ n = 5,\ r = 0.8471 \qquad (10.38)$$

$$k'' = -0.124\sigma + 0.1114;\ r^2 = 0.7623,\ n = 5,\ r = 0.8731 \qquad (10.39)$$

E_a vs. Hammett's constant (σ) for aromatic compounds has the following correlation:

$$E_a = -73.118\sigma + 96.16;\ r^2 = 0.8982,\ n = 4,\ r = 0.9447 \qquad (10.40)$$

10.6 Summary

The QSAR models can be used to estimate the treatability of organic pollutants by SCWO. For two chemical classes such as aliphatic and aromatic compounds, the best correlation exists between the kinetic rate constants and E_{HOMO} descriptor. The QSAR models are compiled in Table 10.13. By analyzing the behavior of the kinetic parameters on molecular descriptors, it is possible to establish a QSAR model for predicting degradation rate constants by the SCWO for organic compounds with similar molecular structure. This analysis may provide an insight into the kinetic mechanism that occurs with this technology.

The SCWO process is an emerging technology that offers a potentially ultrahigh destruction efficiency of hazardous wastes with high recovery of resources, and it could become one of the most useful processes for the treatment of concentrated wastewater and sludge. The QSAR models for oxidation rate constants for SCWO offer great help to design the most efficient SCWO processes for a given class of hazardous wastes. In addition,

TABLE 10.13

QSAR Models between Kinetic Parameters vs. Molecular Descriptors

Chemical Class	Kinetic Parameters	n	E_{HOMO}	n	E_{LUMO}	n	Hammett's Constant	n	Taft's Constant
Aliphatic									
	First kinetic rate constants	4	0.9575	5	0.8685	—	—	—	—
	Activation energy	5	0.9988	5	0.9857	—	—	4	0.8452
Aromatic									
	First kinetic rate	4	0.9198	6	0.7834	5	0.7176	—	—
	Pseudo first-order kinetic rate constants	5	0.9962	5	0.8767	5	0.7709	—	—
	Activation energy	8	0.9676	5	0.8355	4	0.8982	—	—

— No correlation was performed.

molecular descriptors provide the ideal tools for predicting the behavior of kinetic mechanisms of organic compounds treated by SCWO technology.

Molecular descriptors may provide fundamental and scientific explanations of variations in the activity mechanism due to the structure of molecules. QSAR models can provide insight into kinetic mechanisms of organic compounds, and some of the design uncertainty associated with SCWO kinetics can be reduced. By using appropriate molecular descriptors, then, the results of QSAR models for SCWO could be used to determine the oxidation rate constants of organic compounds of structure similar to the training set chemicals of the QSAR model without experiments.

References

Albright, T.A., Burdett, J.K., and Whangbo, M.-H., *Orbital Interactions in Chemistry,* John Wiley & Sons, New York, 1985.

Barner, H.E. et al., Supercritical water oxidation: an emerging technology, *J. Haz. Mat.,* 31, 1–17, 1992.

Beziat, J.C. et al., Catalyst wet air oxidation of wastewater, *J. Stud. Surf. Catalysis,* 110, 615–622, 1997.

Blaney, C.A., Li, L., Gloyna, E.F., and Hossain, F., Supercritical water oxidation of pulp and paper mill sludge as an alternative to incineration, in *Innovations in Supercritical Fluids,* Hutchenson, K.W. and Foster, N.R., Eds., ACS Symp. Ser. 608, 444–455, ACS, Washington, D.C., 1995.

Boock, L.T. et al., Hydrolysis and oxidation in supercritical water, *Endeavour,* 17(4), 180–185, 1993.

Brezonik, P.L., Principles of linear free energy and structure–activity relationships and their applications to the fate of chemicals in aquatic systems, in *Chemical Kinetic Dynamic in Aquatic Systems,* Lewis Publishers, Boca Raton, FL, 1994, pp. 113–143.

Chen, P., Li, L., and Gloyna, E.F., Simulation of a concentric-tube reactor for supercritical water oxidation, *ACS Symp. Ser.: Innovation in Supercritical Fluids,* 608(24), 348–363, 1995.

Ding, Z.Y. et al., Catalytic oxidation in supercritical water, *Indust. Eng. Chem. Res.,* 35, 3257–3279, 1996.

Ding, Z.Y. et al., Supercritical water oxidation of NH_3 over a MnO_2/CeO_2 catalyst, *Indust. Eng. Chem. Res.,* 37, 1707–1716, 1998.

Fleming, I., *Frontier Orbitals and Organic Chemical Reactions.* John Wiley & Sons, New York, 1976.

Frank, E.U., Fluids at high pressures and temperatures, *Pure Appl. Chem.,* 59(1), 25–34, 1987.

Gloyna, E.F., U.S. EPA Engineering Bulletin, Risk Reduction Engineering Laboratory, Cincinnati, OH, EPA/540/S-92/006, 1992.

Gloyna, E.F. and Li, L., Supercritical water oxidation: and engineering update, *Waste Manage.,* 36, 379–394, 1993.

Gloyna, E.F. and Li, L., *Supercritical Water Oxidation Model Development for Selected EPA Priority Pollutants: Project Summary,* EPA/600/SR-95/080, U.S. Environmental Protection Agency, Washington, D.C., 1995, pp. 1-4.

Gopalan, S. and Savage, P.E., Reaction mechanism for phenol oxidation in supercritical water, *J. Phys. Chem.*, 98, 2646, 1994.

Gopalan, S. and Savage, P.E., Reaction network for phenol oxidation in supercritical water: a comprehensive quantitative model, *AIChE J.*, 41, 1864–1873, 1995.

Goto, M. et al., Decomposition of municipal sludge by supercritical water oxidation, *J. Chem. Eng. Jpn.*, 30(5), 813–818, 1997.

Hansch, C. and Leo, A., *Exploring QSAR: Fundamentals and Applications in Chemistry and Biology*, American Chemical Society, Washington, D.C., 1995.

Hansch, C., Leo, A., and Hoekman, D., *Exploring QSAR: Hydrophobic, Electronic, and Steric Constants*, American Chemical Society, Washington, D.C., 1995.

Hauptmann, E.G. et al., Strategies for treatment mixtures of bleach plant effluents and waste water treatment sludges by supercritical water oxidation, preprints of papers to be presented at the annual meeting, 80(B), B71–B77, 1994.

Helling, R.K. and Tester, J.W., *Energy and Fuels*, 1, 417, 1987.

Helling, R.K. and Tester, J.W., Oxidation of simple compounds and mixtures in supercritical water: carbon monoxide, ammonia, and ethanol, *Environ. Sci. Technol.*, 22(1), 1319–1324, 1988.

Holgate, R.H. and Tester, J.W., Oxidation of hydrogen and carbon monoxide in sub- and supercritical water: reaction kinetics, pathways, and water-density effects. 2. Elementary reaction modeling, *J. Phys. Chem. Technol.*, 98, 810–822, 1994.

Hout, R., Pietro, W.J., and Hehre, W., *A Pictorial Approach to Molecular Structure and Reactivity*, John Wiley & Sons, New York, 1984.

Huang, C.P. et al., Advanced chemical oxidation: its present role and potential future in hazardous waste treatment, *Waste Manage.*, 13, 361–377, 1993.

Kirts, R.E. and Beller, J.M., Supercritical water oxidation of organic wastes, 1995. www.Clean.rti.org.

Koo, M.L., Wong, K., and Lee, C.H., New reactor system for supercritical water oxidation and its application on phenol destruction, *Chem. Eng. Sci.*, 52(7), 1201–1214, 1997.

Krajnc, M. and Levec, J., On the kinetics of phenol oxidation in supercritical water, *AIChE J.*, 42(7), 1977–1984, 1996.

Krajnc, M. and Levec, J., The role of catalyst in SCWO of acetic acid, *Appl. Catalysis B*, 13, 93–103, 1997a.

Krajnc, M. and Levec, J., Oxidation of phenol over transition-metal oxide catalyst in supercritical water, *Indust. Eng. Chem. Res.*, 36, 32439–3445, 1997b.

Kriksunov, L.B. and MacDonald, D.D., Understanding chemical conditions in supercritical water oxidation systems, *ASME Heat Transfer Div.*, 317(2), 271–279, 1995a.

Kriksunov, L.B. and MacDonald, D.D., Corrosion testing and prediction in SCWO environments, *ASME Heat Transfer Div.*, 317(2), 281–288, 1995b.

Lee, D.S. and Gloyna, E.F., Supercritical water oxidation microreactor system, paper presented at *Water Poll. Control Fed. Specialty Conf.* New Orleans, LA, April 17–19, 1988.

Lee, D.S. et al., Hydrothermal decomposition and oxidation of *p*-nitroaniline in supercritical water, *J. Hazardous Mater.*, 56, 247–256, 1997.

Li, L. and Egiebor, N.O., Feedstream preheating effect on supercritical water oxidation of dissolved organics, *Energy Fuels*, 8(5), 1126–1130, 1994.

Li, L., Chen, P., and Gloyna, E.F., Kinetic model for wet oxidation of organic compounds in subcritical and supercritical water, *ACS Symp. Ser.*, 514, 305–313, 1992.

Li, L., Chen, P., and Gloyna, E.F., Generalized kinetic model for wet oxidation of organic compounds, *AIChE J.*, 37(11), 1687–1697, 1993a.

Li, L. et al., Kinetic model for wet oxidation of organic compounds in subcritical and supercritical water, *Supercritical Fluid Eng. Sci.*, C24, 305–313, 1993b.

Lin, C.C., Smith, F.R., Ichikawa, N., Baba, T., and Itow, M., Decomposition of hydrogen peroxide in aqueous solutions at elevated temperatures, *Int. J. Chem. Kinet.*, 23, 971, 1991.

Lin, K.S., Wang, H.P., and Li, M.C., Oxidation of 2,4-dichlorophenol in supercritical water, *Chemosphere*, 36(9), 2075–2083, 1998.

La Roche, L.H., Weber, M., and Christian, T., Design rules for the wallcooled hydrothermal burner (WHB), *Chem. Eng. Technol.*, 20, 208–211, 1997.

Margolis, E., *Bonding and Structure*, Appleton Century Crofts, New York, 1968.

Martino, C.J. and Savage, P.E., Supercritical water oxidation kinetics, products, and pathways for CH_3 and CHO-substituted phenols, *Indust. Eng. Chem. Res.*, 36, 1391–1400, 1997.

McBrayer, R.N. and Griffit, J.W., Supercritical water oxidation analysis, 1997. www.jumpnet.com.

Minok, K., Lee, W.K., and Lee, C.H., New reactor system for supercritical water oxidation and its application on phenol destruction, *Chem. Eng. Sci.*, 52(7), 1201–1214, 1997.

Modell, M., Supercritical water oxidation, in *Standard Handbook of Hazardous Waste Treatment and Disposal*, McGraw-Hill, New York, 1989, pp. 153–168.

Modell, M. and Mayr, S., Supercritical water oxidation of aqueous waste, *Proc. Water Conf. (OIWCEQ)*, 56, 479–488, 1995.

Modell, M., Gaudet, G.G., Simson, M., Hong, G.T., and Biemann, K., Supercritical water testing reveals new process holds promise, *Solid Wastes Manag.*, 25, 26, 1982.

Modell, M., Larson, J., and Sobczynski, S.F., Supercritical water oxidation of pulp mill sludges, *Tappi J.*, 75(6), 195–202, 1992.

Muller M. and Klein, W., Estimating atmospheric degradation process by SARs, *Sci. Total Environ.*, 109/110, 261–273, 1991.

Peijnenburg, W.J.G.M., QSARs for predicting biotic and abiotic reductive transformation rate constants of halogenated hydrocarbons in anoxic sediment systems, *Sci. Total Environ.*, 109/110, 283–300, 1991.

Reinhard, M. and Drefahl, A., *Handbook for Estimating Physicochemical Properties of Organic Compounds*. John Wiley & Sons, New York, 1999.

Rice, S.F., Destruction of representative Navy wastes using supercritical water oxidation, *Sandia Rep.*, SAND94-8203.UC-402, 1–35, 1994.

Rice, S.F., *Supercritical Water Oxidation Reactor*, Sandia National Laboratories, 1995. www.ca.sandia.gov.

Rice, S.F., Oxidation rates of common organic compounds in supercritical water oxidation, *J. Hazardous Mater.*, 59(2), 261–278, 1998.

Roberts, D.W., Application of QSAR to biodegradation of linear alkylbenzene sulphonate (LAS) isomers and homologues, *Sci. Total Environ.*, 109/110, 301–306, 1991.

Sako, T. et al., Decomposition of dioxins in fly ash with supercritical water oxidation, *J. Chem. Eng. Jpn.*, 30(4), 744–747, 1997.

Savage, P.E., Gopalan, S., Mizan, T.I., Martino, C.J., and Brock, E.E., Reactions at supercritical conditions: fundamentals and applications, *AIChE J.*, 41, 1723–1778, 1995.

Shanableh, A. and Gloyna, E.F., Supercritical water oxidation-wastewaters and sludges, *Water Sci. Technol.*, 23, 389–398, 1991.

Stadig, W.P., Supercritical water oxidation achieves 99.99% waste destruction, *Chem. Process.*, 58(8), 389–398, 1995.

Steeper, R.R., Methane and methanol oxidation in supercritical water: chemical kinetics and hydrothermal flame studies, *Sandia Rep.*, Sand96-8208.UC-1409, 1–150, 1996.

Takahashi, Y. and Wydeven, T., Subcritical and supercritical water oxidation of cell model wastes, *Adv. Space Res.*, 8, 99–110, 1989.

Tester, J.W. et al., Revised global kinetic measurements of methanol oxidation in supercritical water, *Indust. Eng. Chem. Res.*, 32(1), 236–239, 1993a.

Tester, J.W. et al., Supercritical water oxidation technology, *Emerging Technol. Hazardous Waste Manage. III*, 3, 37–75, 1993b.

Thomason, T.B. and Modell, M, *Hazardous Waste*, 1, 453, 1984.

Thornton, T.D. and Savage, P.E., Kinetics of phenol oxidation in supercritical water, *AIChE J.*, 38, 321, 1992a.

Thornton, T.D. and Savage, P.E. Phenol oxidation pathways in supercritical water, *Ind. Eng. Chem. Res.*, 31, 2451, 1992b.

Tufano, V., A multi-step kinetic model for phenol oxidation in high pressure water, *Chem. Eng. Technol.*, 16, 186–190.

USEPA, Supercritical water oxidation, *Eng. Bull.*, EPA/540/S-92/006, 1–7, 1992.

USEPA, 1995. *Supercritical Water Oxidation Model Development for Selected EPA Priority Pollutants*: *Project Summary*, EPA/600/SR/080, U.S. Environmental Protection Agency, Washington, D.C., 1995, 4 pp.

Vaca, L., Development of QSAR Models for Oxidation Rate Constants in Supercritical Water Oxidation, M.S. thesis, Florida International University, Miami, 1999.

Warren, H.J., Radom, L., Schleyer, P.R., and Pople, J.A., *Ab Initio Molecular Orbital Theory*, John Wiley & Sons, New York, 1986.

Webley, P.A. and Tester, J.W., *Energy & Fuels*, 5, 411, 1991.

Webley, P.A., Tester, J.W., and Holgate, H.R., *Ind. Eng. Chem. Res.*, 30, 1745, 1991.

Wightman, T.J., Studies in supercritical wet air oxidation, M.S. thesis, University of California at Berkeley, 1981.

11

Sonolysis

11.1 Introduction

The sonochemical effects produced by sonolysis are due to the phenomenon of cavitation, which is the nucleation of bubbles in a liquid under the influence of ultrasound. Sonolysis is based on the fundamental concepts and theory involved in sonochemistry; the historical perspective of sonochemistry in Table 11.1 provides an insight into the discovery and understanding of fundamental processes in sonolysis. When a liquid of relatively high vapor pressure and dynamic tensile strength (such as water) is exposed to high-frequency ultrasonic waves (a few to several hundred kilohertz), acoustic cavitation in the liquid will occur. The cavitation process includes the formation, growth, and implosive collapse of small gas bubbles. Cavitation by ultrasound is accompanied by high temperature (2000 to 2500 K) and high pressure (hundreds of atmospheres), which are responsible for the degradation of organic pollutants. Therefore, sonolysis can degrade organic pollutants to CO_2 and H_2O or convert them to compounds that are less harmful than the original compounds. The degradation of organic pollutants may result from combustion, supercritical water oxidation, and oxidation by radicals such as hydroxyl radicals and hydrogen radicals. Sonolysis was found to be efficient and economical to decontaminate industrial organics before they are discharged into aquatic ecosystems. Therefore, the applications of ultrasound to destroy organic pollutants have increased significantly in the past decade.

11.2 Fundamental Processes in Sonochemistry

11.2.1 Physical Processes

Sonochemistry is defined as the chemical effects produced by ultrasonic waves. Ultrasound, with frequencies roughly between 15 kHz and 10 MHz, has a drastic effect on chemical reactions. It is the most important

TABLE 11.1

Historical Perspective of Sonochemistry

Year	Historical Event
1867	Early observations of cavitation by Tomlinson and Gernez
1880	Discovery of the piezoelectric effect
1883	Earliest ultrasonic transducer by Galton
1895	Cavitation as phenomenon recognized and investigated on propeller blades
1917	First mathematical model for cavitational collapse predicting enormous local temperatures and pressures by Rayleigh
1927	Publication of first paper on chemical effects of ultrasound (Richards and Loomis, 1927)
1933–1935	Observation of sonoluminescence effects
1933	Reports on reductions in viscosity of polymer solutions by ultrasound
1943	First patent on cleaning by ultrasound (German Patent No. 733.470)
1944	First patent on emulsification by ultrasound (Swiss Patent No. 394.390)
1950s	Intensification of cavitation and ultrasound research; increasing number of applications using ultrasound
1950	Effect of ultrasound on chemical reactions involving metals (Renaud, 1950)
1950	Hot-spot model by Noltingk and Neppiras
1953	First review on the effects of ultrasound (Barnartt, 1953)
1963	Introduction of plastic welding
1970s	Renaissance of sonochemistry research
1980s	Growing research on sonochemical effects
1986	First international meeting on sonochemistry
1990	Foundation meeting of the European Society of Sonochemistry

Source: Adapted from the European Society of Sonochemistry.

phenomenon that produces a sonochemical effect on chemical reactions. This phenomenon proceeds as follows. A sound wave impinging on a solution is merely a cyclic succession of compression and expansion phases imparted by mechanical vibration. During the solution expansion phase, small vapor-filled bubbles are formed due to weak points in the solution, primarily at trapped gas pockets on particulate surfaces. These bubbles grow and contract in response to the expansion and compression phases of the cycle, respectively. Because the surface area of the bubble is greater during the expansion phase than during the compression phase, growth of the bubble is greater than the contraction, resulting in an increase in the average bubble size over many cycles. Over time, the bubble reaches a critical size depending on the ultrasonic frequency, whereupon the pressure of the vapor within the bubble cannot withstand the external pressure of the surrounding solution. The result is a violent collapse of the bubble, with high-velocity jets of solution shooting into the interior. As a result, extremely high temperatures and pressures are produced in and near the bubble.

The use of ultrasound provides a unique means of integrating energy and matter and differs from traditional sources such as light, heat, or ionizing radiation in duration, pressure, and energy per molecule. Ultrasound provides a form of energy for the modification of chemical reactions; an ultrasound wave produces its effect via the generation and collapse of cavitation bubbles during the rarefaction and compression cycles of the wave when the liquid structure is torn apart. Ultrasonic enhancement has been used for some heterogeneous chemical reactions. The cavitation bubbles generated contain the vapor from the solvent and solute. When they collapse, these vapors are subjected to enormous changes in temperature and pressure; under such extreme conditions, the solvent and solute molecules suffer fragmentation and generate small species, including reactive free radicals; they may further undergo some secondary reactions. Although ultrasound has a broad range of industrial applications, its potential as a water and wastewater treatment alternative has not been explored fully.

It is estimated that 4×10^8 bubbles/s/m^3 are produced. The bubbles are on the order of 10 to 200 μm in diameter, and they are short lived, with a lifetime near 10 μs; therefore, the bulk characteristics of the solution remain relatively unaffected, but the implosion of the bubble causes enormous local effects. For example, the temperature of the vapor within the bubble has been estimated to reach as high as 5000 K with a concomitant pressure near 1000 atm. Due to the extremely high temperatures created during the process, a cooling system generally needs to be included in the design of sonolysis reactors. The principal result of these conditions in an aqueous solution is the breakdown of water vapor in the bubble into hydrogen and hydroxyl radicals. This essentially transforms the bubble into a microreactor, where interesting chemistry can take place. If organic species are also present in the water subjected to ultrasonic waves, it is expected that degradation will occur, ultimately to complete mineralization. The extreme conditions created by acoustic cavitation initiate three distinct destruction pathways for organic contaminants: oxidation by hydroxyl radicals, supercritical water oxidation, and pyrolysis. It has been proposed that pyrolytic mechanisms dominate at high solute concentrations while hydroxyl radical attack dominates at low solute concentrations. From the view of pure physics, the effects of sonolysis on aqueous solutions can be described by three fundamental concepts in sonochemistry. The first phenomenon is compression and rarefaction, the second is cavitation, and the third is microstreaming.

11.2.1.1 Compression and Rarefaction

A rapid movement of fluids caused by a variation of sonic pressure subjects the solvent to compression and rarefaction. This movement can be described as a motion that alternatively compresses and stretches the molecular structure within the cavitation process. This rapid movement of fluids is caused by variation of the sonic pressure. Locally, the rarefaction phases of acoustic pressure wave generate gas and microbubbles (Sochard et al., 1998).

11.2.1.2   Cavitation

Cavitation is a three-step process consisting of nucleation, growth, and collapse of gas- or vapor-filled bubbles in a body of liquid. The instantaneous pressure at the center of a collapsing bubble has been theoretically estimated to be about 75,000 psi. The temperature has been similarly estimated to reach a value as high as 13,000°F. Due to this local high temperature and pressure, it has been well recognized that cavitation can enhance the rate of a chemical reaction. Under such extreme conditions, the solvent and solute molecule fragmentizes to generate small pieces such as reactive free radicals. The free radicals may further precede some secondary chemical reactions.

11.2.1.3   Microstreaming

Microstreaming is an event in which large amounts of vibrational energy are put into a small volume with little heating. Furthermore, microstreaming constitutes an unusual type of fluid flow associated with velocity, temperature, and pressure gradients (Laborde et al., 1998).

11.2.1.4   Cavitation Temperatures Probed by EPR

High temperatures generated due to the diffuse energy produce hot spots in the liquid. When well-defined reactions due to ultrasound were studied, Suslick et al. (1986) determined the temperature of the imploding cavity to be 5500°C and the pressure to be around 500 atm according to thermodynamic principles. This short-lived hot spot, with heating and cooling rates greater than 10^9 K/s, is the source of sonochemistry. The reactions that take place at the gas–liquid interface of the bubbles are similar to combustion. The semiclassical model of the temperature dependence of the kinetic isotope effect for H and D atom formation was used to estimate the effective temperature of the hot cavitation regions in which H and D atoms are formed by ultrasound-induced pyrolysis of water molecules (Misik et al., 1995). The collapsing microbubbles, filled with dissolved gas (i.e., argon) and solvent vapor, are the reaction microchambers in which solvent vapor can be pyrolyzed, thus producing radicals that undergo further chemical reactions.

The physical properties of the supercritical fluid differ from those of the bulk liquid. One of the most notable changes is the lower dielectric constant of polar solvents such as water which allows the accumulation of low-polarity solutes at this interface. This explains the crucial role of the hydrophobicity of solutes during reactions in the solution. Thermolysis as well as radical abstraction reactions occur in this region. A temperature of approximately 800 K was determined for the interfacial region surrounding the cavitation bubbles by using the temperature dependence of C–N bond pyrolysis in oxygenated aqueous solutions.

In general, the accuracy of estimating temperatures by measuring the kinetic isotope effect may be influenced by the following factors: (1) the accuracy of the semiclassical model, (2) the lesser sensitivity of the isotope effect at higher temperatures, and (3) the reactions of H (and D) atoms competing with spin trapping.

Although OH radicals are produced by sonolysis of water, the corresponding spin adducts could not be detected using PBN due to the very short half-life of the PBN/OH adduct in aqueous solutions at neutral and slightly acidic pHs. According to the Rice–Herzfeld mechanism, the primary pyrolysis step is cleavage of the weakest bonds in the molecule, such as C–N (~85 kcal/mol) or C–C bonds (~80 kcal/mol).

In an isotope study by Misik et al. (1995), H_2O, D_2O, or a 1:1 mixture of both (1.7 mL) containing the spin trap was added to a Pyrex test tube, which was fixed in the center of a sonication bath with a frequency of 50 kHz. The temperature of the coupling water was 25°C. The sample was sealed with a rubber septum and bubbled with argon through a Teflon tube attached to a fine needle (argon flow rate was 50 mL/min) for 5 min before and during sonication. The time of sonolysis was kept at minimum (typically 45 s) to minimize decay of the spin adduct during sonolysis. After sonication, the electron paramagnetic resonance (EPR) spectrum of the sample was measured. After each experiment, the pHs in the samples were measured and found to be within a range of 6.7 ± 0.3 in all experiments. Immediately after sonication, the samples were transferred to EPR quarts cells, and acquisition of the spectrum typically started within 1 min after sonication. A Varian E9 X-band spectrometer with a 100-kHz modulation frequency and a microwave power of 20 mW was used to record the spectra.

The temperature dependence of the kinetic deuterium isotope effect for the homolytic cleavage of the O–H and O–D bonds of the water molecule was used to estimate the temperature of the region in which this process occurs. The temperature dependence of the kinetic isotope effect has been used previously to study the temperatures of different sonochemical regions in organic liquids. In the semiclassical treatment, quantum mechanical tunneling, which may contribute at lower temperatures, is not considered and the ratio of k_H/k_D for O–H or O–D bond homolytic cleavage is determined internally by the zero-point energy difference of the initial states, and the difference of the zero-point energies of the transition states is neglected. The zero-point energy difference of the ground states (1.24 kcal/mol) was determined from the infrared frequencies of H_2O and D_2O vapor:

$$k_H/k_D = \exp\{1.24 \text{ kcal mol}^{-1}/RT\} \qquad (11.1)$$

The value of k_H/k_D calculated from Equation (11.1) is 8.09, 1.87, and 1.23 at 298 K, 1000 K, and 3000 K, respectively. The intramolecular isotope effect (the k_H/k_D ratio from HOD) is equal to the intermolecular isotope effect within the semiclassical approximation. The equations below show the reactions of radical formation by O–H and O–D bond cleavage and H and D trapping in 1:1 molar mixtures of H_2O and D_2O exposed to ultrasound:

$$XO\text{–}H \rightarrow H + O \tag{11.2}$$

$$H + ST \rightarrow ST/H \tag{11.3}$$

$$XO\text{–}D \rightarrow D + OX \tag{11.4}$$

$$D + ST \rightarrow ST/D \tag{11.5}$$

$$OX + ST \rightarrow ST/OX \tag{11.6}$$

where ST is the spin trap used (PBN, POBN, PYBN, or DMPO) and X is either H or D (in a 1:1 mixture of H_2O and D_2O, 50% of the water molecules are present as HOD). The spin trap method has also been used by Yanagida et al. (1999) to develop a reaction kinetic model of water sonolysis.

11.2.2   Chemical Processes

Although it is generally agreed that the origin of sonochemical effects lies in the bubble collapses and subsequent radical formation, the actual way these collapses achieve a sonochemical effect has been explained by two different main theories. One concept is the electrical theory, which assumes that the extreme conditions associated with collapse are due to an intense electrical field where the collapse is fragmentative. On the other hand, the thermal theory or "hot-spot" theory considers that the collapses are quite adiabatic. Here, the resulting internal pressures and temperatures are so high that vapor molecules dissociate, giving rise to free radical which, when released in the liquid, can react with other species (Sochard et al., 1998).

Water vapor is pyrolyzed to OH radicals and hydrogen atoms, and gas-phase pyrolysis and/or combustion reactions of volatile substances dissolved in water occur. As a result, interfacial regions exist between the
· cavitation bubbles and the bulk solution. Since the temperature in these regions is lower than in the bubbles, a temperature gradient is present in this region. Locally condensed •OH radicals in this region have been reported. Bulk solutions at ambient temperature might undergo reactions of OH radicals or hydrogen atoms that survive migration from the interface. At the same time, the role of supercritical water during cavitation may play an important role in this region.

Supercritical water is a phase of water that exists above its critical temperature and pressure (647 K and 221 atm, respectively). This unique state of water has different density, viscosity, and ionic strength properties than water under ambient conditions. Because the solubility of organic contaminants increases significantly in supercritical water, these organic species are brought into close proximity with the oxidant, usually oxygen from dissolved air. Oxidation rate is therefore several magnitudes higher than wet air oxidation. During sonolysis, it has been proposed that supercritical water is present in a small, thin shell around the bubble. This mode of destruction is expected to be secondary in importance because the fraction of water in the supercritical state is estimated to be on the order of 0.0015 parts out of 100 parts of water. Alternatively, the volume of the gaseous bubble is estimated to be 2×10^4 times greater than the volume of the thin supercritical water shell surrounding the bubble. The value of supercritical water may be limited to increasing the solubility of the organic contaminant near the bubble interface for radical attack. The possible occurrence of supercritical water oxidation in the sonochemical reactor, however, may be one reason to justify fast degradation of organic compounds without O_2.

Pyrolysis is defined as the thermal destruction of a compound in the absence of oxygen. The high temperatures attained within the bubbles are well above the temperatures required to destroy organic materials. This mechanism, however, requires the compound to be present in the vapor phase within the bubble. Compounds with higher vapor pressures will have a higher vapor concentration inside the bubble. It is expected then that pyrolysis will be more prevalent as the vapor pressure of the contaminant increases. During collapse of the bubble, organic species present within the bubble interior would clearly degrade, but because bubble implosion occurs due to the influx of a jet stream of the surrounding liquid it may not be necessary for the organic contaminants to be initially present inside the bubble for degradation to occur. This implosion scenario is analogous to the injection of contaminated liquid directly into the hot reaction zone. Several parameters such as frequency applied have been found to influence the cavitation process. Following are the most important parameters that influence cavitation:

- Frequency
- Intensity
- Solvent
- Bubbled gas
- External temperature and pressure

Strong oxidation as well as reduction reactions have been observed due to generation of H and OH radicals. The main primary chemical process in the sonolysis of water is the thermal dissociation of water to hydrogen atoms and hydroxyl radicals:

$$H_2O \rightarrow H^{\bullet} + {}^{\bullet}OH \qquad (11.7)$$

In the sonolysis of pure water under argon, the formation rates of hydrogen radical and hydrogen peroxide are estimated to be 10.7 and 10.0 μM min^{-1}, respectively. In an oxygen or air atmosphere, hydrogen radicals will react with oxygen as follows:

$$H^{\bullet} + O_2 \rightarrow {}^{\bullet}HO_2 \tag{11.8}$$

$${}^{\bullet}HO_2 + {}^{\bullet}HO_2 \rightarrow O_2 + H_2O_2 \tag{11.9}$$

11.2.2.1 H_2–O_2 Combustion in Cavitation Bubbles

Hart and Henglein (1986) discovered that typical flame reactions occur when ultrasonic waves at intensities sufficient to produce cavitation are propagated through water containing a gas or a mixture of gases. These reactions are brought about by temperatures of several 1000 K that exist in the compression phase of oscillating or collapsing gas bubbles. The yields of such gas-phase reactions in many cases are substantially higher than the yields of reactions occurring in the liquid phase.

In early studies on the formation of hydrogen peroxide by ultrasound in water under various mixtures of oxygen and hydrogen, it has been found that the yield depends on the composition of the mixture in the complex manner. The intermediates during the formation of hydrogen peroxide are free radicals and free atoms, and the question arises whether the radicals can escape from the cavitation bubbles into the bulk solution.

The gas consumption determinations were carried out during sonolysis by direct addition of the H_2O_2 mixture from a syringe to the gas phase to keep gas pressure constant. The rate of H_2O_2 formation as a function of the composition of the gas atmosphere under which the water has been insonated can be observed. Under pure oxygen, H_2O_2 was formed at a rate of 17 μM/min. No H_2O_2 was formed upon insonation under pure hydrogen. The rate of gas consumption is much higher than that of H_2O_2 formation. Therefore, the rate of gas consumption is a function of the composition of the gas atmosphere.

The H_2O_2 combustion into flames is a branched chain reaction, ${}^{\bullet}H$ and ${}^{\bullet}O$ atoms and ${}^{\bullet}OH$ and ${}^{\bullet}HO_2$ radicals are the intermediates, and generally the chains are very long. The combustion in the cavitation bubbles occurs via short chains, at a maximum rate of 220 μM/min. Ozone in oxygen bubbles decomposes at a rate of about 1 mM/min, and nitrous oxide in argon bubbles decomposes at a rate of about 300 μM/min. A rough estimate of the chain length can be obtained using the ratio of the yields of gas consumption and hydrogen peroxide formation. H_2O_2 is formed according to the following reactions:

$$\text{}^\bullet OH + \text{}^\bullet OH \rightarrow H_2O_2 \tag{11.10}$$

$$HO_2^\bullet + HO_2^\bullet \rightarrow H_2O_2 + O_2 \tag{11.11}$$

It must also be taken into consideration that part of the H_2O_2 that reaches the bulk solution is decomposed, as $\text{}^\bullet OH$ and $HO_2^\bullet$ radicals may escape the hot spots and react with H_2O_2 in the bulk solution according to the well-known mechanism:

$$\text{}^\bullet OH + H_2O_2 \rightarrow H_2O + HO_2 \tag{11.12}$$

$$HO_2^\bullet + H_2O_2 \rightarrow H_2O + O_2 + OH \tag{11.13}$$

In addition to these destructive reactions, the radicals may also form H_2O_2 molecules in the bulk solution. The conditions under which the H_2–O_2 combustion occurs in the cavitation bubbles are quite different from those existing in flames. The yields of H_2O_2 and $HO_2^\bullet$ first increase with increasing H_2 concentration as more $\text{}^\bullet H$ atoms are formed and fewer OH radicals are available that could destroy $HO_2^\bullet$ radicals via Equation (11.14):

$$\text{}^\bullet OH + HO_2^\bullet \rightarrow H_2O + O_2 \tag{11.14}$$

This may be due to the fact that the temperature in the compressed cavity bubbles is lower in H_2-saturated water than in O_2, H_2O_2, or air due to the high thermal conductivity of hydrogen. From the solubilities of hydrogen and oxygen, the concentrations of these two gases in the liquid are calculated to be in the molar ratio of 2:1 at an 80:20 composition of the gas atmosphere. The tiny cavitation bubbles are not in thermodynamic equilibrium with the surrounding solution; that is, the gaseous composition does not correspond to that of the gas atmosphere above the insonated liquid. The second maximum of the yield occurs when the cavitation bubbles contain the H_2–O_2 mixture in a 2:1 ratio. In the presence of Fe^{2+} or Cu^{2+} as scavengers for $\text{}^\bullet OH$ and $HO_2^\bullet$, the H_2O_2 yield of 16 μM/min is practically the same as in the absence of these solutes. Although appreciable amounts of $\text{}^\bullet OH$ and $HO_2^\bullet$ radicals are found in the solution, there seems to be no change in the H_2O_2 yield. It thus seems that as much H_2O_2 is destroyed by the radicals in the bulk solution as is formed there.

In the second maximum of the yields, the H_2O_2 yield in the absence of radical scavengers is even greater than in the presence of the scavengers. In this solution, no destruction of H_2O_2 occurs, and this result is understood in terms of the absence of $\text{}^\bullet OH$ radicals. Roughly 50% of the H_2O_2 is formed in the hot spots; the other 50% of H_2O_2 is formed by $HO_2^\bullet$ radicals in the bulk solution.

11.3 Degradation of Organic Pollutants in Aqueous Solutions

Sonolysis of organic pollutants in water involves research on how and what different factors influence the efficiency of sonolysis. For example, it has been proven that the frequency applied is a very important factor influencing the degradation rate: the higher the frequency applied, the higher the resulting removal efficiency. Moreover, current research is concentrated on optimizing the effects of sonolysis. These optimization techniques can be useful for compliance with environmental laws, pollution prevention, and remediation of aqueous wastes. The degradation of different organic compounds by ultrasound and the combination of sonolysis and other advanced oxidation processes such as combining ozonolysis and sonolysis are also very effective.

In the treatment of hazardous wastewater, ultrasonic radiation can decompose water vapor molecules in the bubbles into free radicals such as hydroxyl (OH), hydrogen (H), and hydroperoxyl (HO_2). Evidence for the formation of free radicals by ultrasound in aqueous solution has recently been demonstrated. The hydroxyl radical is particularly reactive with carbon–chlorine and carbon–carbon double bonds and is capable of cleaving the aromatic ring. The primary mechanism is hydroxyl radical oxidation. The severe conditions are enough to break down water vapor within the bubble into hydrogen and hydroxyl radicals, but the highly reactive nature of these radicals prevents a long travel path-length into the solution; therefore, only organic molecules present within the bubble or very near the bubble surface will be destroyed in this fashion. The simultaneous production of the hydrogen radical indicates that reductive pathways may also be available for the destruction of organic pollutants.

Hydrogen peroxide will also be produced by radical combination of two hydroxyl radicals, even though the amount may be too small to be significant. The addition of hydrogen peroxide can increase free-radical concentration in the solution. Local ultrasound intensities can be affected by several factors, such as water level in the ultrasonic tank, position of reaction vessel in the tank, shape of reaction vessel, and solvent level in the reaction vessel. This variation in ultrasound intensity can lead to differences in the progress of the chemical reaction. Hydrogen peroxide and hydrogen gas are now considered to be the principal products formed when the intensity of ultrasonic waves is strong enough to create cavitation propagated through water. The liquid must contain a monoatomic gas such as argon or a diatomic gas such as oxygen for cavitation to occur. Hydrogen atoms, oxygen atoms, hydroxyl radicals, and perhydroxyl radicals are believed to be the intermediates in the production of hydrogen peroxide. The products obtained when water is sonicated have been found to be dependent on the acoustical power, the insonation cell, temperature, external pressure, and dissolved gas present.

11.3.1 Phenol

Petrier et al. (1994) compared sonochemical degradation of phenol at 20 and 487 kHz. The intermediates of the ultrasonic degradation were identified by comparing their retention times through high-performance liquid chromatography (HPLC) and by electron spin resonance (ESR). Hydrogen peroxide concentration has been determined iodometrically. The H_2O_2 determination was not affected by other products that might have formed during the sonochemical reaction. Reactions were performed at 25°C in a cylindrical jacketed glass cell, and the temperature was monitored with a temperature probe immersed in the reacting medium. A 487-kHz wave was emitted, and the system was driven by a high-frequency power supply. The 20-kHz irradiations were carried out with commercial equipment from Branson. The ultrasonic power dissipated into the reactors was estimated by the calorimetric method. The same ultrasonic power (30 W) was delivered at each run.

Exposure of 200 mL of phenol solution to ultrasound at 20 or 487 kHz resulted in a higher rate of loss for the higher frequency. In two cases, hydroquinone (HQ), catechol (CC), and benzoquinone (BQ) were detected as primary intermediates of the degradation process. When the reactor was closed, analysis of the atmosphere showed CO_2 as the only final gaseous product. At 20 kHz, 2% of the theoretical carbon amount was recovered in the gaseous phase after 300 min of treatment; 15% was found for the same irradiation time at 487 kHz. Initial rates of degradation have been determined when the proportion of degraded phenol does not surpass 40% ($k_{20kHz} = 1.12 \times 10^{-6}$ M/min). The rate is dependent on the initial concentration, which reaches a value limit of 1.84×10^{-6} M/min at 20 kHz and 11.6×10^{-6} M/min at 487 kHz. Nevertheless, whatever the theoretical model describing the origins of the molecular activation (thermal and/or electrical), the place where the molecules are brought to an exited state and dissociate is the interior of the bubble of cavitation, which is filled with gas and vapor. In the case of water saturated with air, the first step appears to be cleavage of water and the dioxygen molecule:

$$H_2O \rightarrow {}^\bullet H + {}^\bullet OH \tag{11.15}$$

$$O_2 \rightarrow 2O^\bullet \tag{11.16}$$

Inside the bubble or in the liquid shell surrounding the cavity, these radicals can combine in various ways or react with the gases and vapor present, leading to the detection of H_2O_2:

$$^\bullet H + {}^\bullet OH \rightarrow H_2O \tag{11.17}$$

$$^\bullet OH + {}^\bullet OH \rightarrow O^- + H_2O \tag{11.18}$$

$$^\bullet OH + {}^\bullet OH \rightarrow H_2O_2 \tag{11.19}$$

$$\text{\textbullet OOH} + \text{\textbullet OOH} \rightarrow H_2O_2 + O_2 \qquad (11.20)$$

The main fraction of the H_2O_2 formed during water sonolysis seems to come from the OH and OOH radicals, which combine in the bubble or in the region surrounding the bubble of cavitation in the absence of substrate. The amount of H_2O_2 produced at each of the two frequencies was determined. The concentration of hydrogen peroxide increases linearly vs. time and the rate of formation was found higher at 487 kHz ($4.9 \times 10^{-6}\ M/\text{min}$) than at 20 kHz ($0.75 \times 10^{-6}\ M/\text{min}$). The hydrogen peroxide yield decreases when the phenol concentration increases, assuming that the first step of the phenol degradation results from \textbullet OH radical reaction at a site close to the surface of the bubble. Sonochemical phenol degradation, which proceeds more rapidly at high frequency than at low frequencies, can be related to improved generating rate of \textbullet OH in the solution at higher frequency. The more important cavitational effects occur when the frequency of the ultrasonic wave is equal to the resonance frequency of the bubble. At 487 kHz, when cavitation is located at the gas–liquid interface no formation of particles results from erosion of the emitting surface. To bring water into cavitation requires more energy at 487 kHz than at 20 kHz, and the threshold of cavitation is lower for water saturated with air than degassed water.

As the ultrasonic frequency continued to increase, phenol degradation was most effective at 600 kHz, while little effect was observed at 19.5 kHz (Petrier et al., 1994). The pH in the aqueous solution fell during the ultrasonic irradiation due to the formation of nitrous and nitric acids from nitrogen in the air and dissolved air. The increase of the degradation rate constant (k) was observed at initial pH values in the vicinity of neutrality. The ultrasonic degradation rate of phenol was enhanced by the presence of Fe^{2+}. On the other hand, the degradation rates were little affected if Ag^+, Cu^{2+}, and Ni^{2+} coexisted. The coexistence of ferrous ions seemed to have an effect similar to that of Fenton's reagent, which is to act as an effective oxidant of a wide variety of organic substances because of the formation of hydrogen peroxide during insonation.

Enterazi et al. (2003) described the degradation of phenol in a cylindrical ultrasonic apparatus operating at 35 kHz. The reaction rates were compared to those obtained from devices operating at 20 and 500 kHz. The rate of phenol destruction was higher at 500 kHz than at 35 or 20 kHz; however, the addition of hydrogen peroxide and copper sulfate proceeded more efficiently at 35 kHz, with 30% faster reaction times compared to operation at 500 kHz. The intermediate organic compounds degraded much faster at 35 kHz in the presence of hydrogen peroxide and copper sulfate than at 500 kHz.

11.3.2 Monochlorophenols

Serpone at al. (1994) studied three chorophenols under pulse sonolytic conditions (frequency, 20 kHz; power, 50 W) in air-equilibrated aqueous media. These phenols were totally transformed to dechlorinated, hydroxylated intermediate products via first-order kinetics in about 10 hr for 2-CPOH and 3-CPOH and about 15 hr for 4-CPOH; rate constants for the disappearance of these phenols were $(4.8 \pm 0.4) \times 10^{-3}$/min, $(4.8 \pm 0.5) \times 10^{-3}$/min, and $(3.3 \pm 0.2) \times 10^{-3}$/min, respectively, for approximately 80 μM initial concentration.

The kinetics show two regimes: a low-concentration regime that is zero order in $[CPOH]_l$ and a second regime at higher concentrations where the rate displays saturation-type kinetics reminiscent of Langmuir-type behavior in solid/gas systems. It suggests that the reaction takes place in the solution bulk at low concentrations of chlorophenol, while at the higher concentrations the reaction occurs predominantly at the gas bubble–liquid interface. Chlorophenols are decomposed and dechlorinated almost quantitatively to form hydroxylated aromatic intermediate products; subsequently, species with fewer carbon atoms remained undetectable under these conditions.

Ultrasonic irradiation of aqueous solutions of the chlorophenols was carried out with a Vibra Cell Model VC-250 direct immersion ultrasonic horn (Sonics & Materials; Newtown, CT) operated at a frequency of 20 kHz with a constant power output of ~50 W (the actual insonation power at the solution was 49.5 W, and the power density was 52.1 W/cm^2). Reactions were done in a glass sonication cell (4.4 cm i.d. by 10 cm), similar to the one described by Suslick (1988). The temporal course of the sonochemical processes was monitored by HPLC.

Ultrasonic irradiation (~50 W/cm^2) of a 100-mL air-equilibrated aqueous solution of 4-chlorophenol resulted in the first-order disappearance of the phenol, accompanied after a 1-hr delay by the first-order growth of Cl$^-$. The pH of the isonated solution dropped gradually from the initial value of 5.1 to 3.5 after 11 hr. Sonolysis of the aqueous solution of 3-chlorophenol showed an induction period of ~90 min following which its concentration decreased via first-order kinetics. The pH of the insonated solution of 2-chlorophenol decreased at first to 4.9 and then recovered to its near initial value until 9 hr of insonation, when it dropped abruptly to pH 4.4 and remained constant. The initial drop in pH that occurred during the induction period was also observed for the first-order disappearance of the phenol. It therefore suggests that there are various possible sites where reactions may occur in sonochemistry.

Combined EPR and spin-trapping studies showed that solvent vapor and ambient gases (e.g., air) decompose to atoms or free radicals in the gaseous bubble interior. Water vapor is thermally dissociated into •OH radicals, H• atoms, and •O atoms. The latter interconvert with •OH radicals at the high pressures in the cavity and recombine in the cooler interfacial region to form O_2 and H_2O_2. The power dependence in the sonolytic transformation of a phenolic aqueous solution was found to be the first order.

Monochlorophenols represent an important class of environmental water pollutants of moderate toxicity to mammalian and aquatic life; they possess relatively strong organoleptic effects, with taste thresholds of 0.1 μg/L (ppb). Their principal sources are the natural degradation of chlorinated herbicides (e.g., chlorophenoxyacetic acids), chlorination of phenolic substances in waste effluents, and chlorine treatment of drinking water. Of the three chlorophenols examined here, 3-chlorophenol exhibited the greatest resistance to sonolysis.

Except for the cavity interior, the possible location for oxidation of these substrates by direct or sensitized photolysis, flash photolysis, ultraviolet/peroxide, or irradiated semiconductor (SC) particulates suggests that the sonochemical oxidation process physically mirrors the heterogeneous photocatalytic process such as UV/TiO_2.

11.3.3   2-Chlorophenol

Ku et al. (1997) detected no formation of H_2O_2 during the sonication process in pure water and suggested that the decomposition of 2-chlorophenol due to oxidation of H_2O_2 is negligible. Many researchers have found that the decomposition rate by sonolysis decreased with increasing solution pH values. The decomposition of 2-chlorophenol may take place in the gaseous (bubbles), film, and bulk solution regions by either pyrolysis or free radical attack. Because the temperature in the gaseous region can increase up to 5000 K, pyrolysis of the organics most likely happens in the gaseous region rather than in the film region (about 1000 K). Free-radical attack of 2-chlorophenol is assumed to be minimal in the bulk solution region because the free radicals generated in the cavitation bubbles are barely transferred to the room-temperature bulk solution region through the much higher temperature film region. For neutral and acidic solutions, molecular 2-chlorophenol species in the bulk solution region diffuse into the film region, and part of the molecular species may even evaporate into the gaseous (bubbles) region from the gas–liquid interface.

Depending upon the speciation of an organic pollutant at various pHs, degradation may take place at different regions of the microbubbles. This is because the solubility or vapor pressure varies greatly according to the neutral molecule or ionized species. At neutral and acidic solutions (pH < 8), the molecular species predominates and the ionic species is predominant in the alkaline solution for pH greater than 9. These two 2-chlorophenol species may possess different reaction behaviors during the sonication process. For neutral and acidic solutions, molecular 2-chlorophenol species in the bulk solution region diffuse into the film region, and part of the molecular species may even evaporate into the gaseous (bubble) region from the gas–liquid interface. Thus, the overall decomposition of 2-chlorophenol can be attributed to the pyrolysis, and free-radical attack occurs in both the gaseous and film regions. On the other hand, the ionic species of 2-chlorophenol predominating in the alkaline solutions cannot evaporate into the gaseous region, leaving the decomposition of 2-chlorophenol to occur only in the film region, therefore reducing the decomposition rate. These reactions are illustrated in Figure 11.1. The gradual decrease of the concentration of organic intermediates after certain reaction times and the increase of CO_2 concentration suggest that the complete mineralization may occur via the formation of organic intermediates.

The presence of dissolved oxygen in aqueous solution was reported to play a very important role in the generation of highly oxidative hydroxyl free radicals; therefore, the free-radical attack in both gaseous and film regions is strongly influenced by the dissolved gas presented in aqueous solutions. Under uncontrolled conditions, the temperature of solution might

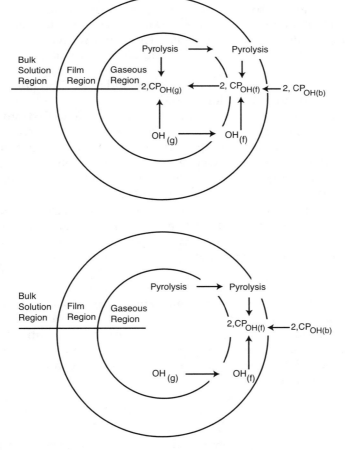

FIGURE 11.1
Reaction regions in the microbubble at various pHs. (From Ku, Y. et al., *Water Res.*, 31(4), 929–935, 1997. With permission.)

be increased during sonication because of the transformation of pressure into heat. Many researchers have mentioned very controversial results regarding the effect of solution temperature on the decomposition of organics by sonication.

11.3.4 Chlorinated C_1 and C_2 Volatile Organic Compounds

Bhatnagar and Cheung (1991) investigated sonochemical destruction of chlorinated C_1 and C_2 volatile organic compounds in dilute aqueous solution. Methylene chloride (CH_2Cl_2), chloroform ($CHCl_3$), carbon tetrachloride (CCl_4), 1-2-dichloroethane ($C_2H_4Cl_2$), trichloroethylene (C_2HCl_4), and perchloroethylene (C_2Cl_4) were exposed to 20-kHz ultrasound in a batch reactor to determine the efficacy of ultrasonic process in the destruction of undesirable compounds in water. Aqueous mixtures of two or more volatile compounds were also sonicated to investigate the potential of sonication in more realistic situations and to study the kinetics of individual halomethanes in the presence of other halomethanes. The initial concentration of VOCs ranged from 350 to 50 mg/L in water and decreased with time as a result of sonication.

When dilute aqueous solution of C_1 compounds and C_2 compounds were exposed to ultrasound under ambient conditions, the concentration decreased exponentially vs. sonication time according to first-order kinetics. pH control must be adjusted for optimizing the rate of sonochemical destruction for chlorinated compounds. The values of the first-order rate constants for the compounds were essentially unchanged by the presence of other reacting species. The rate constant for CH_2Cl_2 in the mixture was 0.0345 and 0.0458/min for CCl_4. Sufficient cavitation bubbles and free radicals were produced to oxidize all of the volatile organic components of the mixture. Conventionally, the reaction rate changing with temperature can be described by the Arrhenius equation. The reaction could take place inside the cavitation bubble, where the temperature and pressure are so high that the solute molecule breaks down, or in the bulk liquid phase, where the free radicals generated as a result of high-intensity ultrasonic waves oxidize the target molecule. In the absence of any solute, these radicals lead to the generation of H_2O_2. If the reaction takes place in the cavitation bubbles and all the experimental conditions such as reaction vessel size, temperature, power supplied, and pressure remain unchanged, the destruction of the compound can be expressed by the first-order rate equation:

$$-d[C_xH_yCl_z]/dt = k[C_xH_yCl_z] \qquad (11.21)$$

where $C_xH_yCl_z$ is the target compound. A fit to the experimental data for all the C_1 compounds indicates first-order kinetics. The rate constants are listed in Table 11.2.

TABLE 11.2

First-Order Rate Constants for Sonochemical Destruction of VOCs

Compound	k (min^{-1})	Estimated Error (min^{-1})	Norm/N_{data}	P_{vap} (20°C) (Torr)
	0.033	0.002	2.080	348.900
$CHCl_3$	0.043	0.005	3.500	160.000
CCl_4	0.043	0.002	0.380	90.000
1,2-Dichloroacetic acid	0.021	0.003	2.480	64.000
Trichloroethylene	0.021	0.001	0.632	57.800
Perchloroethylene	0.026	0.004	1.086	20.000
1,1,1-Trichloroethane	0.046	0.003	1.067	100.000

Source: Bhatnagar, A., and Cheung H.M., *Environ. Sci. Technol.*, 26, 1481, 1991. With permission.

It seems that the cavities enclose a vapor of the solute because of the high vapor pressure of these compounds. The primary reaction pathway for these compounds appears to be the thermal dissociation in the cavities. The activation energy required to cleave the bond is provided by the high temperature and pressure in the cavitation bubbles. This leads to the generation of radicals such as hydroxyl radical, peroxide radical, and hydrogen radical. These radicals then diffuse to the bulk liquid phase, where they initiate secondary oxidation reactions. The solute molecule then breaks down as a result of free-radical attack. The oxidation of target molecules by free radicals in the bulk liquid phase under normal operating pressures and temperatures can be presented by a second-order rate equation:

$$-d[C]dt = k_1[C][OH] + k_2[C][H] + k_3[C][H_2O_2] \qquad (11.22)$$

where C is the concentration of the target compound.

The concentration of free radicals in the equation is a function of the power input by sonication. C_2 compounds probably disintegrate by both the pyrolysis type of reaction in the cavitation bubble and free-radical attack in the liquid phase. Physical operating conditions such as steady-state temperature and initial pH of the solution were found to have little effect upon the destruction rate of the compound. The simplicity and flexibility along with the high efficiency of destruction indicate the potential of a sonochemical-based process to become a competitive technology for water treatment.

11.3.5 Pentachlorophenate

The ultrasonic wave effect on the degradation of pentachlorophenate (PCP) at 530 kHz was studied by Petrier et al. (1992). PCP was chosen as a model because it belongs to the class of aromatic halides that have to be removed from wastewater. It is still a widely used compound for wood treatment and exhibits high toxicity. The system is operated at 530 kHz with a 20-W high-frequency electrical power source. Ultrasonic intensity determined by a calorimetric method was 1.06 W/cm^2. PCP (0.5 M) was prepared as stock concentrated solution from equal molar pentachlorophenol and NaOH. Experiments were performed at pH = 7 in 10^{-3}-M phosphate buffer medium with 10^{-4} M PCP solutions. The irradiated volume (100 mL) was continuously bubbled with gas through a gas dispersion fritted disk at a 25-mL/min flow rate.

When PCP solution (10^{-4} M) under continuous air bubbling is subjected to ultrasound effects, the characteristic absorption bands decrease and the treatment leads to a complex mixture of products. Carbon–chlorine bonds are rapidly cleaved, and after a 150-min sonication time, 90% of the chlorine is recovered in the solution as chloride ions. PCP transformation in aerated solution occurs together with nitrite and nitrate formation. Carbon dioxide is a product of PCP degradation, and it has long been recognized as an inhibitor for sonochemical reactions.

Sonochemical reactions are strongly affected by ambient gas because the temperature inside the collapsing bubble is in close relationship with the polytropic ratio (C_p/C_v) and the thermal conductivity of the gas. In addition, reactions with gases such as O_2, N_2, and CO_2 are directly affected by the high temperatures reached during the collapse of the bubbles. As a consequence, reactive species available for PCP degradation and their rate of production will depend on the nature of the gas.

Under argon bubbling, the degradation is faster than under air or oxygen; no PCP is detected after 50-min sonication. From these experiments, it was postulated that CO_2 is transformed to CO as observed in high-temperature chemistry because of the high temperature inside the cavitation bubble.

11.3.6   *para*-Nitrophenol

Kotronarou et al. (1991) detected temperatures on the order of 2000 K at the gas/liquid interfacial region. Sonochemical reactions are characterized by the simultaneous occurrence of pyrolysis and radical reactions, especially at high solute concentrations. Any volatile solute will participate in the former reactions because of its presence inside the bubbles during the oscillations or collapse of the cavities. In the solvent layer surrounding the hot bubble, both combustion and free radical reactions are possible.

The ultrasonic irradiation of aqueous solution of PNP was carried out with a Branson 200 sonifier that was operated at 20 KHz and with an output of electrical power of 84 W. Reactions were performed in a stainless-steel, continuous-flow reaction cell operated in the batch mode. All reactions were carried out with air-saturated solutions, and the concentration of p-benzoquinone (p-BQ) was determined spectrophotometrically. Hydrogen peroxide was determined iodometrically and fluorometrically. Kinetic experiments were continued after cleaning the cell thoroughly and filling with fresh solution. Loss of PNP or its degradation products by absorption to the HPLC filters was not observed.

The concentration of PNP decreased exponentially with sonication time. The ultrasonic degradation of PNP followed apparent first-order kinetics with the first-order rate constant of $k_1 = 3.7 \times 10^{-4}$/s. H_2O_2 was produced during the decay of PNP by sonication. The decrease of H_2O_2 during the post-irradiation period can be attributed to the reaction of H_2O_2 with NO_2^- to yield NO_3^-. Henglein and Gutierrez (1990) showed that H_2O_2 was generated by the action of ultrasonic waves on pure water. Under these experimental conditions, sonolysis of the solvent alone led to a linear increase of $[H_2O_2]$ vs. irradiation time with a rate constant of 4×10^{-8} (M/s). At longer irradiation times, the observed concentrations of H_2O_2 were lower in PNP solutions than in pure water. The initial rate of formation of H_2O_2 was the same as in pure water, regardless of the PNP concentration.

The sonolysis of pure water produced a linear increase of $[H^+]$ as a function of time with an apparent zero-order rate constant of 8.3×10^{-9} (M/min). On the other hand, in sonicated PNP solutions the increase of $[H^+]$ appeared to be first order with apparent first-order rate constant of $k_1 = 3 \times 10^{-4}$/s. PNP decayed exponentially with time at all pH values. The apparent first-order rate constant decreased from $k_1 = 3.67 \times 10^{-4}$/s at $pH_i = 5$ to $k_1 = 2.0 \times 10^{-4}$/s at $pH_i = 8$. At $pH_i > 10$, the apparent first-order rate constant increased slightly because of the slow thermal reaction between PNP and the products of the base-catalyzed decomposition of H_2O_2. Sonication of air-saturated solutions not exposed to air produces the same results as those obtained with air-equilibrated solutions.

The action of ultrasound on PNP in aqueous solutions resulted in its degradation with the formation of NO_2^-, NO_3^-, and H^+ as primary products and HCO_2^-, $C_2O_4^{2-}$, 4-NC, and H_2O_2 as secondary products. Several reaction mechanisms can be postulated to account for these observations. Because PNP boils at $T > 166°C$, high-temperature reactions of PNP vapor inside the cavitating bubbles can be neglected due to its low vapor pressure. In aerated solutions, the peroxide radicals formed will recombine to generate H_2O_2:

$$2HO_2^{\bullet} \rightarrow H_2O_2 + O_2 \tag{11.23}$$

For every pair of $^{\bullet}OH$ radicals that react with PNP, one molecule of H_2O_2 is produced. This means that, in PNP solutions, the same amount of peroxide is produced from hydroxyl radical reactions as by sonolysis of pure water. In the absence of PNP, $^{\bullet}OH$ recombines to form hydrogen peroxide:

$$2^{\bullet}OH \rightarrow H_2O_2 \tag{11.24}$$

Hydrogen atoms are known to react with PNP, but the rate constant of this reaction has not been measured. PNP decreases exponentially with sonication time, while PNP attacked by $^\bullet$OH in homogenous solution follows zero-order kinetics. Extensive investigations of high-temperature reactions of $^\bullet$OH with aromatic compounds have demonstrated that at $T > 400$ K the main reaction pathway is hydrogen atom abstraction instead of hydroxyl radical addition to the aromatic ring.

11.3.7 *para*-Nitrophenyl Acetate

Hua et al. (1995) proposed a supercritical water region in addition to two reaction regions such as the gas phase in the center of a collapsing cavitation bubble and a thin shell of superheated liquid surrounding the vapor phase. Chemical transformations are initiated predominantly by pyrolysis at the bubble interface or in the gas phase and attack by hydroxyl radicals generated from the decomposition of water. Depending on its physical properties, a molecule can simultaneously or sequentially react in both the gas and interfacial liquid regions.

p-Nitrophenol (p-NP) is degraded completely by sonolysis to yield short-chain carboxylic acids. Intermediate products resulting from both hydroxyl radical attack and thermal bond cleavage are detected. Hydroxyl radical attack on p-NP is thought to occur in a region of the bubble interface with $T < 440$ K, while pyrolysis occurs in a hotter interfacial region with an average temperature of 900 K.

Supercritical water (SCW) provides an additional phase for chemical reaction. This phase of water exists above the critical temperature (T_c) of 647 K and the critical pressure (P_c) of 221 bar and has physical characteristics between those of a gas and a liquid.

The irradiation horn was immersed (3 cm below the surface) into the sample solution. The average power delivered to the aqueous phase was 115 W, which corresponds to an intensity of approximately 96 W/cm^2. Solutions of 100-μM p-NP were adjusted to pH 4.8 to 5.2 with phosphoric acid, and sonolysis reactions were performed in a modified stainless-steel cell on a total volume of 25 mL. The reaction chamber was stirred with a magnetic stirring bar and mixing motor. Compressed air was blown continuously through the converter to minimize changes in temperature of the piezoelectric crystal. pH was adjusted to the range of 4.8 to 5.2 to ensure that all of the p-NP (pK_a = 7.16 at I = 0.05 M) was in the neutral form such that it would preferentially partition to the gas bubble interfaces. The reaction solution was sparged for 15 min at a flow rate of 10 to 15 mL/min with the appropriate gas. Constant temperature was maintained.

The reaction rate constant for carbon–nitrogen bond cleavage in nitrobenzene has been determined to be:

$$k = A \exp\{-E_a/RT\} \tag{11.25}$$

where $A = 1.9 \times 10^{15}$ s and $E_a/R = 33{,}026$ K. Using these values for A and E_a/R, the effective temperature of the cavitation bubbles can be estimated. The relative temperature of bubble collapse can be adjusted by saturating the solution with gases characterized by substantially different specific heats and thermal conductivities. An important factor controlling bubble collapse temperature is the polytropic factor, K, of the saturating gas. From the knowledge of K we can estimate the maximum temperature during bubble collapse from Equation (11.25).

The value of K is associated with the amount of heat released from the gas inside the bubble during adiabatic compression. As K increases, the heat released upon bubble collapse also increases. Additional physicochemical properties that may influence the temperature attained during bubble collapse include thermal conductivity (L) and gas solubility. A low thermal conductivity favors high collapse temperatures because the heat of collapse will dissipate less quickly from the cavitation site. A gas with both a low thermal conductivity and high water solubility should yield the highest temperature upon cavitational bubble collapse. The sonolytic degradation of p-NP has been found to be a first-order reaction for all gases. Therefore, the effective average temperatures at the interface of the collapsing bubbles for each gas can be estimated.

To analyze the overall rate of reaction, where E_a is the intrinsic activation energy, A is the preexponential factor, $A = k_B T/h$, in which $R = 8.31$ J/mol, $k_B = 1.38 \times 10^{-23}$ J/K, and $h = 6.626 \times 10^{-34}$ J/s, E_a and A are determined experimentally from kinetic data. Given that the interface between the hot gas in the bubble and the surrounding cooler liquid is hydrophobic, organic compounds will partition to that region much more effectively than ions. Over a pH range of 3 to 8, the overall first-order rate constants are essentially invariant in a sonicated solution. This pH dependency is consistent with the tendency for water in a dense supercritical state to have a higher ion activity product relative to normal-phase water; a higher value for water equilibrium constant (K_w) should result in a higher concentration of both OH$^-$ and H$^+$ for a given set of conditions.

Flint and Suslick (1991) and Seghal and Wang (1989) clearly demonstrated that temperature and pressure within a collapsing cavitation bubble exceed the critical point of water, on the basis of previously estimated temperatures within a collapsed bubble and a smaller layer of surrounding liquid. However, no experimental data are available for the density of nuclei or actual cavitation bubbles in water during ultrasonic irradiation or SCW accelerated chemical reactions.

Ultrasonically irradiated solution of p-NPA exhibits an observed rate constant that is enhanced by about two orders of magnitude in comparison to the same hydrolysis under ambient conditions at 25°C. Observations in unsonified solutions suggest that the reaction rate enhancement occurs at the cavitation–bubble interface. Both the enhanced hydrolysis rate and its pH independence may be explained by the existence of SCW around the collapsing cavitation bubbles. The apparent decrease of $S^=$ in the presence of ultrasound suggests that the rate enhancement might be due to physical changes in the microscopic environment surrounding the substrate.

Supercritical water represents a potentially important component of sonochemistry, in addition to the free-radical reactions and thermal/pyrolytic effects. Because the reaction occurs at or close to the bubble/water interface, compounds more hydrophobic than *p*-NPA are expected to exhibit even higher hydrolysis rate enhancements. Finally, the existence of the supercritical phase in an ultrasonically irradiated solution suggests a modification of the conventional view of the reactive area at the cavitation site. This region is normally considered to consist of two discrete phases: a high-temperature, low-density gas phase and a more condensed, lower temperature liquid shell.

11.3.8   Nitrobenzene

Sonochemical reactions of nitrotoluenes in the presence of potassium permanganate oxidize the –CH_3 group to –$COOH$. *p*-Nitrophenol degraded primarily by denitration during pyrolysis, with free-radical attack being of minor importance, to produce NO_2^-, NO_3^-, benzoquinone, hydroquinone, 4-nitrocatechol, formate, and oxalate. Sonolytic by-products from 2,4,6-trinitrotoluene (TNT) suggest that the first step is hydroxyl radical attack on the methyl group to form 2,4,6-trinitrobenzoic acid. Other products include trinitrobenzene, hydroxylated TNT derivatives, acetate, formate, glycolate, oxalate, CO_2, NO_3^-, and NO_2^-.

Cropek and Kemme (1997/1998) studied the destruction of nitrobenzene (NB) and dinitrobenzene (DNB). Aqueous solutions of these two species were sonolytically treated under two different frequencies. Instrumentation for monitoring the progress of degradation included one or more of the following: a total organic carbon instrument, a liquid chromatograph with a UV/visible absorbance detector, and a gas chromatograph/mass spectrometer (GC/MS).

DNB is more resistant to sonolysis than NB. Because the production of by-products does not track the disappearance of DNB, volatility is the primary mode of DNB loss. The poor solvation of DNB by water as well as the large solution surface area allows rapid volatility. Because NB is more soluble in water, it remains available for sonolysis longer and produces a wider variety of by-products than sonolysis of DNB; therefore, organic compounds that are less soluble in water will be more difficult to effectively treat by sonolysis unless steps are taken to increase the presence of the contaminant in the irradiation zone. It is also noted that a higher irradiation frequency is more efficient for sonolysis. More intermediates are observed at shorter irradiation time than with a lower ultrasonic frequency. In addition, a different subset of intermediates was detected, indicating alternate sonolytic pathways are followed.

The work performed on DNB and NB illustrates two main problems with the use of ultrasound for contaminant degradation. First, under the given experimental conditions, the process is slow. Therefore, the fate of future applied research may rest in the ability to show favorable comparisons to other treatment processes in terms of both cost and efficiency. Second, as mentioned above, the agitation produced by ultrasonic irradiation initiates more volatilization than degradation.

11.3.9 Hydroxybenzoic Acid

Nagata et al. (1996) studied the decomposition of aqueous solutions of hydroxybenzoic acid (HBA) by ultrasound irradiation under atmospheric air and under argon. Their study showed that the kinetics followed first-order kinetics during initial stages. Based on the finding that the decomposition of 3-HBA is completely inhibited by the addition of 0.1-mM t-BuOH, which is an effective scavenger of hydroxyl radicals. Therefore, the main degradation mechanism is the reaction of 3-HBA with hydroxyl radicals produced during sonolysis. Unlike volatile and hydrophobic organics, the thermal decomposition of HBA in cavitation bubbles and the interfacial region was suggested to have a smaller effect on the overall degradation process. Under air, oxygen molecules participate in the decomposition in cavitation bubbles and the interfacial region; however, under argon atmosphere the degradation mainly proceeds via reactions with OH radicals in the bulk solution.

Table 11.3 lists the pseudo first-order rate constants for the decomposition of hydroxybenzoic acids. It shows that the ratio of the mean initial decomposition rate of HBAs under argon, $R(HBA)_{Ar}$ to that under air, $R(HBA)_{air}$ is equal to $(3.0 + 4.9 + 5.1)/(2.7 + 3.4 + 3.1) = 1.4$ (see Table 11.3). The ratio of the rate of ${}^{\bullet}OH$ radical formation under argon to that under air was estimated to be $R({}^{\bullet}OH)_{Ar}/R({}^{\bullet}OH)_{air} = 20/15 = 1.3$ and is close to the value of $R(HBA)_{Ar}/R(HBA)_{air}$ which suggests that the decomposition of HBAs is mainly caused by ${}^{\bullet}OH$ radicals, and oxygen molecules have little effect on the decomposition.

Sonolytic decomposition of chlorophenol in water was enhanced in the presence of Fe(II), assuming that the Fenton oxidations occur:

$$H_2O_2 + Fe\ (II) \rightarrow {}^{\bullet}OH + OH^- + Fe\ (III) \qquad (11.26)$$

$${}^{\bullet}OH + chlorophenols \rightarrow Cl^- + CO_2 + H_2O \qquad (11.27)$$

By analogy, it may be possible that an appropriate amount of Fe(II) addition will increase the efficiency of decomposition of HBAs by ultrasound.

TABLE 11.3

Pseudo First-Order Rate Constants for Decomposition of Hydroxybenzoic Acids

	Argon		Air	
Compound	$10^2 \times k$ (min^{-1})	R^2	$10^2 \times k$ (min^{-1})	R^2
2-Hydroxybenzoic acid	3.0	0.9981	2.7	0.9875
3-Hydroxybenzoic acid	4.9	0.9997	3.4	0.9972
4-Hydroxybenzoic acid	5.1	0.9999	3.1	0.9988
3,4-Dihydroxybenzoic acid	1.9	0.9760	1.9	0.9835
GA	2.6	0.9875	5.5	0.9955
TA	6.4	0.9997	16.4	0.9869

Note: R is the correlation coefficient.

Source: Nagata, Y. et al., *Environ. Sci. Technol.*, 30, 1133–1138, 1996. With permission.

11.3.10 Chlorinated Hydrocarbons

Cheung et al. (1991) investigated sonication of 1,1,1-trichloroethane and trichloroethylene by using a W-385 Ultrasonicator and a 2-L glass reaction vessel equipped with a stainless-steel cooling coil capable of delivering 475 W of ultrasound energy. The pH in the reactor decreased rapidly in all cases, indicating the probable formation of HCl from the chlorinated reactants. The pH vs. sonication time follows the first-order kinetics. The first-order rate constants of 1,1,1-trichloroethane and trichloroethylene are (3.93×10^{-2}) and (2.7×10^{-3}) min^{-1}, respectively. The pH results for carbon tetrachloride are particularly interesting as it contains no hydrogen. The results suggest that it is likely that the hydrogen required for HCl formation is obtained from the water. The rate of the HCl formation seems to peak at approximately pH 4. It appears that pH may be useful in driving the reaction toward HCl as the final chlorinated product. The GC/MS results indicate that no other chlorinated organics are present in the sonication of the methylene chloride.

11.3.11 Chloroform

Hua and Hoffmann (1996) used a Branson ultrasonic cleaning tank (model ATH 610-6) and power generator (model EMA 30-6) to study the destruction of chloroform. The total amount of chloroform in each bottle was 8 mg. After 2 hr of exposure to ultrasound, the remaining chloroform and chloride concentrations were measured by gas chromatography and ion chromatography. The chloroform concentration in the gas phase was estimated by Henry's law, with a Henry's constant of 0.150 at 24.8°C. The average recovery was 99.6%, with a standard deviation of 3.2%. After 2 hr of ultrasonic treatment, the remaining amount of chloroform in solution varied over a wide range, from 0.9 mg to 6.8 mg, because of the existence of methanol, which could inhibit free-radical reactions.

Hydrogen peroxide can serve as an initiator. The optimal concentrations for H_2O_2 are those with a molar ratio of H_2O_2 to $CHCl_3$ between 30 and 50. Addition of Fe^{++} will increase the reaction efficiency due to Fenton's reaction:

$$Fe^{2+} + H_2O_2 \rightarrow Fe^{3+} + OH^- + {}^{\bullet}OH \tag{11.28}$$

This ferrous ion could be considered a catalyst for the decomposition of hydrogen peroxide, which reached a maximum at pH of 3.5. The first-order reaction had a rate constant of 0.0177/min. Therefore, the ultrasound/H_2O_2 process is quite effective for dechlorination of chloroform; however, a 50:1 ratio of H_2O_2 to $CHCl_3$ is quite high and unacceptable. Much effort should be put toward developing a better reactor design to reduce reaction time and/ or more efficient catalysts to decrease the amount of hydrogen peroxide applied. The kinetics and mechanisms of CCl_4 degradation have been developed using the mass balance of chlorine atoms during CCl_4 sonolysis (Hua and Hoffmann, 1996). The decomposition of CCl_4 during ultrasonic irradiation is significant, with first-order rate constants ranging from 10^{-2} to 10^{-3}/s. The Cl mass balance for the ultrasonic irradiation of CCl_4 in aqueous solution can be written as follows:

$$\text{Chlorine atom yield} = [Cl^-] + [HOCl] + 4[C_2Cl] + 6[C_2Cl_6] + 4[CCl_4]_f$$
$$+ 4([CCl_4]_i - [CCl_4]_f) \tag{11.29}$$

A typical chlorine balance was >70%, with Cl^- found to be the dominant product. Hexachloroethane and tetrachloroethylene were the only organic by-products detected and were not present when the initial CCl_4 concentration was low (19.5 µM). A peak concentration of 2.8 µM was observed after 15 min of sonication. The chromatographic retention times of the peaks in each sample were in agreement within 0.36 s. In the absence of ultrasonic irradiation, *p*-NP does not degrade in a saturated solution of CCl_4; however, the addition of CCl_4 during sonication of *p*-NP in an Ar-saturated aqueous solution enhanced the rate constant of *p*-NP degradation by a factor of 4.5 compared to sonication without CCl_4. Degradation during sonolysis appears to be as follows:

$$CCl_4 \rightarrow {}^{\bullet}CCl_3 + {}^{\bullet}Cl \tag{11.30}$$

$${}^{\bullet}CCl_3 \rightarrow {}^{\bullet}CCl_2 + {}^{\bullet}Cl \tag{11.31}$$

Formation of dichlorocarbene (${}^{\bullet}CCl_2$) is also thought to occur by the simultaneous formation of two chlorine atoms:

$$CCl_4 \rightarrow {}^{\bullet}CCl_2 + Cl_2 \tag{11.32}$$

In the presence of oxidizing species, the trichloromethyl radical can act as a scavenger of hydroxyl radicals:

$$^{\bullet}CCl_3 + {}^{\bullet}OH \rightarrow HOCCl_3 \tag{11.33}$$

Sufficient quantities of trichloromethyl radical are formed. Therefore, recombination and radical scavenging occur in parallel. Ozone can react directly with solute molecules and also decomposes in the gas phase to yield oxygen atoms and hydroxyl radical as follows:

$$O_3 \rightarrow O_2 + O \tag{11.34}$$

$$^{\bullet}O + H_2O \rightarrow 2^{\bullet}OH \tag{11.35}$$

Aqueous ozone decomposition in solution is strongly influenced by pH and results in a variety of reactive species, including peroxide radical and hydrogen peroxide. During the decomposition of C_2Cl_4 by O_3, arrangement of molozonide to the ozonide and subsequent hydrolysis yields phosogene and hydrogen peroxide. The reactive pathways of CCl_4 in a sonicated, aqueous solution are in agreement with those observed during sonolysis of the pure liquid. Multicomponent waste streams are a more realistic matrix to consider for practical treatment situations. The sonication of a mixture of CCl_4 and *p*-NP results in enhancement of *p*-NP degradation and demonstrates that ultrasound is not limited to single-solute solutions. Because of the variety of reaction pathways and reactive regions that occur during cavitation, competition between different solutes is minimized. The effect of CCl_4 in a mixed wastestream is particularly interesting because it releases a residual oxidant that can continue to attack other refractory molecules in solution after the ultrasonic irradiation is halted.

11.3.12 CFC 11 and CFC 113

Cheung and Kurup (1994) constructed a batch reactor with a 35-mL working volume and a circulating reactor with a 250-mL working volume. This unit produces 20-kHz ultrasound with up to 475 W maximum power. Also included in the system was a closed reactor body into which the ultrasonicator horn was threaded with provision being made for temperature control via a cooling jacket and for headspace analysis. Periodic calibrations were performed for chlorofluorocarbon (CFC)-11 using solutions in which the CFC concentrations were set via careful volumetric measurements and relatively large volumes to ensure accuracy. The solutions used were prepared by adding the requisite amount of the CFC to 2 L of deionized, distilled water. The temperature and pH were monitored online for the circulating reactor. The pH fell from its initial value of 7.4 to 5.4 by the end of run, indicating the formation of acidic species. The final reported k_1 values were the first-order rate constants for the rate of chlorofluorocarbon destruction scaled by the power per volume delivered by the ultrasonicator. Less than 5% of the original CFC-11 was volatized, with the majority of the material being destroyed in the liquid phase. This is consistent with similar measurements made on other originated hydrocarbons.

Runs at 10°C for both CFC-11 and CFC-113 were conducted in the circulating reactor system to examine temperature effects. In both cases, the apparent first-order rate constant declined slightly. The decline was somewhat lower for CFC-113 than for CFC-11, which may even have remained unchanged. A sonochemical approach has the advantage of not requiring any transference of the target molecule from an aqueous phase as would be required in combustion or catalytic oxidation.

11.3.13 Parathion

Mills and Hoffmann (1992) investigated ultrasonic degradation of parathion. Parathion (*O,O*-diethyl *O-p*-nitrophenyl triphosphate) is a major pesticide used in large quantities worldwide. Organophosphate esters such as parathion have been used as alternatives to DDT and other chlorinated hydrocarbon pesticides; however, the organophosphate esters are not rapidly degraded in natural waters. At 20°C and pH 7.4, parathion has a hydrolytic half-life of 108 days and its toxic metabolite, paraoxon, has a similar half-life of 144 days. Ultrasonic irradiation of 25 mL of parathion-saturated, deionized water solution was conducted in a water-jacketed, stainless-steel cell with a Branson 200 sonifier operating at 20 kHz and ~75 W/cm². The temperature of the sonicated solution was kept constant at 30°C. All sonolytic reactions were carried out in air-saturated solution. The concentration of the parathion hydrolysis product *p*-nitrophenol (PNP) was determined in alkaline solution with a Shimadzu MPS-2000 UV/visible spectrophotometer.

Sulfate, nitrite, nitrate, *p*-nitrophenol, phosphate, and oxalate were identified as products of parathion sonolysis. The maximum solubility of parathion in water is 82 μM (24 μg/mL) at 25°C. After 2 hr of sonolysis, [SO_4^{2-}] was found to be 82 μM. All the initial parathion was degraded in <2 hr at 30°C under sonolysis at 20 kHz with an intensity of 75 W/cm². In addition, the [SO_4^{2-}] increased linearly with sonolysis time of a zero-order rate constant of $k_0 = 0.68$ μM/min. Because the reactants were unbuffered, the pH of the solution changed during sonication. After 30 min of sonication, the pH dropped from 6.1 to 4.1 and remained close to that value. The observed decrease in pH is consistent with the formation of H^+ during ultrasonic irradiation of PNP under similar experimental conditions. Over a broad pH range, *p*-nitrophenol has been found to be the major hydrolysis product of parathion. When parathion is exposed to electrohydraulic cavitation during sonolysis, both the P=S and the P–C bonds are broken to produce SO_4^{2-} and PNP. The chemical effects of ultrasonic irradiation are a direct result of acoustic cavitation. Sound waves traveling through water with frequencies greater than 15 kHz force the growth and subsequent collapse of small bubbles of gas in response to the passage of expansion and compression waves. The greater coupling occurs when the natural resonance frequency of the bubbles equals the ultrasonic frequency.

Parathion appears to undergo thermal decomposition in the hot interfacial region in the collapsing cavitation bubbles and degrades secondarily via reaction with •OH radicals. Pyrolysis in the interfacial region is predominant at high solute concentrations, while at low solute concentrations free-radical reactions are likely to predominate. In the bulk solution, the chemical reaction pathways are similar to those observed in aqueous radiation chemistry. Sonolysis is relatively inefficient with respect to total input energy. Only a small portion of total energy supplied to the direct immersion horn system results in useful free-radical reactions.

The same energy input dispersed over a broader area results in significant enhancement of reaction rates and energy utilization efficiency. The low energy utilization efficiency may limit the use of direct probe sonolysis to special applications such as groundwater remediation or for low-flow pretreatment of hazardous industrial wastes. However, sonolysis does not require the addition of chemical additives to achieve a viable degradation rate.

11.3.14 Hydantoin Chemicals

Hydantoin compounds are biorefractory compounds. The structure of the parent compound consists of a five-member ring containing two nitrogen atoms and two carbonyl groups. These compounds are polar and nonvolatile and have higher water solubility in solution, acting as weak acids due to ionization. Hydantoin compounds can be found in the stripped-gas liquor stream at a coal gasification plant. Gas produced at such a plant is initially cooled and scrubbed in a spray washer. As stripped gas liquor is recycled for cooling purposes, the increasing concentration of hydantoin compounds in this cooling water can affect the speciation and aquatic chemical behavior of metal ions in the water. However, existing techniques for the treatment of industrial effluent such as ultrafiltration and activated carbon adsorption simply transfer biorefractory compounds from one phase to another without ultimately destroying these compounds.

The reaction solution (10 mM in a 50-mM conical flask) was sonicated in a Pulsatron 330 ultrasonic cleaning bath (38 kHz) filled with 5 L of water by Schwikkard et al. (1990). The reaction was maintained at 25°C by circulating the water in the ultrasonic bath through cooling coils in a cold water bath. The ultrasonic bath was characterized by calculating its power output, by measuring the initial temperature rise of the water in the ultrasonic bath, and by measuring the formation of 10 mM hydrogen peroxide in water in the reaction vessel over 24 hr. The sonochemical degradation of 50 mg/M

1-methylhydantoin over 24 hr was studied with either 332 mg/M hydrogen peroxide or oxygen by placing an equivalent solution in a constant temperature bath maintained at 25°C by Schwikkard et al. (1990).

Hydrogen peroxide in water in the reaction vessel (a maximum of 3.5 mg/M over 25 hr) indicated that the ultrasonic bath was sufficiently powerful to produce hydroxyl radicals (due to ultrasonic cavitation). The degradation of 1-methylhydantoin over 24 hr increased from 30 to 62% with the addition of hydrogen peroxide. Ultrasonic cavitation leads to the formation of hydroxyl radicals; hence, 1-methylhydantoin is degraded through the oxidation reaction with hydroxyl radicals. No degradation of 1-methylhydantoin occurred in the control without hydrogen peroxide added, indicating that ultrasound supplied sufficient energy to the system to overcome the required activation energy of the hydroxyl radical reaction.

11.3.15 Methanol

Buttner et al. (1991) studied the sonolytic effects in water–methanol mixtures under argon and oxygen conditions. Formaldehyde and peroxide have been identified as the main products. In addition, products similar to the ones formed during the pyrolysis or combustion of methanol have been observed with relative yields changing with the composition of the mixture (see Table 11.4).

In pure water, hydrogen was formed at a rate of 22 μM/min, and the hydrogen yield strongly increased with increasing methanol concentration up to a maximum at 10 vol%. $^{\bullet}CH_3$ and $^{\bullet}CH_2OH$ radicals have been detected in the sonolysis of water–methanol mixtures by an ESR spin-trapping method, with maximum yields also occurring at a methanol concentration close to 10 vol%. (Under oxygen and at low methanol concentrations, the main product was CO_2, which indicates that the combustion of methanol vapor is almost complete under these conditions:

TABLE 11.4

Relative Yields in the Pyrolysis and Sonolysis of Methanol

Reaction Product	Pyrolysis at 1160 K	Sonolysis	
		10 vol%	40 vol%
H_2	100	100	100
CO	32	30	39
CH_2O	16	27	78
CH_4	0.9	16	39
C_2H_4	—	2.0	2.0
C_2H_6	—	1.8	1.7

Source: Buttner, J. et al., *J. Phys. Chem.*, 95, 1528–1530, 1991. With permission.

$$CH_3OH + 1.5O_2 \rightarrow CO_2 + 2H_2O \qquad (11.36)$$

Incomplete combustion of methanol can be described by the following over-all processes:

$$CH_3OH + O_2 \rightarrow CH_2O + H_2O_2 \qquad (11.37)$$

$$CH_3OH + 2O_2 \rightarrow CO + 2H_2O_2 \qquad (11.38)$$

$$CH_3OH + 1.5O_2 \rightarrow HCOOH + H_2O_2 \qquad (11.39)$$

The maximal hydrogen peroxide yield occurs in a 40-vol% methanol solution irradiated under argon–oxygen atmospheres. This behavior is quite different from previous observations of H_2O_2 yield in pure water sonicated under various argon–oxygen mixtures, where a maximum of 30 vol% O_2 was observed due to cavitation bubbles. The increase in the yield of hydrogen peroxide with increasing oxygen concentration, therefore, cannot be attributed to a change in temperature on the bubbles but to an increase in the efficiency of methanol combustion. Sonochemical yields in a pure organic liquid are much smaller than in the sonolysis of pure water; however, in certain mixtures of the two liquids, much higher yields than those resulting from the sonolysis of the components can be observed. The phenomenon can be understood in terms of the hot-spot theory of sonochemical reactions, according to which temperature reached in the adiabatic compression phase of cavitation bubbles depends on the heat content of the gases and vapors in the bubbles.

11.3.16 Polymer and Iodide

Henglein and Gutierrez (1990) reported that redox reactions are initiated by free radicals formed by the thermal decomposition of solvent or solute molecules in compressed cavitation bubbles. Temperatures of several thousand Kelvin can be reached in the adiabatic compression phase of such bubbles. On the other hand, main-chain degradation is caused by hydrodynamic shear forces that appear in the vicinity of oscillating or collapsing cavitation bubbles. Redox reactions occur with appreciable yields only in the presence of a mono- or diatomic gas; the yield of macromolecule degradation depends little on the nature of the gas and can still be large in the presence of apolyatomic gases such as N_2O. In other words, the degradation process depends much less on the cavitational conditions than the free-radical reactions.

When a polymer such as polyacrylamide is exposed to ultrasound, the decrease in chain length of the polymer can be a measure of the viscosity of the solution. In the case of continuous irradiation, the approximation is based on the formula:

$$n = n_0(1/1 + k_d^0 t_0) \qquad (11.40)$$

where n is viscosity, n_0 is the relative viscosity before irradiation, and k_d^0 is the degradation constant for continuous irradiation. Because of the linear relationship, one viscosity determination after irradiation is sufficient to obtain the degradation constant. Similarly, for pulse irradiation, we have:

$$n = n_0(1/1 + k_d t) \qquad (11.41)$$

where n is viscosity, n_0 is the degradation constant, and t is the irradiation time for pulse irradiation.

In the study by Henglein and Gutierrez, a commercial 20-kHz sound generator (KLN System 582; horn diameter, 14 mm) was also used. The solution (20 mL) was irradiated in a beaker (30-mm i.d.), with the distance between the horn and the bottom of the vessel being 24 mm.

The experiment was carried out with a 0.2-M KI solution in the absence and presence of poly(acrylamide), with both high- and low-molecular-weight samples being used. Under all conditions, the maximum rate of iodine showed a rather steep decrease at higher intensities. The maximum had been observed previously and was attributed to a change in cavitation conditions. The bubbles combine at high intensities to form large bubbles in which adiabatic compression does not produce temperatures as high as in the smaller, resonating bubbles. In the presence of high-molecular-weight polymer, the viscosity of the solution was strongly increased; however, the iodine yield was slightly greater in the presence of polyacrylamide, the effect being practically the same for samples of high and low molecular weight. This result can be attributed to a chemical interference of the polymers with the oxidation of iodide. The degradation and also the thermal decomposition of the polymer produce organic peroxy radicals.

The maximum iodine yield is shifted to lower intensities with decreasing pulse length because it takes time to activate the liquid (to form gas bubbles). If this time is shorter than the pulse length, chemical effects do not occur. The specific rate of degradation under continuous irradiation reaches its maximum at 100 W. The degradation of the polymer occurs relatively effi-ciently under conditions where the gas bubbles coalesce. Collapsing bubbles still produce strong shear forces in the liquid even under conditions where adiabatic compression temperature is not very high.

The decrease in polymer degradation was less pronounced than that of the iodine oxidation. High temperatures are not required to generate effec-tive shear forces in the vicinity of the cavitation bubbles; however, the effects of the two kinds of sonication cannot be simply compared by their intensities. Polymer degradation is very much favored under sonication conditions where the gas bubbles combine. In such conditions, a certain number of free radicals, which may oxidize organic compounds or form H_2O_2, are generated.

11.3.17 Hydrogen Sulfide

Flosdorf and Chambers (1933) reported that metal sulfides were oxidized in the presence of audible sound (1 to 15 kHz) while investigating the bactericidal action of audible sound; however, Schmitt et al. (1929) were the first researchers to observe the rapid oxidation of dissolved H_2S gas to colloidal sulfur during sonication at 750 kHz with a 250-W power source. They reported that an increase in the total pressure of the system (P_{02}) led to higher oxidation rates up to a limiting critical pressure. This critical pressure depended on the amount of dissolved H_2S gas and the intensity of irradiation. The primary oxidation product was found to be elemental sulfur. The overall reaction was thought to proceed via reactions of HS^- with $^•OH$ radicals, $HO_2^•$ radicals, or H_2O_2.

Sound waves with frequencies higher then 16 kHz traveling through a liquid can force the growth and subsequent collapse of small bubbles in response to the passage of expansion and compression waves. The radius of the bubble prior to collapse can be estimated to be on the order of several hundred micrometers (400 to 500 μm). The time scale for the collapse of the bubble is <100 ns, and the time scale for heat diffusion in the surrounding liquid is a few microseconds with a corresponding diffusion length of a few hundred nanometers. Water vapor present in the cavities undergoes thermal dissociation to give hydrogen atoms ($^•H$) and hydroxyl radicals ($^•OH$). Solutes present near the bubble/water interface can also undergo thermal decomposition. Secondary reactions take place in the liquid phase between solute molecules and radical species, mainly $^•OH$ in the case of aqueous solutions, escaping from the gas phase into solution. $^•OH$ is a powerful and efficient chemical oxidant in both the gas and liquid phases. Reactions of OH with inorganic and organic substrates are often near the diffusion-controlled rate.

The ultrasonic irradiation of aqueous sulfide solutions was conducted with a Branson 200 sonifier operating at 20 kHz by Kotronarou et al. (1992). Reactions were performed in a 50-mL water-jacketed, stainless-steel cell from Sonics & Materials. The temperature inside the reaction vessel was kept constant at 25°C. All irradiations were carried out in air-saturated solutions at $t = 0$. A Haake A80 water circulating and temperature-controlling system was used for temperature control. Hydrogen peroxide was analyzed fluorometrically. Deionized water was used to prepare all the solutions.

Ultrasonic irradiation of 25-mL solutions of bivalent sulfur at pH 10 (borate, I = 0.06 M) resulted in a linear decrease of [S(–II)] with sonication time. A post-irradiation oxidation of sulfide to sulfate was observed and was attributed to the subsequent thermal reaction of HS^- with H_2O_2, which was formed during the sonolysis of water. In the control experiments, a linear increase of $[H_2O_2]$ vs. sonication time was observed with a zero-order rate constant of 2.52 μM/min. As expected, only SO_4^{2-} was present in the H_2O_2-treated samples. Therefore, the fraction of the total sulfide, $[S(-II)]$ that is in the form of HS^- increases with increasing pH. At pH around neutral, 50% of the total $S(-II)$ is present in the form of HS^-, and at pH > 9 practically all $S(-II)$ is present in the form of HS^-. The results in alkaline solution are in good agreement with the results from radiation chemistry. The higher SO_4^{2-} / HS^- ratios observed can be explained by further oxidation of the reaction intermediates by H_2O_2. Following is the proposed rate law for the $S(-II)$ by H_2O_2:

$$d[S(-II)/dt = k_1[H_2S][H_2O_2] + k_2[HS^-][H_2O_2] \qquad (11.42)$$

where $k_1 = 0.5$ M/s and $k_2 = 29$ M/s. Based on this expression, the half-life for $[S(-II)]$ at $[H_2O_2] = 20$ μM and at pH 10 is 1.2 min. In alkaline solutions, the main product of $S(-II)$ oxidation by H_2O_2 is SO_4^{2-}. It is therefore reasonable to assume that ultrasonic oxidation of $S(-II)$ in alkaline solutions is the result of the direct reaction of HS^- with $^\bullet OH$. The generation of $^\bullet OH$ during water sonolysis follows the second-order reaction. The dependence of the zero-order rate on $[S(-II)]_i$ is similar to what has been reported for the ultrasonic oxidation of ferrous sulfate. The observed decrease in the apparent zero-order rate constant at pH > 10 can be partly explained by the dissociation of OH in alkaline solutions:

$$^\bullet OH + OH^- \underset{k_b}{\overset{k_f}{\rightleftarrows}} O^{\bullet -} + H_2O \qquad (11.43)$$

where $k_f = 1.2 \times 10^{10}$ M/s and $k_b = 9.3 \times 10^7$ M/s.

O^- reacts more slowly with the same substrates than $^\bullet OH$. The thermal decomposition of H_2S occurs within the collapse cavitation bubbles or within the gas–liquid interface. $S(-II)$ is destroyed via two different pathways during sonication: oxidation by $^\bullet OH$ and thermal decomposition.

Reaction of HS^- with $^\bullet OH$ is the main pathway for the oxidation of $S(-II)$ at pH >10. When O_2 is present, the rate of $S(-II)$ oxidation increases linearly with initial sulfide concentration. At pH < 8.5, thermal decomposition of H_2S within or near collapsing cavitation bubbles becomes the important pathway. The formation of free radicals upon sonication has been traditionally used to define the "chemicoacoustic" or "sonoacoustic" efficiency of ultrasonic irradiation. The sonolytic efficiency would then depend on the chemical compound of interest and could be higher than the chemicoacoustic efficiency. The sonoacoustic efficiency of the system analyzed here is of the same order of magnitude as those reported by other investigators working under different sonication conditions. The yields of water sonolysis depended only on the total energy deposited and were independent of the frequency and intensity distribution of the ultrasonic field.

In summary, chlorinated compounds such as carbon tetrachloride, trichloroethylene, chloroform, pentachlorophenate, methylene chloride, and chlorofluorocarbons have high vapor pressures and tend to concentrate within the bubbles, allowing fast degradation to chloride ions, CO_2, and other inorganic species. The degradation mechanism is pyrolysis rather than hydroxyl radical attack. Phenolic compounds can be mineralized with benzoquinone, hydroquinone, and catechol as the organic intermediates and oxalic, maleic, acetic, formic, and propanoic acids as the short-chain polar intermediates. Other similar species include methyl phenols, hydroxyanisoles, resorcinol, and chlorophenols. The sonolytic degradation of other chemical species includes dextran, sodium hypochlorite, iodide and bromide, arylalkanes, and glyceraldehyde. All the species can be completely mineralized if sufficient frequency and energy are provided.

11.4 Engineering Application

The unusual microenvironments created during acoustic cavitation permit a wide variety of uses for ultrasound. These applications include homogenization and cell disruption in the biological field, dissolution and mixing in chemistry, soldering and welding in metal working, degreasing and emulsification in industry, and polymerization and depolymerization. Novel chemical applications result from the catalytic effects of sonochemistry, including the creation of new compounds (for example, new iron carbonyl compounds), new and more effective catalysts, and greater rates and yields on chemical reactions. Over the past several years, ultrasonic irradiation has been applied to degrade a variety of organic pollutants. Specifically, the technology is applicable for aqueous media for the treatment of hazardous organic pollutants.

In general, ultrasonic irradiation in environmental remediation is primarily applied to:

- Volatile organic compounds
- Phenolic compounds
- Pesticides
- Dyes

Using ultrasonic irradiation as a treatment technology for the degradation of organic pollutants in aqueous solutions offers several benefits. One important advantage is the equipment used in the ultrasonic processes. The equipment is generally of simplistic design and is very easy to apply and maintain (Vinodgopal et al., 1998). The cost is also a factor that has been proven advantageous in some cases when compared to other treatment technologies.

The following beneficial sonochemical effects can be expected when using sonolysis as an advanced oxidation process:

- Decrease of reaction time and/or increase of yield
- Use of less forcing conditions (e.g., lower reaction temperature)
- Possible switching of reaction pathway
- Use of less or no phase-transfer catalysts
- Forced reactions with gaseous products by degassing
- Use of crude or technical reagents
- Activation of metals and solids
- Reduction of any induction periods
- Enhancement of the reactivity of reagents or catalysts
- Generation of useful reactive species

Ultrasonic degradation of organic pollutants in aqueous solution depends strongly on the nature of the organics. For example, hydrophobic and volatile molecules such as tetrachloromethane and TCE are degraded mainly by direct thermal decomposition in the cavitation bubbles (Drijvers et al., 1999). The escape of volatile organic compounds from the reaction chamber requires special attention in many of the processes using sonolysis. Studies have shown that the technology will work more efficiently in aqueous solutions where the organic pollutants are well soluble in water. Sonolysis is a relatively new technology, and many of the applications have still taken place only at an experimental level; therefore, field applications should be a priority to prove the true effectiveness of the technology.

Ultrasonic irradiation investigations are focused on further optimizing the processes involved in sonolysis. Future work in sonolysis should study the effects of different experimental parameters on the kinetics and mechanisms of destruction. First, the effects of solution temperature should be studied. Maximum cavitation occurs in water at 35°C, and it is expected that this would also increase the degradation rate. Conversely, for some processes, increasing the temperature actually decreases the rate. Lack of a prominent temperature effect would reduce the energy required to heat or cool the solution in full-scale treatment systems. Second, the effect of sparge gas should be studied. Use of noble gases increases the cavitation temperature due to both a decrease in the kinetic degrees of freedom of the gas molecule where energy can be lost and a lower thermal conductivity may prevent heat from dissipating from the bubble.

References

Barnartt, *Q. Rev.*, 7, 84, 1953.

Bhatnagar, A. and Cheung H.M., Sonochemical destruction of chlorinated Cl and C2 volatile organic compounds in dilute aqueous solution, *Environ. Sci. Technol.*, 26, 1481–1486, 1991.

Buttner, J., Gutierrez, M., and Henglein, A., Sonolysis of water–methanol mixtures, *J. Phys. Chem.*, 95, 1528–1530, 1991.

Cheung, H.M. and Kurup, S., Sonochemical destruction of CFC11 and CFC113 in dilute aqueous solution, *Environ. Sci. Technol.*, 28, 1619–1622, 1994.

Cheung, H.M., Bhatnagar, A., and Jansen, G., Sonochemical destruction of chlorinated hydrocarbons in dilute aqueous solution, *Environ. Sci. Technol.*, 25, 1510–1512, 1991.

Drijvers, D., Van Langenhove, H., and Beckers M., Decomposition of phenol and trichloroethylene by the ultrasound/H_2O_2/CuO process, *Water Res.*, 33(5), 1187–1194, 1999.

Entezari, M.H., Petrier, C., and Devidal, P., TI sonochemical degradation of phenol in water: a comparison of classical equipment with a new cylindrical reactor, *Ultrason. Sonochem.*, 10(2), 103–108, 2003.

Flint, E.B. and Suslick, K.S., *Science*, 253, 1397, 1991.

Flosdorf, E.W. and Chambers, L.A., *J. Am. Chem. Soc.*, 55, 305, 1933.

Hart, E.J. and Henglein, A., *J. Phys. Chem.*, 90, 5992, 1986.

Hart, E.J. and Henglein, A., Sonochemistry of aqueous solutions: H_2–O_2 combustion in cavitation bubbles, *J. Phys. Chem.*, 91, 3654–3656, 1991.

Henglein, A. and Gutierrez, M., Effects of continuous and pulsed ultrasound: a comparative study of polymer degradation and iodide oxidation, *J. Phys. Chem.*, 94, 5169–5172, 1990.

Hua I. and Hoffmann M.R., Kinetics and mechanism of the sonolytic degradation of CCl_4 intermediates and by-products, *Environ. Sci. Technol.*, 30, 864–871, 1996.

Hua, I. and Hoffmann, M.R., Optimization of ultrasonic irradiation as an advanced oxidation technology, *Environ. Sci. Technol.*, 31(8), 2237–2243, 1997.

Hua I., Hochemer, R.H., and Hoffmann, M.R., Sonolytic hydrolysis of *p*-nitrophenyl acetate: the role of supercritical water, *J. Phys. Chem.*, 99, 2335–2342, 1995.

Kotronarou, A., Mills, G., and Hoffmann M.R., Ultrasonic irradiation of *p*-nitrophenol in aqueous solution, *J. Phys. Chem.*, 95, 3630–3638, 1991.

Kotronarou, A., Mills, G., and Hoffmann, M.R., Oxidation of hydrogen sulfide in aqueous solution by ultrasonic irradiation, *Environ. Sci. Technol.*, 26, 2420–2428, 1992.

Ku, Y., Chen, K.-Y., and Lee, K.-C., Ultrasonic destruction of 2-chlorophenol in aqueous solution, *Water Res.*, 31(4), 929–935, 1997.

Laborde, J.-L., Bouyer, C., Caltagirone, J.-P., and Gerard, A., Acoustic cavitation field prediction at low and high frequency ultrasounds, *Ultrasonics*, 36, 581–587, 1998.

Misik, V., Miyoshi, N., and Riesz, P., EPR spin-trapping study of the sonolysis of H2O/D2O mixtures: probing the temperatures of cavitation regions, *J. Phys Chem.*, 99, 3605–3611, 1995.

Nagata, Y. et al., Decomposition of hydroxybenzoic and humic acids in water by ultrasonic irradiation, *Environ. Sci. Technol.*, 30, 1133–1138, 1996.

Petrier, C., Micolle, M., Merlin, G., Luche, J., and Reverdy, G., Characteristics of pentachlorophenate degradation in aqueous solution by means of ultrasound, *Environ. Sci. Technol.*, 26, 1639–1642, 1992.

Petrier, C., Lamy, M., Francony, A., Benahcene, A., David, B., Renaudin, V., and Gondrexon, N., Sonochemical degradation of phenol in dilute aqueous solution: comparison of the reaction rates at 20 and 487 kHz, *J. Am. Chem. Soc.*, 98(41), 10514–10520, 1994.

Renaud, *Bull. Soc. Chim. Fr.*, 1044, 1950.

Richards and Loomis, *J. Am. Chem. Soc.*, 49, 3086, 1927.

Schmitt, F.O., Johnson, C.H., and Olson, A.R., *J. Am. Chem. Soc.*, 51, 370, 1929.

Schwikkard, G.W., Winship, S.J., and Buckley, C.A., Sonochemical Degradation of Hydantoin Chemicals, Project Report, Department of Chemical Engineering, University of Natal, Republic of South Africa, 1990.

Sehgal, C.M. and Wang, S.Y., *J. Am. Chem. Soc.*, 103, 6606, 1989.

Serpone, N., Terzian, R., Hidaka, H., and Pelizzetti, E., Ultrasonic-induced dehalogenation and oxidation of 2-, 3-, and 4-chlorophenols in air-equilibrated aqueous media: similarities with irradiated semiconductor particles, *J. Phys. Chem.*, 98, 2634–2640, 1994.

Sochard, S., Wilhelm, A.-M., and Delmas, H., Gas-vapor bubble dynamics and homogenous sonochemistry, *Chem. Eng. Sci.*, 53(2), 239–254, 1998.

Suslick, K.S., *Ultrasound: Its Chemical, Physical, and Biological Effects*, VCH Publishers, New York, 1988.

Suslick, K.S. and Hammerton, D.A., The site of sono-chemical reaction transactions on ultrasonics, *Ferroelectrics and Frequency Control*, 33, 143–150, 1986.

Suslick, K.S., Hammerton, D.A., and Cline, D.E., *J. Am. Chem. Soc.*, 108, 5641, 1986.

Vinodgopal, K., Peller, J., Makogon, O., and Kamat, P.V., Ultrasonic mineralization of a reactive textile azo dye, Remazol Black B, *Water Res.*, 32(12), 3646–3650, 1998.

Yanagida, H., Masubuchi, Y., Minagawa, K., Ogata, T., Takimoto, J.I., and Koyama, K., A reaction kinetics model of water sonolysis in the presence of a spin trap, *Ultrasonics Sonochemistry*, 5, 133–139, 1999.

12

High-Energy Electron Beam

12.1 Introduction

Electron beams have been in commercial use since the 1950s. Early applications involved the cross-linking of polyethylene film and wire insulation. Later, the process was extended to include sterilization of medical supplies, rubber vulcanization, disinfection of wastewater, food preservation, and many more applications. In solving environmental problems, the process has been shown to be efficient for the destruction of several classes of hazardous organic compounds as well as the inactivation of total coliform and bacteria in sewage sludge (Kurucz et al., 1991). It can be applied to treat wastewater from numerous industries such as the food, health, pharmaceutical, pulp and paper, and textile sectors. The groups of compounds from these wastewaters may contain benzene; substituted benzenes such as toluene, *m*-xylene, and *o*-xylene; phenol; halogenated ethenes such as trichloroethylene (TCE) and tetrachloroethylene (PCE); halogenated methanes such as trihalomethanes (THMs); carbon tetrachloride; and methylene chloride. Table 12.1 shows the removal efficiency of these organic compounds in aqueous solution by high-energy electron beam, as a function of solute concentration, absorbed dose, pH, and scavengers in potable water and raw and secondary wastewaters. This innovative treatment process is not only limited to simple toxic organic chemicals but is also applicable to complex mixtures of organic pollutants under varying water quality.

12.2 Chemistry of Aqueous Electrons

12.2.1 Formation of Radical Species

The irradiation of pure water with high-energy electrons generated by an accelerator results in the rapid formation of electronically excited states and/or free radicals in 10^{-14} to 10^{-9} s. These reactive species will react with organic

TABLE 12.1

Summary of Overall Removal Efficiencies for Various Organic Compounds
Tested at the Miami Electron Beam Research Facility

Compound	% Removal	Required Dose (krad)
Chloroform	83–99	586–650
Bromodichloromethane	>99	80
Dibromochloromethane	>99	80
Bromoform	>99	80
Carbon tetrachloride	>99	80
Trichloroethylene (TCE)	>99	57–500
Tetrachloroethylene (PCE)	>99	241–500
trans-1,2-Dichloroethane	93	800
		800
1.1 – Dichloroethene	>99	
		800
1.2 – Dichloroethene	60	800
Hexachloroethane	>99	800
1,1,1-Trichloroethane	89	650
1,1,2,2-Tetrachloroethane	88	650
Methylene chloride	77	800
Benzene	>99	49–650
Toluene	97	45–650
Chlorobenzene	97	650
Ethylbenzene	92	650
1,2-Dichlorobenzene	88	650
1,3-Dichlorobenzene	86	650
1,4-Dichlorobenzene	84	650
l,m-Xylene	91	650
o-Xylene	92	650
Dieldrin	>99	800
Total phenol	88	37–800
TNT[a]	40	800–1700
DEMP[a]	90	150–780
DMMP[a]	90	220–950

[a] Bench-scale studies using ^{60}Co.

Source: Kurucz, C.N. et al., *Radiat. Phys. Chem.*, 45(2), 299–308, 1995. With permission.

and inorganic compounds in aqueous solutions. The reactions result in either oxidation or reduction of compounds and lead to the formation of end products such as CO_2, H_2O, and inorganic salts. After 10^{-12} s, an ionization species such as e_{aq}^-, $OH^{\bullet}$, $H_3O_{aq}^+$ and molecular fragments resulting from dissociation of excited-state molecules are in thermal equilibrium with the medium. After 10^{-7} s of irradiation, the radiolysis products begin to diffuse, resulting in a fraction of them reacting to form radical products. Because the energy required to produce a chemical change is only a few electron-volts (eV) per molecule, a high-energy electron is capable of initiating several thousand reactions. These reactions can be summarized by the following equation:

$$H_2O \xrightarrow{e^-} [2.7]^\bullet OH + [2.6]\, e_{aq}^- + [0.6]^\bullet H + [2.7]H_3O^+ + [0.45]H_2 + [0.7]H_2O_2$$

$$(12.1)$$

Removal efficiency by high-energy electron beam is usually expressed by G values. The G value is defined as the number of excited states, radicals or other products formed or lost in a system when 100 eV of energy is absorbed (Kurucz et al., 1995b). The G values for each species is shown in the brackets in Equation (12.1). Removal efficiency of solutes can be quantitatively expressed in terms of two constants, G_D and k'. G_D describes the percent removal of solute at a given dose. It is defined by the disappearance of the solute in aqueous solution and is determined experimentally using the following equation (Kurucz et al., 1991):

$$G_D = [\Delta C \bullet N_A / D(6.24 \times 10^{17})]$$

$$(12.2)$$

where ΔC is the difference in organic solute concentration (M) at dose D and zero dose after irradiation; D is the dose in krad; N_A is Avogadro's number (6.02×10^{23}); and 6.24×10^{17} is the conversion factor from moles to 100 eVL^{-1}.

12.2.2 Hydroxyl Radical

Of the chemical species formed in Equation (12.1), the most reactive are the oxidizing species such as hydroxyl radical ($^\bullet OH$) and the reducing species such as aqueous electron (e_{aq}^-) and hydrogen atom ($^\bullet H$). The concentration of the reactive radicals formed in the irradiated solutions can be determined according to the fact that 1 krad of irradiation adsorbed will form 1.04 M reactive species when G equals 1 for a solute. Thus, for a G of 2.7, the concentration of $^\bullet OH$ from Equation (12.1) is 0.28 mM at an absorbed dose of 100 krad. Based on these calculations, the total reactive radical species concentration formed usually ranges from 0.40 to 2.23 mM for $^\bullet OH$. The presence of both oxidizing and reducing species is unique to this process and distinguishes it from other treatment processes.

Hydroxyl radicals, $^\bullet OH$, can undergo several types of reactions with chemical species in aqueous solution. The types of reactions that are likely to occur are hydroxylation, hydrogen abstraction, electron transfer, and radical–radical recombination. Hydroxylation reaction occurs readily with aromatic and unsaturated aliphatic compounds, which result in the formation of hydroxylated radicals:

$$^\bullet OH + CH_2 = CH_2 \rightarrow HOCH_2 - CH_2^\bullet$$

$$(12.3)$$

Hydrogen abstraction occurs with saturated molecules such as aldehydes and ketones:

$$\text{·OH} + CH_3\text{--}CO\text{-}CH_3 \rightarrow \text{·}CH_2COCH_3 + H_2O \qquad (12.4)$$

When hydroxyl radical is reacting with inorganic species, electron transfer will occur after aqueous solutions are irradiated with high energy electrons. For example, halogen ions (X^-) will react with hydroxyl radical as follows:

$$\text{·OH} + X^- \rightarrow X^\bullet + OH \qquad (12.5)$$

$$X^\bullet + X^- \rightarrow X_2^- \qquad (12.6)$$

12.2.3   Hydrogen Peroxide

When two hydroxyl radicals recombine, hydrogen peroxide (H_2O_2) is formed:

$$\text{·OH} + \text{·OH} \rightarrow H_2O_2 \qquad (12.7)$$

H_2O_2 can also be generated in oxygenated aqueous solutions by the reactions of e_{aq}^- and ·H with O_2. The reactions result in the formation of reduced oxygen, the superoxide radical ion, and/or the conjugate acid:

$$e_{aq}^- + O_2 \rightarrow O_2 \qquad (12.8)$$

$$H + O_2 \rightarrow HO_2^\bullet \qquad (12.9)$$

These reactions lead to the formation of H_2O_2:

$$2O_2^- + 2H^+ \rightarrow H_2O_2 + O_2 \qquad (12.10)$$

$$HO_2^- \rightarrow H_2O_2 + O_2 \qquad (12.11)$$

A further possible reaction that may result is:

$$e_{aq}^- + H_2O_2 \rightarrow \text{·OH} + OH^- \qquad (12.12)$$

This reaction occurs with a second rate constant of 1.2 to 1.4×10^{10} M^{-1} s^{-1} and suggests that the addition of H_2O_2 would increase the OH· concentration, which would increase the removal efficiency of organic pollutants.

12.2.4   Aqueous Electron

Aqueous electron, e_{aq}^-, is a powerful reducing reagent with an $E° = 2.7$ V. It reacts with many hazardous halogenated and nonhalogenated organic com-

TABLE 12.2

Second-Order Rate Constants (M^{-1} s^{-1}) of Selected Compounds
for Reactivity with Aqueous Electrons

Compound	$k_{e_{aq}^-}$ (M^{-1} s^{-1})
Benzene	9.0×10^6
Carbon tetrachloride	1.6×10^{10}
Chlorobenzene	5.0×10^8
Chloroform	3.0×10^{10}
1,2-Dichlorobenzene	4.7×10^9
1,3-Dichlorobenzene	5.2×10^9
1,4-Dichlorobenzene	5.0×10^9
trans-1,2-Dichloroethylene	7.5×10^9
Nitrobenzene	3.7×10^{10}
Phenol	2.0×10^7
Tetrachloroethylene (PCE)	1.3×10^{10}
Toluene	1.4×10^7
Trichloroethylene (TCE)	1.9×10^9
Vinyl chloride	2.5×10^8

Source: Buxton, G.V. et al., *J. Phys. Chem. Ref. Data*, 17, 513–886, 1988. With permission.

pounds during their removal from aqueous solutions. The reaction of an aqueous electron, e_{aq}^-, is a single electron transfer process:

$$e_{aq}^- + RCl \rightarrow R^\bullet + Cl^- \tag{12.13}$$

The rate constants of organic pollutants with e_{aq}^- are listed in Table 12.2.

12.2.5 Hydrogen Radical

Hydrogen radical ($^\bullet$H) accounts for approximately 10% of the total free radical concentration in irradiated water. $^\bullet$H undergoes two general types of reactions with organic compounds:

- Hydrogen addition:

$$^\bullet H + C_6H_6 \rightarrow C_6H_7 \tag{12.14}$$

- Hydrogen abstraction:

$$^\bullet H + CH_3OH \rightarrow H_2 + {}^\bullet CH_2OH \tag{12.15}$$

The second-order rate constant of $^\bullet$H with common radical scavengers is relatively small. However, it is large enough to account for the removal of organic compounds. The high-energy electron beam process is the only process that produces these radicals. Table 12.3 summarizes the reaction rate constants of major organic pollutants with $^\bullet$OH, e_{aq}^-, and $^\bullet$H.

TABLE 12.3

Rate Constants of Organic Pollutants (M^{-1} s^{-1})

Organic Chemicals	•OH	e^-_{aq}	•H
Alkyl-substituted benzene			
m-Xylene	7.5×10^9	—	2.6×10^9
Ethylbenzene	7.5×10^9	—	—
o-Xylene	6.7×10^9	—	2.0×10^9
Toluene	3.0×10^9	1.4×10^7	2.6×10^9
Aromatic compound			
Benzene	7.8×10^9	9.0×10^6	9.1×10^8
Phenol	6.6×10^9	2.0×10^7	1.7×10^9
Chlorinated and brominated aliphatic compounds			
Tetrachloroethylene	2.8×10^9	$1.3 \times 101°$	—
trans-1,2-Dichloroethylene	6.2×10^9	7.5×10^9	—
Trichloroethylene	4.0×10^9	1.9×10^9	—
Chlorinated benzene compounds			
Chlorobenzene	5.5×10^9	5.0×108	1.4×10^9
1,3-Dichlorobenzene	—	5.2×10^9	—
1,4-Dichlorobenzene	—	5.0×10^9	—
1,2-Dichlorobenzene	—	4.7×1^9	—
Chlorinated ethene			
Vinyl chloride	1.2×10^{10}	2.5×10^8	—
Chlorinated methanes			
Chloroform	$\sim5.5 \times 10^6$	3.0×10^{10}	1.1×10^7
Carbon tetrachloride	—	1.6×10^{10}	3.8×10^7
Methylene chloride	9.0×10^7	6.3×10^9	4.0×10^6
Heterocyclic aromatic compounds			
Pyridine	3.1×10^9	1.0×10^9	7.8×10^8
Methylated phenols			
p-Cresol	1.2×10^{10}	4.2×10^7	—
o-Cresol	1.1×10^{10}	—	—
Nitroaromatic compounds			
Nitrobenzene	3.9×10^9	3.7×10^{10}	1.0×10^9

Source: Buxton, G.V. et al., *J. Phys. Chem. Ref. Data*, 17, 513–886, 1988. With permission.

12.3 Irradiation of Toxic Organic Chemicals in Aqueous Solutions

High-energy electron beam irradiation can be used to effectively remove and/or destroy toxic organic chemicals in aqueous solutions by oxidation or reduction. It has been used to treat many compounds in water treatment (trihalomethanes) (Cooper et al., 1993b), groundwater contamination (chlorinated methanes and ethenes) (Kurucz et al., 1993), and hydrocarbons leaking from underground storage tanks (benzene and substituted benzenes) (Nickelsen et al., 1992a,b), as well as many other hazardous organic chemicals that are regulated.

To gain a better understanding of the electron beam process it is necessary to study the kinetics at various solute concentrations, the pH effects, absorbed doses, and the effect of scavengers in various water qualities. Water quality plays an important role in the removal efficiency of toxic chemicals and has led to experiments investigating the destruction of selected organic compounds suspended in different water matrices. Many studies have been conducted in pure aqueous solutions and have not taken into account the presence of inorganic and organic matter that naturally exist in water and may affect removal rates of the solute in question by interacting with the transient reactive species formed during irradiation. In the presence and absence of known scavengers, such as methanol, oxygen, bicarbonate/carbonate ions, and dissolved organic carbon, the degradation rates change significantly. By examining the rate constants for each transient species with a known scavenger and comparing it with the rate constant of the same transient species with the solute to be removed in a specific water quality, it is possible to predict which will be the preferred reaction, thereby determining the removal efficiency of the solute.

The following sections discuss the research results from Cooper and his colleagues at the Miami High-Energy Electron Beam Facility. In a typical experiment, chemicals are injected into different water quality streams at various concentrations, the pH is varied, and samples of the solution are collected and analyzed after it has passed through the beam at various absorbed doses.

12.3.1   Saturated Halogenated Methanes

Halogenated methylenes, including carbon tetrachloride, methylene chloride, and four trihalomethanes (THMs), namely, chloroform, bromodichloromethane, dibromochloromethane, and bromoform, have been studied. The formation of THMs in drinking water results from chlorination during disinfection. THMs in drinking water are classified as probable human carcinogens. Carbon tetrachloride is used in the production of refrigerator fluid, propellants for aerosol cans, and other end uses and is found in groundwater; it can cause serious health effects and is now regulated by the U.S. Environmental Protection Agency at a maximum contaminant level (MCL) of 5 µg/L. Methylene chloride is used widely in industry as a process solvent in the manufacture of photographic films and pharmaceuticals and as an agent in foam and paint-stripping operations. If consumed, methylene chloride can affect the central nervous system and is therefore regulated at an MCL of 5 µg/L in the United States.

Chloroform removal was found to be dependent on the water quality. In potable water, at two solute concentrations of 75 and 750 µg/L, 99% removal was observed at 800 krads (Cooper et al., 1993a). With the similar initial concentrations and irradiation doses in secondary and raw wastewaters, the removal efficiency was 85 and 90%, respectively. However, for

99% bromoform removal, the effect of water quality was insignificant when the initial concentration was between 100 and 1500 µg/L and the irradiation dose was above 300 krads. Three experiments were carried out in potable water of three different compositions: (1) the addition of individual THMs, (2) a mixture of THMs, and (3) the addition of chlorine to form THMs in the water. In each case, G_D decreased as the relative percentage of bromide and the dosage increased; in other words, at the low concentration, removal is relatively independent of solute concentration. However, at the highest concentration, an increased dose was required to meet the same solute removal. A possible explanation for this is that radical–radical recombination of the OH$^{\bullet}$ and the e_{aq}^- increases with increasing dosage; therefore, the relative concentration of the reactive species available for the THMs is lowered. Thus, the factors that will most affect the removal efficiency of THMs are those that affect the concentration of e_{aq}^- in solution, because THM removal depends more on the concentration of the e_{aq}^- than the $^{\bullet}$OH. This was the general trend noted among other low-molecular-weight organic pollutants, such as CCl_4 and methylene chloride.

Two studies were conducted for CCl_4 removal (Cooper, 1993a). One was conducted with three different CCl_4 concentrations at one pH, and the other was conducted with three different pHs at one solute concentration. Removal efficiency was similar to that of the THMs. pH was found not to be a contributing factor in the removal of CCl_4 by high-energy electron beam irradiation. When the pH of water was adjusted to make it potable, the bicarbonate/carbonate equilibrium was disturbed. As the pH increased, so did the alkalinity, resulting in increased scavenging of the $^{\bullet}$OH, thus reducing the removal efficiency of CCl_4 as well as the THMs.

Similar experiments for methylene chloride (CH_2Cl_2) removal were performed. The results were similar to those for THM removal. Due to the high solubility of CH_2Cl_2, the experiments were carried out at higher concentrations. Such concentrations would require a higher dose to meet treatment objectives in the low microgram per liter range.

High-energy electron beam degradation of halogenated saturated methanes led to the formation of by-products (Mak et al., 1997). To determine which by-products may be formed, the mechanism for the destruction of the compound was investigated. For example, the irradiation of chloroform ($CHCl_3$) and the formation of by-products have been studied by Mak et al. (1997). The formation of oxidized organic compounds such as formaldehyde and formic acid has been observed. However, when chloroform is irradiated and no halogenated compounds have been detected.

A mechanism for the decomposition of $0.07\text{-}M$ $CHCl_3$ at low irradiation dosages is proposed in Equation (12.16) to Equation (12.29), with or without dissolved oxygen. The bimolecular rate constant, k, is in M^{-1} s^{-1}. The reaction equations were found to be in agreement with experimental data (Dickson et al., 1986):

$$e_{aq}^- + CHCl_3 \rightarrow Cl^- + {}^\bullet CHCl_2; \; k = 3.0 \times 10^{10} \qquad (12.16)$$

$$^\bullet H + CHCl_3 \rightarrow H_2 + {}^\bullet CCl_3; \; k = 2.4 \times 10^6 \qquad (12.17)$$

$$\rightarrow HCl + {}^\bullet CHCl_2; \; k = 1.2 \times 10^7 \qquad (12.18)$$

$$^\bullet OH + CHCl_3 \rightarrow H_2O + {}^\bullet CCl_3; \; k = 5.0 \times 10^6 \qquad (12.19)$$

$$^\bullet COOH + {\cdot}COOH \rightarrow HOOC{-}COOH \qquad (12.20)$$

$$\rightarrow HCOOH + CO_2 \qquad (12.21)$$

$$^\bullet CHO + HCCl_3 + H_2O \rightarrow {}^\bullet CHCl_2 + HCOOH + HCl \qquad (12.22)$$

At the higher solute concentration of 0.07 M $CHCl_3$, several radical–radical recombination reactions were inferred from the by-product analysis:

$$H^\bullet + {}^\bullet CHCl_2 \rightarrow CH_2Cl_2 \qquad (12.23)$$

$$^\bullet CHCl_2 + {}^\bullet CHCl_2 \rightarrow CHCl_2CHCl_2 \qquad (12.24)$$

$$^\bullet CCl_3 + {}^\bullet CHCl_2 \rightarrow CCl_3CHCl_2 \qquad (12.25)$$

$$^\bullet CCl_3 + {}^\bullet CCl_3 \rightarrow CCl_3CCl_3 \qquad (12.26)$$

$$^\bullet H + {}^\bullet CHO \rightarrow HCHO \qquad (12.27)$$

In solutions with high oxygen concentrations, the following reactions appear to play an important role in the decomposition of $CHCl_3$:

$$^\bullet CHCl_2 + O_2 \rightarrow {}^\bullet O_2CHCl \qquad (12.28)$$

$$^\bullet CCl_3 + O_2 \rightarrow {}^\bullet O_2 CCl_3 \qquad (12.29)$$

At high organic concentration and high radiation dosage, the removal of $CHCl_3$ was inhibited at high dosages when the oxygen concentration was depleted. This observation suggests that radical–radical recombination has occurred, thus allowing halogenated by-products to remain. In the experiments conducted by Mak et al. (1997), formaldehyde was the only aldehyde observed. No halogenated reaction by-products such as haloacetic acids or ketones were observed after irradiation.

The relative percent removal of $CHCl_3$ can be calculated from the bimolecular reaction rate constants and the G values for the reaction of $CHCl_3$ with the three transient species ${}^\bullet OH$, e_{aq}^-, and ${}^\bullet H$. These calculations indicate that the e_{aq}^- initiates greater than 99% of the removal of $CHCl_3$, whereas ${}^\bullet OH$ is responsible for less than 0.2% and ${}^\bullet H$ for less than 0.01%. Therefore, the factors that will affect the removal efficiency of $CHCl_3$ from solution are those that affect the e_{aq}^- concentration.

12.3.2   Unsaturated Halogenated Ethenes

The removal of TCE and PCE from potable water and secondary and raw wastewater was studied by full-scale experiments at both low and high solute concentrations (Cooper et al., 1993a). TCE removal was found to be dependent on both ${}^\bullet OH$ and e_{aq}^-, with second-order rate constants of 4.0×10^9 and 1.9×10^9 M^{-1} s^{-1} (Buxton et al., 1988), respectively. In contrast, PCE removal is dependent primarily on the e_{aq}^- at 1.3×10^{10} M^{-1} s^{-1}. Greater than 90% removal was obtained for both TCE and PCE at both concentrations in all waters (Cooper et al., 1993a); however, PCE was not as effectively removed as was TCE in solution of equal solute concentration and water quality. The reason for this may be related to the number of chlorine atoms present in both compounds. Overall, the removal rates of TCE and PCE were best in potable water and about the same for secondary and raw wastewater. Although the percent removal at lower concentration was higher in all waters than at higher concentrations, greater than 90% removal was observed for both TCE and PCE in all waters.

Irradiation was also successful in the decomposition of THMs to chloride and bromide ions (Cooper et al., 1993b). Toxic organic by-products such as haloacetic acid, aldehyde, ketones, or halogenated organic compounds were formed after irradiation. It has also been proven effective in the destruction of halogenated ethenes such as TCE and PCE. Removal rates decreased 20-fold in the presence of methanol as opposed to its absence. Aldehydes and formic acid were found when low solute concentrations of TCE and PCE were irradiated; however, at high concentrations no more than 5% formic acid was found. Complete conversion of organic chlorine to chloride ion can be achieved.

The G_D decreased in all three waters as the radiation dose increased (Cooper et al., 1992a-c). As the solute concentration was increased tenfold, the G_D also increased tenfold. When G_D was compared at each radiation dose, the results were remarkably similar regardless of water quality. At high concentration, the G_D value for TCE was higher in secondary wastewater than in potable water. This may be due to higher influent solute concentration in the secondary wastewater. Although water quality is a factor in removal efficiency, it does not appear to be as important when compared to other treatment processes.

In the presence of 3.3-mM methanol, which is a ${}^\bullet OH$ scavenger, the removal efficiency of the solutes was reduced up to 20-fold when compared to solution with no methanol; however, it is not clear why the removal of

PCE by the $^\bullet$OH is so dramatically reduced when PCE removal primarily depends on the e_{aq}^- more than $^\bullet$OH.

The proposed reaction mechanism for the destruction of aqueous solutions of TCE or PCE predicts the formation of stable oxidized polar organic compounds. These compounds consist of acids, aldehydes, and possibly haloacetic acids. Three possible mechanisms have been proposed for the formation of by-products due to the irradiation of aqueous solutions containing TCE and PCE. The first is for the formation of formaldehyde, acetaldehyde, and glyoxal, which are formed at a concentration of approximately two orders of magnitude less than the influent solute concentration. Second, the formation of formic acid decreased with increasing radiation dose. The formic acid concentration was found to be higher for PCE than TCE. These results are most probably due to the slower reaction rate constants of PCE with e_{aq}^- and $^\bullet$OH, compared to TCE. The third possible reaction is the formation of haloacetic acids when TCE and $^\bullet$OH react. The mechanism of decomposition of PCE by $^\bullet$OH is shown in Equation (12.30) to Equation (12.42) (Dickson et al., 1986):

$$CCl_2 = CCl_2 + {}^\bullet OH \rightarrow HOCCl_2\text{--}CCl_2{}^\bullet \tag{12.30}$$

$$HOCCl_2\text{--}CCl_2{}^\bullet \rightarrow COCl\text{--}CCl_2{}^\bullet + H^+ + Cl^- \tag{12.31}$$

$$COCl\text{--}CCL_2{}^\bullet + O_2 \rightarrow COCl\text{--}CCl_2OO^\bullet \tag{12.32}$$

$$2COCl\text{--}CCl_2OO \rightarrow O_2 + 2COCl\text{--}CCl_2O^\bullet \tag{12.33}$$

$$COCl\text{--}CCl_2O^\bullet \rightarrow {}^\bullet COCl + CCl_2O \tag{12.34}$$

$$CCl_2O + H_2O \rightarrow CO_2 + 2Cl^- + 2H^+ \tag{12.35}$$

$${}^\bullet COCl + H_2O \rightarrow {}\cdot COO^- + 2H^+ + Cl^- \tag{12.36}$$

$${}^\bullet COO^- + O_2 \rightarrow {}^\bullet O_2{}^- + CO_2 \tag{12.37}$$

$$2COCl\text{--}CCl_2 \rightarrow COCl\text{--}C_2Cl_4 - COCl \tag{12.38}$$

$$COCl\text{--}C_2Cl_4 - COCl + H_2O \rightarrow HOOC\text{--}C_2Cl_4\text{--}COOH + 2Cl^- \tag{12.39}$$

$$COCl\text{--}CCl_2 + {}^\bullet COO \rightarrow COCl\text{--}CCl_2\text{--}COO^\bullet \tag{12.40}$$

$$COCl\text{--}CCl_2\text{--}COO^\bullet + H_2O \rightarrow HOOC\text{--}CCl_2\text{--}COOH + Cl^- \tag{12.41}$$

$$2{}^\bullet COO^- + 2H^+ \rightarrow HOOC\text{--}COOH \tag{12.42}$$

Once again the principal reaction products at high absorbed doses are the more oxidized organic aldehydes and acids.

12.3.3 Substituted Benzenes

The maximum contaminant level (MCL) of benzene is 0.5 mg/L for solid waste and wastewater and 0.005 mg/L in drinking water. Toluene, *m*-xylene, and *o*-xylene, which are all alkyl-substituted benzenes, are also regulated in drinking water due to their carcinogenic nature; therefore, their removal or destruction is imperative not only for the population's health but also for the safety of the environment. Mixtures of these three compounds, plus benzene, in potable water and secondary wastewater effluent were studied using continuous high-energy electron beam irradiation in the absence and presence of methanol (Nickelsen et al., 1992a,b). Three factors were studied in the removal efficiency of these compounds: water quality, solute concentration, and dose. All four compounds when added at a low concentration and at a dose of 787 krad were removed to below detection limits in potable water, whereas in secondary wastewater 90 to 96% removal was observed (Nickelsen et al., 1992a,b). At a concentration 20 times higher and at the same dose of 787 krad, removal efficiency was a little above the detection limit — for example, 93% for benzene in both potable water and secondary wastewater. Similar results were observed for all other compounds studied in potable water and secondary wastewater. The dose required to remove 90% of the initial toluene, *m*-xylene, and *o*-xylene concentrations in potable water was about 1.5-fold lower than the dose required to remove similar concentrations in secondary wastewater. At the higher concentrations, the dose required for benzene in secondary wastewater was higher when compared to potable water at both low and high concentrations. During the removal of benzene and its alkyl-substituted derivatives, phenol was detected as the reaction by-product. Hydroxyl radical was primarily responsible for their disappearance; however, water quality was shown to affect removal rates due to dissolved organic carbon (DOC) at higher organic concentrations.

The G_D values at all doses for both potable and secondary wastewater were similar at low concentrations. At high solute concentrations and only at the lowest dose, the G_D values in potable water were higher when compared with secondary wastewater. At the two higher doses, the G_D values were similar in both waters, with the exception of benzene, where the G_D value was higher in potable water. The G_D values for the higher concentration were one order of magnitude higher than those observed for low concentrations. The removal of these compounds follows the first order with respect to solute concentration.

The effect of oxygen, bicarbonate/carbonate, and methanol on removal efficiency of these compounds was also studied. All four compounds relied primarily on •OH for removal efficiency, with the exception of toluene (for which 16% of its removal depended on H•), and the removal rate increased with increasing dose because less scavenging effect occurs at the higher doses. The oxygen content of both waters was approximately 3.7 mg/L. Both e_{aq}^- and •H rapidly reduce O_2 to form O_2^- with second-order rate constants of 1.9×10^{10} and 2.1×10^{10} M^{-1} s^{-1}, respectively, so dissolved oxygen was not expected to affect removal efficiency except in the case of toluene. In the presence of methanol, which is a •OH scavenger, all four compounds were expected to have a reduction in removal efficiency. Several by-products have been identified upon the irradiation of benzene; its alkyl-substituted derivatives are shown in Table 12.4, which also shows the concentration of these intermediates.

Phenols are initial by-products in the decomposition of these compounds. At low absorbed doses, the phenol concentration increased and then decreased below influent concentration at high absorbed doses, which suggests that the high-energy electron beam process can also be used to degrade phenols to simpler compounds.

After hydroxylation of benzene by hydroxyl radical, the intermediate radical may decompose to phenol. Figure 12.1 shows the reaction pathways of benzene: phenol will form upon irradiation of aqueous solutions of solute, involving reaction of the compounds with •OH, forming hydroxycyclohexadienyl radicals. Hydroxycyclohexadienyl radicals generate phenol by disproportionation. Reaction of toluene, *m*-xylene, and *o*-xylene with •OH would yield alkyl-substituted phenols by similar mechanisms. Glyoxal is formed upon irradiation of benzene, which first forms hydroxycyclohexadienyl radicals and then phenol or combines with oxygen to form an unstable hydroxy hydroperoxide. Subsequent water elimination and ring opening lead to the formation of mucondialdehyde, which is further oxidized to muconic acid. Continued oxidative processes ultimately lead to the formation of glyoxal.

12.3.4   Phenol

Phenols and substituted phenols are found naturally occurring in natural waters at low concentrations; however, their high concentrations due to industrial and agricultural waste may be toxic, especially to aquatic life. Experiments were conducted on a large scale to determine the effect of solute concentration, absorbed dose, total alkalinity, and total solids on the removal of phenol. Three solute concentrations (10.6, 106, and 531 µmol/L) were used over a pH range of 5 to 9 in the presence or absence of 3% Kaolin clay. As much as 94.2% phenol removal could be attributed to •OH, 5.4% to •H, and 0.4% by e_{aq}^- (Nickelsen et al., 1992a,b). Because •OH is the primary reacting species, hydroxylated compounds through •OH were expected to be the major by-products. As with all compounds studied with electron beam technology, the greatest percent removal was observed at the lowest initial solute concentration (10.6 µmol/L) and increased with increasing dose. The percent

TABLE 12.4

Hydroxylated Reaction By-Products

	Detection limit (10⁻⁶ M)	Dose (krad)	pH 5	pH 7	pH 9
		Benzene			
Phenol (10⁻⁶ *M*)	0.01	0	bmdl	bmdl	bmdl
		50	9.95	11.88	12.50
		100	5.09	7.52	9.50
		200	bmdl	1.53	4.10
Hydroquinone (10⁻⁶ *M*)	0.10	0	bmdl	bmdl	bmdl
		50	7.63	6.30	2.80
		100	2.83	2.50	1.20
		200	1.39	bmdl	bmdl
Catechol (10⁻⁶ *M*)	0.05	0	bmdl	bmdl	bmdl
		50	2.53	2.96	0.71
		100	3.21	2.01	bmdl
		200	1.00	0.89	bmdl
Resorcinol (10⁻⁶ *M*)	1.00	0	bmdl	bmdl	bmdl
		50	3.04	5.92	bmdl
		100	bmdl	bmdl	bmdl
		200	bmdl	bmdl	bmdl
		Toluene			
0-Cresol (10⁻⁶ *M*)	0.01	0	bmdl	bmdl	bmdl
		50	4.08	3.55	6.70
		100	0.70	0.40	2.70
		200	bmdl	bmdl	0.02

bmdl = below method detection limit.
Source: Nickelsen, M.G. et al., *Water Res.*, 28(5), 1227–1237, 1992a. With permission.

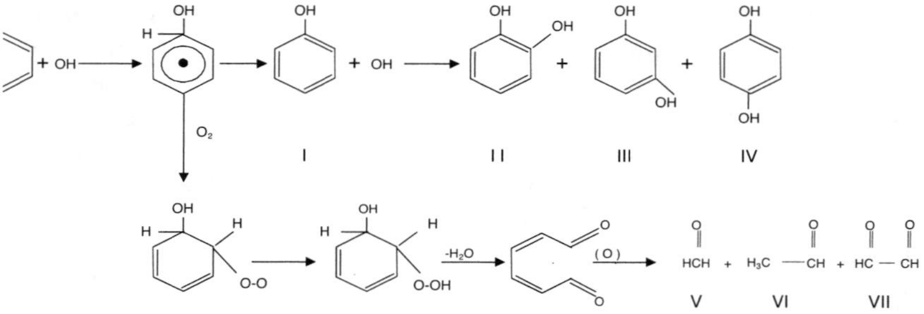

FIGURE 12.1
Proposed mechanism for the destruction of benzene. Identified dose-dependent reaction by-products include phenol (I), catechol (II), and resorcinol (III). (From Nickelsen, M.G. et al., *Water Res.*, 28(5), 1227–1237, 1992. With permission.)

removal decreased with increasing pH from 5 to 9 due to increasing bicarbonate/carbonate ion concentrations. The 3% solids as clay appeared not to have any effect on removal efficiency; therefore, removal efficiency was greatest at low pH and low bicarbonate/carbonate ion concentrations (Lin et al., 1995). At high phenol concentrations (~950 μmol/L), recirculation of the waste stream was necessary to destroy all of the phenol and its reaction by-products.

Removal efficiency of phenol can also be expressed in terms of G_D values. As the pH increases, the G_D value decreases. At any given solute concentration and pH, the removal rate decreases with increasing dose, suggesting that the radical–radical recombination reaction involving $^\bullet$OH, e^-_{aq}, and $^\bullet$H also increased.

The mechanism for the breakdown of phenol after hydroxylation of benzene can be seen in Figure 12.1. Several dihydroxy-substituted phenol derivatives were observed upon irradiation: hydroquinone, catechol, and trace amounts of resorcinol (Nickelsen et al., 1993a,b). However, hydroquinone and catechol were both formed at the highest concentration at a low absorbed dose and their concentration decreased with increasing radiation to below detection limits at a dose of 300 krad. At a pH of 9 and a dose of 100 krad, dihydroxy-substituted phenol concentrations increased. When the pH was 5 and 7, the maximum concentrations of these products were found at 50 krads, suggesting a more efficient removal at a lower pH.

During irradiation of phenol, aldehydes and formic acid were formed, indicating ring rupturing of phenol and further oxidation and mineralization in some cases. At low irradiation doses and low substrate concentrations, phenol reacts with $^\bullet$OH to form hydroquinone, catechols, and resorcinol. For all compounds studied, the G_D value and the dose required for 90% removal of the initial solute concentration increased as the concentration of organic solute increased. Other compounds formed were formaldehyde, acetaldehyde, and glyoxal, which accounted for less than 1% of the initial concentration. The formation of aldehydes indicates that phenol undergoes ring cleavage. Formic acid formation was also noted; its concentration increased with radiation dose and it accounted for less than 10% of the initial phenol concentration.

12.3.5 Disinfection of Sewage Sludge

High-energy electron beams can be used for the disinfection and inactivation of fecal coliform and bacteria and to a lesser extent viruses in sewage sludge. Inactivation of total coliform and bacteria occurred at a dose of 3 kGy, and at 4 kGy eggs are destroyed (Kurucz et al., 1991). The electron beam can also reduce human enteric viruses by one or two orders of magnitude at a dose of about 4 kGy. The size of an organism may also play a role in its removal or inactivation. For example, the larger the organism, the more effective the removal process by irradiation.

The effect of high-energy electrons on selected microorganisms has been studied. For example, eggs of *Ascaris suum* and *Shistoma mansoni* were inactivated at a very low dose (Kurucz et al., 1991). *S. mansoni* was totally eliminated at a dose less than 700 Gy. For 90% inactivation, a higher dose was required in the presence of sludge, suggesting that sludge slows down the deactivation process. Complete sterilization was found to be below 1 kGy for most organisms (Kurucz et al., 1991).

Water quality plays a role in the inactivation procedure for coliform. For instance, when the effluent was chlorinated, inactivation was fast (Kurucz et al., 1991). Once the dose exceeded 1.5 kGy, resistance to inactivation was observed. In raw and nonchlorinated secondary effluent, inactivation of the total bacteria population was observed. Four to five logs of coliform microorganisms and total bacteria occurred between 6 and 7 kGy (Kurucz et al., 1991). Experiments were carried out to study the inactivation of fecal coliform microorganisms in raw wastewater and chlorinated and nonchlorinated secondary wastewater effluents. At a dose of 1.3 kGy, fecal coliform in chlorinated secondary wastewater was reduced by four units of logarithm to the base ten of activation; above 1.3 kGy, no further increase of removal was observed, suggesting that bacteria are resistant at this dose. In nonchlorinated secondary wastewater, fecal coliform required a higher dose of 6.5 kGy for five log units of inactivation. In the chlorinated secondary wastewater, the presence of chlorine may have created a stressful environment for the organism, thereby allowing it to be inactivated with ease and thus accounting for the lower dosage.

Fecal organism inactivation was also studied in nonchlorinated secondary wastewater and raw wastewater by Kurucz et al. (1991). Fecal organisms are absorbed in a dose-dependent manner, and four to five log units of inactivation can be expected as the dose approaches 6 kGy. The total bacterial flora was inactivated at approximately the same level as the coliform for each absorbed dose, as opposed to other disinfectant methods. The fact that the total fecal organisms are inactivated at approximately the same rate indicates that high-energy electrons are indiscriminate disinfectants.

The mechanism of inactivation of bacteria by high-energy electron beam radiation is thought to be indirect due to the hydrogen peroxide residual monitored throughout the experiment. The reactive species formed by the irradiation of water react with the molecular constituents of the microorganisms. Because these species have a very short half-life, it is believed that the electrons from the beam may also affect the inactivation process.

12.3.6 Estimation of Removal Efficiency of Organic Pollutants

The relative importance of the three transient reactive species in regard to the removal efficiency of some organic compounds in natural waters can be estimated. According to the rate constants listed in Table 12.3, the removal efficiency of organic pollutants in pure water in the absence of radical scavengers such as oxygen, bicarbonate/carbonate, etc. can be calculated. Assuming the removal reactions are the ones between the transient species $^{\bullet}OH$, e_{aq}^{-}, and $^{\bullet}H$ and the organic chemical of interest (R), then the overall removal of any solute can be described by the following kinetic expression:

$$-d[R]t/dt = k_1[R][OH^{\bullet}] + k_2[R][\, e_{aq}^{-}\,] + k_3[R][H^{\bullet}] \qquad (12.43)$$

where k_1, k_2, and k_3 are the respective second-order rate constants. The relative concentrations of each of the three reactive species in pure aqueous solution are expressed as G values; therefore, at any dose, the product of the second-order rate constant of a given solute and the G value of the respective reactive species is a proportionality constant, k'_1, k'_2, and k'_3. The relative solute removal of each species can be determined from:

$$\text{Relative removal } [R] = k'_1 / \Sigma\big(k'_1, k'_2, k'_3\big) \times 100 \qquad (12.44)$$

Summing these proportionality constants provides an estimated value for the relative solute removal:

$$\text{Relative solute removal} = k'_1[R] \qquad (12.45)$$

For example, in Table 12.3, when the removal rates of one compound from four different groups (chloroform, TCE, benzene, and toluene) are examined, the following assumptions can be made. The removal of $CHCl_3$ depends primarily on the e_{aq}^{-}, whereas TCE removal is achieved by both e_{aq}^{-} and $^{\bullet}OH$. The removal of benzene depends almost entirely on $^{\bullet}OH$, while a significant amount of toluene is removed by $^{\bullet}H$ as well as $^{\bullet}OH$. The removal rate of chloroform appears to be faster by one order of magnitude than benzene, TCE, and finally toluene. Specific substituent groups may enhance or slow down the removal process; for instance, the presence of the nitro group of nitrobenzene may help speed up a reaction with the e_{aq}^{-}.

The degradation rates of two compounds in the same group, such as CE and PCE, are dependent on the number of chlorine atoms that each molecule contains. PCE, with four chlorine atoms, has the higher rate constant and is faster than TCE of three chlorine atoms. Because chlorine is a good leaving group, the more chlorine atoms a compound has the faster the degradation will be. Halide ions are relatively stable and are very weak bases, thereby making them excellent leaving groups.

12.3.7   Radical Scavenger Effect

12.3.7.1   Methanol

Methanol is the primary scavenger of ·OH and is naturally occurring in water. In all the experiments conducted, 3.3-mM methanol was used as a carrier of the organic solute of interest. Methanol was found to react with ·OH and to a lesser extent with e^-_{aq} and ·H with second-order rate constants of 9.7×10^8, 1.0×10^4, and 2.6×10^6 M^{-1} s^{-1}, respectively. From the rate constants it can be seen that methanol scavenges primarily the ·OH and the ·H radicals.

12.3.7.2   Bicarbonate/Carbonate Ions

Bicarbonate and carbonate ions are common ·OH scavengers that exist in equilibrium in natural waters and react with ·OH with second-order rate constants of 8.5×10^6 and 3.9×10^8 M^{-1} s^{-1}, respectively. Of the waters tested, potable water had a pH of about 7 and secondary wastewater a pH of about 9. As the pH increased, ·OH scavenging increased due to the presence of increased alkalinity species, bicarbonate and carbonate ions, thereby reducing removal efficiency. The effect of these ions on ·OH can be predicted according to the reactions:

$$·OH + HCO_3^- = H_2O + CO_3^= \tag{12.46}$$

$$·OH + CO_3^{2-} = OH^- + CO·_3^- \tag{12.47}$$

Although the alkalinities of the secondary wastewater tested were fivefold higher than for potable water, the ·OH scavenging was approximately 2.5 times higher in the potable water based on the HCO_3^-/CO_3^{2-} equilibrium; therefore, high removal efficiency results from either a low pH and/or a low bicarbonate/carbonate ion concentration.

12.3.7.3   Dissolved Organic Carbon

Dissolved organic carbon (DOC) is another common component of natural waters. In secondary wastewater, the DOC concentration is sixfold higher than potable water, thereby requiring a larger dose of radiation due to its scavenging effect to achieve the required removal efficiency. Only indirect studies for the reaction of ·OH with DOC have been conducted, while scavenging of e^-_{aq} and ·H by DOC has not been studied.

12.3.7.4 Oxygen

Oxygen is reduced rapidly by both e^-_{aq} and $\cdot H$ to form O_2^- with second-order rate constants of 1.9×10^{10} and 2.1×10^{10} M^{-1} s^{-1}, respectively. For example, in the presence of 3.7 mg/L of oxygen, 35% of these two reactive species would be removed at a dose of 100 krad as opposed to 5% at a dose of 800 krad.

12.4 QSAR Models

Substituents have a profound effect on the electrophilic substitution rate constants of organic compounds reacting with e^-_{aq} (Anbar and Hart, 1964). Table 12.5 lists the rate constants for substituted benzenes, toluene, and phenols. Four orders of magnitude of rate constants may vary from 4×10^6 M^{-1} s^{-1} for phenol to 3×10^{10} M^{-1} s^{-1} for nitrobenzene. To gain insight into the reaction mechanisms involving e^-_{aq} , Anbar and Hart (1964) applied Hammett's equation to the rate constants and the substituent constants. η is defined as the ratio of rate constants of substituted compounds vs. nonsubstituted reference compounds such as benzene, toluene, and phenol as follows:

$$\eta = \log \frac{k}{k_{\mathrm{H}}} = \log \frac{k_{C_6H_5X+e^-_{aq}}}{k_{C_6H_6+e^-_{aq}}} \tag{12.48}$$

When η values are plotted against the σ_{para} values, Hammett's equation is satisfied for all the substituents except for bromine and iodine, as shown in Figure 12.2. The ρ value is 4.8, which is within the range of ρ values reported for a variety of substituted reactions such as nitration, bromination, and hydrolysis (Stock and Brown, 1956). Therefore, Hammett's equation for substituted benzenes reaction with e^-_{aq} can be expressed as:

$$\eta = \log \frac{k}{k_{\mathrm{H}}} = \log \frac{k_{C_6H_5X+e^-_{aq}}}{k_{C_6H_6+e^-_{aq}}} = 4.8\sigma_{para} \tag{12.49}$$

Because the ρ value is positive, negatively charged carbon ions are considered to be the primary transition state complex (TSC). The TSC will dissociate to a substituted phenol radical and a stable anion. It may also be neutralized by the toluene, resulting in the addition of a proton, H^+ (Arai and Dorfman, 1964); therefore, e^-_{aq} is considered to interact with the π-orbital of the ring as in electrophilic substitution rather than to affect electron distribution and polarizability of a certain substituent.

TABLE 12.5

Specific Rate Constants and Relative Rates of Reaction of Substituted Benzene, Toluene, and Phenols

Substituent X	C_6H_6X		$p\text{-}XC_6H_4CH_3$		$m\text{-}XC_6H_4OH$	
	$k_{C_6H_6X+e^-_{aq}}$ $(M^{-1}\,s^{-1})$	$\mathrm{Log}\left(\dfrac{k_{C_6H_6X+e^-_{aq}}}{k_{C_6H_6+e^-_{aq}}}\right)$ (η)	$k_{XC_6H_4CH_3+e^-_{aq}}$ $(M^{-1}\,s^{-1})$	$\mathrm{Log}\left(\dfrac{k_{XC_6H_4CH_3+e^-_{aq}}}{k_{C_6H_4CH_3+e^-_{aq}}}\right)$ (η)	$k_{XC_6H_4OH+e^-_{aq}}$ $(M^{-1}\,s^{-1})$	$\mathrm{Log}\left(\dfrac{k_{XC_6H_4OH+e^-_{aq}}}{k_{C_6H_4OH+e^-_{aq}}}\right)$ (η)
H	1.4×10^7	0.00	1.2×10^7	0.00	4×10^6	0.00
CH_3	1.2×10^7	$-.07 \pm 0.07$	1.2×10^7	0.00	4×10^6	0.00
OH	4.0×10^6	$-.54 \pm 0.20$	1.2×10^7	0.00	4×10^6	0.00
F	6.0×10^7	$.63 \pm 0.08$	1.2×10^7	0.00	2.0×10^8	1.70
Cl	5.0×10^8	1.55 ± 0.05	4.5×10^8	1.57	5.0×10^8	2.10
Br	4.3×10^9	2.52 ± 0.08	4.5×10^8	1.57	2.7×10^9	2.83
I	1.2×10^{10}	2.93 ± 0.06	1.3×10^{10}	3.03	2.7×10^9	2.83
CN	1.6×10^{10}	3.06 ± 0.06	1.4×10^{10}	3.07	4.8×10^9	3.08
COO-	3.1×10^9	2.35 ± 0.05	3.6×10^9	2.48	1.06×10^9	2.42
SO_3	4.0×10^9	2.46 ± 0.05	1.65×10^9	2.14	1.06×10^9	2.42
NO_2	3.0×10^{10}	3.33 ± 0.05	1.9×10^{10}	3.22	2.5×10^{10}	3.80
CF_3	1.8×10^9	2.11 ± 0.05	1.9×10^{10}	3.22	2.5×10^{10}	3.80
CH_2OH	1.3×10^8	0.97 ± 0.05	1.9×10^{10}	3.22	2.5×10^{10}	3.80
SH	4.7×10^7	0.53 ± 0.05	1.9×10^{10}	3.22	2.5×10^{10}	3.80
$CONH_2$	1.7×10^{10}	3.08 ± 0.05	1.9×10^{10}	3.22	2.5×10^{10}	3.80
SO_2NH_2	1.6×10^{10}	3.06 ± 0.05	1.9×10^{10}	3.22	2.5×10^{10}	3.80

Source: Anbar, M. and Hart, E.J., *J. Am. Chem. Soc.*, 86, 5633–5636, 1964. With permission.

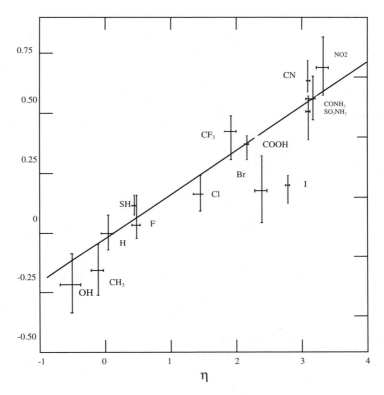

FIGURE 12.2

η as a function of σ of Hammett's equation; $\eta = \log\left(k_{C_6H_6X} + e_{aq}^- / k_{C_6H_6} + e_{aq}^-\right)$. (From Anbar, M. and Hart, E.J., *J. Am. Chem. Soc.*, 86, 5633–5636, 1964. With permission.)

To further understand the mechanistic difference between aromatic substitution and the e_{aq}^- reactions, reaction rate constants of disubstituted benzenes reacting with e_{aq}^- were studied. Specifically, substituted benzoic acid was studied, and one substituent (–COOH) was kept constant. Table 12.6 lists the rate constants of substituted benzoic acid reacting with e_{aq}^-. It also calculates η, which is defined as the following:

$$\eta_{C_6H_5COO} = \log \frac{k_{XC_6H_4COO^-}}{k_{C_6H_5COO^-}} \tag{12.50}$$

when $\eta_{C_6H_4COO}$ is plotted against $\eta_{C_6H_6}$. An excellent straight line with a slope of 6.5 was obtained (Figure 12.3).

The above correlation gives the following equation between the relative rate constants for substituted benzoic acid and substituted benzenes reacting with e_{aq}^-:

TABLE 12.6

Specific Rate Constants and Relative Rates of Reaction of *para*-Substituted Benzoic Acids

Substituent	$k_{pXC_6H_4COO^-}$ $(M^{-1}\,s^{-1})$	$\text{Log}\left(k_{XC_6H_4COO^-} / k_{C_6H5COO^-}\right)$ $(M^{-1}\,s^{-1})$
H	3.1×10^9	0.00
O⁻	4.0×10^8	−0.89
NH$_2$	2.1×10^9	−0.17
CH$_3$	3.0×10^9	−0.01
F	3.8×10^9	0.09
Cl	6.0×10^9	0.29
Br	7.7×10^9	0.40
I	9.1×10^9	0.47
CX	1.0×10^{10}	0.51
COO⁻	7.3×10^9	0.37

Source: Anbar, M. and Hart, E.J., *J. Am. Chem. Soc.*, 86, 5633–5636, 1964. With permission.

$$\eta_{C_6H_6} = 6.5\eta_{C_6H_4COO} \tag{12.51}$$

Therefore, the ρ value of the benzoic acid series is equal to 4.8/6.5 = 0.74. Hammett's equation for benzoic acid is:

$$\log \frac{k_{XC_6H_4COOH}}{k_{C_6H_5COOH}} = 0.74\sigma \tag{12.52}$$

The above equation suggests that the benzoic acid series are much more insensitive to substitution effects due to the negatively charged benzoic functional group [–COO⁻]. In other words, it is much more difficult to add an additional electron to benzoic series than to benzene series; furthermore, the higher the rate constant of the nonsubstituted parent compound, the lower the η values of the series will be. This is because the more reactive series have the rate constants approaching the diffusion-controlled limit of $10^{10}\,M^{-1}\,s^{-1}$, where a further increase in rate cannot occur.

Table 12.7 presents the relative rates of reaction of nonsubstituted aromatic compounds at the *meta*, *para*, and *ortho* positions. The compounds may be divided into two groups:

- Group A includes *meta*- and *para*-substituted compounds of comparable reactivity.
- Group B includes *meta*-substituted compounds having higher rate constants than *para*-substituted compounds (Table 12.8).

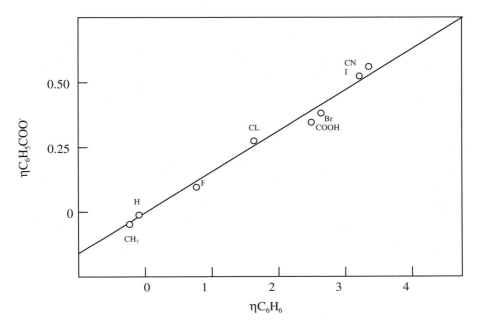

FIGURE 12.3
$\eta C_6H_5COO^-$ as a function of ηC_6H_6, where $\eta C_6H_5COO^-=\log\ \{\kappa XC_6H_4COO^-+e_{aq^-},/$
$\kappa C_6H_6COO^-+e_{aq^-}\}\ \eta C_6H_6=\log\ \{\kappa C_6H_5X+e_{aq^-}/\kappa C_6H_6+e_{aq^-}\}$ (From Anbar, M. and Hart, E.J., *J. Am. Chem. Soc.*, 86, 5633–5636, 1964. With permission.)

Table 12.8 suggests that one of substituents in Group B increases the π-electron density of the ring by resonance, while it withdraws electrons by induction. As a result, k_{meta} is greater than k_{para}. When three rate constants at the *meta*, *para*, and *ortho* positions are compared (Table 12.9), two distinct groups can again be identified:

- Group A includes compounds that follow a decreasing order of rate constants: $k_{ortho} > k_{meta} > k_{para}$. For this group, the neighboring electron-withdrawing groups partially neutralize the electron-donating capacity of O⁻.

- Group B includes all nonsubstituted benzenes that show $k_{ortho} < k_{meta}$ and $k_{para} = k_{meta}$. This occurrence is referred to as the *ortho effect*, which has nothing to do with steric hindrance. The ortho effect results from a mutual inductive effect of two electron-withdrawing groups, which is enhanced by the negative charge on one of them. As a result of the mutual inductive effect, their ability to withdraw electrons from the π-orbital of the aromatic nucleus decreases (Taft, 1956).

TABLE 12.7

The Relative Rates of Reaction of Disubstituted Aromatic Compounds

Substituents		$k_oXC_6H_4Y$, M^{-1} sec^{-1}	$k_mXC_6H_4Y$, M^{-1} sec^{-1}	$k_pXC_6H_4Y$, M^{-1} sec^{-1}	log (k_{meta} / k_{para})	log (k_{ortho}/ k_{meta})
X	Y					
COO$^-$	CH2	2.7 X 10^8	2.6 X 10^9	3.0 X 10^9	− 0.06	− 1.03
COO$^-$	F	3.1 X 10^9	6.7 X 10^9	3.8 X 10^9	0.25	− 0.30
COO$^-$	Cl	1.2 X 10^9	5.5 X 10^9	6.0 X 10^9	− 0.04	− 0.66
COO$^-$	I	4.6 X 10^9	1.3 X 10^{10}	9.1 X 10^9	0.16	− 0.93
COO$^-$	O$^-$	3.2 X 10^9	1.1 X 10^9	4.0 X 10^8	0.42	0.48
F	O$^-$	3.4 X 10^8	2.0 X 10^8	1.2 X 10^8	0.22	0.23
Cl	O$^-$	2.0 X 10^8	5.0 X 10^8	6.4 X 10^8	− 0.10	− 0.40
Br	O$^-$	1.9 X 10^9	2.7 X 10^9	2.9 X 10^9	− 0.03	− 0.15
CN	O$^-$	8.2 X 10^9	4.8 X 10^9	2.0 X 10^9	0.38	0.23
NO$_2$	O$^-$	2.0 X 10^{10}	2.5 X 10^{10}	2.5 X 10^{10}	0.00	− 0.10
Cl	Cl	4.7 X 10^9	5.2 X 10^9	5.0 X 10^9	0.01	− 0.04

Source: Anbar, M. and Hart, E.J., *J. Am. Chem. Soc.*, 86, 5633–5636, 1964. With permission.

TABLE 12.8

Resonance Interaction of Two Substituents

Group A	COO$^-$CH$_3$	$k_{meta} = k_{para}$
	CL$-$O$^-$	
	Br$-$O$^-$	
	NO$_2$$-O^-$	
	Cl$-$Cl	
Group B	COO$^-$F	$k_{meta} > k_{para}$
	COO$^-$I	
	COO$^-$O$^-$	
	CN$-$O$^-$	
	F$-$O$^-$	

TABLE 12.9

Ortho Effect of Two Substituents

Group A	COO$^-$O$^-$	$k_{ortho} > k_{meta} > k_{para}$
	F$-$O$^-$	
	CNO$-$	
Group B	COO$^-$CH$_2$	$k_{ortho} > k_{meta} = k_{para}$
	COO$^-$F	
	COO$^-$Cl	
	COO$^-$I	
	NO$_2$ $-$O$^-$	
	Cl$-$Cl	

12.5 Engineering Applications

In summary, high-energy electrons produced by electron beam accelerators are effective in removing toxic organic chemicals from potable water and secondary and raw wastewater. This is indeed a unique process that is able to treat solutions that contain up to 10% solids and can effectively and efficiently destroy many organic compounds in sludge. In all the compounds tested, the presence of suspended solids (3% kaolin clay) had no significant effect on overall removal (Cooper et al., 1992b). pH and temperature also play little or no role in removal efficiency. Also, this is the only available advanced oxidation process that produces both e_{aq}^- and $^\bullet OH$ in aqueous solutions. $^\bullet H$ may react with solutes of interest and further increase the removal efficiency of toxic compounds. Finally, a low solute concentration has no effect on removal rates; however, at high concentrations, a higher dose is required to achieve the same removal efficiency as that for a lower concentration due to radical–radical recombination as well as scavengers.

As with all advanced oxidation technologies, a number of advantages and disadvantages are associated with the high-energy electron beam process. The numerous advantages of high-energy electron beam make it suitable to be used as an ultimate treatment process for hazardous organic chemicals. For example, this is the only available technology that produces approximately equal concentrations of the reducing species, e_{aq}^- and $^\bullet H$, and the oxidizing species, $^\bullet OH$, at the same time. The process is not selective in the destruction of organic chemicals and allows for simultaneous treatment of complex mixtures of hazardous chemicals such as halogenated aliphatic and nonhalogenated aromatic compounds.

Reactions of organics are very rapid and occur in less than a second, thereby allowing a continuous flow-through system with good process flexibility at full scale when flows vary with time. The absorption dosage of radiation as a function of the penetration depth is shown in Figure 12.4. Clearly, the penetration depth increases with the applied voltage of electrons. Because an electron is able to penetrate as deep as 1 cm, a continuous flow reactor can be used.

The high-energy electron beam process is pH independent in the range of 3 to 11; therefore, any changes in pH over time will not adversely affect the treatment efficiency. If necessary, a lowering of pH can be done as a pretreatment to remove alkalinity, which may contain $^\bullet OH$ scavengers such as bicarbonic and carbonic ions in natural waters. Because water quality is important, pretreatment may decrease operation cost.

The process can effectively treat aqueous streams, soil slurry, sediments, and sludge. It is temperature independent, meaning variations in water temperatures have no effect on the treatment efficiency. No organic sludge is produced, and the target organic chemicals are either mineralized or broken down into low-molecular-weight organic compounds. Although, the

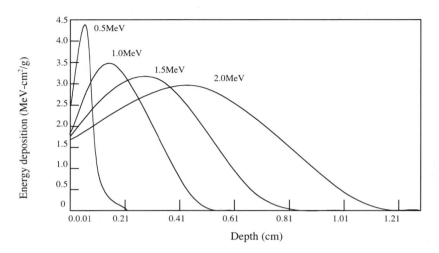

FIGURE 12.4
Energy deposition in pure water at various electron energies. (From Kurucz, C.N. et al., *Radiat. Phys. Chem.*, 45(2), 299–308, 1995. With permission.)

process cannot remove heavy metals, the formation of an inorganic precipitate was observed when hazardous waste leachate was studied.

The process produces no air emissions. Because this is an aqueous-based technology, neither NO_x nor SO_x is produced. The delivery system is closed so no volatilization of toxic organic compounds occurs. The electron beam can be used efficiently and effectively as pretreatment for biological remediation. The electron beam can break complex organic compounds, making them suitable for microbiological degradation.

The major disadvantage to this process is its high cost; however, it shows promise as it is still in its infancy. The cost of treatment using high-energy electron technology depends on the dose required to obtain the desired detoxification, the volume of waste to be treated, the size of the treatment facility, the time utilization of the facility, and the manner in which capital recovery is handled. For example, Kurucz et al. (1990, 1995a) summarized capital and operating costs based upon their experiments conducted at the 1.5-MeV Miami Electron Beam Research Facility (MEBRF), located at the Virginia Key Wastewater Treatment Plant in Miami, FL. A breakdown of its cost is shown in Table 12.10.

The estimated capital requirements represent an approximate 5% annual inflation of the total price actually paid for the Miami facility. Changes in the amortization period or interest rate can significantly affect annual capital charges. Also note that no direct costs for supervision and overhead are included.

The MEBRF can be operated over a range of flow rates from 60 to 160 gpm. The maximum absorbed dose delivered by the electron beam depends on the flow rate; an application requiring a dose of 1290 krad would limit

TABLE 12.10

Costs for 1.5-MeV and 50-ma Electron Beam System

		Capital costs ($1000s)
Installed beam		$1850
Support facility[a]		$500
Total		$2350
Amortization	(10 years ~ 15%) – $466/year	
	(20 years ~ 15%) - $374/year	
		Operating costs (per hour)
Operator		$20.00
Power		$10.50
Water		$ 2.50
Maintenance		$ 8.00
Total		$41.0

[a] Shielding, delivery system, etc.
Source: Kurucz, C. et al., in *Proc. of the 45th Industrial Waste Conference*, Purdue University, West Lafayette, IN, 1990, pp. 539–545. With permission.

the flow to 60 gpm. A lower dose of about 500 krad, such as that required for TCE, would allow treatment at 160 gpm. Modifications in the waste stream delivery and changes in the beam operating system can be made to reduce the need for a full-time operator, which would help lower the cost of the process.

A 160-gpm flow would incur operating costs of approximately $2.5/1000 gal, while a 2100-gpm flow would cost $0.25/1000 gal. If a similar comparison to other processes were made, the operating cost of an ultraviolet/peroxidation process, regardless of the flow, would be $2.59/1000 gal, and the total cost for a wet oxidation process would be $200/1000 gal at high concentrations. Incineration was estimated to cost $76/1000 gal to $517/1000 gal. Transportable electron beam units will be significantly higher in cost than permanent facilities due to transportation costs, increased maintenance, and shorter useful life. Although the treatment costs for this process are highly dependent on the required dose and flow rate, it is competitive with alternative treatment methods. Total treatment costs are estimated to be approximately 0.4 to 0.5¢/L (1.5 to 2.0¢/gal), assuming that one pass through the beam at a dose of 800 krad will be sufficient to treat to the desired level.

References

Anbar, M. and Hart, E.J., Reactivity of aromatic compounds toward hydrated electrons, *J. Am. Chem. Soc.*, 86, 5633–5636, 1964.

Arai, S. and Dorfman, L.M., *J. Chem. Phys.*, 41, 2190, 1964.

Buxton, G.V., Greenstock, C.L., Helman, W.P., and Ross, A.B., Critical review of rate constants of reactions of hydrated electrons, hydrogen atoms and hydroxyl radicals ($OH^{\bullet}/O^{\bullet-}$) in aqueous solution, *J. Phys. Chem. Ref. Data*, 17, 513–886, 1988.

Cooper, W.J., Nickelsen, M.G., Meacham, D.E., Kurucz, C.N., and Waite, T.D., High energy electron beam irradiation: an advanced oxidation process for the treatment of aqueous based organic hazardous wastes, *Water Pollut. Res.*, 27(1), 69–95, 1992a.

Cooper, W.J., Kurucz, C.N., and Waite, T.D., An overview of the use of high energy electron beam irradiation for the destruction of toxic organic chemicals and waters containing solids, in *Emerging Technologies for Hazardous Waste Management*, 1992b.

Cooper, W.J., Nickelsen, E., Meacham, D.E., Cadavid, M.G., Kurucz, C.N., and Waite, T., High-energy electron beam irradiation: an innovative process for the treatment of aqueous based organic hazardous wastes, *J. Environ. Sci. Health*, A27(1), 219–244, 1992c.

Cooper, W.J., Cadavid, E., Nickelsen, M.G., Lin, K., Ford, D.B., Kurucz, C.N., and Waite, T.D., The removal of tri- (TCE) and tetrachloroethylene (Tetra-CE) from aqueous solution using high-energy electrons, *J. Air Waste Manage.*, 43, 1358–1366, 1993a.

Cooper, W.J., Cadavid, E., Nickelsen, M.G., Lin, K., Kurucz, C.N., and Waite, T.D., Removing THMs from drinking water using high-energy electron-beam irradiation, *J. Am. Water Works Assoc.*, 85(9), 106, 1993b.

Dickson, L.W., Lopata, V.J., Toft-Hall, A., Kremers, W., and Singh, A., Radiolytic removal of trihalomethanes from water, in *Proc. from the 6th Symp. on Radiation Chemistry*, 1986, pp. 173-182.

Hart, E.J., Sheffield, G., and Thomas, J.K., Rate constants of hydrated electron reactions with organic compounds, *J. Phys. Chem.*, 68(6), 1271–1274, 1964.

Kurucz, C., Cooper, W.J., Nickelsen, M.G., and Waite, T.D., Full-scale electron beam treatment of hazardous waste, in *Proc. of the 45th Industrial Waste Conference*, Purdue University, West Lafayette, IN, 1990, pp. 539–545.

Kurucz, C.N., Waite, T.D., Cooper, W.J., and Nickelsen, M.G., High energy electron beam irradiation of water, wastewater and sludge, *Adv. Nuclear Sci. Technol.*, 22, 1–43, 1991.

Kurucz, C.N., Waite, T.D., and Cooper, W.J., The Miami electron beam research facility: a large scale wastewater treatment application, *Radiat. Phys. Chem.*, 45(2), 299–308, 1995a.

Kurucz, C.N., Waite, T.D., and Nickelsen, M.G., Empirical models for estimating the destruction of toxic organic compounds utilizing electron beam irradiation at full scale, *Radiat. Phys. Chem.*, 45(5), 805–816, 1995b.

Lin K.J., Cooper, W.J., Nickelsen, M.G., Kurucz, C., and Waite, T.D. Decomposition of aqueous solutions of phenol using high energy electron beam irradiation — a large-scale study, *Appl. Radiat. Isotopes*, 46(12), 1307–1316, 1995.

Mak, F.T., Zele, S.R., Cooper, W.J., Nickelsen, M.G., Kurucz, C.N., and Waite, T.D., Kinetic modelling of carbon tetrachloride, chloroform and methylene chloride removal from aqueous solution using the electron beam process, *Water Res.*, 31(2), 219–228, 1997.

Nickelsen, M.G., Cooper, W.J., Lin, K., Kurucz, C.N., and Waite, T.D., High-energy electron beam generation of oxidants for the treatment of benzene and toluene in the presence of radical scavengers, *Water Res.*, 28(5), 1227–1237, 1992a.

Nickelsen, M.G., Cooper, W.J., Kurucz, C.N., and Waite, T.D., Removal of benzene and selected alkyl-substituted benzenes from aqueous solution utilizing continuous high-energy electron irradiation, *Environ. Sci. Technol.*, 26, 144, 1992b.

Sehested, K., Coffitzen, H., Christensen, H.C., and Hart, E.J., Rates of reaction of O~, OH, and H with methylated benzenes in aqueous solution: optical spectra of radicals, *J. Phys. Chem.*, 79(I), 1975.

Shaukat, F., Kurucz, C.N., Cooper, W.J., and Waite, T.D., Disinfection of wastewaters: high-energy electron vs. gamma irradiation, *Water Res.*, 27(7), 1177–1184, 1991.

Spinks, J.W.T. and Woods, R.J., *An Introduction to Radiation Chemistry*, 3rd ed., John Wiley & Sons, New York, 1990.

Stock, L.M. and Brown, H.C., *Adv. Phys. Org. Chem.*, 1, 36, 1962.

Taft, R.W., Jr., in *Steric Effect in Organic Chemistry*, Newton, M.S., Ed., Wiley, New York, 1956.

13

Zero-Valent Iron

13.1 Introduction

The use of zero-valent iron to reduce chlorinated hydrocarbons was first reported in patent literature by Sweeny and Fischer in 1972; however, Sweeny and Fischer never published their studies in peer reviewed journals, so their work was overlooked by the research community (Gillham and O'Hannesin, 1994). In the late 1980s, Reynolds observed that organics disappeared from the iron pipes used in his study on the corrosion of PVC and iron pipes by water contaminated with organics. Several years later, Gillham realized the potential of using the reduction ability of zero-valent iron for practical purposes and holds several patents for the application of zero-valent iron degradation of organic compounds (Wilson, 1995).

Pump-and-treat and impermeable confinement are frequently used to degrade halogenated and nonhalogenated hydrocarbons in groundwater; however, these remediation techniques have limitations. For example, pump-and-treat only transfers contaminants to another media such as air stripping or activated carbon. In addition, the discharge of large volumes of water and the production of secondary waste may be costly. Also, the hydraulic characteristics of the aquifer may be adversely affected. Permits are required for discharge, and groundwater rights have to be purchased for the disposal of large volumes of treated groundwater, which may result in excessive operating costs (Cantrell and Kaplan, 1997). The use of a common alternative, biological degradation, has increased for remediation, but gaining an understanding of the biochemical pathways and associated by-products involved, as well as developing effective strains of bacteria and managing the population of bacteria, can be difficult and the process has not yet been well defined (Gillham and O'Hannesin, 1994).

Zero-valent iron is a promising *in situ* remediation technology for the degradation of many common pollutants, as it is comparatively inexpensive, does not restrict land use, and requires no energy for operating. Zero-valent iron has been successfully utilized to destroy trichloroethenes, chromate, chlorinated organics, and mixed wastes. It is capable of reducing and dehalogenating a wide variety of halogenated hydrocarbons over wide concentration ranges (Hardy and Gillham, 1996). In addition, iron is nontoxic and inexpensive (Gillham and O'Hannesin, 1994). It is easily oxidized by organic compounds, thereby reducing the contaminant without an additional reactant. In addition to being highly reductive to many halogenated hydrocarbons, zero-valent iron can reduce highly mobile oxycations (UO_2^{2+}) and oxyanions (CrO_4^{2-}, MoO_4^{2-}, TcO_4^-) into insoluble forms (Cantrell and Kaplan, 1997).

This chapter presents the theory and application of zero-valent iron and includes the relevant *in situ* chemical/physical processes. To illustrate these *in situ* technologies, the basic mechanisms of adsorption reduction and oxidation processes are discussed for *in situ* treatment of (1) organic pollutants, (2) heavy metals, and (3) mixtures of organic and inorganic pollutants. The history of zero-valent iron, current applications, mechanisms and kinetics of the system, system improvements, and advantages and disadvantages for zero-valent iron are also discussed.

13.2 Fundamental Theory

The reactions in zero-valent iron are heterogeneous due to the strong dependence of the reaction rate on the surface area of the iron (Burris et al., 1995). The surface reaction proceeds in four steps. First, the reactant undergoes mass transport from the groundwater to the iron surface (Matheson and Tratnyek, 1994). Second, the contaminant is absorbed onto the surface of the iron, where the chemical reaction occurs. Third, the reaction products desorb from the surface, which allows the site to become available for another reaction (Burris et al., 1996a). Finally, the products of the reaction return to the groundwater. Rate limitation could occur at any step. Where it may not be the sole limitation, mass transport plays an essential role in the kinetics of dechlorination (Matheson and Tratnyek, 1994).

Essentially, reduction of hazardous wastes by zero-valent iron is due to the beneficial corrosion of iron. This process takes advantage of the chemical reaction that occurs when iron is oxidized. The contaminant is the oxidant (Fairweather, 1996), while zero-valent iron is a strong reductant capable of dehalogenating several halogenated hydrocarbons (Kaplan et al., 1996). Commercial-grade iron and industrial scrap iron are sufficient to reduce chlorinated solvents (Matheson and Tratnyek, 1994). Although iron is actually consumed during the reaction, it remains effective for a long period of time. For example, 1 kg of iron can dechlorinate chloromethane at a concentration of 1 mg/L and sufficiently treat 0.5 million liters of water (Gillham and O'Hannesin, 1994).

The reductive reaction is slow under anaerobic conditions, because iron may be oxidized by oxygen. Chlorinated contaminants possess an oxidizing potential similar to that for oxygen (Tratnyek, 1996). When strong oxidizing compounds such as chlorinated contaminants are not present, the iron is spontaneously corroded in water (Matheson and Tratnyek, 1994).

13.2.1 Thermodynamics

The reaction during reduction of organic pollutants by Fe^0 has two parts. The anodic half-reaction produces Fe^{2+} from Fe^0 and causes the iron metal to corrode (Agrawal and Tratnyek, 1996):

$$Fe^0 + 2H^+ \longleftrightarrow Fe^{2+} + H_2 \qquad (13.1)$$

The cathodic half-reaction varies with the reactivity of the available electron acceptors, such as H^+ and H_2O in aqueous solutions. If conditions are aerobic, the cathodic half reaction makes O_2, which is the electron acceptor, and H_2 will not be produced (Agrawal and Tratnyek, 1996):

$$Fe^0 + 2H_2O \longleftrightarrow Fe^{2+} + H_2 + 2OH^- \text{ (anaerobic corrosion)} \qquad (13.2)$$

$$Fe^0 + 2H_2O + O_2 \longleftrightarrow 2Fe^{2+} + 4OH^- \text{ (aerobic conditions)} \qquad (13.3)$$

The redox pair formed from oxidizing the zero-valent iron has a reduction potential of -0.440 V; therefore, zero-valent iron can reduce hydrogen ions, carbonate, sulfate, nitrate, and oxygen, in addition to alkyl halides (Matheson and Tratnyek, 1994). Both Equation (13.2) and Equation (13.3) cause the pH to increase.

Zero-valent iron and organic substrate can react with a net result of iron oxidation and reduction of the substrate. In such a reaction, iron acts as a reducing agent:

$$Fe^0 \rightarrow Fe^{2+} + 2e^- \qquad (13.4)$$

The Pourbaix diagram shown in Figure 13.1 illustrates the thermodynamic stability of iron species in aqueous solutions of a few organic substrates. The relative position of each substrate shows that the reaction between iron and the corresponding organic is thermodynamically favorable. Three elementary reactions involved in the reductive dechlorination of organic compounds are shown in Figure 13.2.

13.2.2 Kinetics

While the reaction thermodynamics is important, the reaction kinetics is equally important in designing zero-valent iron system; furthermore, the

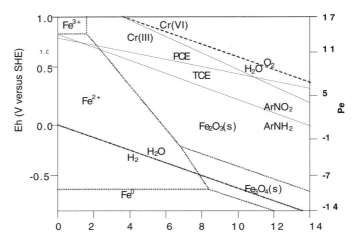

FIGURE 13.1

Eh–pH diagram (or Pourbaix diagram) showing equilibria with water, iron, and common environmental contaminants including perchloroethene (PCE), nitrobenzene (ArNO$_2$), and chromate (Cr[VI]). Hematite (α-Fe$_2$O$_3$) and magnetite (Fe$_3$O$_4$) are assumed to be the controlling phases for iron speciation. The stability lines for the reduction of nitrobenzene (ArNO$_2$) to (ArNH$_2$), Cr(VI) to Cr(III), and PCE to TCE are superimposed to show the instability of Fe^0 in the presence of these contaminants. (From Scherer, M.M. et al., *CRC Crit. Rev. Environ. Sci. Technol.*, 30(3), 363–411, 2000. With permission.)

reaction is a heterogeneous reaction. The kinetics of heterogeneous reactions involving Fe^0 is determined by the following factors:

- The reaction rate constants from Equation (13.1) to Equation (13.3)
- Physical processes on the surface of the catalyst (reducing agent), including its properties
- Mass transfer limitations and diffusion effects
- Sorption/desorption processes involving the substrate and availability of active reaction sites on the iron surface
- Fluid flow characteristics, including velocity, flow regime

The mass transfer limitations have been shown to be less significant in regard to the kinetics of chlorinated aliphatics based on relatively slow rates of degradation (Scherer et al., 2000). On the other hand, nitroaromatics and azo dyes have higher reduction rates under which the diffusion and mass transfer effects may become reaction rate-limiting factors (Agrawal and Tratnyek, 1996). Scherer et al. (2000) identified three steps that may impose limitations on the reduction rates. The formation of precursor complex on active metal sites can be rate limited by the number of reaction sites. Burris et al. (1996) suggested that the hydrophobicity of the contaminant may significantly affect the sorption rate of substrates. Scherer et al. (2000) illus-

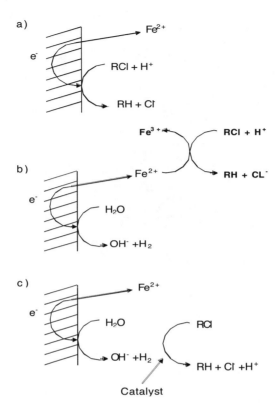

FIGURE 13.2
Scheme showing proposed pathways for reductive dehalogenation in Fe^0–H_2O systems: (A) direct electron transfer from iron metal at the metal surface; (B) reduction by Fe^{2+}, which results from corrosion of metal; (C) catalyzed hydrogenolysis by the H_2 that is formed by reduction of H_2O during anaerobic corrosion. Stoichiometries are shown. (From Matheson, L.J. and Tratnyek, P.G., *Environ. Sci. Technol.*, 28, 2045–2053, 1994. With permission.)

trated that the transfer of electrons from the surface of the reducing agent to the substrate could affect the rates by the three major mechanisms shown in Figure 13.3:

- Electron transfer from bare iron metal exposed by pitting of the oxide layer, while the pitting mechanism involves localized corrosion and possible catalytic dissolution pathways
- Electron transfer from conduction bands in the oxide layer
- Electron transfer from adsorbed or lattice Fe(II) surface area, expressing reduction of a sorbed or lattice surface site

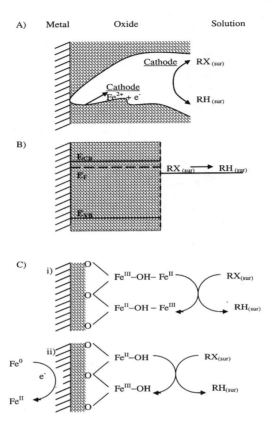

FIGURE 13.3
Conceptual models of electron transfer (ET) mechanisms at Fe^0–oxide–water interface: (A) ET from bare iron metal exposed by pitting of the oxide layer; (B) ET from conduction bands in the oxide layer; (C) ET from adsorbed or lattice Fe(II) surface sites. (From Scherer, M.M. et al., *CRC Crit. Rev. Environ. Sci. Technol.*, 30(3), 363–411, 2000. With permission.)

The physical properties of the iron metal are, therefore, important factors associated with mass transfer limitations of the electron transfer processes and reaction-limiting steps.

13.2.3 Adsorption

As a contaminant moves through soil and groundwater, chemical processes will affect both contaminant concentration and overall hydrogeochemistry (Schoonen, 1998) of the system. Different adsorption mechanisms cause pollutants to adsorb onto the soil, volatilize, precipitate, and be part of the oxidation–reduction processes. Adsorption is loosely described as a process in which chemicals partition from a solution phase into or onto the surfaces of solid-phase materials. Adsorption at particle surfaces tends to retard contaminant movement in soil and groundwater.

The three types of adsorption are (1) physical, (2) chemical, and (3) exchange adsorption. Especially important to the success of *in situ* treatment by Fe^0 are the soil characteristics, which affect soil sorptive behavior such as mineralogy, permeability, porosity texture, surface qualities, and pH. Physical adsorption is due to van der Waal's forces between molecules where the adsorbed molecule is not fixed on the solid surface but is free to move over the surface and may condense and form several superimposed layers. An important characteristic of physical adsorption is its reversibility. On the other hand, chemical adsorption is a result of much stronger forces with a layer forming, usually of one molecule thickness, where the molecules do not move. It is normally not reversible and must be removed by heat. The exchange adsorption and ion exchange process involves adsorption by electrical attraction between the adsorbate and the surface (Rulkens, 1998).

Adsorption may occur in a combination of three possible mechanisms: hydrophobic expulsion, electrostatic attraction, and complexation. Most nonpolar compounds, such as various organics, adsorb by this mechanism, and the degree of partitioning is correlated to the octanol/water partitioning coefficient, K_{ow}. Polar substrates such as various metals sorb via electrostatic attraction and complexation. Table 13.1 shows the typical sorption mechanisms and typical examples.

Sorption isotherm curves are graphical relationships showing the partitioning between solid and liquid form where mass adsorbed per unit mass of dry solids (*S*) is plotted against the concentration (*C*) of the constituent in solution. *K* is the sorption equilibrium constant; *N* is a constant describing the intensity of sorption. The linear sorption isotherm can be expressed as follows:

$$S = K_d C^N \qquad (13.5)$$

TABLE 13.1

Sorption Mechanisms in Soils

Mechanism	Other Terminology	Examples
Hydrophobic expulsion	Partitioning	Nonpolar organics (e.g., PCBs, PAHs)
Electrostatic attraction	Outer-sphere	Some anions (e.g., NO^{3-})
	Nonspecific	Alkali and alkaline earth metals (Ba^{2+},
	Physisorption	Ca^{2+})
	Physical	
	Ion exchange	
Complexation reaction	Inner sphere	Transition metals (e.g., Cu^{2+}, Pb^{2+},
	Specific	CrO_4^{2-})
	Chemisorption	
	Chemical	
	Ligand exchange	

Source: Adapted from Scherer, M.M. et al., *CRC Crit. Rev. Environ. Sci. Technol.*, 30(3), 363–411, 2000.

The Freundlich isotherm can be described as:

$$S = K_d \, C^{1/N} \tag{13.6}$$

The Langmuir isotherm is used to describe single-layer adsorption based on the concept that a solid surface possesses a finite number of sorption sites. When these active sites are filled, the site will no longer sorb solute from the solution:

$$S = K_{\alpha d} \cdot \Gamma_{max} \tag{13.7}$$

where α = absorption constant related to the binding energy (L/mg), and Γ_{max} = maximum amount that can be adsorbed by the solid (mg/kg).

Because adsorption isotherms are equilibrium equations, the rate at which the material is adsorbed has to be studied in terms of chemical affinities, pH, solubility, hydrophobicity, and many other physical and chemical characteristics.

Because organic nonpolar compounds have stronger attraction to organic matter than to mineral content, the amount of adsorption of an organic contaminant is more dependent on the organic content of the soil. The adsorption partition coefficient is generally used to determine this adsorption amount, as it is empirically related to the organic fraction of the soil (f_{oc}), and the normalized partition coefficient K_{oc} can be expressed as follows:

$$K_p = K_{oc} f_{oc} \tag{13.8}$$

The amount of adsorption is also dependent on the moisture content of the soil, as water competes for adsorption sites, as illustrated in Figure 13.4.

In a zero-valent iron system, pollutants will be retained at the Fe surface. Movement of metals into other environmental compartments (i.e., groundwater, surface water, or the atmosphere) should be minimal as long as the retention capacity of Fe is not exceeded. The extent of movement of a pollutant in the zero-valent iron system is intimately related to the solution pH, the surface chemistry of the Fe, the specific properties of pollutants, and the associated waste matrix. The retention mechanisms for pollutants include adsorption of the pollutant by the Fe surfaces and precipitation. The retention of cationic metals by Fe has been correlated with Fe properties such as pH, redox potential, surface area, cation exchange capacity, organic matter content, clay content, iron and manganese oxide content, and carbonate content. Anion retention has also been correlated with pH, iron and manganese oxide content, and redox potential. In addition to Fe properties, consideration must be given to the type of pollutants, their concentration, the competing ions, and the complexing ligands. Transport of metals associated with various wastes may be enhanced due to the following reasons (Puls et al., 1995):

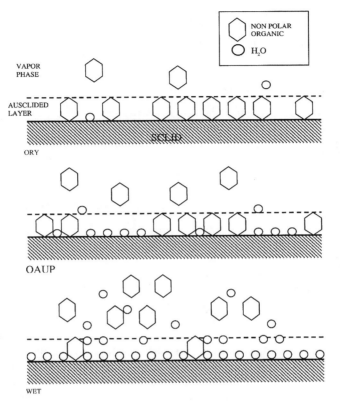

FIGURE 13.4
Adsorption mechanism for nonpolar organic species. (Adapted from Semer, R. and Reddy, K.R., *J. Haz. Mat.*, 57, 209–230, 1998.)

- Transportation caused by metal association with mobile colloidal size particles
- Formation of metal organic and inorganic complexes that do not sorb to soil solid surfaces
- Competition with other constituents of waste, both organic and inorganic, for sorption sites
- Deceased availability of surface sites caused by the presence of a complex waste matrix

The available surface area of the iron has the greatest effect on the reaction rate (Johnson et al., 1996). The iron is commonly treated with hydrochloric acid prior to the experiment to accelerate dechlorination and improve reproducibility. By cleaning the metal surface, the passive oxide layer is broken off, the surface area is increased by etching and pitting, and the density of highly reactive sites is increased (Agrawal and Tratnyek, 1996). Because of the wide range of waste characteristics and various ways in which metals can be adsorbed to the Fe surface, the extent of pollutant retention by a soil is specific to the type of site, soil, and waste involved.

13.2.4 Halogenated Hydrocarbons

Dechlorination is a surface reaction with the zero-valent iron serving as the electron donor. When there is a proton donor, such as water, chlorinated compounds will be dehalogenated. The reaction kinetics depends upon the mass transfer to the surface of the iron, the available surface area, and the condition of the surface. The reaction is pseudo first order, and direct contact with the surface of the iron is required for degradation to take place (Gillham and O'Hannesin, 1994). The basic equation for dechlorination by iron metal is as follows:

$$Fe^0 + RX + H^+ \longleftrightarrow Fe^{2+} + RH + Cl^- \qquad (13.9)$$

The reduction potentials for various alkyl halides range from +0.5 to +1.5 V; therefore, when Fe^0 serves as an electron donor, the reaction is thermodynamically favorable. Because three reductants are present in the treatment system (Fe^0, H_2, and Fe^{2+}), three possible pathways exist. Equation (13.9) represents the oxidation of Fe^0 by reduction of a halogenated compound. In the second pathway, the ferrous iron behaves as a reductant, as represented in Equation (13.10). This reaction is relatively slow because the ability to reduce a pollutant by ferrous iron is dependent on the speciation ferrous ions, which is determined by the ligands present in the system. The third possible pathway, Equation (13.11), is dehalogenation by hydrogen. This reaction does not occur easily without a catalyst. In addition, if hydrogen levels become too high, corrosion is inhibited (Matheson and Tratnyek, 1994):

$$2Fe^{2+} + RX + H^+ \rightarrow 2Fe^{3+} + RH + X^- \qquad (13.10)$$

$$H_2 + RX \rightarrow RH + H^+ + X^- \qquad (13.11)$$

If all three pathways are not possible, then reactions will be limited. Additional limitations may occur if the reaction is aerobic because Fe^{3+} could be produced by further oxidation of Fe^{2+} and cause precipitation of iron oxides (Helland et al., 1995). The end products such as ferrous chloride and ferric oxide are not capable of reducing chlorinated compounds (Gillham and O'Hannesin, 1994). Burris et al. (1995) state that Fe^{2+} and H_2 do not have an effect on degradation. Hardy and Gillham (1996) have suggested the possibility that degradation may be due to a catalytic reaction utilizing hydrogen produced from the reduction of water. In developing and improving the performance of the zero-valent iron technique in the field, detail mechanisms are important and critical for a specific site. The dominant process is the oxidation of the iron metal (Matheson and Tratnyek, 1994). The reduction of the halogenated hydrocarbon involves the transfer of an electron and occurs at the iron surface. A carbon-centered radical, $R^\bullet$, is formed as:

$$RX + e^- \rightarrow R^\bullet + X^- \tag{13.12}$$

Then, a second electron transfer occurs, and the radical is protonated as in Equation (13.13).

$$R^\bullet + H^+ + e^- \rightarrow {}^\bullet RH \tag{13.13}$$

The result is dehalogenation, and during this step Fe^{2+} is formed by dissolving into solution. Hydrocarbon formation also adds perplexity to mechanism determination. Because C_1 to C_5 hydrocarbons have been found in iron and water systems, the hydrocarbons formed may be due to reduction of aqueous CO_2 by zero-valent iron. Therefore, iron behaves as both a reactant, by corroding and supplying electrons, and as a catalyst, by promoting the formation of hydrocarbons. Ten hydrocarbons were identified up to C_5; also, iron pretreated with H_2 formed hydrocarbons with longer chain lengths. The hydrocarbon concentration and the growth in chain length increase with time; therefore, if the hydrocarbons are not desorbed, the production of the hydrocarbons could be rate limiting in the dechlorination of chlorinated organics. Although this does not determine the mechanism by which iron metal removes halogens from halogenated organics, it does show that the product distribution, carbon mass balance, and reaction rate may be affected by catalysts (Hardy and Gillham, 1996).

For modeling purposes, zones are constructed in the column tests performed. A subsurface barrier will have three zones: upgradient (area before the barrier), iron-bearing gradient (the barrier itself), and downgradient (area after the barrier). In the upgradient zone, the Fe^0 is oxidized to Fe^{2+} and Fe^{3+} by O_2. The reaction increases the pH, and precipitation of the iron oxides is initiated. The precipitation could reduce the permeability of the barrier and surface area of the iron, thus reducing the reaction rate. This also causes the iron-bearing zone and down gradient zone to be anoxic (Tratnyek, 1996). This production of precipitates has been observed regularly in the laboratory, but the degree to which the effect occurs in the field is under much debate. For these reasons, buffers such as pyrite, which lowers the pH, are used.

The reactivity of the iron with the contaminants determines the feasibility and the design of the site. Many factors contribute to the reaction rate. For example, the oxide layer that forms on the iron affects the surface of the iron and inhibits further corrosion, so this layer needs to be minimized. Fortunately, iron possesses a "porous and incoherent" nature with oxide film, and iron usually exhibits satisfactory degradation rates over a long period of time (Tratnyek, 1996).

Because reaction rates vary widely among contaminants, Fe barrier design should be dependent on the least reactive contaminant. Some contaminants are quickly reduced, and others react too slowly for consideration of remediation by iron. Although much data exist regarding degradation rates, quantitative predictions concerning the reaction rates cannot be made because many of the experiments are highly inconsistent in accounting for the effect of pH, surface area of the iron, mixing rate, and other experimental variables. Comparisons are of better quality if the rate constants are normalized. For example, when rate constants are normalized to the surface area of iron (k_{SA}), the specific rate constant varies by only one order of magnitude for individual halocarbons. Upon further correlation, the k_{SA} has indicated that dechlorination is more rapid for saturated carbon than unsaturated carbons, and reaction rates are faster for perhalogenated compounds. By obtaining this quality information, the amount of iron required to obtain a decrease in contaminant concentration by a magnitude of three can be calculated (Tratnyek, 1996). Also, quantitative structure–activity relationships (QSARs) could be developed for such prediction of iron required for different organic pollutants.

13.3 Degradation of Hazardous Wastes

13.3.1 Organic Pollutants

Table 13.2 presents organic classes that have been successfully degraded by zero-valent iron.

13.3.1.1 Unsaturated Halogenated Compounds

Trichloroethylene (TCE) and perchloroethylene (PCE) require cleavage of the carbon–halogen bonds. Two methods of cleavage are β-elimination by dehydrohalogenation, as shown in Equation (13.14), and nucleophilic substitution by either water or hydrogenolysis in Equation (13.15). The proposed pathways for reduction of chloroethylenes by zero-valent iron are as follows:

$$RX = RX + 2e^- \rightarrow R \equiv R + 2X^- \tag{13.14}$$

$$RX + 2e^- + H^+ \rightarrow RH + X^- \tag{13.15}$$

The reduction of the triple bond may form an olefin (Equation 13.15) or an alkane (Equation 13.16) (Burris et al., 1995):

$$R \equiv R + 2e^- + 2H^+ \rightarrow RH + RH \tag{13.16}$$

TABLE 13.2

Organic Contaminants Treated by Fe^0

Methanes	Tetrachloromethane
	Trichloromethane
	Dichloromethane
Ethanes	Hexachloroethane
	1,1,1-Trichloroethane
	1,1,2-Trichloroethane
	1,1-Dichloroethane
Ethenes	Tetrachloroethene
	Trichloroethene
	cis-1,2-Dichloroethene
	trans-1,2-Dichloroethene
	1,1-Dichloroethene
	Vinyl chloride
Propanes	1,2,3-Trichloropropane
	1,2-Dichloropropane
Aromatics	Benzene
	Toluene
	Ethylbenzene
Other	Hexachlorobutadiene
	1,2-Dibromoethane
	Freon 113
	N-nitrosodimethylamine

Source: USEPA, EPA/600/R-98/125, U.S. Environmental Protection Agency, Washington, D.C., 1998.

$$R \equiv R + 4e^- + 4H^+ \rightarrow RH_2 - RH_2 \qquad (13.17)$$

The overall reaction as shown in Equation (13.17) for the degradation of the TCE produces ethene and ethane at a ratio of 2:1:

$$C_2HCl_3 + 3H^+ + 6e^- \rightarrow C_2H_4 + 3Cl^- \qquad (13.18)$$

Ethene and ethane account for 80% of the mass of the hydrocarbons identified as products. Trace amounts of methane and acetylene are also produced (Orth and Gillham, 1996). The reduction of PCE forms *cis*-1,2-dichloroethylene (DCE), *trans*-1,2-DCE, 1,1-DCE, vinyl chloride, ethylene, dichloroacetylene, acetylene, ethene, ethane, chloroacetylene, methane, and several alkenes ranging from C_3 to C_6. The trace amounts of dichloroethylene and vinyl chloride formed during the reduction of PCE and TCE are further reduced (Burris et al., 1995). Reaction rates vary with substrate, chemical, and microbiological conditions. Selected $t_{1/2}$ values are provided in Table 13.3.

TABLE 13.3

$t_{1/2}$ Values for Selected Halogenated Aliphatics

Compound	$t_{1/2}$ (min)
Perchloroethylene	17.9
Tetrachloroethene	—
Trichloroethene	13.6
1,1-Dichloroethene	40.0
trans-1,2-Dichloroethene	55.0
cis-1,2-Dichloroethene	432.0
Vinyl chloride	374.0

Source: Gillham, R.W. and O'Hannesin, S.F., *Ground Water*, 32(6), 958–967, 1994. With permission.

13.3.1.2 Saturated Halogenated Compounds

Helland et al. (1995) compared the degradation rates of halogenated aliphatic compounds by zero-valent iron in both anoxic and oxic reductions. The reactions are as follows:

$$2Fe^0 \rightarrow 2Fe^{2+} + 4e^- \tag{13.19}$$

$$3H_2O \rightarrow 3H^+ + 3OH^- \tag{13.20}$$

$$2H^+ + 2e^- \rightarrow H_2 \tag{13.21}$$

$$2Fe^0 + 3H_2O + X\text{-}Cl \rightarrow 2Fe^{2+} + 3OH^- + H_2 + X\text{-}H + Cl^- \tag{13.22}$$

When water dissociation is taken into consideration, the reaction becomes (Gillham and O'Hannesin, 1994):

$$Fe^0 + H_2O + X\text{–}Cl \rightarrow Fe^{2+} + OH^- + X\text{–}H + Cl^- \tag{13.23}$$

If no dissociation of water occurs, the following reaction will take place:

$$X\text{–}Cl + H^+ + 2e^- \rightarrow X\text{–}H + Cl^- \tag{13.24}$$

The reaction proceeds until each chlorine ion is removed. For example, carbon tetrachloride would be reduced to chloroform, then to methylene chloride, and finally to methane (the reduction of methylene chloride takes several months, however). No degradation products other than the parent compounds were found; therefore, degradation is simple, reductive dechlorination, with the zero-valent iron serving as an electron donor. The reaction was pseudo first-order and the reaction constant, k, decreased with each additional dehalogenation step (Gillham and O'Hannesin, 1994).

Gillham and O'Hannesin (1994) reported that the degradation rates increased as the surface area of the iron increased with respect to the volume of the solution. Also, degradation declined with decreasing degree of chlorination. Dichloroethene is the only compound in Table 13.4 not degraded by the zero-valent iron process.

Saturated hydrocarbons such as 1,2-dibromo-3-chloropropane (DBCP) have been used as soil nematicides. DBCP has contaminated the groundwater due to its extensive use. Although banned in 1977, concentrations in some parts of the United States exceed the maximum contaminant level of 0.2 μg/L. The dehalogenation of DBCP has two possible pathways: (1) a series of three hydrodehalogenation reactions requiring a proton and two electrons for each reaction or (2) the formation of a transitional state of propene followed by hydrogenation of the double bond, with the product being propane in both mechanisms. The only intermediate formed is propene, with a reaction rate constant of 0.28 ± 0.03 min^{-1} (Siantar et al., 1996). Table 13.5 shows the effect of pH on the pseudo first-order rate constants and half-lives of DBCP with or without mixing. Table 13.6 indicates that sulfate and nitrite ions do not significantly affect the pseudo first-order rate constants and half-lives of DBCP.

13.3.1.3 Polychlorobiphenyls

Degradation of polychlorobiphenyls (PCBs) by zero-valent iron requires temperatures of 400°C (Grittini et al., 1995). At 400°C, PCBs are reduced to biphenyls. For further degradation of the biphenyl compound, temperatures have to exceed 500°C (Chuang et al., 1995). Under normal temperatures, zero-valent iron has little effect on PCBs.

TABLE 13.4

$t_{1/2}$ Values for Sample Halogenated Aliphatics

Compound	$t_{1/2}$ (hr)
Tetrachloromethane	0.25
Trichloromethane	33.0
Tribromomethane	0.24
Dichloromethane	No decline
Hexachloroethane	0.13
1,1,2,2-Tetrachloroethane	19.2
1,1,1,2-Tetrachloroethane	4.4
1,1,1-Trichloroethane	5.3

Source: Gillham, R.W. and O'Hannesin, S.F., *Ground Water*, 32(6), 958–967, 1994. With permission.

TABLE 13.5

Effect of Ions on Pseudo First-Order Rate Constants and Half-Lives of DBCP
Transformation Using 0.1-*M* HEPES Buffer Solution at pH 7, Mixing at 400 rpm,
and 22°C

Aqueous Phase	pH	k (hr⁻¹) (No Shaking)	k (hr⁻¹) (Mild Shaking)	$t_{1/2}$ (hr) (No Shaking)	$t_{1/2}$ (hr) (Mild Shaking)
Milli Q™ water	7	0.099	0.132	6.99	5.23
0.1-M MES	6	0.341	0.482	2.03	1.44
0.1-M HEPES	7	0.393	0.513	1.76	1.35
0.1-M Tricine	8	0.259	—	2.67	—
0.1-M CHES	9	—	—	—	—

Source: Siantar, D.P. et al., *Water Res.*, 30(10), 2315–2322, 1996. With permission.

TABLE 13.6

Effect of Sulfate and Nitrite Ions on Pseudo First-Order Rate Constants and
Half-Lives of DBCP Transformation

Concentration	k (min⁻¹)	$t_{1/2}$ (min)
No sulfate or nitrite added	0.300	2.31
Sulfate concentration (mg/L)		
3.7	0.273	2.54
19.8	0.266	2.61
27.6	0.270	2.57
Nitrite concentration (mg/L)		
4.3	0.255	2.72

Source: Siantar, D.P. et al., *Water Res.*, 30(10), 2315–2322, 1996. With permission.

13.3.1.4 *Nitroaromatic Compounds*

Nitroaromatic compounds (NACs) are one of the widespread contaminants
in the environments. Sources of NACs are numerous; they originate from
insecticides, herbicides, explosives, pharmaceuticals, feedstock, and chemicals for dyes (Agrawal and Tratnyek, 1996). Under anaerobic conditions, the
dominant action is nitro reduction by zero-valent iron to the amine. Other
pathways do exist, such as the formation of azo and azoxy compounds,
which is followed by the reduction of azo compounds to form amines. Also,
in addition to the possibility of azo and azoxy compounds, phenylhydrox-
ylamine may be an additional intermediate (Agrawal and Tratnyek, 1996).
Nitrobenzene reduction forms the amine aniline. Known for its corrosion
inhibition properties, aniline cannot be further reduced by iron. Additionally,
it interferes with the mass transport of the contaminant to the surface of the
iron. The overall reaction is as follows:

$$ArNO_2 + 3Fe^0 + 6H^+ \rightarrow ArNH_2 + 3Fe^{2+} + 2H_2O \qquad (13.25)$$

The elementary steps are shown in Equation (13.26) to Equation (13.28):

$$ArNO_2 + Fe^0 + 2H^+ \xrightarrow{k_1} ArNO + 3Fe^{2+} + H_2O \qquad (13.26)$$

$$ArNO + Fe^0 + 2H^+ \xrightarrow{k_2} ArNHOH + Fe^{2+} \qquad (13.27)$$

$$ArNHOH + Fe^0 + 2H^+ \xrightarrow{k_3} ArNH_2 + 3Fe^{2+} + 2H_2O \qquad (13.28)$$

During the reduction of NACs, changes in corrosion and precipitation of iron products were not significant. As long as the diffusivities of the different substitutions on the benzene were similar, the reaction rate did not vary significantly (Agrawal and Tratnyek, 1996). The rate constants are given in Table 13.7.

Because nitrobenzene degradation is faster in batch experimental systems than in column studies, a mass-transfer limitation exists; therefore, when determining the effectiveness of *in situ* groundwater treatment systems, hydraulics, mass transfer, and reaction kinetics should be taken into consideration (Burris et al., 1996).

Weber (1996) performed additional experimental studies on 4-aminoazobenzene (4-AAB). The compound was chosen based on the hypothesis that azo compounds capable of reduction by zero-valent iron and an amino group would allow for determination of surface-reaction occurrence. The compound 4-AAB reduces and forms aniline. The reduction of 4-AAB was performed without cleaning the iron with hydrochloric acid, which is a standard method in most experiments. Complete loss of 4-AAB occurred in 2 hr. When the iron was washed with hydrochloric acid, the reaction occurred so quickly that aniline formation could not be measured. An additional experiment with bound 4-AAB was conducted to determine if the reaction

TABLE 13.7

Pseudo First-Order Rate Constant of Substrate Reduction

Substrate	k (min^{-1})
Nitrobenzene	0.0339
Nitrosobenzene	0.0339
1,3-Dinitrobenzene	0.0339
4-Chloronitrobenzene	0.0336
4-Nitroanisole	0.0327
4-Nitrotoluene	0.0335
2,4,6-Trinitrotoluene	0.0330
Parathion	0.0250

Source: Agrawal, A. and Tratnyek, P.G., *Environ. Sci. Technol.*, 30, 153–160, 1996. With permission.

occurred at the iron surface. Because no degradation of 4-AAB occurred, the reaction appears to be surface mediated.

When NACs are reduced, aromatic amines are formed and are of significant toxicological concern. Zero-valent iron cannot accomplish this degradation; therefore, remediation has to extend beyond the reduction of the parent NAC (Agrawal and Tratnyek, 1996). Two methods of further reduction are biodegradation and enzyme-catalyzed coupling reactions. Biodegradation can degrade the NACs further to mineralization, and the reaction is more rapid for aromatic amines than for the biodegradation of the original NAC compound (Burris et al., 1996b); however, biodegradation of NACs can be difficult and can produce toxic metabolites; therefore, further reduction of the NACs can also be accomplished by enzyme-catalyzed coupling reactions, which integrates the amine into organic matter (Monsef et al., 1997). Both methods could be combined with nitro reduction by Fe^0 to treat NAC contamination (Agrawal and Tratnyek, 1996).

13.3.1.5   Nitrates and Nitrites

The presence of nitrates and nitrites in groundwater is of growing concern as it poses serious health risks. The typical sources of nitrate are nitrogen fertilizers, septic tanks, and animal wastes. Excessive nitrates present in groundwater can be reduced by microorganisms to nitrite, which is harmful to human health and animals, agriculture, and the environment. Nitrites and secondary amines can react to form nitroamines, which cause serious health effects, including cancer. Such adverse effects with increasing nitrate contamination in groundwater draw attention to treatment and removal of nitrate and nitrite contamination in groundwater. Physicochemical removal by ion exchange resins does not destroy nitrates, frequent resin regeneration by salt produces much brine, and biological degradation by denitrifying microorganisms is often slow and incomplete. The process requires expensive maintenance due to production of excessive biomass.

The feasibility of chemical degradation of nitrate by reducing agents has been investigated for application in treatment of contaminated groundwater by Horold et al. (1993a,b). The transformation of aqueous nitrate into benign products was studied using Fe^0 in a pH-buffered anaerobic aqueous medium from groundwater. Their results showed that 75 to 85% of the nitrate was reduced to ammonia at room temperature within an hour in an acidic pH of 3. The chemical denitrification by organic and inorganic reductants and catalysts showed a 55% reduction in nitrate within 48 hr with Fe^0 powder in an anaerobic medium at 85°C. A rapid catalytic reduction of nitrite with hydrogen gas using Pd metal supported on alumina and of nitrate with hydrogen gas using a Cu–Pd bimetallic catalyst supported on alumina have been demonstrated. This technique seemed to be efficient at the temperature and pH ranges of natural groundwater and was recommended for an *ex situ* treatment of groundwater.

To gain insight into kinetics, reaction pathways, and reaction end products, laboratory investigations were performed by Rahman and Agrawal (1997). Sodium nitrate and sodium nitrite were selected as model pollutants. Reactions were carried out at room temperature in the dark with untreated [Fe^0] = 69.4 g/L. In some cases, Fe^0 was treated with 10% HCl (v/v) for nearly 2 min and then washed with deionized water four to six times prior to reaction. The nitrate and nitrite stock solution was nearly 0.16 mM, and mixing was achieved at 40 rpm.

Experiments with sodium nitrate showed rapid reduction by Fe^0, with nitrite as an intermediate and ammonia as final product. Iron acts as an electron donor and the reduction is coupled with metal corrosion (Equation 13.9). The reduction reaction in the model system was found to proceed in two sequential steps (Equation 13.30 and Equation 13.31), and the overall reaction was represented as follows:

$$Fe^0 \leftrightarrow Fe^{2+} + 2e^- \qquad (13.29)$$

$$NO_3^- + 2H^+ + 2e^- \rightarrow NO_2^- + H_2O \qquad (13.30)$$

$$NO_2^- + 8H^+ + 6e^- \rightarrow NH_4 + 2H_2O \qquad (13.31)$$

$$4Fe^0 + NO_3^- + 10H^+ \rightarrow 4Fe^{2+} + NH_4 + 3H_2O \qquad (13.32)$$

In investigations into nitrate reduction with Fe^0, the concentration of nitrate decreased through time with concurrent increase in nitrite concentrations. As no nitrite was injected, the appearance of nitrite and the increase in its concentration with concurrent decreases in nitrate concentration throughout the experiment suggested reduction of nitrate (Equation 13.30) by zero-valent iron (Rahman et al., 1997). The transformation reaction of nitrate to nitrite was found to be first-order in substrate concentration, and the reaction rate constant was obtained. These reduction experiments were performed with untreated as well as acid-treated Fe^0 turnings. Experiments were also performed to investigate if only nitrite was reduced with Fe^0 metal under identical conditions. Results showed that nitrite can also be reduced by acid treated or untreated Fe^0 metal (Rahman and Agrawal 1997). The transformation reaction (Equation 13.31), similar to nitrate reduction, was found to be first order in substrate concentration and, unlike nitrate, no by-products of nitrite reduction were evident in ion chromatography. This suggested that either reduced products (N_2 gas) were lost in gaseous forms or they may be present in solution as NH_4^+ cation. The [NH_4^+] was estimated in a batch system after several days as the final reduction product and indicated a mass balance of nearly 80%. The hypothesized reaction (Equation 13.35) consisted of the following steps:

$$NO_2^- + 4H^+ + 3e^- \rightarrow 0.5\ N_2\ (gas) + 2H_2O \qquad (13.33)$$

$$0.5N_2 + 4H^+ + 3e^- \rightarrow NH_4^+ \qquad (13.34)$$

$$NO_2^- + 8H^+ + 6e^- \rightarrow NH_4^+ + 2H_2O \qquad (13.35)$$

Thus, it was observed that the first-order rate constants (k_1) for nitrate reduction by untreated Fe^0 increase due to the pretreatment of iron metal with HCl; however, observed increases in the rate constant for nitrite reduction have been relatively small under similar acid pretreatment conditions. During the first 12 hr, the rate constant for nitrate reduction showed a gradual decline, and this decline seems to have been clearly influenced by the presence of chloride. The reaction rate constants for nitrate and nitrite reduction by untreated Fe^0 turnings are directly dependent on the concentration of Fe^0 used, ranging between 69.4 and 208.2 g/L; thus k_1 and k_2 increase linearly with increases in the surface area of the untreated iron. Table 13.8 demonstrates that acid-treated Fe^0 is more reactive than its untreated counterpart.

13.3.2 Reduction of Heavy Metals

Anions or oxyanions of arsenic, selenium, chromium, technetium, and antimony are important groundwater contaminants and may occur under natural groundwater conditions. Indirect precipitation of inorganic cations results from the reduction of an anion-forming species, usually sulfate. Sulfate reduction generates hydrogen sulfide, which combines with metals to form relatively insoluble metal sulfide precipitates. Many heavy metals are treatable using this approach, including Ag, Cd, Co, Cu, Fe, Ni, Pb, and Zn, as listed in Table 13.9 (Waybrant et al., 1998). Column experiments conducted using a range of organic substrates demonstrated the potential to remove a

TABLE 13.8

Reduction Kinetics of Nitrate and Nitrite with an Fe^0 System

Fe^0 (g/L)	Treatment	Rate Constant	Half-Life (hr)	R^2	Conditions
Nitrate reduction					
69.4	Untreated	0.0168	41.3	0.99	
69.4	Acid treated	0.2592	2.67	0.97	
Nitrate reduction in presence of chloride					
69.4	Acid treated	0.0348	19.9	0.99	$[Cl^-] = 7\ \mu M$
69.4	Acid treated	0.089	7.78	1.00	$[Cl^-] = 70\ \mu M$
Nitrite reduction					
69.4	Untreated	0.0963	7.2	0.98	
69.4	Acid treated	0.1864	3.72	0.99	

Source: USEPA, EPA/600/R-98/125, U.S. Environmental Protection Agency, Washington, D.C., 1998.

TABLE 13.9

Inorganic Contaminants Treated

Trace metals	Chromium
	Nickel
	Lead
	Uranium
	Technetium
	Iron
	Manganese
	Selenium
	Copper
	Cobalt
	Cadmium
	Zinc
Anion	Sulfate
contaminants	Nitrate
	Phosphate
	Arsenic

Source: Waybrant, K.R. et al., *Environ. Sci. Technol.*, 32(13), 1972–1979, 1998. With permission.

range of dissolved metals at groundwater velocities similar to those observed at sites of groundwater contamination. A field-scale reactive barrier for the treatment of acid mine drainage and removal of dissolved Ni was installed in 1995 at the Nickel Rim mine site near Sudbury, Ontario. It was composed of municipal compost, leaf compost, and wood chips. Monitoring of the reactive barrier indicates continued removal of the acid-generating capacity of the groundwater flowing through the permeable reactive barrier (PRB) and decreases in dissolved Ni concentrations from up to 10 mg/L to <0.1 mg/L within the PRB.

The mechanism of adsorption implies attachment of the chemical to reactive sites on mineral surfaces. These sites usually result from an excess either positive or negative charge on the surfaces. These surface charges can be constant (fixed) due to ion substitutions in the mineral matrix (isomorphous substitution), variable with pH, or a mixture of both. In addition, the adsorption can result from either inner-sphere or outer-sphere complexation. Inner-sphere complexation is due to actual covalent and ionic chemical bond formation. In outer-sphere complexation, adsorption results from ion-pair bonding due to electrostatic forces; hydration water separates the solvated ion from the surface. Many metal oxides and some clay minerals have net surface charges that vary with pH due to the proportion of protonated vs. deprotonated surface sites. Among these variably charged materials are the iron oxyhydroxides (rusts) that result when zero-valent iron corrodes. These materials are very significant to adsorption of both inorganic and organic charged solution species (ionic species). The charges of both the surface and the solution ion control whether adsorption will occur or whether the surface and the ion will repulse one another. The pH of zero-point of charge, or pH_{zpc}, is the pH at which negatively and positively charged surface sites exist in approximately equal numbers on the mineral. Above pH_{zpc}, the surface will have a net negative charge, enhancing cation adsorption; below pH_{zpc}, the surface will have a net positive charge, enhancing anion adsorption.

13.3.2.1   Chromium

Elemental iron, iron-bearing oxyhydroxides and iron-bearing aluminosilicate minerals have been observed to promote the reduction and precipitation of Cr(VI). The overall reactions for the reduction of Cr(VI) by Fe^0 and the subsequent precipitation of Cr(III) and Fe(III) oxyhydroxides are:

$$CrO_4^{2-} + Fe^0 + 8H^+ \rightarrow Fe^{3+} + Cr^{3+} + 4H_2O \tag{13.36}$$

$$(1-x)Fe^{3+} + (x)Cr^{3+} + 2H_2 \rightarrow Fe_{(1-x)} Cr_x OOH_x + 3H^+ \tag{13.37}$$

The main physical processes leading to the above reactions that occur with PRB treatment are sorption, dissolution, and precipitation (USEPA, 1998).

13.3.2.2   Arsenic

Manning et al. (2002) applied zero-valent iron for treatment of arsenic (III) and (V) and investigated their reactions with two Fe^0 materials, their iron oxide corrosion products, and several model iron oxides. Different species of arsenate (As(V)) and arsenite (As(III)) were treated. By applying x-ray diffraction, scanning electron microscopy/energy-dispersive spectrometry (SEM/ED), x-ray absorption spectroscopy, and high-performance liquid chromatography (HPLC)/hydride generation atomic absorption spectrometry, a number of corrosion products of Fe^0 were identified, including lepidocrocite (γ-FeOOH), magnetite (Fe_3O_4), and maghemite (γ-Fe_3O_4). The results indicated that Fe(II) oxidation was an intermediate step in the Fe^0 corrosion process. Under aerobic conditions, the Fe^0 corrosion reaction does not reduce As(V) to As(III) but the As(III) has been oxidized to As(V). The oxidation of As(III) was also caused by magnetite and hematite minerals, indicating that the formation of certain iron oxides during Fe^0 corrosion favors the As(V) species as a final reaction product. It was hypothesized that water reduction with subsequent release of OH$^-$ to solution on the surface of corroding Fe^0 may also promote the oxidation of As(III) to As(V). The existence of inner-sphere bidentate As(III) and As(V) complexes has been assumed to have significance in the iron corrosion reaction. Thus, *in situ* treatment of groundwater containing As(III) and As(V) has been demonstrated.

13.3.2.3 Uranium

Cui and Spahiu (2002) investigated the reduction of uranium(VI) on corroded iron under anoxic conditions and demonstrated that U(VI) can be treated by Fe^0. The anoxic conditions were attained by flushing with a gas mixture composed of 99.97% Ar and 0.03% CO_2 through the test vessel. The test vessel contained an oxygen trap and synthetic groundwater (10 mM NaCl and 2 mM HCO^{3-}). A corrosion product of dark green formed on the iron surface after 3 months of corrosion. The chemical was identified as carbonate green rust — (Fe_4Fe_2III)–Fe–$II(OH)CO_3$. The iron foil that reacted in a solution (10 ppm U(VI), 10 mM NaCl, and 2 mM HCO_3) for 3 months was analyzed by SEM-EDS. The study showed the formation of (1) an uneven layer of carbonate green rust (1 to 5 μm thick) on the iron surface; (2) thin (0.3 μm) uranium-rich layer; and 3) UO_2 crystals (3 to 5 μm) on the thin uranium layer. The experimental results proved that the U(VI) removal capacity of metal iron is not hindered by formation of a layer of carbonate green rust on the iron. Tests with cast iron and pure iron indicate that they have similar U(VI) removal capacities. At the end of the experiment, the uranium concentrations approached the solubility of UO_2 (solid), 10^{-8} M.

13.3.2.4 Mercury

Mercury represents a serious environmental risk, and the study of removal of mercury from wastewater has received considerable attention in recent years. Mercury concentration was usually reduced by deposition on a cathode with high surface area. Removal of mercury is studied using extended surface electrolysis which reduces the level of mercury to below acceptable concentrations of 0.01 ppm in wastes by employing a Swiss roll cell with a cadmium-coated, stainless-steel cathode. An industrial cell with a fluidized bed electrode has also been studied. Graphite, as an efficient porous electrode, has been used to remove traces of mercuric ions form aqueous electrolyte solutions. In order to apply the electrochemical method for some effluents, it is necessary to use sodium hypochlorite to convert elemental mercury and less soluble mercury compounds to water-soluble mercuric–chloride complex ions.

Grau and Bisang (1995) investigated the removal of mercury form wastewater containing chloride ions through contact deposition using a less expensive reducing agent such as iron, instead of electrical energy. Due to the complexation of mercuric ions, mercuric compounds are highly soluble in aqueous chloride solutions. The following equilibrium reactions must be considered:

$$Hg^{2+} + Cl^- \leftrightarrow HgCl^+; \; K_1 = 5.6 \times 10^6 \qquad (13.38)$$

$$Hg^{2+} + 2Cl^- \leftrightarrow HgCl_{2(aq)}; \; K_2 = 1.7 \times 10^{13} \qquad (13.39)$$

$$Hg^{2+} + 2Cl^- \leftrightarrow HgCl_3^-; K_3 = 1.2 \times 10^{14} \qquad (13.40)$$

$$Hg^{2+} + 4Cl^- \leftrightarrow HgCl_4^{2-}; K_4 = 1.2 \times 10^{15} \qquad (13.41)$$

When simplified, the predominant reaction during the deposition of mercury from a solution containing chloride ions in high concentration is as follows:

$$HgCl_4^{2-} + 2e^- \rightarrow Hg(l) + 4Cl^- \qquad (13.42)$$

Reaction (13.42) has a reversible electrode potential under standard conditions of 0.4033 V. This potential was calculated from standard Gibbs energy data, and its value indicates that iron can be used as the reducing agent.

$$Fe \rightarrow Fe^{2+} + 2e^- \qquad (13.43)$$

This equation has a standard electrode potential of –0.447 V. Thus, the solution containing mercuric and chloride ions in contact with iron forms a battery. The reduction of the complex ions to metallic mercury is the cathodic reaction. The dissolution of iron is the anodic reaction. The overall reaction in the battery is given by the addition of Equation (13.42) and Equation (13.43). Due to the high value of its reversible cell voltage under standard conditions (0.85 V), it is expected that a very low equilibrium concentration of the complex ion can be achieved.

13.3.3   Reduction of Inorganic Pollutants

13.3.3.1   Chlorine

Water from the wastewater treatment plants of paper mills, power plants, etc., contains high chlorine residues in aqueous media, which causes environmental concern. Several methods have been used for dechlorination, including granular activated carbon, hydrogen peroxide, sodium thiosulfate, ammonia, sodium sulfite, and metabisulfite. In addition, ferrous sulfate heptahydrate has also been proposed for the removal of chlorine residues.

The methods using sulfur(IV) species have some disadvantages as pH decreases, such as handling difficulties and high cost. Removal of chlorine residues with ammonia and chloro-aminines have harmful effects on aquatic environments and have other unpleasant properties, such as obnoxious odor, dechlorination odor, etc. Therefore, a method for reduction of chlorine to chloride using metallic iron in chlorine solutions has been studied by Özdemïr and Tufekcï (1997). Chlorine solutions were prepared from chlorine obtained either by NaCl electrolysis or commercially. The chlorine solution in water was mixed at 20°C in a temperature-controlled bath. The experiments were carried out at pHs of 4, 5, 6, 7, 8, and 9. The diameter of granular metallic iron was about 0.2 to 0.5 mm.

The possible reactions occurring during removal of chlorine residues in aqueous media by metallic iron can be described as follows:

$$2Fe + 2HOCl + 2H_2O \leftrightarrow 2Fe(OH)_2 + 2H^+ + 2Cl^- \quad (13.44)$$

$$2Fe(OH)_2 + HOCl + H_2O \leftrightarrow 2Fe(OH)_3 + H^+ + Cl^- \quad (13.45)$$

Addition of Equation (13.44) and Equation (13.45) gives the following equation:

$$2Fe + 3HOCl + 3H_2O \leftrightarrow 2Fe(OH)_3 + 3Cl^- + 3H^+ \quad (13.46)$$

At higher pH, Reaction (13.44) and Reaction (13.45) occur as follows:

$$2Fe + 2OCl^- + 2H_2O \leftrightarrow 2Fe(OH)_2 + 2Cl^- \quad (13.47)$$

$$2Fe(OH)_2 + OCl^- + H_2O \leftrightarrow 2Fe(OH)_3 + Cl^- \quad (13.48)$$

Thus, in this case, addition of Equation (13.47) and Equation (13.48) gives Equation (13.49):

$$2Fe + 3OCl^- + 3H_2O \leftrightarrow 2Fe(OH)_3 + 3Cl^- \quad (13.49)$$

Ferric hydroxide can be quantitatively determined. The rate of removed chlorine residues was approximately 100%. Table 13.10 and Table 13.11 show

TABLE 13.10

Removal of Chlorine Residues and Amount of Chloride in pH Range of 4 to 6

Contact Time (min)	[HOCl + OCl⁻] (Concentration, mg/L)			[Cl⁻] (Concentration, mg/L)			% Removal		
	pH 4	pH5	pH6	pH 4	pH5	pH6	pH 4	pH5	pH6
0	210	292	292	203	278	278	0	0	9
5	62	96.6	111	295	411	400	71	67	62
10	2.2	5.8	2.9	336	473	456	99	98	90
15	0	0	0	338	479	473	100	100	99
20	—	—	—	—	—	475	—	—	100

Source: Özdemïr, M. and Tüfekcï, M., *Water Res.*, 31(2), 343, 1997. With permission.

TABLE 13.11

Removal of Chlorine Residues and Amount of Chloride in pH Range of 7 to 9

Contact Time (min)	[HOCl + OCl⁻] (Concentration, mg/L)			[Cl⁻] (Concentration, mg/L)			% Removal		
	pH 4	pH 5	pH 6	pH 4	pH 5	pH 6	pH 4	pH 5	pH 6
0	360	234	234	322	225	224	0	0	0
5	175	178	181.4	448	263	266	51	24	22.5
10	79	152	128	514	282	298	78	35	45
15	35	102	93.6	544	314	322	90	56	60
20	15	58.5	64.4	556	343	346	96	75	72.5
25	3.0	38	46.8	564	355	354	99.5	84	80

Source: Özdemïr, M. and Tüfekcï, M., *Water Res.*, 31(2), 343, 1997. With permission.

that when the pH is lower than 7, the reaction between granular metallic iron and chlorine residues has a high rate. When the pH is greater than 7, the reaction rate is lower than those at lower pHs. The higher reaction rates at low pHs arose from increasing concentrations of HOCl. At pH of 7, the composition of hypochloride solution was 75.2% HOCl and 24.8% OCl⁻, whereas at pH 9 the composition was 2.9% HOCl and 97.1% OCl⁻.

When granular metallic iron smaller than 0.5 mm was used, an optimum reaction rate was obtained. The ratio of metallic iron to hypochlorite concentration was approximately 1.1-fold in terms of the stoichiometric ratio. The contact time of granular metallic iron with chlorinated water was approximately 25 min; thus, at pH less than 7, the chlorine residues were totally removed within 25 min from the beginning of the reaction.

Because experiments were performed in nitrogen-free water, NH_2Cl, $NHCl_2$, and NCl_3 did not exist. Although it is thought that metallic iron can reduce nitrogen-combined chlorine species, further studies are required to find out if metallic iron can reduce them. The only drawback of this method is the earthy odor and a suspension from ferric hydroxide; this disadvantage can be overcome by precipitation within 15 to 20 min followed by filtration through an ordinary sand filter to remove the earthy odor.

13.4 QSAR Models

To analyze the reactivity of organics in systems using zero-valent iron, ΔH_f (standard heat of formation) and E_{LUMO} (energy of the lowest unoccupied molecular orbital) were computed using the PM3 Hamiltonian method. Reductive dechlorination rate constants for five chlorobenzenes in the presence of Pd/Fe as catalyst were determined experimentally. Table 13.12 lists the rate constants (Lin et al., 2003).

Based on the computed ΔH_f, E_{LUMO} values, and the kinetic rates, linear free-energy relationships (LFERs) for the dechlorination rate constants were developed by using a partial least squares (PLS) regression. Using this model, the reaction rate constants can be expressed as:

$$\text{Log } k = -1.63 + 1.46 \times 10^{-3}\Delta H_f - 7.69 \times 10^{-1}E_{LUMO}$$

$$n = 8, \ Q^2_{CUM} = 0.879, \ r = 0.954, \ p < 0.001 \tag{13.50}$$

where k stands for the dechlorination rate constants, ΔH_f is the standard heat of formation, and E_{LUMO} is the energy of the lowest unoccupied molecular orbital. The Q^2_{CUM} value of the model is 0.879, indicating robustness and predictive power of the model. Table 13.13 shows the observed and predicted rate constants.

E_{LUMO} measures the ability of a molecule to accept electrons. Compounds with low E_{LUMO} tend to accept electrons easily. Thus, the coefficient of E_{LUMO} is negative. Theoretically a compound with more negative ΔH_f or lower ΔH_f is more stable (Fried et al., 1977); therefore, it is reasonable that the coefficient of ΔH_f in Equation (13.50) is positive, as shown in Table 13.13. Thus, the higher the ΔH_f value is, the more unstable or reactive the chlorinated compounds will be. As a result, the log k values are greater. LFER analysis on dechlorination by Fe^0 of chlorinated aliphatic compounds, E_{LUMO}, and ΔH_f have been confirmed to be more significant molecular descriptors than other

TABLE 13.12

Observed First-Order Rate Constants of Reductive Dehalogenation

No.	Compounds	k (min⁻¹)	r	n
1	Chlorobenzene	0.0243	0.927	9
2	1,2-Dichlorobenzene	0.0369	0.941	8
3	1,3-Dichlorobenzene	0.0435	0.978	8
4	1,4-Dichlorobenzene	0.0438	0.976	8
5	1,2,4-Trichlorobenzene	0.0485	0.947	7
6	2-Chlorophenal	0.0215	0.962	8
7	3-Chlorophenal	0.0165	0.986	11
8	4-Chlorophenal	0.0112	0.953	9

Source: Liu, Y. et al., *Chemosphere*, 50, 1275–1279, 2003. With permission.

TABLE 13.13

Observed and Predicted First-Order Rate Constants of Reductive Dehalogenation
Using Computed ΔH_f and E_{LUMO}

No.	Compounds	ΔH_f (kJ)	E_{LUMO} (eV)	Log k (Observed)	Log k (Predicted)	SE	Residual[a]
1	Chlorobenzene	69.441	−9.388	−1.61	−1.57	±0.03	−0.04
2	1,2-Dichlorobenzene	46.269	−9.295	−1.43	−1.43	±0.03	0.00
3	1,3-Dichlorobenzene	42.337	−9.421	−1.36	−1.42	±0.03	0.06
4	1,4-Dichlorobenzene	41.990	−9.235	−1.36	−1.38	±0.04	0.02
5	1,2,4-Trichlorobenzene	19.755	−9.241	−1.31	−1.26	±0.05	−0.05
6	2-Chlorophenal	−18.158	−9.210	−1.67	−1.78	±0.04	0.11
7	3-Chlorophenal	−19.340	−9.236	−1.78	−1.80	±0.04	0.01
8	4-Chlorophenal	−19.129	−9.009	−1.95	−1.84	±0.04	−0.11

[a] Residual is the difference between observed and predicted log k values; SE is the standard error for the predicted value.

Source: Liu, Y. et al., *Chemosphere*, 50, 1275–1279, 2003. With permission.

molecular descriptors (Chen et al., 2002). Figure 13.5 shows the observed
and predicted values of the rate constant. With the exception of points 6 (2-
chlorophenal) and 8 (4-chlorophenal), the model has shown that an LFER
model based on ΔH_f and E_{LUMO} can be used to predict the reactivity of the
chlorinated benzenes in a zero-valent system. Points 6 and 8 have slightly
higher residuals (0.11 vs. 0.05), which could deviate slightly from the main
trend.

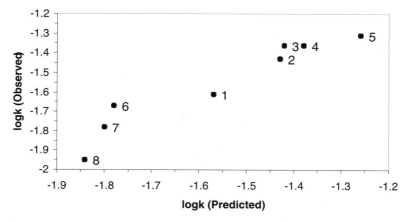

FIGURE 13.5

Plot of observed log k values vs. those predicted by the PLS model. (From Liu, Y. et al.,
Chemosphere, 50, 1275–1279, 2003. With permission.)

13.5 Engineering Applications

One of the most attractive features for permeable reactive iron barriers (PRBs) is the low operating costs in comparison to alternative treatment technologies. In addition, the efficiency of PRBs is suggested to exceed 5 to 10 years (USEPA, 1998), which additionally contributes to efficient operation. PRBs have been constructed in two basic design configurations: funnel-and-gate and continuous PRB.

13.5.1 Continuous and Funnel-and-Gate PRBs

The most important PRB design parameter is the residence-time determined by groundwater velocity through the PRB. In addition, a number of other design features must be considered in order to achieve a cost-efficient solution. Two basic configurations are currently being used for full-scale field applications: continuous PRBs or funnel-and-gate designs. Both are similar in terms of the use of heterogeneous reactions in a porous media to reduce contaminants (Burris et al., 1996b). Trench-and-fill barriers are installed by digging a trench in the path of the contaminant flow and filling the trench with materials that will either precipitate or destroy the contaminant. The material chosen depends on the type of contaminant, remediation objectives, and site conditions. The materials should be inexpensive and have enough hydraulic conductivity to allow groundwater to flow through the barrier. Possible materials include minerals with chelates and quaternary amines, organic peat, titanium hydroxide, sawdust, and iron. If the objective is to destroy the contaminants, sufficient retention time should be designed so that the reactions can occur completely. If a reduction in the concentration is desired, the barrier will retain the contaminants for release at a slower rate or contain the contaminants for subsequent excavation (Kaplan et al., 1996). Zero-valent iron is an optimum choice for these purposes. The construction costs are dependent on the depth, width, and saturated thickness of the plume, as well as on the residence time. In addition, the reactive materials and the funnel materials (if required), will significantly contribute to the overall cost. Other important factors include the need to dewater during excavation, the costs of groundwater and soil disposal, and health and safety factors. The U.S. Environmental Protection Agency (EPA) has proposed the design criteria as shown in Table 13.14. The table also compares some of the characteristics of continuous PRB with the funnel-and-gate technique.

13.5.1.1 Characteristics of Reactive Media

In general terms, the reactive media for a PRB should be benign to subsurface environments and should not inhibit bioactivity. In addition, the reactive media should be efficient for a sufficiently long period to reduce costs. The

TABLE 13.14

Summary of Design Features for Continuous PRB and Funnel-and-Gate

Continuous PRBs	Funnel-and-Gate Systems

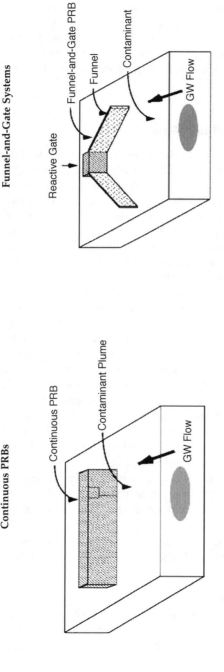

A trench has been excavated and backfilled with granular Fe.

As long as the hydraulic conductivity of the aquifer is less than that of the PRB, underflow of contaminated groundwater should not occur. Continuous PRBs have relatively minimal impact on the natural groundwater flow conditions at a site.

The continuous design allows the water to pass through the barrier under its natural gradient and at its natural flow velocity. The groundwater flow velocity through the PRB will be very similar to the velocity in the aquifer.

Low permeability funnels direct groundwater toward permeable treatment zones or gates.

The permeability of the reactive material in the gated zone must be equal to or greater than the aquifer permeability to minimize flow restrictions. This funnel is placed to encompass and direct the flow of contaminated water to a "gate" or "gates" containing a permeable zone of granular Fe^0 or other reactive material.

Due to directing a large cross-sectional area of water through the much smaller cross-sectional area of the gate, groundwater velocities within the gate will be higher than those resulting from the natural gradient.

The funnel portion of the design is engineered to completely encompass the path of the contaminant plume, and the overall design must prevent the contaminant plume from flowing around the barrier in any direction.

It would seem unlikely that the total volume of water in a high flux system, directed by a high-surface-area funnel, could be infiltrated through the much smaller surface area of the gate and aquifer material interface that contacts the downgradient side of the gate.

Recent variations include the use of backfilled caissons; media-filled, hollow-vibrating beams; and emplaced reaction vessels. The funnel typically consists of sheet pilings, slurry walls, or some other material and is preferably keyed into an impermeable layer (clay, bedrock) to prevent contaminant underflow. Particular care is required in designing and constructing the connection between the impermeable funnel section and the permeable gate section in order to avoid bypass of contaminated ground water.

Sheet pilings can be driven to delineate the sides and ends of the media gate. The interior of this construction can then be dewatered and excavated to make room for the reactive material (Gillham and Burris, 1992). For the emplaced reaction vessels, contaminated upgradient waters are directed into the subsurface vessel which contains reactive media, and the treated water is discharged through a pipe that extends downgradient through the impermeable wall.

The gate shape may be controlled by construction techniques, but most commonly they have been rectangular. The more typical rectangular or box-shaped gate can be built by driving temporary sheet pilings and/or building removable subsurface walls within which the reactive materials are placed.

Continuous PRB only needs to cover an area comparable to the cross-sectional area of the plume. It is built to a depth that extends over the vertical and horizontal dimensions of the contaminant plume, as a safety factor, and is filled with granular iron or some other reactive material. The upgradient and downgradient surface areas of the aquifer material contacting the PRB will be approximately the same, minimizing disruption in the natural groundwater flow relative to current funnel-and-gate designs.

Trench emplacement involves trenching and filling in the fully saturated zone due to problems of immediate water intrusion and potential collapse of the trench walls. Unless subsurface constructions (sheetpile walls, trench filling with biodegradable slurries such as guar, etc.) are utilized, the aquifer materials must be removed and the reactive media emplaced nearly simultaneously. This rapid process of aquifer material removal and reactive media emplacement is possible using continuous trenching devices.

In an actual field implementation, more iron may be required by a continuous PRB using current commercial-scale installation techniques (trenching), particularly if the plume is very broad and requires a relatively long PRB. Some limiting thickness must be maintained to ensure PRB integrity throughout its volume. The use of jetting, as an emplacement method rather than trenching, could potentially mitigate this problem for thin continuous PRBs.

A drawback of this type of PRB installation is difficulty tracking the volume of reactive material actually emplaced during the construction.

Source: U.S. EPA, Remediation Technologies Development Forum (RTDF): Permeable Reactive Barrier Technologies for Contamination Remediation, EPA/600/R-98/125, U.S. EPA, Washington, D.C., 1998.

reactive material should not significantly affect the plume by excessively low hydraulic conductivity. The material should be comparatively uniform with only small fluctuations of permeability across the longitudinal dimensions and across the depth of the wall. Iron metal and zero-valent iron powder are the most common reactive media, and they can be obtained as scrap iron in granular form.

13.5.1.2   Types of Reactive Media

A number of modifications have been proposed by the EPA, including injection of slurries, hydrofracturing, and driving mandrels. The design process involves careful consideration of a number of issues as follows:

- Nature of the contaminant
- Long-term availability of reactive media
- System hydrology and plume boundaries
- Groundwater/contaminant residence time per unit thickness of reactive media
- Monitoring for compliance and performance
- Sodium dithionite Fe(III) → Fe(II)
- *In situ* electrodes to supply hydrogen as an electron donor to subsurface
- Bioenvironment

In the field, a porous iron wall is placed in the path of the contaminated groundwater. As water flows through, contaminants are transformed into harmless substances. In the funnel-and-gate system, the groundwater is directed toward the gate that is placed perpendicular to the water flow, and the gate removes contaminants while allowing water and other constituents to flow through freely (Cantrell and Kaplan, 1997). Currently, the process works to depths of 50 ft, but some researchers say that depths of at least 100 ft should be possible; however, the cost effectiveness at that depth is an issue (Fairweather, 1996). In addition to *in situ* methods, the use of zero-valent iron in above-ground reactors has been considered, but further understanding of the mechanism behind the reaction is needed.

Before one of the zero-valent iron treatment processes can be used at a site, extensive batch tests must be performed. Then, column tests are performed for several months to predict performance based on site conditions. Following pilot tests that take 6 months to a year, full-scale operation may begin. Thorough research of the site has to be conducted. For the treatment to be successful, the precise water velocity, depth of contamination, and soil matrix have to be known (Wilson, 1995). Errors in these calculations would cause the treatment to be ineffective in contaminant removal. From these results, the size of the barrier is determined by the least reactive contaminant.

As of 1996, two full-scale field tests were in operation in the U.S. They are located in Sunnyvale, CA. The Sunnyvale sites chose zero-valent iron usage primarily because of scheduling of the project. The soils around the site consisted of clay that could be used for a funnel-and-gate system. To achieve a safety factor of five, a $60 \times 40 \times 25$-ft gate was installed using 90 tons of iron. Costing $80,000 to $100,000, the system should work for 30 years and require no maintenance. Two more projects are pending in Somersworth, NH, and Elizabeth City, NC (Fairweather, 1996). Intense monitoring is being conducted; data collection is essential to improve the system and resolve unanswered questions (Fairweather, 1996).

In Canada, the first field demonstration involved decontaminating groundwater of trichloroethene and tetrachloroethene at Canadian Forces Base in Borden, Ontario. A permeable barrier made of scrap iron filings was installed in the path of the water that was contaminated with chlorinated solvents. For years following, the groundwater tested has been free of chlorinated contaminants (Tratnyek, 1996). The Geomatrix consulting firm pursued the zero-valent iron process and installed two walls of 300 ft and 235 ft, respectively, and a reactive gate that was 40 ft wide by 4 ft thick and 13 ft deep. During construction, 220 tons of iron filings were used (Fairweather, 1996). The water velocity rate of 1 ft per a day means that the water travels through the wall in 4 days (Wilson, 1995). Both existing contaminants and contaminants produced from degradation are reduced. The site is continuously monitored and meets all regulatory requirements. The results have been excellent and approved by the government. Due to the success of remediation by zero-valent iron, the original lease was terminated and a new lessee has developed the land into a parking lot. At an initial construction cost of $770,000, an additional $1.3 million is expected for monitoring and replacing the wall every 10 years (Fairweather, 1996). A portion of the cost is due to extensive monitoring of the site for research purposes; therefore, operating remediation sites in the future should not be as costly. Even the cost of the iron, at $400 to $450 a ton, should decrease as more suppliers become available. The initial cost of the project may be greater than a simple pump-and-treat, but it should pay off in a few years (Fairweather, 1996). In addition, when making cost-to-benefit comparisons, other methods do not consider the fact that the site can be reused.

13.5.2 Monitoring

For permeable reactive barriers, the monitoring effort must consider the contaminant that exceeds the groundwater standards. In addition, monitoring of degradation products from reductive dehalogenation reactions or other contaminant transformation products may be required. Other groundwater parameters such as pH, alkalinity, specific conductance, and major compositions with respect to cations and anions must be included as well. Therefore, it is necessary to develop a network of monitoring wells, which are important factors that ensure adequate assessment of system performance.

13.5.2.1 Planning the Monitoring Effort

The EPA has developed recommendations with respect to sampling and monitoring the performance of groundwater (Table 13.15). Development of a quality assurance project plan (QAPP) is recommended prior to initiating groundwater sampling. Specifications of a QAPP are summarized in Table 13.15. In general, the QAPP must consider a monitoring plan that includes data collection methods, analytical methods, and sample handling. Thus, the monitoring plan has been defined by the EPA with the following objectives in mind:

- Sampling requirements
- Pre-sampling activities
- Sample collection
- *In situ* or field analyses and equipment calibration
- Sample preservation and handling
- Equipment decontamination
- Chain-of-custody control and records management
- Analytical procedures and quantification limits for both laboratory and field methods
- Field and laboratory quality assurance/quality control
- Evaluation of data quality
- Health and safety

TABLE 13.15

Recommended Groundwater Sampling and Monitoring Efforts Relevant to PRBs

Application	U.S. EPA Document
Quality assurance project plan (QAPP)	*Quality Assurance Plan* (EPA/600/9-89/087)
Groundwater monitoring	*Test Methods for Evaluating Solid Waste* (USEPA 1986c; commonly known as SW-846)
Groundwater sampling	*A Workshop Summary* (EPA/600/R-94/205)
Subsurface characterization and monitoring techniques	*A Desk Reference Guide* (EPA/R-93/003b), Vol. 1: *Solids and Ground Water*; Vol. 2: *The Vadose Zone, Field Screening and Analytical Methods*

13.5.2.2 Compliance Monitoring

For compliance sampling purposes, the EPA recommends low-flow sampling methods, with flow rates between 0.1 to 0.5 L/min, depending on site-specific hydrogeologic conditions. The most important consideration in designing the flow rates is to minimize stress to the hydrogeologic system. Low-flow sampling techniques allow the compliance wells to be placed in close proximity to the reactive wall, as the reduced purge volume will decrease the chance of mixing waters from unintended sources and zones. Continuous measurement of the operation parameters can be accomplished by using instruments that utilize in-line flow cells. The more common groundwater compliance sampling parameters include the following:

- Volatile organics (VOAs or VOCs)
- Dissolved gases and total organic carbon (TOC)
- Semivolatile organics (SMVs or SVOCs)
- Metals and cyanide
- Major water-quality cations and anions
- Radionuclides

13.5.2.3 Performance Monitoring

It is important to evaluate or confirm that the desired degradation or transformation of the target contaminant species is occurring as the plume moves through the reaction zone. In addition to those degradation products that may also be regulated contaminants, it may also be desirable to analyze for non-contaminant species to confirm the contaminant transformation.

A number of geochemical parameters can provide additional information about the performance of permeable iron barriers. The monitoring program can consider measuring the geochemical quantities in order to analyze the performance of the PRB. Some of the parameters indicate that iron corrosion reactions are occurring and additionally provide some indication of the extent of precipitate formation. The corrosion effects can be monitored by the increase in pH and generation of free Fe^{2+} and H_2. Corrosion processes also decrease the redox status of the aqueous environment, and levels to less than -400 mV vs. the standard hydrogen electrode (SHE) can be expected. Similarly to dehalogenation, the key indicator parameters for chromate reduction by Fe^0 are pH increase and ferrous iron formation and may also include the loss of dissolved Cr from solution accompanied by the formation of a mixed Fe–Cr (oxy)hydroxide mineral-phase solid solution. Increases of the pH cause a shift in the dissolved carbonate equilibria with decreasing concentrations of carbonic acid and bicarbonate species in solution and increasing concentrations of carbonate ions. Changes in alkalinity, ferrous iron, and calcium can be

TABLE 13.16

Parameters Recommended by the U.S.
EPA for Geochemical Analysis

Parameter	Indication
pH	Corrosion
Eh	Corrosion
Alkalinity	Precipitation, corrosion
Dissolved oxygen	Corrosion
Total dissolved sulfide	Corrosion
Ferrous iron	Precipitation, corrosion
Dissolved hydrogen	Corrosion

used to evaluate the potential for precipitation of calcite and siderite minerals. The EPA has suggested monitoring a number of geochemical parameters, as shown in Table 13.16.

13.5.2.4 Microbial Characterization

In general, under conditions of suitable pH and substrate availability, iron PRB may promote the activity of iron- and sulfate-reducing bacteria and the methanogens. Enhanced activity could influence zero-valent iron reductive dehalogenation reactions through favorable impacts on redox potential or the iron surfaces or through direct microbial transformations of the target compounds; however, this enhanced activity may result in biofouling of the treatment zone. Systems dependent upon biological remediation must be analyzed for the microbial activity and its effect on the PRB performance. Table 13.17 lists the recommended analytical methods for microbial activity by the U.S. EPA.

TABLE 13.17

Methods Recommended by the U.S. EPA for Analysis of Microbial Activity

Type of Analysis	Recommended Method
Microbial characterization	Epifluorescence microscopy
Microbial identification and enumeration	Scanning electron microscopy
Bacterial populations upgradient, downgradient, and within the wall	Fatty acid methyl ester (FAME) analysis

13.5.3 Engineering Improvement

Remediation by the use of zero-valent iron offers a wide range of advantages, such as the reduction of many halogenated compounds, tremendous cost savings, *in situ* remediation, and site reuse. Most importantly, the contaminants are effectively degraded on-site, not transported elsewhere to be stored or transferred to another media. Typically saving taxpayers $75 million per site, the economic benefits offer a savings of up to 50% of average cleanup costs (Fairweather, 1996). Although the initial cost is equivalent to or slightly higher than traditional remediation (such as pump-and-treat), the zero-valent iron process works without above-ground structures, operation costs, and energy supply requirements, Furthermore, because the remediation takes place in subsurface, the land can be utilized for other purposes (Fairweather, 1996). Although other metals may be effective in the reduction of groundwater contaminants, iron is nontoxic, readily available, and inexpensive by comparison.

Zero-valent iron remediation is not without issues and concerns. Because it is a new technology, long-term performance data have not been collected. The effects of precipitates, depth of operation, and life of the iron walls have not been completely determined. Due to increases in the pH during the reaction, precipitates may inhibit degradation by clogging the iron. Thus far, this occurrence has only been seen in the laboratory and not in the field. This may be attributed to the fact that the formation of precipitates may be due to excess oxygen at the upper level of subsurface, while the subsurface environment is very anaerobic (Fairweather, 1996). But, further field research must be completed before a conclusion is drawn. Also, site conditions play an integral role in the effectiveness of the treatment. For example, the Intersil site had a 65-ft layer of clay that aided in containment. Another issue is the depth at which the technique is cost effective and efficient for use. Currently, zero-valent iron has been used at depths of from 50 to 75 ft. More studies need to be done to determine the required site conditions. Finally, the mechanism of the chemical reaction has not been fully explained. Therefore, understanding the degradation processes is essential to understanding the system.

One concern of the use of zero-valent iron is the accumulation of precipitates on the iron wall. The hydrology of the barrier and the contaminated site could be significantly altered by the process of reductive dehalogenation. The process generates ferric hydroxides and oroxyhydroxides, which increase the buildup of precipitates. The formation of iron(oxy)-hydroxides, iron-sulfide minerals, calcite, and siderite as potential products may negatively affect the iron reactivity. Based on precipitate buildup, the wall permeability may significantly deteriorate, thus affecting the hydrogeological conditions in the PRB zone. The EPA has suggested the use of the methods shown in Table 13.18 for analysis of the buildup.

Instead of wall replacement, another process by which to rejuvenate the wall is desired (Fairweather, 1996). One method of avoiding wall replacement

TABLE 13.18

Methods Recommended by the U.S. EPA for Analysis of
Precipitate Buildup

Recommended Methods to Analyze Precipitate Formation
Scanning electron microscopy with energy dispersive x-ray analysis
Auger spectroscopy
Laser Raman spectroscopy
Conventional wet chemical extractions
X-ray photoelectron spectroscopy

is colloidal injections. Injecting Fe^0 colloid suspensions into the subsurface can be accomplished through the use of injection and extraction wells. This approach is advantageous because it is less expensive than installing a barrier. The porosity is not reduced, and the flows remain stable and flow freely. The barrier method is effective for approximately 30 years before requiring replacement. With colloid suspensions, subsequent injections would accomplish rejuvenation of the iron (Cantrell and Kaplan, 1997). The colloidal suspension is installed by digging three wells in a line. The first well is used to inject colloids, and the second well withdraws groundwater, which pulls the colloids toward the second well and forms a barrier between the two wells. After the colloids are in place, the second well becomes an injection well for colloids, and the third well extracts groundwater to complete the formation of the colloid barrier. The thickness of the barrier is determined by the flow rate, composition of the groundwater, and kinetics between the contaminants and the colloids. The advantages of installing only a series of temporary wells include the lack of a large excavation, a reduction in cost, and an increase in the depth to which remediation occurs. Also, if a problem arises as the groundwater is monitored, injections can be made instead of excavations.

Several governing factors have become evident from experiments; for example, the concentration of the colloid injection and the injection rate determine the extent to which the barrier can be retained, and the removal efficiency of the colloids decreases as the colloids increase in the sediments produced from the reactions that occur. Also, when injected, the zero-valent iron is dispersed evenly throughout, and flow rates required to move the iron colloids through the column are higher than the flow rates found in the surrounding groundwater; therefore, once the barrier is in place, the groundwater flow rates will not move the colloids, thus keeping contaminants in place. However, limitations do exist such as difficulty in validating the integrity of the barrier. If low-porosity soils and particles greater than micron sized are used or if too much of the colloid is injected, then permeability is greatly reduced and the flow of the groundwater is inhibited (Kaplan et al., 1996).

Some products produced from the reduction reaction require further treatment. Zero-valent iron enhanced with other technologies is effective in achieving results. Most promising is bimetallic treatment by the addition of nickel, copper, zinc, or palladium. For substances that are not affected by the reductive properties of iron, the combination makes resistant compounds such as *cis*-1,2-dichloroethene and PCBs reduce to safe, nontoxic products rapidly (Grittini, 1995). Also, biodegradation could be used not only to reduce amines formed from the reduction of nitro aromatic compounds but also to break down iron precipitates that may form.

Developed by the U.S. Department of Energy and several corporations, the "lasagna" approach uses electrokinetics to guide contaminated groundwater through low-permeability zones to treatment zones that contain iron (Tratnyek, 1996). Other improvements that increase the reaction rate of the system include the removal of the oxide layer by physical and chemical means. Thus far, this application works better in the laboratory than in the field, but further research is being completed.

Zero-valent iron can also be used for heavy-metal removal. Applications for cadmium and chromate removal have already been shown to be successful. When combined with chloride ions, iron has been shown to be a simple and inexpensive method to remove mercury from wastewater (Grau and Bisang, 1995). Nitrates also degrade in the presence of zero-valent iron, but the application of treating nitrate-contaminated water has not been extensively studied (Siantar et al., 1996).

13.6 Summary

The mechanism of the reaction of zero-valent iron with organic contaminants is the oxidation of the iron with the reduction of the organic compound. The reaction is pseudo first order, and direct contact is required. Direct contact of the target pollutants with the iron surface is critical to ensure the success of the remediation process. Comparison of reaction rates is difficult because the experiments have varied in the methods used. Normalization of reaction rates allows for better comparisons. Results show that nitro reduction by Fe^0 is faster than dehalogenation (Agrawal and Tratnyek, 1996). Within the halogenated compounds, the reaction rate is the greatest for saturated and perhalogenated organic oxidants, especially when water is present (Matheson and Tratnyek, 1994). It is hoped that as more data are collected results can be normalized for better comparison. Zero-valent iron can be used to treat a wide range of organic and heavy metal contamination. Its economic advantages make zero-valent iron a worthy candidate for consideration as a treatment process. However, before implementation can occur, conditions of the site must be thoroughly assessed for the treatment to be effective. Also, a full understanding of pathways of reduction and compounds or intermediates produced is needed for development of an optimal design (Roberts et al., 1996).

References

Agrawal, A. and Tratnyek, P.G., Reduction of nitro aromatic compounds by zero-valent iron metal, *Environ. Sci. Technol.*, 30, 153–160, 1996.

Burris, D.R., Campbell, T.J., and Manoranjan, V.S., Sorption of trichloroethylene and tetrachloroethylene in a batch reactive metallic iron–water system, *Environ. Sci. Technol.*, 29, 2850–2855, 1995.

Burris, D.R., Hatfield, K., and Wolfe, N.L., Laboratory experiments with heterogeneous reactions in mixed porous media, *J. Environ. Eng.*, 122(8), 685–691, 1996.

Campbell, T.J., Burris, D.R., Roberts, A.L., and Wells, J.R., Trichloroethylene and tetrachloroethylene reduction in a metallic iron–water–vapor batch system, *Environ. Toxicol. Chem.*, 16(4), 625–630, 1997.

Cantrell, K.J. and Kaplan, D.I., Zero-valent iron colloid emplacement in sand columns, *J. Environ. Eng.*, 123(5), 499–505, 1997.

Cantrell, K.J., Kaplan, D.I., and Wietsma, T.W., Zero-valent iron for the *in situ* remediation of selected metals in groundwater, *J. Hazardous Mater.*, 42, 201–212, 1995.

Chen, J.W., Pei, J., Quan, X., Zhao, Y.Z., and Chen, S., Schramm, K.W., Kettrup, A., Linear free energy relationships on rate constants for dechlorination by zero-valent iron, *SAR and QSAR Environ. Res.*, 13, 597–606, 2002.

Chuang, F., Larson, R.A., and Wessman, M.S., Zero-valent iron-promoted dechlorination of polychlorinated biphenyls, *Environ. Sci. Technol.*, 29, 2460–2463, 1995.

Cui, D.Q. and Spahiu, K., BP 623: the reduction of U(VI) on corroded iron under anoxic conditions, *Radiochim. Acta*, 90(9–11), 623–628, 2002.

Fairweather, V., When toxics meet metal, *Civil Eng.*, May, 44–48, 1996.

Fried, V., Hameka, H.F., Blukis, U. *Physical Chemistry,* Macmillan Publishing Co., New York, 1977.

Gillham, R.W. and O'Hannesin, S.F., Enhanced degradation of halogenated aliphatics by zero-valent iron, *Ground Water*, 32(6), 958–967, 1994.

Gojkovic, S.L., Zecevic, S.K., and Drazic, D.M., Hydrogen peroxide oxidation on passive iron in alkaline solutions, *Electrochim. Acta*, 37(10), 1845–1850, 1992.

Grau, J.M. and Bisang, J.M., Removal and recovery of mercury from chloride solutions by contact deposition of iron felt, *J. Chem. Tech. Biotechnol.*, 62, 153–158, 1995.

Grittini, C., Malcomson, M., Fernando, Q., and Korte, N., Rapid dechlorination of polychlorinated biphenyls on surface of a Pd/Fe bimetallic system, *Environ. Sci. Technol.*, 29(11), 2898–2900, 1995.

Hardy, L.I. and Gillham, R.W., Formation of hydrocarbons from the reduction of aqueous CO_2 by zero-valent iron, *Environ. Sci. Technol.*, 30, 57–65, 1996.

Heijman, C.G., Grieder, E., Holliger, C., and Schwarzenbach, R.P., Reduction of nitroaromatic compounds coupled to microbial iron reduction in laboratory aquifer columns, *Environ. Sci. Technol.*, 29, 755–783, 1995.

Helland, B.R., Alvarez, P.J.J., and Schnoor, J.L., Reductive dechlorination of carbon tetrachloride with elemental iron, *J. Hazardous Mater.*, 41, 205–216, 1995.

Horold, S., Tacke, T., and Vorlop, K.D., Catalytical removal of nitrate and nitrite from drinking water 1. Screening for hydrogenation catalysts and influence of reaction conditions on activity and selectivity, *Environ Technol.*, 14(10), 931–939, 1993a.

Horold, S., Vorlop, K.D., Tacke, T., et al., Development of catalysts for a selective nitrate and nitrite removal from drinking water, *Catal. Today*, 17(1–2), 21–30, 1993b.

Johnson, T.L., Scherer, M.M., and Tratnyek, P.G., Kinetics of halogenated organic compound degradation by iron metal, *Environ. Sci. Technol.*, 30, 2634–2640, 1996.

Kaplan, D.I., Cantrell, K.J., Wietsma, T.W., and Potter, M.A., Retention of zero-valent iron colloids by sand columns: application to chemical barrier formation, *J. Environ. Qual.*, 25, 1086–1094, 1996.

Kliger, G.A., Shiukin, A.N., Glebov, L.S., Mikaya, A.I., Loktev, S.M., and Zaikin, V.G., Mechanism of the hydroamination of alcohols and carbonyl compounds on a fused iron catalyst, *Acad. Sci. USSR*, 26(8), 1923–1926, 1990.

Ku, Y. and Chen, C., Removal of chelated copper from wastewaters by iron cementation, *Indust. Eng. Chem. Resources*, 31, 1111–1115, 1992.

Liu, Y., Yang, F., Chen, J., Gao, L., and Chen, G., Linear free energy relationship for dechlorination of aromatic chlorides by Pd/Fe, *Chemosphere*, 50, 1275–1279, 2003.

Manning, B.A., Hunt, M.L., Amrhein, C., and Yarmoff, J.A., Arsenic(III) and arsenic(V) reactions with zero-valent iron corrosion products, *Environ. Sci. Technol.*, 36(24), 5455–5461, 2002.

Matheson, L.J. and Tratnyek, P.G., Reductive dehalogenation of chlorinated methanes by iron metal, *Environ. Sci. Technol.*, 28, 2045–2053, 1994.

Monsef, H.R., Michels, D.A., Brewtra, J.K., Biswas, N., and Taylor, K.E. (1997). Removal of nitroaromatic compounds from water through combined zero valent metal reduction and enzyme-based oxidative coupling reactions, *I&EC Special Symp. Am. Chem. Soc.*, 1–3, 1997.

Muftikian, R., Nebesny, K., Fernando, Q., and Korte, N., A method for the rapid dechlorination of low molecular weight chlorinated hydrocarbons in water, *Water Resources*, 29(10), 2434–2439, 1995.

Muftikian, R., Nebesny, K., Fernando, Q., and Korte, N., X-ray photoelectron spectra of the palladium–iron bimetallic surface used for the rapid dechlorination of chlorinated organic environmental contaminants, *Environ. Sci. Technol.*, 30, 3593–3596, 1996.

Orth, W.S. and Gillham, R.W., Dechlorination of trichloroethene in aqueous solution using Fe^0, *Environ. Sci. Technol.*, 30, 66-71, 1996.

Özdemïr, M. and Tüfekcï, M., Removal of chlorine residues in aqueous media by metallic iron, *Water Res.*, 31(2), 343–345, 1997.

Powell, R.M., Puls, R.W., Hightower, S.K., and Sabatini, D.A., Coupled iron corrosion and chromate reduction mechanisms for subsurface remediation, *Environ. Sci. Technol.*, 29, 1913–1922, 1995.

Puls, R.W., Powell, R.M., and Paul, C.J., In situ remediation of ground water contaminated with chromate and chlorinated solvents using zero-valent iron: a field study, *The 209th ACS National Meeting, Division of Environmental Chemistry*, 35(1), 788–791, 1995.

Rahman, A. and Agrawal, A., Reduction of nitrate and nitrite by iron metal: implications for ground water remediation. Abstract Paper. *Am. Chem. Soc., Environ.* Part 1, 213, 175, April 13, 1997.

Roberts, A.L., Totten, L.A., Arnold, W.A., Burris, D.R., and Campbell, T.J., Reductive elimination of chlorinated ethylenes by zero-valent metals, *Environ. Sci. Technol.*, 30, 2654–2659, 1996.

Rulkens, W.H. et al., Remediation of polluted soil and sediment: perspectives and failures, *Water Sci. Technol.*, 37(8), 27–35, 1998.

Sawyer, D.T., Sobkowiak, A., and Matsushita, T., Metal [MLx; M=Fe, Cu, Co, Mn]/ hydroperoxide-induced activation of dioxygen for the oxygenation of hydrocarbons: oxygenated Fenton chemistry, *Acc. Chem. Res.*, 29(9), 409–416, 1996.

Scherer, M.M., Richter, S., Valentine, R.L., and Alvarez, P.J.J., Chemistry and microbiology of permeable reactive barriers for *in situ* groundwater clean up, *CRC Crit. Rev. Environ. Sci. Technol.*, 30(3), 363–411, 2000.

Schoonen, M.A. et al., An introduction to geocatalysis, *J. Geochem. Explor.*, 62, 201–215, 1998.

Semer, R. and Reddy, K.R., Mechanisms controlling toluene removal from saturated soils during in-situ air sparging, *J. Haz. Mat.*, 57, 209–230, 1998.

Siantar, D.P., Schreier, C.G., Chou, C., and Reinhard, M., Treatment of 1,2-dibromo-3-chloropropane and nitrate-contaminated water with zero-valent iron or hydrogen/palladium catalysts, *Water Res.*, 30(10), 2315–2322, 1996.

Tomiyasu, T., Sakamoto, H., and Yonehara, N., Catalytic determination of iron by a fixed-time method using the oxidation reaction of chlorpromazine with hydrogen peroxide, *Analyt. Sci.*, 12, 507–509, 1996.

Tratneyek, G., Putting corrosion to use: remediating contaminated groundwater with zero-valence metals, *Chem. Indust.*, July, 499-503, 1996.

USEPA, *Remediation Technologies Development Forum (RTDF): Permeable Reactive Barrier Technologies for Contamination Remediation*, EPA/600/R-98/125, U.S. Environmental Protection Agency, Washington, D.C., 1998.

Warren, K.D., Arnold, R.G., Bishop, T.L., Lindholm, L.C., and Betterton, E.A., Kinetics and mechanism of reductive dehalogenation of carbon tetrachloride using zero-valence metals, *J. Hazardous Mater.*, 41, 217–227, 1995.

Waybrant, K.R., Blowes, D.W., and Ptacek, C.J., Selection of reactive mixtures for use in permeable walls for treatment of mine drainage, *Environ. Sci. Technol.*, 32(13), 1972–1979, 1998.

Weber, E.J., Iron-mediated reductive transformations: investigation of reaction mechanism, *Environ. Sci. Technol.*, 30(2), 716–719, 1996.

Wilson, E.K., Zero-valent metals provide possible solution of groundwater problems, *Chem. Eng. News*, July 3, 1995, pp. 19–23.

Zielinski, R.E. and Grow, D.T., An iron catalyst for CVD of methane on carbon fibers, *Carbon*, 30(2), 295–299, 1992.

14

Combinations of Advanced Oxidation Processes

14.1 Introduction

As advanced oxidation processes (AOPs) continue to enter the commercial market, treatment efficiency becomes an important factor for selecting a process with better cost efficiency. To analyze the cost efficiency, comparisons between selected AOPs have been reported in recent years. A number of variables, including initial conditions, pH, oxidant, and catalyst concentrations, have been analyzed in combination with kinetic parameters of the process. The kinetics of each process are described by the rate constants and/or the reaction half times. These investigations have been conducted under identical experimental conditions so that treatment efficiency can be compared under the same conditions. In addition, combinations of various AOPs have also been analyzed with respect to cost efficiency. This comparison was carried out by combining various oxidants such as oxygen, ozone, and hydrogen peroxide (H_2O_2) with catalysts such as ultraviolet (UV), Fe^{2+}, and TiO_2. In general, synergic effects exist when two oxidation systems were combined. This synergy was reflected in a marked increase in the free-radical reaction pathway. The effect was also observed to increase with the complexity of the oxidation systems used. The only exception arose in the comparison of Fenton's reagent system with a combined Fenton's reagent/ozone system, in which case there seems to be interference of ozone with Fenton's reagent through oxidation of Fe^{2+}, which reduces the amount of Fe^{2+} to decompose H_2O_2 into hydroxyl radicals.

14.2 Fundamental Theory

The destruction kinetics of organic pollutants may be inhibited in wastewaters containing complex compounds. Under such conditions, the combination of UV, O_3, and H_2O_2 systems could be a better process because hydroxyl radicals are generated by several mechanisms. As a result, $UV/O_3/H_2O_2$ is less affected by color and turbidity in wastewater than UV/O_3 or UV/H_2O_2 system. They are also applicable over a wider pH range when compared to UV/H_2O_2 systems. Figure 14.1 shows a proposed mechanistic pathway that exhibits a network of free-radical chain reactions involving $HO_2^{\bullet}/O_2$, $HO_3^{\bullet}/O_3^{\bullet-}$, and $OH^{\bullet}/O^{\bullet-}$ in the generation and consumption of $OH^{\bullet}$ as proposed by Hong et al. (1996). This reaction network was a revised version based upon the reaction pathways formulated by Staehelin and Hoigne (1985) and Peyton and Glaze (1988).

Kinetic rate constants of various reactions in the reaction pathways are listed in Table 14.1. The model includes (1) dark reactions such as the decomposition of O_3 in water and the interaction of O_3 and $H_2O_2^{\bullet}$, and (2) UV-assisted actions such as the photolysis of O_3 and H_2O_2. In aqueous media, the free-radical chain reactions are initiated either by the k_1 step or by the k_7 followed by the k_6 step. When H_2O_2 is added to ozonated water (i.e., peroxone), the k_7 step that produces HO_2^- becomes inconsequential, and a large amount of H_2O_2 is available through the addition of the peroxone system. Next, the k_6 step becomes the predominant super-oxide- and ozonide-producing pathway, an important chain-initiation reaction. This is evident when the reaction rates of the k_1 and k_6 steps are compared with a typical batch dose of H_2O_2 (e.g., 10 to 200 mg/L). For applied $[O_3] = 2 \times 10^{-5}$ M (1 mg/L) and $[H_2O_2]_T = 1.5 \times 10^{-3}$ M (50 mg/L) at pH 7 (Hong et al., 1996):

$$k_1[O_3][OH^-] = 70(2 \times 10^{-5})(10^{-7}) = 1.4 \times 10^{-10} \ M^{-1} \ s^{-1} \qquad (14.1)$$

$$k_6[O_3][HO_2^-] = k_6[O_3][H_2O_2]_T ([H^+]/K[H_2O_2] + 1 + K[H_2O_2]/[H^+])^{-1}$$
$$= (2.8 \times 10^6)(2 \times 10^{-5})(1.5 \times 10^{-3})(2.5 \times 10^{-5}) = 2.1 \times 10^{-6} \ M^{-1} \ s^{-1} \quad (14.2)$$

Therefore, the enhancement of the peroxone system is due to faster chain initiation. In addition, when a large amount of H_2O_2 is added, the scavenging of $OH^{\bullet}$ by H_2O_2 (k_5 step) may overtake the O_3 (k_4 step). For example, $[O_3] = 1$ mg/L and $[H_2O_2]_T = 50$ mg/L at pH 7:

$$k_4[O_3][OH^{\bullet}] = (1.1 \times 10^8)(2 \times 10^{-5})[OH^{\bullet}] = (2.2 \times 10^3)[OH^{\bullet}] \ M^{-1} \ s^{-1}$$
$$(14.3)$$

$$k_5[H_2O_2][OH^{\bullet}] \approx k_5[H_2O_2]_T[OH^{\bullet}] = (2.7 \times 10^7)(1.5 \times 10^{-3})[OH^{\bullet}]$$
$$= (4.0 \times 10^4)[OH^{\bullet}] \ M^{-1} \ s^{-1} \qquad (14.4)$$

The model shows that UV illumination introduces additional and more productive $OH^{\bullet}$ generation pathways through three different mechanisms: (1) direct photolysis of O_3, (2) photolysis of O_3 to produce H_2O_2, and (3) photolysis of the formed H_2O_2. The contribution of each process to the

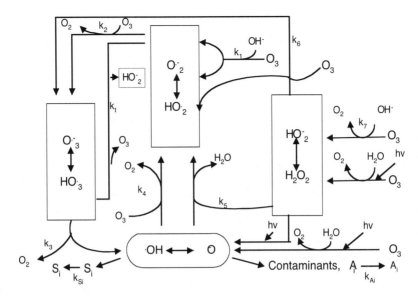

FIGURE 14.1
Free-radical chain reaction of $UV/O_3/H_2O_2$ system. (From Hong, A. et al., *J. Environ. Eng.*, 122(1), 58–62, 1996. With permission.)

generation of hydroxyl radical will be presented in detail in Section 14.4. The reaction mechanism in Figure 14.1 demonstrates that pH is an important parameter. Due to the fact that some anions such as HO_2^- react faster than their conjugate acids, pH will influence the reaction kinetics and thus the steady-state concentrations of various intermediates. As a result, the treatment efficiency can be greatly impacted by pH. The acid–base equilibria of various reacting species must be considered, as shown in Table 14.1 (Hong et al., 1996). The relative amounts of UV illumination, O_3, H_2O_2, and scavengers (S_i) such as HCO_3^- are critical in determining the level of active $OH^\bullet$ radical in the system. In UV-illuminated reactions, contaminant degradation rates are typically observed to be much faster than dark reactions (Zappi et al., 1993).

14.3 Process Description

From the late 1980s to the early 1990s, Ultrox International, a commercial manufacturer, has demonstrated the efficacy of ultraviolet-light-enhanced oxidation at sites belonging to the Department of Defense (DOD) and the Superfund sites (Zeff and Barich, 1992). Figure 14.2 illustrates a flow diagram of the Ultrox UV/oxidation treatment system. It shows that two different oxidants are used in the process. Ozone is generated from air and hydrogen

TABLE 14.1

Rate and Equilibrium Constants for Various Reactions in the $UV/O_3/H_2O_2$ Process

Reaction	Constant	Ref.
$O_3 + OH^- = O_3 \cdot + HO_2{}^\bullet$	$k_1 = 70\ M^{-1}\ s^{-1}$	Buhler et al. (1984)
$O_2{}^{\bullet-} + O_3 = O_3{}^\bullet + O_2$	$k_2 = 1.6 \times 10^9\ M^{-1}\ s^{-1}$	Buhler et al. (1984)
$HO_3 = H^\bullet + O_2$	$k_3 = 1.1 \times 10^5\ M^{-1}\ s^{-1}$	Buhler et al. (1984)
$OH^\bullet + O_3 = HO_2{}^\bullet + O_2$	$k_4 = 1.1 \times 10^8\ M^{-1}\ s^{-1}$	Sehested et al. (1984)
$OH^\bullet + H_2O_2 = HO_2{}^\bullet + H_2O$	$k_5 = 2.7 \times 10^7\ M^{-1}\ s^{-1}$	Christensen et al. (1982)
$HO_2{}^- + O_3{}^\bullet = HO_2{}^\bullet + O_3{}^\bullet$	$k_6 = 2.8 \times 10^6\ M^{-1}\ s^{-1}$	Staehelin and Hoigne (1982)
$O_3 + OH^- = HO_2{}^\bullet + O_2$	$k_7 = 48\ M^{-1}\ s^{-1}$	Bahnemann and Hart (1982)
$OH^\bullet + A_i = A'_i + OH$	$k_{A_i}\ ;\ A_i = \text{contaminant } I$	—
$OH^\bullet + HCO_3 = HCO_3{}^\bullet + OH$	$k'_1 = 1.5 \times 10^7\ M^{-1}\ s^{-1}$	Staehelin and Hoigne (1982)
$OH^\bullet + CO_3{}^{2-} = CO_3{}^{\bullet-} + OH^-$	$k'_2 = 4.2 \times 10^8\ M^{-1}\ s^{-1}$	Staehelin and Hoigne (1982)
$O_2{}^{\bullet-} + HO_3{}^\bullet = O_3 + HO_2{}^-$	$k_t = 10^{10}\ M^{-1}\ s^{-1}$	Staehelin and Hoigne (1982)
$O_3 + H_2O + h\nu = 2^\bullet OH + O_2$	$\nu\alpha\phi_1 I_{0\lambda}^{O_3}$	—
$O_3 + H_2O + h\nu = H_2O_2 + O_2$	$\nu\alpha\phi'_1 I_{0\lambda}^{O_3}$	—
$H_2O_2 + h\nu = 2^\bullet OH$	$\nu\alpha\phi_2 I_{0\lambda 2}^{H_2O_2}$	—
$H_2O_2 = H^+ + HO_2{}^-$	$pK^{H_2O_2} = 11.6$	Saucre et al. (1984)
$HO_2{}^\bullet = H^+ + HO_2{}^{\bullet-}$	$pK = 4.8$	Bielski et al. (1985)
$HO_3{}^\bullet = H^+ + HO_3{}^{\bullet-}$	$pK = 6.2$	Buhler et al. (1984)
$OH^\bullet = H^+ + O^{\bullet-}$	$pK = 11.8$	Weeks and Rabani (1966)

Source: From Hong, A. et al., *J. Environ. Eng.* 122(1), 58–62, 1966. With permission.

peroxide from the feed tank; therefore, the process can be operated in any combinations, such as UV/O_3, UV/H_2O_2, and $UV/O_3/H_2O_2$.

Figure 14.3 presents an engineering drawing of the process. Standard equipment designs is used in all of these installations. Reactor size varies from 300 to 4800 gal, and reactors are fabricated from stainless steel. Ozone generators range from 21 to 140 lb/day. The ozone is dispersed through porous, stainless-steel diffusers. The number of diffusers required depends on the organic compound being oxidized and the degree of removal required. The UV lamps are enclosed within quartz tubes for easy replacement and are mounted vertically within the reactor (Zeff and Barich, 1992). By utilizing a combined $UV/O_3/H_2O_2$ process, the Ultrox oxidation systems proved to be efficient in removing volatile organic compounds (VOCs), benzene, toluene, xylene, hydrazines, phenols, chlorophenols,

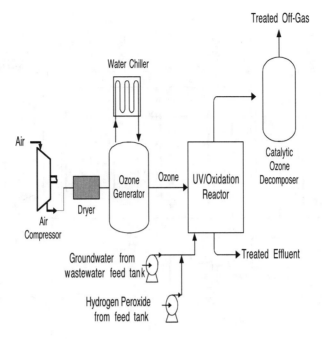

FIGURE 14.2
Flow diagram of Ultrox system. (From USEPA, Ultrox International Ultraviolet Radiation/ Oxidation Technology—Applications Analysis Report, EPA/540/A5-89/012, September 1990.)

dioxanes, polychlorinated biphenyls (PCBs), and pesticides present in wastewaters and groundwaters.

Figure 14.4 is the three-dimensional view of the process. The figure illustrates the relative water head at different stages. Because of flexible design, Ultrox UV/oxidation treatment systems have a number of advantages: (1) very few moving parts; (2) operation at low pressure; (3) minimum maintenance; (4) full-time or intermittent operation in either a continuous or batch treatment mode; (5) use of efficient, low-temperature, and long-life UV lamps; and (6) use of a microprocessor to control and automate the treatment process (Zeff and Barich, 1992).

14.4 Degradation of Organic Pollutants

14.4.1 Phenol

Phenol is degraded faster by ozone processes under basic pH than acidic pH because the contribution of hydroxyl radicals increases with pH. Table 14.2 lists the degradation efficiency of phenol under various experimental

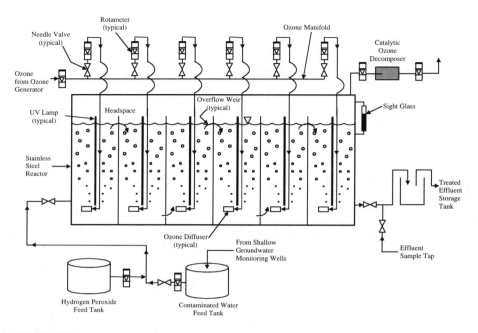

FIGURE 14.3

Flow diagram of Ultrox system. (From USEPA, Ultrox International Ultraviolet Radiation/ Oxidation Technology—Applications Analysis Report, EPA/540/A5-89/012, September 1990.)

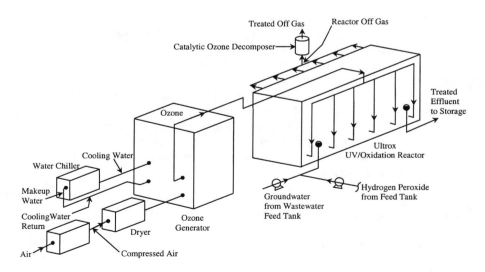

FIGURE 14.4

Isometric view of Ultrox system. (From USEPA, Ultrox International Ultraviolet Radiation/ Oxidation Technology—Applications Analysis Report, EPA/540/A5-89/012, September 1990.)

TABLE 14.2

Phenol Degradation in H_2O_2 Oxidation Systems as a Function of Time, pH, and Initial Reactant Concentrations

AOPs	pH	$C_{H_2O_2}$ (mM)	$C_{Fe(II)}$ (mM)	C_{TiO_2} (g/L)	Time of Treatment (min)	Phenol Degradation (%)
O_3	5.7–3	0	0	0	80	85.4
O_3	7.2 buffered	0	0	0	80	90.0
O_3	9.4 buffered	0	0	0	80	100
O_3/H_2O_2	5–3.4	6.8	0	0	80	80.6
O_3/H_2O_2	5–3.4	534	0	0	80	58.3
O_3/H_2O_2	6.8 buffered	0.62	0	0	80	90.4
O_3/H_2O_2	6.8 buffered	6.2	0	0	80	93.4
O_3/H_2O_2	6.8 buffered	31	0	0	80	86.9
O_3/H_2O_2	6.8 buffered	78	0	0	80	86.5
O_3/H_2O_2	6.8 buffered	155	0	0	80	77.7
O_3/H_2O_2	9.3 buffered	6.2	0	0	80	92.5
O_3/H_2O_2	9.3 buffered	31	0	0	80	88.8
UV	4.4 – 4	0	0	0	30	24.2
UV	6.8 buffered	0	0	0	30	14.0
UV	11.5 buffered	0	0	0	30	5.0
UV/H_2O_2	5–3.8	—	0	0	30	24.2
UV/H_2O_2	(3.5–2.5)		0	0	30	87.1
UV/H_2O_2	(3.2–2.3)		0	0	30	90.6
UV/C	(3.1–2.3)		0	0	30	89.8
UV/O_3	5.2–3	0	0	0	80	80.9
UV/O_3	6.9 buffered	0	0	0	80	92.6
UV/O_3	9.4 buffered	0	0	0	80	91.9

Source: Esplugas, S. et al., *Water Res.*, 36, 1034–1042, 2002. With permission.

variable conditions. At neutral pH and at low H_2O_2 concentrations, H_2O_2 improved ozonation slightly but showed an inhibitory effect at concentrations higher than 6.2 mM. The same effect was observed for the combination of $UV/O_3/H_2O_2$, and the limiting H_2O_2 concentration was found to be 0.07 mM. Under UV illumination alone, the best conditions were found at pH 5 without any buffer. The degradation rate increased considerably when H_2O_2 was used; nevertheless, the initial H_2O_2 concentration exerted little influence on the range used. In photocatalysis, the degradation rate increased with the catalyst concentration up to a value of 0.5 g/L. From that point, the rate was almost constant. In Fenton's reaction, the limiting factor was the amount of hydrogen peroxide. The higher the amount of H_2O_2, the faster the degradation will be. Also, as the concentration of Fe(II) ion is increased, the degradation rate is improved.

14.4.1.1 Comparison of Pseudo First-Order Kinetic Constant

The kinetic parameters chosen for comparison are rate constants and $t_{1/2}$. Radiation influences and the effect of reactor design are usually identical when these kinetic data are compared between the various AOPs tested. The values for pseudo first-order kinetics and half-lives for various processes are given in Table 14.3. In most cases, the values of $t_{3/4}$ are equal to two times those of $t_{1/2}$; therefore, the reactions obey a first-order kinetics. Figure 14.5. shows that Fenton's reagent has the largest rate constant, e.g., approximately 40 times higher than UV alone, followed by UV/H_2O_2 and O_3 in terms of the pseudo first-order kinetic constants. Clearly, UV alone has the lowest kinetic rate constant of 0.528 hr^{-1}.

14.4.1.2 Cost Estimation

The economic evaluation of treatment processes is very important in selecting an AOP from different available systems. The overall costs are represented by the sum of the capital, operating, and maintenance costs (Table

TABLE 14.3

First-Order Rate Constants $t_{1/2}$ and $t_{3/4}$ for Phenol Degradation under Various AOPs

Process	k (hr^{-1})	$t_{1/2}$ (hr)	$t_{3/4}$ (hr)
O_3/H_2O_2	2.13	0.325	0.63
UV/O_3	3.14	0.221	0.417
O_3	4.42	0.157	0.317
$UV/O_3/H_2O_2$	4.17	0.166	0.333
UV/H_2O_2	6.26	0.111	0.383
UV	0.528	1.31	3.33
Photo catalysis	0.582	1.19	2.47
Fenton	22.2	0.0312	0.067

Source: Esplugas, S. et al., *Water Res.,* 36, 1034–1042, 2002. With permission.

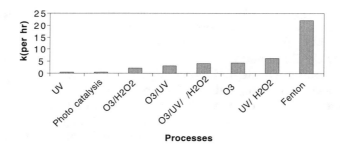

FIGURE 14.5

Comparison of the rate constants for selected AOPs. All processes have been approximately first-order with respect to substrate concentration. (Data from Esplugas, S. et al., *Water Res.,* 36, 1034–1042, 2002. With permission.)

14.4). An estimation of costs has been made in this section; however, it should be pointed out that costs could considerably decrease for photocatalytic treatments when solar light is used. Figure 14.6 indicates that UV alone has the maximum cost because it has the lowest kinetic rate constant. Fenton's reagent has reasonably lower costs and the highest rate constant. Different AOPs (ozone and its combinations, photocatalysis and UV/H_2O_2, photocatalysis and Fenton's reagent) have been compared in terms of the degradation of phenol in aqueous solution. In UV processes (UV, UV/H_2O_2, and photocatalysis), the degradation rate produced by the UV/H_2O_2 process was almost five times higher than photocatalysis and UV alone. Fenton's reagent showed the fastest degradation rate, 40 times higher than the UV process and photocatalysis and five times higher than ozonation. Nevertheless, the degradation rates and lower costs obtained with ozonation make it the most appealing choice for phenol degradation.

14.4.2 *para*-Hydroxybenzoic Acid

para-Hydroxybenzoic acid is a very common pollutant in a variety of industrial wastewaters (olive oil and distillation industries). Because it is

TABLE 14.4

Analysis of Costs Associated with Various AOPs

Process	k (hr^{-1})	Cost ($/kg)
UV	0.528	172.2
O_3/H_2O_2	2.13	2.71
UV/O_3/	3.14	9.28
UV/O_3/H_2O_2	4.17	7.12
O_3	4.42	0.81
Fenton	22.2	3.92

Source: Esplugas, S. et al., *Water Res.*, 36, 1034–1042, 2002. With permission.

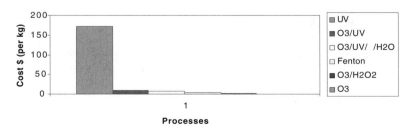

FIGURE 14.6
Cost analysis of selected AOPs. (Data from Esplugas, S. et al., *Water Res.*, 36, 1034–1042, 2002. With permission.)

Physicochemical Treatment of Hazardous Wastes

highly toxic and refractory to anaerobic biological treatment, it was chosen for the comparison as being representative of phenolic acids, and its degradation efficiency and kinetics are compared for several AOPs. The first stage was to compare the 12 oxidation processes applied to the destruction of *p*-hydroxy-benzoic acid. Table 14.5 lists the semi-reaction times and the conversions attained at 5 and 10 min of reaction time. The kinetic parameters of different oxidation processes are also shown in Table 14.5. Figure 14.7 demonstrates that Fenton's reagent has the highest kinetic rates followed by ozonation and then UV irradiation. To study the improvement by different AOPs, the various oxidation systems are divided into three groups corresponding to the three basic oxidation processes from which they derive: UV irradiation, ozonation, and Fenton's reagent.

TABLE 14.5

Oxidation Rate Constants for Various AOPs

Oxidation Process	k (min^{-1})	$t_{1/2}$ (min)
$UV/O_3/H_2O_2/Fe^{2+}$	0.46510067	1.49
$UV/H_2O_2/Fe^{2+}$	0.33157895	2.09
$O_3/H_2O_2/Fe^{2+}$	0.231	3.00
$UV/O_3/H_2O_2$	0.20086957	3.45
H_2O_2/Fe^{2+}	0.17111111	4.05
UV/H_2O_2	0.154	4.50
O_3/H_2O_2	0.1125	6.16
UV/O_3	0.10862069	6.38
O_3/Fe^{2+}	0.10358744	6.69
O_3	0.08640898	8.02
UV/TiO_2	0.06861386	10.1
UV	0.03705882	18.7

Source: Data from Beltan-Heredia, J. et al. *Chemosphere*, 42, 351–359, 2001.

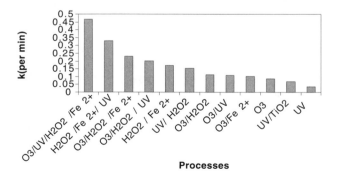

FIGURE 14.7

Selected destruction rate constants for *p*-hydroxybenzoic acid. (Data from Beltan-Heredia, J. et al. *Chemosphere*, 42, 351–359, 2001.)

14.4.2.1 Oxidation Processes Using UV Radiation

The overall reaction process consists of three contributions: direct oxidation by UV irradiation (photolysis), direct oxidation by ozone (ozonation), and oxidation by free radicals:

$$\text{Substrates} + h\text{v} \rightarrow \text{Products} \tag{14.5}$$

$$\text{Substrates} + O_3 \rightarrow \text{Products} \tag{14.6}$$

$$\text{Substrates} + OH^{\bullet} \rightarrow \text{Products} \tag{14.7}$$

Direct oxidation of organic compounds by hydrogen peroxide in the entire pH range is negligible. At pH 5, ozone reacts with the compound directly and not via free radicals (Staehelin and Hoigne, 1985). Figure 14.8 shows that the order of efficiency in the oxidation systems under the UV irradiation can be described as:

$$UV < UV/TiO_2 < UV/O_3 < UV/H_2O_2 < UV/H_2O_2/O_3$$

The pseudo first-order kinetic constants (k_t) of p-hydroxybenzoic acid are given in Table 14.6. UV/H_2O_2 has a higher rate constant than that of UV/O_3 because the former requires only 0.5 peroxide molecule and 0.5 photon, whereas the latter process requires 1.5 ozone molecules and 0.5 photon. This is shown in Equation (14.8) and Equation (14.9):

$$H_2O_2 \xrightarrow{h\text{v}} 2OH^{\bullet} \tag{14.8}$$

$$3O_3 \xrightarrow{h\text{v}} O_2 + {}^{\bullet}O \tag{14.9}$$

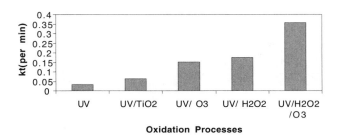

FIGURE 14.8
Oxidation rates for processes involving UV irradiation. (Data from Beltan-Heredia, J. et al. *Chemosphere*, 42, 351–359, 2001.)

TABLE 14.6

Oxidation Rate Constants for Various AOPs

Oxidation Process	k_t (min^{-1})
UV	0.032
UV/TiO$_2$	0.063
UV/O$_3$	0.151
UV/H$_2$O$_2$	0.176
UV/O$_3$/H$_2$O$_2$	0.358

Source: Data from Beltan-Heredia, J. et al. *Chemosphere,* 42, 351–359, 2001.

Because H$_2$O$_2$ can be infinitely soluble in water, there is no limitation on mass transfer; however, O$_3$ has a lower solubility in water and has mass transfer limitations. Thus, the rate constant for UV/H$_2$O$_2$/O$_3$ is the greatest, followed by that for UV/H$_2$O$_2$ and then UV/O$_3$.

14.4.2.2 AOPs Using Ozone

The pseudo first-order kinetic rate constants for processes based on the application of ozone are given in Table 14.7. Figure 14.9 presents the pseudo first-order rate constants of *p*-hydroxybenzoic degradation. The degradation rates follow the increasing order:

$$O_3 < O_3/H_2O_2 < O_3/H_2O_2/Fe^{2+} < O_3/Fe^{2+} < UV/O_3 < UV/O_3/H_2O_2 < UV/O_3/H_2O_2/Fe^{2+}$$

14.4.2.3 AOPs Using Fenton's Reagent

In regard to the degree of conversion of *p*-hydroxybenzoic acid at 5 and 10 min of reaction time, the degradation efficient of AOPs using Fenton's reagent follows the increasing order:

TABLE 14.7

Oxidation Rate Constants for Various AOPs

Oxidation Process	k_t (min^{-1})	$t_{1/2}$ (min)
O$_3$	0.100	6.93
O$_3$/Fe^{2+}	0.135	0.828
O$_3$/H$_2$O$_2$	0.122	5.680
UV/O$_3$	0.151	4.589
O$_3$/H$_2$O$_2$/Fe^{2+}	0.128	5.414
UV/O$_3$/H$_2$O$_2$	0.358	1.935
UV/O$_3$/H$_2$O$_2$/Fe^{2+}	0.869	0.797

Source: Data from Beltan-Heredia, J. et al. *Chemosphere,* 42, 351–359, 2001.

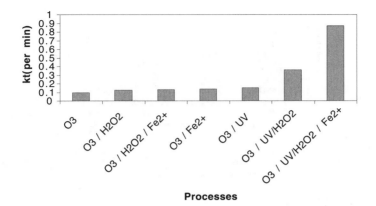

FIGURE 14.9
Oxidation rates for processes using ozone. (Data from Beltan-Heredia, J. et al. *Chemosphere*, 42, 351–359, 2001.)

$$Fe^{2+}/H_2O_2/O_3 < Fe^{2+}/H_2O_2 < Fe^{2+}/UV/H_2O_2 < Fe^{2+}/H_2O_2/UV/O_3$$

Table 14.8 shows the rate constants and reaction half times for oxidation of *p*-hydroxybenzoic acid by these AOPs. The results are plotted in Figure 14.10. Figure 14.10 indicates that the addition of ozone has neither synergic nor additive effects (Fenton's reagent + ozone). This may be because ozone interferes with the action of Fenton's reagent, probably by oxidizing Fe^{2+} to Fe^{3+}. The combined effect is faster at the beginning (when Fe^{2+} concentration is high), but is eventually overtaken by Fenton's reagent. The last two oxidation systems in Table 14.8 show the synergic effects of adding ozone to the photo–Fenton's reagent system due to the notable increase of degradation rates. The order of reactivity determined for the oxidation of *p*-hydroxybenzoic acid in aqueous solution is as follows:

$$UV < UV/TiO_2 < O_3 < O_3/Fe^{2+} = O_3/H_2O_2 < UV/O_3 < UV/H_2O_2 = Fe^{2+}/H_2O_2/O_3$$

$$< Fe^{2+}/H_2O_2 < UV/O_3/H_2O_2 < Fe^{2+}/UV/H_2O_2 < UV/O_3/H_2O_2/Fe^{2+}$$

14.4.3 Chlorophenols

Chlorophenols (CPs) constitute a group of organic substances that are introduced into the environment as a result of several human activities (e.g., water disinfection, waste incineration, uncontrolled use of pesticides and herbicides). The rate constants for the global processes and the efficiency of the various AOPs tested are compared in the following sections.

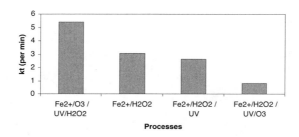

FIGURE 14.10

p-Hydroxybenzoic oxidation rate constants for various Fenton's reaction systems. (Data from Beltan-Heredia, J. et al. *Chemosphere*, 42, 351–359, 2001.)

TABLE 14.8

Oxidation Rate Constants for Various AOPs

Oxidation Process	$t_{1/2}$ (min)	k_t (min^{-1})
$UV/O_3/H_2O_2/Fe^{2+}$	5.414063	0.128
H_2O_2/Fe^{2+}	3.052863	0.227
$UV/H_2O_2/Fe^{2+}$	2.634981	0.263
$UV/O_3/H_2O_2/Fe^{2+}$	0.797468	0.869

Source: Data from Beltan-Heredia, J. et al. *Chemosphere*, 42, 351–359, 2001.

14.4.3.1 Comparison of Various AOPs

The decomposition of all four selected CPs (4-CP, 2,4-DCP, 2,4,6-TCP, 2,3,4,6-TeCP) was achieved by utilizing the following oxidants: UV radiation, Fenton's reagent, and ozone. The results are tabulated in Table 14.9. Figure 14.11 shows the degradation rate of chlorophenols by UV/H_2O_2. Figure 14.12 suggests that 4-CP was rapidly degraded by Fenton's reagent, and decreasing rates were obtained for 2,4-DCP, 2,4,6-TCP, and 2,3,4,6-TeCP. This suggests that the increase of chlorine atoms in a chlorophenol molecule decreases the susceptibility of aromatic rings to attack by the hydroxyl radicals gen-

TABLE 14.9

Comparison of Rate Constants for Various AOPs

Compound	Ultraviolet		Fenton		O_3 at pH = 2		O_3 at pH = 9	
	$k_p \times 10^3$ (min^{-1})	$t_{1/2}$ (min)	$k_p \times 10^3$ (min^{-1})	$t_{1/2}$ (min)	$k_p \times 10^3$ (min^{-1})	$t_{1/2}$ (min)	$k_p \times 10^3$ (min^{-1})	$t_{1/2}$ (min)
4-CP	564	1.1	1877	0.4	17	38.5	239	3.4
2-4-DCP	38	17.5	209	2.4	24	30.4	315	3.3
2,4,6-TCP	26	25.2	98	5.1	44	20.6	314	3.1
2,3,4,6-TeCP	21	30.6	9	49.5	94	10.6	415	1.9

Source: Data from Benitez, F.J. et al., *Chemosphere*, 41, 1271–1277, 2000.

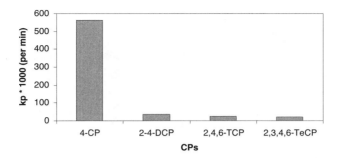

FIGURE 14.11
Decomposition rates for selected chlorophenols in UV/H_2O_2 oxidation system. (Data from Benitez, F.J. et al., *Chemosphere*, 41, 1271–1277, 2000.)

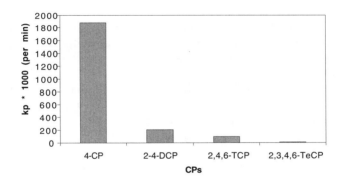

FIGURE 14.12
Decomposition rates for selected chlorophenols in the Fenton's reaction system. (Data from Benitez, F.J. et al., *Chemosphere*, 41, 1271–1277, 2000.)

erated by UV radiation. This can be explained by the electronically excited states of polychlorinated phenols generated during photochemical treatments (Burrows et al., 1998). In these excited states, the CP molecules undergo intramolecular transformations and stabilize with different electron distributions, followed by decomposition to radical or molecular products. A higher degree of chlorine substitution withdraws electrons from the aromatic rings. As a result, electrophilic addition of hydroxyl radical to the ring is not favored. The formation of an excited state or stabilization of the intermediate state becomes more difficult to oxidize than non-chlorine substituted phenols.

Figure 14.13 shows that the degradation by ozone alone is much faster at pH 9 than at pH 2. In addition, the greater the number of chlorine substituents contained in a chlorophenol molecule, the higher the degradation rate is. The figure indicates that the oxidation rates are not the same as in the photodegradation process. In UV/O_3, hydroxyl radicals usually attack the aromatic ring at the sites that are not occupied by chlorine atoms; therefore,

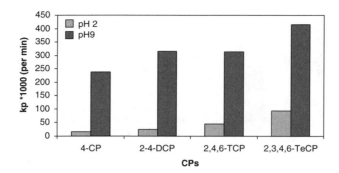

FIGURE 14.13
Effect of pH on chlorophenol oxidation rates at various pH values. (Data from Benitez, F.J. et al., *Chemosphere*, 41, 1271–1277, 2000.)

hydroxylation is the first elementary step and precedes dissociation of chlorine atoms (Tang and Huang, 1996). So, the increase in number of chlorine atoms in the aromatic ring usually decreases the reactivity toward hydroxyl radicals. However, at pH 2 and 9, hydroxyl radical is not the dominant species in ozonation alone. Therefore, oxidation by molecular ozone contributes mostly to the CP degradation. The trend in k_p values is reversed and k_p increases with increasing substituent chlorine atoms. The rate of decomposition increases in the following order as shown in Figure 14.13:

$$2,3,4,6\text{-TeCP} < 2,4,6\text{-TCP} < 2\text{-4-DCP} < 4\text{-CP}$$

The pathways proposed by Chen and Ni (2001) to describe the complete ozonation of 2–chlorophenol were hydroxylation, dechlorination, and ring cleavage. According to this mechanism, the presence of more atoms of chlorine enhances the dechlorination step; therefore, the degradation is faster. Degradation efficiency of CPs can be compared for combinations of UV radiation + H_2O_2, UV radiation + Fenton's reagent, and UV radiation + ozone as described in the following sections.

14.4.3.2   UV/H_2O_2 System

Photodegradation experiments with the four selected CPs in the presence of hydrogen peroxide were conducted at 25°C and pH 2 by Benitez et al. (2000). The degradation rates achieved were the same as those obtained for the single photochemical process. The degradation rates decreased when the number of chlorine substituents increased, as shown in Table 14.10. Figure 14.14 shows the combined process constant (k_t) and radical rate constant (k_r). It can be seen that these rate constants have moderately higher values than those obtained in the single photodecomposition process due to hydroxyl radical reactions.

TABLE 14.10

Oxidation Rate Constants for Various AOPs

Compound	$k_t \times 10^3$ min^{-1}	$k_r \times 10^3$ min^{-1}
4-CP	601	36
2-4-DCP	44	6.9
2,4,6-TCP	33	7.2
2,3,4,6-TeCP	29	7.9

Source: Data from Benitez, F.J. et al., *Chemosphere*, 41, 1271–1277, 2000.

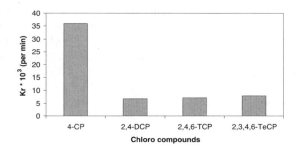

FIGURE 14.14
Combined process constant (k_t) and radical rate constant (k_r) for UV/H$_2$O$_2$ process. (Data from Benitez, F.J. et al., *Chemosphere*, 41, 1271–1277, 2000.)

TABLE 14.11

Oxidation Rate Constants for Various AOPs

Compound	$k_t \times 10^3$ min^{-1}	$k_r \times 10^3$ min^{-1}	$t_{1/2}$ (min)
4-CP	642	79	0.9
2,4-DCP	88	50	7.4
2,4,6-TCP	78	52	8.5
2,3,4,6-TeCP	58	37	12.0

Source: Data from Benitez, F.J. et al., *Chemosphere*, 41, 1271–1277, 2000.

14.4.3.3   Photo–Fenton's Reagent System

Decomposition experiments for these CPs listed in Table 14.11 were carried out by the simultaneous action of UV radiation and Fenton's reagent (Benitez et al., 2000). Table 14.11 shows the first-order rate constants and half-lives. During the photo–Fenton's reagent reaction, the single photodecomposition rate constant, k_t, decreased as the number of chlorine substituents increased. In addition, combined rate constants, k_t, are much greater than the radical reaction constants, k_r. Therefore, this confirms the additional contribution of the radical reaction due to generation of the hydroxyl radicals by Fenton's reagent and by UV/H_2O_2. Figure 14.15 shows the radical decomposition rate constants for the photo–Fenton's reagent reaction. Figure 14.16 plots the combined rate constants for the photo–Fenton's reagent system. It show that the overall rate constants, k_t, increased much more than the radical constants, k_r.

14.4.3.4 UV/O_3 System

Finally, degradation experiments with these four CPs by the combined process UV/O_3 were conducted at 25°C and pH 2 by Benitez et al. (2000). Table 14.12 shows the rate constants, k_t, and $t_{1/2}$ determined through kinetic studies. Figure 14.17 shows that the rate constants decreased significantly when chlorine substituents were increased from one to two. However, the rate constants remained relatively the same for 2-, 3-, and 4-chlorine-substituted phenols. Comparing these results to those of the single photodecomposition or pH 2 ozonation processes, this combination accelerates the decomposition rate, with an extremely high rate constant (k_t) for 4-CP and lower rates for the other CPs. The remaining CPs (2-4-DCP, 2,4,6-TCP, 2,3,4,6-TeCP)

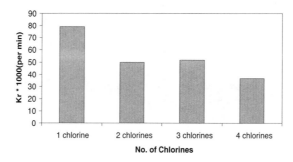

FIGURE 14.15
Radical decomposition of CP by photo–Fenton system. (Data from Benitez, F.J. et al., *Chemosphere*, 41, 1271–1277, 2000.)

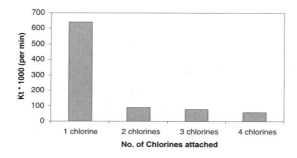

FIGURE 14.16
Combined process constant (k_t) for photo–Fenton system. (Data from Benitez, F.J. et al., *Chemosphere*, 41, 1271–1277, 2000.)

TABLE 14.12

Rate Constants and $t_{1/2}$ for Chlorophenols

Compound	$k_t \times 10^3$ min^{-1}	$t_{1/2}$ (min)
4-CP	644	1.3
2-4-DCP	65	15.6
2,4,6-TCP	68	11.8
2,3,4,6-TeCP	71	9.7

Source: Data from Benitez, F.J. et al., *Chemosphere*, 41, 1271–1277, 2000.

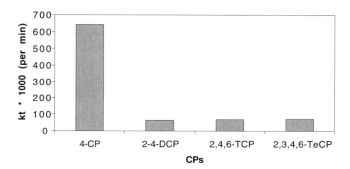

FIGURE 14.17
Combined process constant (k_t) and radical rate constant (k_r) for UV/ozone system. (Data from Benitez, F.J. et al., *Chemosphere*, 41, 1271–1277, 2000.)

presented decreasing rate constants when the substituent chlorine atoms increased. During the decomposition by either UV radiation or Fenton's reagent, 4-CP was the most rapidly degraded and 2,3,4,6-TeCP exhibited the lowest rate, while in the single ozonation process the sequence of degradation was the reverse: 2,3,4,6-TeCP > 2,4,6-TCP > 2-4-DCP > 4-CP, with a clear increase in the degradation rate when the pH was increased from 2 to 9. Therefore, ozone attacks CPs with a different mechanism than do OH• radicals. The degradation is enhanced due to generation of hydroxyl radicals by UV and high pH. This improvement is moderate in the UV/H$_2$O$_2$ system and significant in the UV/Fenton's reagent system. In the UV/O$_3$ system, the contribution of the radical reaction to the global reaction is higher than that of direct photolysis.

14.4.3.5   pH Effect on the Ozone Oxidation of Chlorophenols

Takahashi et al. (1994) showed the pH effect on the rate constant for 2-CP and 4-CP. Taking an approximate value for pH vs. k, Table 14.13 can be obtained. Figure 14.18 indicates that as pH increased, the degradation rate of 2- and 4-CP increased. In the temperature range from 287 to 318 K, the reaction temperature will barely have any effect on the observed degradation

TABLE 14.13

Rate Constants at Various pH Values

p H	2	4	5	7	9	11	2.5	4.8	7.5	8.4	9.5
k	0.06	0.04	0.08	0.05	0.1	0.13	0.04	0.045	0.14	0.12	0.13

Source: Data from Boncz, M.A. et al., *Water Sci. Technol.*, 35(4), 65–72, 1997.

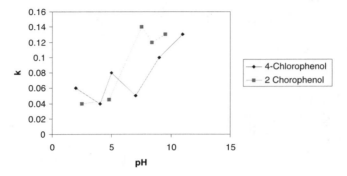

FIGURE 14.18
Rates of formation of 2- and 4-chlorophenol by-products. (From Boncz, M.A. et al., *Water Sci. Technol.*, 35(4), 65–72, 1997. With permission.)

rates of *para*- and *ortho*-chlorophenols. The reactivity of chlorophenolate anion is higher than that of chlorophenol itself, as can be seen from the comparison between the oxidation of 2–CP and 4-CP at increasing pH values.

14.4.4 Reactive Dyes

14.4.4.1 Ozone Treatment

Ozone reacts with commercial dyes found in textile wastewater through two different pathways — namely, direct molecular and indirect radical chain reactions. The ozonation pathway strongly depends upon the quality of the wastewater to be treated (i.e., pH and concentration of initiators, promoters, and scavengers in the reacting medium). In all the cases, almost instantaneous decolorization was observed by Alaton et al. (2002). Table 14.14 shows that appreciably faster removal rates at UV_{280} and at pH 7 were achieved than at pH 11 and 3. Initial decolorization rates (r_{id}) and the first-order UV_{280} removal kinetics ($k_{UV_{280}}$) were obtained after 1 hr of ozonation and are summarized in the table. Table 14.14 indicates that color removal could be simply achieved via both reaction pathways of ozone, whereas the reaction pH had to be at least 7 to enhance ozone decomposition for higher overall UV_{280} and TOC reduction. Table 14.14 also shows that the initial decolorization rates (r_{id}; 1/m.min) were almost identical. The overall UV_{280} removal increased from 70.7% at pH 3 to 93.1% at pH 7 and remained almost unchanged (92.4%) at pH 11. This observation indicated that color removal

TABLE 14.14

Decolorization Rate Constant at Different pH Values

Reaction pH	r_{id} (1/m.min)	TOC Removal (%)	$k_{UV_{280}}$ (1/min)
3.0	11.5	17.3	0.15
7.0	11.2	27.8	0.33
11.0	11.4	15.9	0.28

Source: Alaton, I.A. et al., *Water Res.*, 36, 1143–1154, 2002. With permission.

could be achieved by both reaction pathways (direct molecular and indirect radical change types of reactions) of ozone, whereas the reaction pH had to be at least 7 to enhance ozone decomposition for higher overall UV_{280} and TOC reduction. The highest overall TOC removal rate was found as 28% at pH 7, when the radical reaction became effective; at the same time, the inhibiting effect of carbonate ions was not very pronounced. The effect of the applied specific ozone dose was evaluated in separate experiments where the specific ozone input rate was varied from 544 to 2970 mg/(L.hr) and the reaction was carried out at pH 7, as shown in Figure 14.19 (Alaton et al., 2002).

Table 14.15 shows the values for H_2O_2/UV-C oxidation of the reactive dye bath effluent in the photochemical (UV-C) reactor in the presence of 680 mg/L H_2O_2 at varying pH values. Table 14.15 shows that k_d and $k_{UV_{280}}$ values increased by factors of 2.7 and 2.6, respectively, by elevating the pH from 3 to 7. Overall $k_{UV_{280}}$ removal efficiency was not seriously affected by the reac-

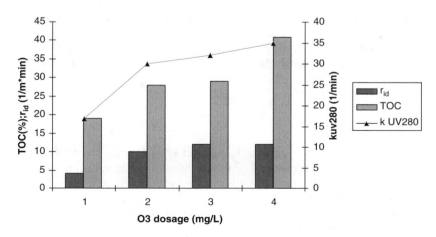

FIGURE 14.19

TOC remaining as a function of ozone dosage. O_3 dosage was 544, 1310, 2340, and 2970 mg/L for 1, 2, 3, and 4. (From Alaton, I.A. et al., *Water Res.*, 36, 1143–1154, 2002. With permission.)

TABLE 14.15

Effect of pH on the Decolorization Rate Constants

Reaction pH	k_d (1/min)	r_{id} (1/m.min)	$k_{UV_{280}}$ (1/min)	TOC Removal (%)
3.0	0.14	5.5	0.05	30.4
7.0	0.36	6.8	0.13	13.9
11.0	0.07	0.9	0.04	1.1

Source: Alaton, I.A. et al., *Water Res.*, 36, 1143–1154, 2002. With permission.

tion pH and varied between 87% and 91% after 1 hr of treatment. Increasing the pH to 11 inhibited the color removal and the $k_{UV_{280}}$ rates decreased drastically. A completely different kinetic trend was observed for the total organic carbon (TOC) reduction rate; overall TOC removal efficiency dropped from 30% at pH 3 to 14% at pH 11 (average TOC removal, 1%). This inhibition of TOC reduction at pH 11 was a consequence of the OH• scavenging effect of carbonate anions present in the simulated reactive dye bath effluent in the form of Na_2CO_3 (= 867 mg/L).

Figure 14.20 shows that the removal rates increased considerably with increasing initial H_2O_2 concentrations from 0 to 680 mg/L (20 mM). A maximum value for color removal was reached at a H_2O_2 dose of around 680 mg/L (k_d = 0.36 1/min; $k_{UV_{280}}$ = 0.13 1/min). Increasing the H_2O_2 concentration further from 680 to 1360 mg/L (40 mM) resulted in a significant inhibition of color removal; however, the overall TOC removal efficiency continued to increase, reaching its highest value (14.6%) in the presence of 40-mM (1360 mg/L) H_2O_2. Thus, the presence of excess H_2O_2 may lower the treatment efficiency of AOP. Increasing the pH of reactive dye bath effluents had a more detrimental effect upon the treatment performance of the UV/H_2O_2 oxidation system than on the ozonation process. The applied H_2O_2 dose and pH have to be in the optimal range to maximize the treatment performance of the AOP.

14.4.4.2   *UV/TiO₂*

Table 14.16 provides values of the k_d, r_{id}, and $k_{UV_{280}}$ together with the overall percent TOC removal efficiencies obtained for 1-hr photocatalytic treatment of 15 times diluted reactive dye bath effluent at various reaction pH values (Alaton et al., 2002). At the end of the photocatalytic oxidation period, overall color removal ranged between 89.5 and 94.6% at varying pH values. In terms of TOC removal, pH 3 appeared to be the optimal condition for the UV/TiO₂ system.

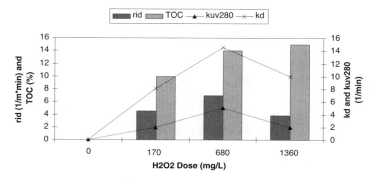

FIGURE 14.20

TOC remaining as a function of the initial concentration of H_2O_2. Reaction pH = 7.0; treatment time = 1 hr. (From Alaton, I.A. et al., *Water Res.*, 36, 1143–1154, 2002. With permission.)

TABLE 14.16

Effect of pH on the UV/TiO_2: Treatment of Simulated Reactive Dyebath Effluent (TiO_2 dose = 1000 mg/L)

Reaction pH	k_d (1/min)	r_{id} (1/m.min)	$k_{UV_{280}}$ (1/min)	TOC Removal (%)
3.0	0.04	0.24	0.02	12.5
7.0	0.05	0.31	0.02	10.3
9.0	0.04	0.28	0.02	9.0
11.0	0.02	0.15	0.01	0.0

Source: Alaton, I.A. et al., *Water Res.*, 36, 1143–1154, 2002. With permission.

14.4.5 1,3,5-Trichlorobenzene (TCB) and Pentanoic Acid (PA)

TCB is widely present in the chemical industry as a by-product of pesticide manufacturing (Masten et al., 1996). Pentanoic acid was used as an •OH radical probe due to its low reactivity with molecular ozone and high reactivity with hydroxyl radicals generated from the decomposition of ozone. pH, humic acid, and bicarbonate may impact the efficiency of oxidation of TCB and PA. Natural water was simulated by adding humic substances or bicarbonate to deionized water. The degradation rate was assumed to be first order with respect to the concentration of the target chemicals by using a continuous-flow, stirred-tank reactor (CFSTR) (Masten et al., 1996). Following is the expression for rate constants:

$$k = \frac{1}{\theta}\left(\frac{[C]_0 - [C]}{[C]}\right) \tag{14.10}$$

where:

k is the rate constant.

$[C]_0$ is the initial concentration.

$[C]$ is the target concentration.

θ is the retention time.

Figure 14.21 illustrates the effect of humic acid or carbonate on the efficiency of the oxidation of TCB. The rate of TCB degradation decreased at both 1.6 and 10 mg/L humic acid. It appears that humic acid scavenges •OH and other radicals responsible for the degradation of TCB. Furthermore, the presence of humic acid is likely to reduce the transmission of UV light, thereby decreasing the rate of ozone decomposition through OH radicals (Masten et al., 1996).

For bicarbonate, the rate of TCB degradation decreases with increasing bicarbonate concentration (as shown in Figure 14.21). This is due to the scavenging of OH radicals by bicarbonate. The degradation of the target chemical was assumed to be first order with respect to light intensity, $[O_3]$, and [•OH]. The observed first-order rate constant, k, is given by the following expression:

$$k = k_{O_3}\left[O_3\right] + k_{photo}\phi I + k_{OH\bullet}\left[OH^\bullet\right] \tag{14.11}$$

where:

I is the light intensity.

k_{O_3} is $2.3\ M^{-1}\ s^{-1}$.

$k_{photo}\phi$ is $0.0013\ W^{-1}\ s^{-1}$.

Equation (14.11) demonstrates that O_3, photons, and •OH are all involved in the degradation of TCB, subject to the effect of pH, bicarbonate, and humic acids (Masten et al., 1996). Table 14.17 provides a comparison of the effect of pH on the efficiency of oxidation of pentanoic acid using O_3, UV/O_3, O_3/H_2O_2, and $UV/O_3/H_2O_2$ (Masten et al., 1996).

14.4.6 Polycyclic Aromatic Hydrocarbons (PAHs)

PAHs are formed as the by-products of incomplete combustion of fossil fuels. These compounds have been identified in many emission sources, such as vehicle exhausts; power plants; chemical, coke, and oil-shale industries; and municipal sewage (Trapido et al., 1995). Some PAHs are known to be carcinogens. PAHs have been observed to be degraded by ozone treatment in aqueous media. The degradation kinetics of five PAHs — anthracene,

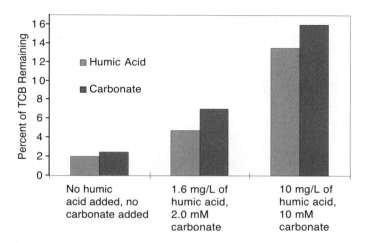

FIGURE 14.21
Degradation of TCB under the influence of humic acid and bicarbonate. (From Masten, S. et al., *J. Hazardous Waste Hazardous Mater.*, 13(2), 265–284, 1996. With permission.)

TABLE 14.17

Comparison of the Percentage of PA Remaining under Various Systems

Process	pH	PA Remaining (%)
Ozone	2.2	92.2
	7.0	15.8
	11.5	31.7
Ozone/UV	2.3	27.0
	6.8	13.6
	11.0	16.6
Ozone/H_2O_2	2.3	80.9
	6.8	8.3
	12.2	22.6
Ozone/H_2O_2/UV	8.2	13.2

Note: Input ozone concentration is 127 ± 3 μM; hydrogen peroxide dosage is 0 or 60 μM.

Source: Masten, S. and Davies, S., *Environ. Sci. Technol.*, 28(4), 180–185, 1994. With permission.

phenanthrene, pyrene, fluoranthene, and benzopyrene — by UV/O_3/H_2O_2 are discussed in this section. Figure 14.22 shows that 90% removal can be achieved within 40 min in the degradation of procatechuic acid by the UV/O_3/H_2O_2 process.

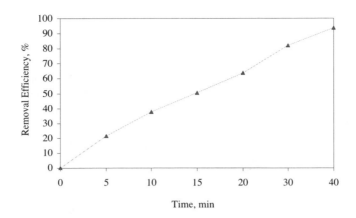

FIGURE 14.22
Degradation of procatechuic acid in UV/ozone/H_2O_2 treatment. (From Trapido, M. and Veressinina, Y., *Environ. Technol.*, 16, 729–740, 1995. With permission.)

14.4.6.1 Anthracene

During ozonation, no differences in the degradation rates of anthracene in acidic and neutral media have been observed. In basic media, the degradation of anthracene proceeds remarkably slowly (Trapido et al., 1995). A comparison of the half-lives of anthracene is provided in Table 14.18.

14.4.6.2 Pyrene

The results obtained in AOP-treatment of pyrene in aqueous solutions were quite similar to those of anthracene and phenanthrene. For pyrene, ozonation has been more effective at lower pH values than at neutral pH values. The half-lives of pyrene at pH 3, 7, and 9 are 17, 24, and 42 s, respectively; however, the half-life of pyrene increased in the series of $O_3 < UV/O_3 < O_3/H_2O < UV/O_3/H_2O_2$ (Trapido et al., 1995). The UV degradation of pyrene is quite fast; the half-life is only 69 s (Trapido and Veressinina, 1995). Figure 14.23 illustrates the oxidation kinetics of anthracene and pyrene in neutral media.

TABLE 14.18

Comparison of the Half-Lives of Anthracene under Various Systems

System	Half-Life (s)
O_3/H_2O_2	30
UV/O_3	56
$UV/O_3/H_2O_2$	70
O_3	26

Source: Data from Trapido, M. and Veressinina, Y., *Environ. Technol.*, 16, 729–740, 1995. With permission.

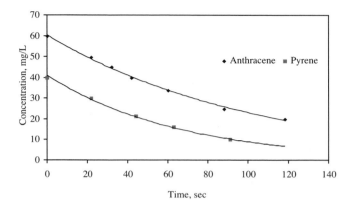

FIGURE 14.23
Oxidation kinetics of anthracene and pyrene in neutral media with UV/ozone/H₂O₂ system.
(From Trapido, M. and Veressinina, Y., *Environ. Technol.*, 16, 729–740, 1995. With permission.)

14.4.6.3 Phenanthrene

AOPs have also been applied for phenanthrene degradation in acidic and
basic media, as shown in Figure 14.24. The results are similar to those
obtained in neutral media. Ozonation alone was shown to be the most
effective for destruction of phenanthrene. Table 14.19 provides the half-lives
of phenanthrene under various AOP systems (Trapido and Veressinina,
1995). Figure 14.24 illustrates the degradation kinetics of phenanthrene with
the UV/O₃/H₂O₂ system.

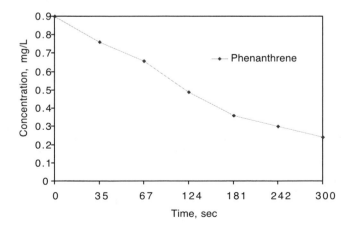

FIGURE 14.24
Degradation kinetics of phenanthrene with UV/ozone/H₂O₂ system. (From Trapido, M. and
Veressinina, Y., *Environ. Technol.*, 16, 729–740, 1995. With permission.)

TABLE 14.19

Half-Life of Phenanthrene by Various AOPs

System	Half-Life (s)
O_3/H_2O_2	58
UV/O_3	60
$UV/O_3/H_2O_2$	135
O_3	56

Source: Data from Trapido, M. and Veressinina, Y., *Environ. Technol.*, 16, 729–740, 1995. With permission.

14.4.6.4 Fluoranthene

Table 14.20 provides the half-lives of fluoranthene at various pH values under ozonation, UV degradation, and AOP systems (Trapido and Veressinina, 1995). The half-lives of fluoranthene in AOPs followed the series $O_3 = UV/O_3 < O_3/H_2O_2 < UV/O_3/H_2O_2$, as shown in Table 14.21.

14.4.6.5 Benzo(a)pyrene

Ozonation of benzo(a)pyrene proceeds more effectively in neutral media and is significantly decelerated in basic media. Table 14.22 shows that the half-lives of benzopyrene in AOPs followed the series $O_3 = UV/O_3 < O_3/H_2O_2 < UV/O_3/H_2O_2$.

TABLE 14.20

Half-Lives of Fluoranthene by Ozonation or UV at Various pH

pH	Half-Life Ozonation	UV Alone
3	134	336
7	78	277
9	357	753

Source: Data from Trapido, M. and Veressinina, Y., *Environ. Technol.*, 16, 729–740, 1995. With permission.

TABLE 14.21

Half-Lives of Fluoranthene by Various AOPs

System	Half-Life (s)
O_3/H_2O_2	86
UV/O_3	86
$UV/O_3/H_2O_2$	165
O_3	78

Source: Data from Trapido, M. and Veressinina, Y., *Environ. Technol.*, 16, 729–740, 1995. With permission.

TABLE 14.22

Half-Lives of Benzo(*a*)pyrene by Ozonation or Ultraviolet at Different pHs

pH	Half-Life	
	Ozonation	UV Alone
3	25	136
7	19	110
9	33	174

Source: Data from Trapido, M. and Veressinina, Y., *Environ. Technol.*, 16, 729–740, 1995. With permission.

14.4.7 Chlorinated Aliphatic Compounds

Zeff and Barich (1992) have illustrated the oxidation of methylene chloride and methanol by various AOP systems. Lewis et al. (USEPA, 1990) completed a series of studies on VOCs with the Ultrox system at the Lorentz Barrel & Drum Superfund site. They concluded that the key difference between ozone and peroxone is the primary oxidation modes: direct oxidation vs. hydroxyl radical oxidation. Ozone/hydrogen peroxide reacts with aromatic compounds; moreover, it also reacts on aliphatic acids with hydroxyl radicals. Due to these conditions, mineralization was slightly higher than with ozone alone. The results of these studies showed that DOC mineralization reached 15 and 18% with ozone alone and with an ozone/peroxide system (peroxone), respectively (applied ozone dose of 6.5 mg/L). The ozone process relies heavily on direct oxidation due to molecular ozone, while peroxone relies primarily on oxidation with hydroxyl radical. In the peroxone process, the ozone residual is short lived because the added peroxide greatly accelerates the ozone decomposition. However, the increased oxidation achieved by the hydroxyl radical greatly outweighs the reduction in direct ozone oxidation because the hydroxyl radical is much more reactive than molecular ozone. The net result is that oxidation is more reactive and much faster in the peroxone process than the ozonation process. Table 14.23 summarizes the key differences between ozone and peroxone as they relate to their application in drinking water treatment. Vermont's VOC treatment facility at IBM showed increasing treatment efficiency with the addition of hydrogen peroxide. The treatment efficiencies increased to an optimal level of 91% and nearly 100% for PCE and TCE at mass ratios of hydrogen peroxide to dissolved ozone of between 1 and 2, as shown in Table 14.24 and Figure 14.25.

14.4.8 Fulvic Acids

Volk et al. (1997) assessed the effects of ozone, ozone/hydrogen peroxide, and catalytic ozone by changes in the organic constituents of a synthetic solution of fulvic acids. Initial dissolved organic carbon (DOC) and

TABLE 14.23

Comparison between Ozone and Peroxone Oxidation

Factor	Ozone	Peroxone
Ozone decomposition rate	Normal decomposition, which produces hydroxyl radical as an intermediate product	Accelerated ozone decomposition, which increases the hydroxyl radical concentration above that of ozone alone
Ozone residual	5 to 10 min	Very short lived due to rapid reaction
Oxidation path	Usually direct aqueous molecular ozone oxidation	Primarily hydroxyl radical oxidation
Ability to oxidize iron and manganese	Excellent	Less effective
Ability to oxidize taste and odor compounds	Variable	Good; hydroxyl radical more reactive than for ozone
Ability to oxidize chlorinated organics	Poor	Good; hydroxyl radical more reactive than for ozone
Disinfection ability	Excellent	Good, but systems can only receive CT credit if they have a measurable ozone residual
Ability to detect residual for disinfection monitoring	Good	Poor; cannot calculate CT value for disinfection credit

Source: USEPA, Ultrox International Ultraviolet Radiation/Oxidation Technology—Applications Analysis Report, EPA/540/A5-89/012, September 1990.

TABLE 14.24

Oxidations of Methylene Chloride and Methanol

Contact Time (min)	Control	Ultraviolet	UV/H_2O_2	O_3/H_2O_2	UV/O_3	UV/O_3/ H_2O_2
Methylene Chloride						
0	100	100	100	100	100	100
15	100	59	46	32	36	19
25	100	42	17	21	16	7.6
Methanol						
0	75	75	75	NDA	75	75
30	75	75	75	NDA	31	1.2

Note: All concentrations are reported in mg/L; NDA = no data available.

Source: USEPA, Ultrox International Ultraviolet Radiation/Oxidation Technology—Applications Analysis Report, EPA/540/A5-89/012, September 1990.

biodegradable dissolved organic carbon (BDOC) concentrations of the fulvic acid solution were 2.84 and 0.23 mg/L, respectively. Oxidation tests were performed according to the ozone method, which provides an assessment of the extent of oxidation. Ozone, ozone/hydrogen peroxide, and catalytic ozonation mineralized 15, 18, and 24% of the initial DOC, respectively

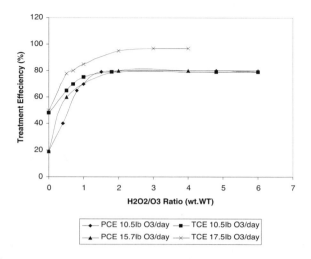

FIGURE 14.25
Effect of hydrogen peroxide on ozone treatment system at ozone production rates of 10.5 and 15.7 lb/day. The *x*-axis shows the hydrogen peroxide-to-ozone mass ratio. The treatment efficiency report is the sum of oxidation and air stripping. (From USEPA, Ultrox International Ultraviolet Radiation/Oxidation Technology—Applications Analysis Report, EPA/540/A5-89/ 012, September 1990.)

(Figure 14.26). The oxidation system that generated the highest BDOC concentration was ozone/hydrogen peroxide, while catalytic ozone produced the lowest concentrations; with ozone doses greater than 3.5 mg/L, BDOC levels were 0.90, 0.80, and 0.60 mg/L for ozone/hydrogen peroxide, ozone, and catalytic ozone, respectively. Catalytic ozone induced oxidation of ozone by-products into CO_2.

14.4.9   Tomato Wastewaters

Beltran et al. (1997) reported the results of oxidation of two tomato wastewaters with ozone combined with hydrogen peroxide or UV radiation (254 nm). The oxidation yields of these systems were compared with those from ozonation alone at similar experimental conditions. It was found that O_3/H_2O_2 oxidation leads to the increase of COD degradation rate (e.g., 86% at pH 6 in tomato wastewaters). It can be observed that an increase in hydrogen peroxide concentration, especially above 10^{-3} M, leads to an increase of the COD and TOC degradation rate, as shown in Figure 14.27. The differences between the oxidation types (O_3 and O_3/H_2O_2) diminish with increases in pH (see Figure 14.28). With distillery wastewaters, the presence of hydrogen peroxide barely increases the oxidation rate; however, the combination of UV/O_3 radiation was the best oxidation method applied because of the improvements achieved in both COD and TOC reduction rates compared to

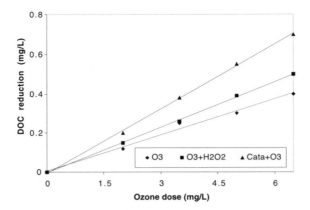

FIGURE 14.26

DOC reduction vs. applied ozone dose for a fulvic acid solution treated with ozone, ozone/hydrogen peroxide, or catalytic ozone. Contact time = 10 min; pH = 7.5; initial DOC = 2.84 mg/L. (From Volk, C. et al., *Water Res.*, 31(3), 650–656, 1997. With permission.)

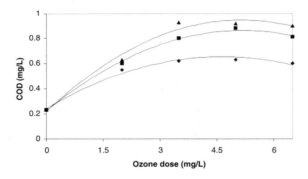

FIGURE 14.27

COD vs. ozone fed in during oxidation of tomato wastewaters with ozone combined with hydrogen peroxide. *Conditions:* average inlet ozone mass rate = 25.3 mg/min; COD_0 = 916 mg O_2/L; T = 18°C; pH = 6.3. Ozonation alone: ◆, C_{H2O2} = 10^{-3} M; ■, C_{H2O2} = 10^{-2} M; ▲, C_{H2O2} = 10^{-1} M. (From Beltran, F.J. et al., *Water Res.*, 31(10), 2415–2428, 1997. With permission.)

those achieved by ozonation alone, regardless of wastewater type treated. As happened to the other oxidation systems, the COD and TOC reductions were higher for tomato wastewater oxidation.

14.5 Hydroxyl Radical Concentrations in AOPs

From the literature, the rate constants between p-hydroxybenzonic acid and ozone, UV photon, and hydroxyl radical are known (Gurol and Nekouinaini, 1984). Therefore, contribution of ozonation, photolysis, and hydroxyl radical

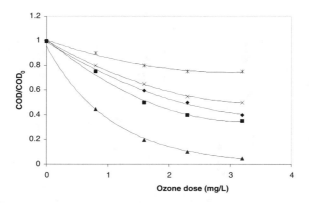

FIGURE 14.28

COD/COD_0 vs. ozone fed in during oxidation of tomato wastewaters with ozone combined with hydrogen peroxide at different pH values. *Conditions:* average inlet ozone mass rate = 25.9 mg/min; COD_0 = 750 mg/L; T = 18°C. Ozonation alone: *, pH = 6.3; ×, pH = 11. Ozone/ hydrogen peroxide (H_2O_2 = 0.02 M): ◆, pH = 6.3; ■, pH = 9.1; ▲, pH = 11. (From Beltran, F.J. et al., *Water Res.*, 31(10), 2415–2428, 1997. With permission.)

to the degradation of p-hydroxybenzonic acid has been estimated by Beltran-Heredia et al. (2001). Figure 14.29 shows a summary of the importance of each of the three reaction routes (direct photolysis, ozonation, and free radicals) in each of the combined oxidation processes. It shows that the combined systems that have the greatest free radical component (>80% of overall oxidation process) are: Fe^{2+}/H_2O_2, $UV/Fe^{2+}/H_2O_2$, $UV/Fe^{2+}/H_2O_2/O_3$, UV/H_2O_2.

Since the reaction constant between benzoic acid such as p-hydroxybenzonic acid with the hydroxyl radical is well documented (Ashton et al., 1995), it has been used as a reference compound in calculating the hydroxyl radical concentration. Using the comparative method, a value was obtained for the reaction constant of the hydroxyl radical with p-hydroxybenzonic acid of 1.63×10^9 $M^{-1}s^{-1}$ at 20°C. Having determined this constant, the next step was to calculate the concentration of hydroxyl radicals for each of the oxidation systems employed. Figure 14.30 shows the values of the hydroxyl radical concentrations for each oxidation system. Since $UV/Fe^{2+}/H_2O_2/O_3$ process has the highest hydroxyl radical concentration, it is not surprising that the system has the highest oxidation efficiency among all the AOPs.

14.6 Conclusions

Extensive research has been conducted on advanced oxidation systems with two operational elements, such as UV/H_2O_2, UV/O_3, and H_2O_2/O_3; however, not much research has been performed on systems with three

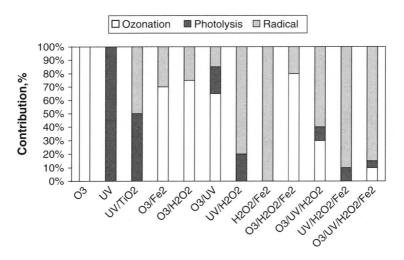

FIGURE 14.29
Contribution of photolysis, ozonation, and radical reaction in each oxidation process. (From Beltran-Heredia, J., Torregrosa, J., Domingues, J.R., and Peres, J.A., *Chemosphere*, 42, 351-359, 2001. With permission.)

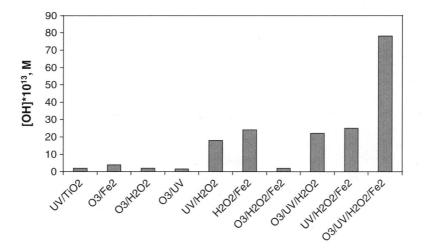

FIGURE 14.30
Values of the hydroxyl radical concentration for each oxidation system. (From Beltran-Heredia, J., Torregrosa, J., Domingues, J.R., and Peres, J.A., *Chemosphere*, 42, 351-359, 2001. With permission.)

operational elements, such as $UV/O_3/H_2O_2$ system. This is due to the increased cost of three different oxidation systems and the limited number of compounds that can be treated effectively. VOCs have been successfully treated by the UV/oxidation process. Compounds that have been oxidized include TCE, methanol, and methyl chloride. Ozonation has been shown to be quite effective for the destruction of PAHs in neutral media. The effect of pH on the oxidation of pollutants played a major role in the AOP systems studied. Using p-hydroxybenzoic acid as the reference compound, $UV/O_3/H_2O_2/Fe^{2+}$ process produces the highest hydroxyl radical concentration of 7.8×10^{-12} M and has the highest degradation efficient among the ten different AOP systems studied by Beltran-Heredia et al. (2001).

References

Alaton, I.A., Balcioglu, I.A., and Bahnemann, D.F., Advanced oxidation of a reactive dyebath effluent: comparison of O_3, H_2O_2/UV-C and TiO_2/UV-A processes, *Water Res.*, 36, 1143–1154, 2002.

Ashton, L., Buxton, G.V., and Stuart, C.R., Benzoic acid, rate constant with hydroxyl radical, *J. Chem. Soc. Faraday Trans.*, 91, 1631–1633, 1995.

Bahnemann, D. and Hart, E.J., Mechanism of the hydroxide ion initiated decomposition of ozone in aqueous solution, *J. Phys. Chem.*, 86(2), 255–259, 1982.

Beltran, F.J., Kolaczkowski, S.T., Crittenden, B.D., and Rivas, F.J., Degradation of *ortho*-chlorophenol with ozone in water, *Trans. I. Chem. Eng.*, 71(B), 57–65, 1993.

Beltran, F.J., Encinar, J.M., and Gonzalez, J.F., Industrial wastewater advanced oxidation. Part 2: Ozone combined with hydrogen peroxide or UV radiation, *Water Res.*, 31(10), 2415–2428, 1997.

Beltran-Heredia, J., Torregrosa, J., Domingguez, J., and Peres, J.A., Comparison of the degradation of p-hydroxybenzoic acid in aqueous solution by several oxidation processes, *Chemosphere*, 42, 351–359, 2001.

Benitez, F.J., Beltan-Heredia, J., Acero, J.L., and Rubio, F.J., Contribution of free radicals to chlorophenols decomposition by several advanced oxidation processes, *Chemosphere*, 41, 1271–1277, 2000.

Bielski, B.H.J., Cabelli, D.E., Arudi, R.L., and Ross, A.B., *J. Phys. Chem. Ref. Data*, 14, 1041–1100, 1985.

Boncz, M.A., Bruning, H., Rulkens, W.H., Sudholter, E.J.R., Harmsen, G.H., and Bijsterbosh, J.W., Kinetic and mechanistic aspects of the oxidation of chlorophenols by ozone, *Water Sci. Technol.*, 35(4), 65–72, 1997.

Buhler, R.E., Staehelin, J., and Hoigne, J. Ozone decomposition in water studied by pulse radiolysis: 1. HO2/O2, and HO3/O3 as intermediates, *J. Phys. Chem.*, 88, 2560–2564, 1984.

Burrows, H.D., Ernestova, L.S., Kemp, T.J., Skurlatov, Y.I., Purmal, A.P., Yermakov, A.N., Kinetics and mechanism of photodegradation of chlorophenols, *Progress Chem. Kinet.*, 23(3), 145–207, 1998.

Christensen, H., Sehested, K., and Corfitzen, H., Reactions of hydroxyl radicals with hydrogen peroxide at ambient and elevated temperatures, *J. Phys. Chem.*, 86(a), 1588–1590, 1982.

Esplugas, S., Gimenez, J., Contreras, S., Pascual, E., and Rodriguez, M., Comparison of different advanced oxidation processes for phenol degradation, *Water Res.*, 36, 1034–1042, 2002.

Gurol, M. and Nekouinaini, S., Kinetics behavior of ozone in aqueous solutions of substituted phenols, *Ind. Eng. Chem. Fundam.*, 23, 54–60, 1984.

Hong, A., Zappi, M., Kuo, C.-H., and Hill, D., Modeling kinetics of illuminated and dark advanced oxidation processes, *J. Environ. Eng.*, 122(1), 58-62, 1996.

Lewis, N., Topudurti, K., Welshans, G., and Foster, R., A field demonstration of the UV/oxidation technology to treat ground water contaminated with VOCs, *J. Air Waste Manage. Assoc.*, 40(4), 540–547, 1990.

Masten, S. and Davies, S., The use of ozonation to degrade organic contaminants in wastewaters, *Environ. Sci. Technol.*, 28(4), 180–185, 1994.

Masten, S., Shu, M., Galbraith, M., and Davies, S., Oxidation of chlorinated benzenes using advanced oxidation processes, *J. Hazardous Waste Hazardous Mater.*, 13(2), 265–284, 1996.

Ni, C.H. and Chen, J.N., Heterogeneous catalytic ozonation of 2-chlorophenol aqueous solution with alumina as a catalyst, *Water Sci. Technol.*, 43(2), 213–220, 2001.

Peyton, R. and Glaze, W. Destruction of pollutants in water with ozone in combination with ultraviolet radiation: 3. Photolysis of aqueous ozone. *Environ. Sci. Technol.* 22, 761–767, 1988.

Saucer, M.L. Jr., Brown, W.G., and Hart, E.J., O(P)atom formation by the photolysis of hydrogen peroxide in alkaline aqueous solutions, *J. Phys. Chem.*, 88(7), 1398–1400, 1984.

Sehested, K., Holcman, J., Bjerbakke, E., and Hart, E., A pulse radiolysis study of the reaction OH + O3 in aqueous medium, *J. Phys. Chem.*, 88(8), 4144–4147, 1984.

Sotelo, J.L., Beltran, F.J., Benitez, F.J., and Beltran-Heredia, J., Henry's law constant for the ozone-water system, *Water Res.*, 23, 1239–1246, 1989.

Staehelin, J. and Hoigne, J., Decomposition of ozone in water: rate of initiation by hydroxide ions and hydrogen peroxide, *Environ. Sci. Technol.*, 16(10), 676–681, 1982.

Staehelin, J. and Hoigne, J., Decomposition of ozone in water in the presence of organic solutes acting as promoters and inhibitors of radical chain reactions, *Environ. Sci. Technol.*, 19, 1206–1213, 1985.

Takahashi, N., Nakai, T., Satoh, Y., and Katoh, Y., Variation of biodegradability of nitrogenous organic compounds by ozonation, *Water Res.*, 28(7), 1563–1570, 1994.

Tang, W.Z. and Huang, C.P., 2,4-Dichlorophenol oxidation kinetics by Fenton's reagent, *Environ. Technol.*, 17, 1371–1378, 1996.

Trapido, M., Veressinina, Y., and Munter, R., Ozonation and advanced oxidation processes of polycyclic aromatic hydrocarbons in aqueous solutions: a kinetic study, *Environ. Technol.*, 16, 729–740, 1995.

USEPA, Ultrox International Ultraviolet Radiation/Oxidation Technology—Applications Analysis Report, EPA/540/A5-89/012, September 1990.

Volk, C., Roche, P., Joret, J.C., and Paillard, H., Comparison of the effect of ozone, ozone/hydrogen peroxide system and catalytic ozone on the biodegradable organic matter of a fulvic acid solution, *Water Res.*, 31(3), 650–656, 1997.

Weeks, J.L. and Rabani, J., The pulse radiolysis of deaerated aqueous carbonate solutions. I: Transient optical spectrum and mechanism; II: pK for OH radicals, *J. Phys. Chem.*, 70(7), 2100–2106, 1966.

Zappi, M.E., Hong, A., and Cerar, R., Treatment of groundwater contaminated with high levels of explosives using traditional and non-traditional advanced oxidation processes, *HMCRI Superfund Conf.*, Washington D.C., 1993.

Zeff, J.D. and Barich, J., UV/oxidation of organic contaminants in ground, waste and leachate waters, *Water Pollut. Res. J. Can.*, 27(1), 139–150, 1992.

Index